T. Mabee
Jan. 1973

SIMULATION

Statistical Foundations and Methodology

This is Volume 92 in
MATHEMATICS IN SCIENCE AND ENGINEERING
A series of monographs and textbooks
Edited by RICHARD BELLMAN, *University of Southern California*

The complete listing of the books in this series is available from the Publisher upon request.

SIMULATION

Statistical Foundations and Methodology

G. ARTHUR MIHRAM

Moore School of Electrical Engineering
University of Pennsylvania
Philadelphia, Pennsylvania

1972

ACADEMIC PRESS New York and London

ACADEMIC PRESS, INC.
111 Fifth Avenue, New York, New York 10003

United Kingdom Edition published by
ACADEMIC PRESS, INC. (LONDON) LTD.
24/28 Oval Road, London NW1

LIBRARY OF CONGRESS CATALOG CARD NUMBER: 72-159609

PRINTED IN THE UNITED STATES OF AMERICA

TO MY PARENTS

With Gratitude for Stressing the Importance of a Prepared Mind

CONTENTS

Chapter 3 **Time Series**

Chapter 4 **The Monte Carlo Method**

Chapter 5 **Modeling**

Chapter 6 **Fundamentals of Simular Experimentation**

Chapter 7 **Experimental Designs for Qualitative Factors**

PREFACE

Man has almost continually accumulated, at an increasing rate, his awareness and knowledge of the world about him. In perspective, it would seem only natural that the stockpiling of information would lead inevitably to a need to systematize and organize this knowledge. Moreover, the need to organize and systematize the *search* for further knowledge has become even more imperative. As a consequence, this apparently insatiable thirst has given rise to the current surge of interest in *systems* and in methods for their facile analysis and description.

The systems approach, at first successful in military, industrial, and managerial contexts, has now become firmly embedded in practically every other discipline which endeavors to explain or predict observable phenomena. The essence of the systems approach is modeling, the effort to mime, and therefore to understand, systems and their components and interactions. The fundamental premise of the systems analyst (or operations researcher, systems engineer, or management scientist, as he is often termed in similar contexts) is that the systems which he analyzes are atomistic; i.e., systems consist of component modules which may be isolated and whose interdependencies and interactions may be, under observation and/or reflection, stipulated.

In studying and analyzing systems, the systems analyst finds himself eventually constrained from carrying his atomistic description of the system to the limit. As is indicated in the first chapter, the realization of this constraint leads to the inevitable acceptance of an uncertainty principle of modeling, a principle which reveals the importance of probabilistic system descriptions in conscientious modeling activities.

Moreover, since the systems of greatest interest evolve with time, models of them are of necessity dynamic in character.

This book is devoted to the dynamic, stochastic, simulation model. However, in the initial chapter, a discussion of model categories reveals that the scope of the book is not so limited as might at first appear. The dynamic, stochastic model of the simular variety is one of our most general types of symbolic models for the representation of systems.

Because probability represents our mechanism for the quantification of uncertainty, the second and third chapters are concerned with a rigorous development of random variables and time series. Prerequisite to an understanding of this material (as well as most of that which follows) is a rudimentary knowledge of the probability calculus (such as the multiplication and addition rules of probability), a knowledge of the differential and integral calculi, and an elementary notion of vector and matrix operations (such as vector and matrix sums and products). Since most complex simulations are implemented on electronic computers, a rudimentary acquaintance with programming will probably be an asset (in terms of concept formation only, however) to students of this book, though the reader will not find herein any computer programs.

The need for such a text became apparent while lecturing at the University of Pennsylvania's Wharton School of Finance and Commerce and its Moore School of Electrical Engineering. The postgraduate students of systems theory in both schools inevitably possess a quite adequate background regarding random variables and their transformations, so that the material in Chapters 2 and 3 constitutes basically a review. Considerable emphasis is placed, however, on techniques for the generation of random (and pseudorandom) numbers and on methods for the generation therefrom of random variables of many distributional families.

Consequently, Sections 2.7, 2.9 become essential aspects of the text. When employed in this mode, emphasis should rightfully be placed on Sections 2.12, 2.14, 2.15, and 3.4, for the material therein provides an essential understanding of the fundamental types of randomness which can frequently be found to exist in systems. Furthermore, the relationships of the multivariate normal distribution to the Pearson's chi-squared, the Student's t, and the Snedecor's F distributions are essential to the later chapters which develop techniques for the (necessarily statistical) analyses of the responses (outputs) emanating from the stochastic model. The exercises at the end of Section 2.15.3 are intended to be especially valuable in this regard.

Chapter 4 provides a basic description of the Monte Carlo method, and describes the relationship of this technique to, and the distinctions of the method from, stochastic modeling. Indeed, the dynamic, stochastic simulation model is shown to be a generator of time series and, therefore, its repeated use constitutes a Monte Carlo experiment with a stochastic process. Applications to stochastic PERT networks, productivity systems, queues, and inventory modeling are indicated.

Chapter 5 returns the reader to the aspects of modeling complex systems. The five stages of model development are delineated, and each is discussed in turn, the statistical bases of each phase becoming consequently more apparent. Apropos of this discussion is a fundamental introduction to, and comparison of, special-purpose, *ad hoc*, simulation programming languages. Those features required by the analyst/investigator are developed in an endeavor to illustrate the importance of these languages to successful computerised modeling.

The five penultimate chapters are concerned with the application of extant statistical methodology to the analysis and comparison of responses emanating from independently seeded encounters with (i.e., iterations of) stochastic models. Fundamental limit laws of probability theory are shown to be applicable to the prediction of the distributional nature of many random variables which are emitted by the stochastic model. Such predictions are frequently possible, as it develops, from reflections regarding the intrinsic behavior mechanisms incorporated within a model's structure.

Taking cognizance of a principium of seeding, developed in Chapter 5, the discussion of the five following chapters underscores the fact that simulation is indeed the manager's laboratory, in which experimentation regarding alternative control mechanisms and system substructures may be tested symbolically before being implemented physically. Such experimentation, just as is that undertaken in the scientist's laboratory, may be quite efficiently organized in accordance with the well-established principles of the statistician's experimental designs. Simular experimentation, then, not only employs factorial experiments and regression analyses, but also may invoke response surface methodology (in order to ascertain the optimal operating conditions of the system under study) and time series analyses (in order to ascertain the dynamic behavior mechanisms of the system as it evolves through time). The applications of these statistical tools to simular experimentation are delineated in the five penultimate chapters in some detail.

The present text does not endeavor to be a source of case studies.

Though mention is made of diverse applications in ecological, societal, economic, and environmental disciplines, the reader seeking such case studies would be referred to the earlier text of Chorafas (1965), *Systems and Simulation*, to the list of bibliographic sources presented at the conclusion of this book, or to manuals and texts prepared by the designers of *ad hoc* simulation languages. Indeed, an ideal supplement to the current book, when employed as a textbook, might be a judicious selection of one or more of these programming manuals. Though not designed as a source of exemplary simulation models, the book does conclude with a chapter describing the likely perspectives of, and the challenges confronting, the simulation methodologists.

Due to its emphasis on the stochastic aspects of modeling, the book may also prove useful as a text for courses in fundamental statistics as well. Though the development of the statistical techniques is, for the most part, quite rigorous, the text may become of value in applied probability or statistics courses in which students have access to a digital computer in order to perform sampling experiments and Monte Carlo analyses, as suggested in the exercises.

Though answers to the many exercises in the text do not appear explicitly as an addendum, every effort has been made either to include the answers within the particular exercises or to ensure that the desired solution is somewhat obvious from the context of the material immediately preceding.

Presumably, a text of this magnitude shall not appear without its share of errors. It is hoped that all are typographical, though the responsibility for any type of error must, in the final analysis, reside with the author. An alphabetized bibliography of referenced sources is provided at the end of the book; sins of omission, if any, shall hopefully be forgiven by those affected.

ACKNOWLEDGMENTS

An expression of appreciation is due the National Science Foundation, whose research grant (Number GK-5289) has provided support for the author's current studies in the design and analysis of computerized simulation models.

Expressions of gratitude are due a number of individuals who, in one way or another, contributed to the book: Bradley Efron, Editor of the *Journal of the American Statistical Association*, gave his kind permission for the extraction of material appearing in Table A.6 of the Appendix; Mrs. Vera Watt, Chief Clerk of the Moore School of Electrical Engineering, supervised much of the typing of the manuscript; and Mr. Robert Lefferts devoted a considerable amount of time and effort in reading through the manuscript and in performing the computations which have yielded Tables A.1–A.5 of the Appendix.

Without doubt the book would not be before the reader were it not for the patience of, and the inspiring atmosphere created by Danielle. The effects of her French upbringing and her Australian citizenship are perhaps more than once evident in the orthography of the text.

"Le hasard ne favorise que les esprits préparés."

LOUIS PASTEUR

Chapter 1

MODELS

"—... *you do not see with your little eyes certain stars of the fiftieth magnitude that I perceive very distinctly; do you conclude from this that these stars do not exist?*
—But, ... I have searched well.
—But, ... you have perceived badly."

—Voltaire

1.1. Prolegomenon

Scientists, since the beginning of their organized search for knowledge, have been engaged in modeling, often without really being totally aware of their occupation with the activity. In a certain sense, they are akin to M. Jourdain, Molière's character (1670) who spoke prose for 40 years without really knowing it. However, it is far removed from our purpose here to belittle scientific activity; the comparison is made strictly to emphasize a rather recent departure in the general conception of the activities of the scientist. Until the present century, the scientist had been viewed as an individual primarily engaged in the search for deterministic laws, any one of which would serve to explain some naturally occurring phenomenon.

Aided by elementary machines, such as the telescope, the microscope, and the pendulum clock, the post-Renaissance European scientists were able to devise formulations that related two or more observable quantities in such a manner that predictions of their interactions and their behavior could be quite accurately made. Especially in classical physics, these formulations inevitably described mathematically the temporal–spatial relationships among physical phenomena, such as Newton's and Boyle's laws. However, in the simultaneously developing anatomical and biological sciences, working descriptions, as opposed to mathematical formulas, were employed in order to depict the underlying nature of observable natural phenomena; to wit, the descriptive theories of the biological roles played by microscopic cells and gametes, which became observable as a result of extended human visual capabilities made possible by the perfected lens. Either of these two scientific approaches, each dedicated to the explanation of naturally occurring phenomena, is a form of modeling, an act of structuring, physically or conceptually, temporal–spatial relationships among physical entities.

With the increased development of scientific knowledge, man acquired an ability to construct better and more refined instruments, or *machines*, for the purpose of extending his naturally endowed powers and capabilities. The machines, in turn, provided scientists with a greater number of natural phenomena to observe, study, and explain; for example, the aforementioned gamete and the discovery of the radioactive elements have led to scientific disciplines of such breadth that their total specifics are not likely known to any single individual of the mid-twentieth century.

The success of these studies and the acceptability of their derived explanations has rested to a large extent upon the credibility of proposed models of the particular phenomena. For example, studies of electromagnetic phenomena led to the development of radio transmitters and receivers, which have led, in their turn, to the recent discoveries of previously unnoticed (and often humanly unnoticeable) characteristics of the stars and the nonilluminating bodies of our universe.

Each acknowledged advance in knowledge or machinery has been accompanied by a model of significantly explicative quality. Newton's models were expressed as mathematical relationships in a presumably extant Euclidean universe; Einstein's models assumed mathematical forms in a presumably Lobachevskian geometry.

Alternatively, popular atomic models of the current century have been of a material nature, being miniature orreries which themselves mime Copernicus' model of our own planetary system. Engineers have successfully employed physical models in their professional activities ever since the inception of their profession. Structural properties of large systems can thus be tested in the small before being implemented in the large. Much of the success in the fields of navigation, architecture, communication, and transportation owes a significant debt of gratitude to the modeling process.

What constitutes modeling—a physical miniature, a set of one or more equations in two or more variables, a pictorial or verbal description of some phenomenon? As a first semantic approximation, modeling may be defined as the *act of mimicry*. Such a broad definition would necessarily encompass puppet shows, statues, scale models, animated cartoons, paintings, equations of motion, and war training exercises, each of which provides a certain degree of mimicry, though of diverse natures and behaviors.

1.2. Categories of Models

Realizing fundamental similarities in these many human activities, Rosenblueth and Wiener (1945) prepared a fundamental categorization of models as they are employed in science:

(*a*) *material models*, conceived of as transformations of original physical objects,
(*b*) *formal models*, defined as symbolic assertions, in logical terms, of idealized, relatively simple situations, the assertions representing the structural properties of the original factual system.

Thus, according to this categorization, scale models and the planetarium become material models because they are spatial transformations of their intended counterparts. Genetic studies of *Drosophila*, when conducted for the purpose of inferring the behavior of populations of mammalian species, also constitute material models because the study constitutes a temporal transformation of genetic behavior; long-term hereditary effects on, say, human populations may then be induced from observations on a few generations of the insects.

Formal models were apparently deemed by Rosenblueth and Wiener to be of a mathematical nature. Subcategories of these formal models were presented as:

(*a*) *open-box models*, predictive models for which, given all inputs, the output could be determined,
(*b*) *closed-box models*, investigative models, designed so as to develop understanding of a system's output under differing input conditions.

In science, the closed-box model has become more and more important as a mechanism for eventually developing an open-box model of the phenomenon being investigated. Indeed, the history of scientific endeavor is leading us to believe that very few models are truly open-box. Newton's laws of gravitation have constituted predictive models for the description of the spatial–temporal relationships of major objects in our solar system; however, the model assumed a closed-box attitude when Adams and Leverrier employed it in the mid-nineteenth century to explain the apparently inconsistent behavior of Uranus, thereby locating Neptune. Furthermore, it is becoming realized that few, if any, truly open-box models exist, for all predictive models seem to possess only a limited (though very useful) preciseness.

Hence, the subcategorization of formal models would seem to be predicated primarily on the purpose to which the model is addressed; both attempt to be explicative in purpose, but the open-box model is designed to be predictive and is usually accompanied by an aura of rectitude, especially so if it adequately fulfills its purpose; whereas, the closed-box model serves an investigative role, and is therefore often considered of a more approximate nature.

With the advent of the electronic computer, modeling has taken a significant growth, finding applications in areas of endeavor not normally associated with the scientific method. Studies of international relationships and national political interactions have been developed using symbolic models, as have examinations of industrial concerns, manufacturing plants, military strategies, service agencies, transportation facilities, communication networks, urban systems, satellite projects, and health systems.

The advantages of the electronic computer for modeling were begun to be realized during the early flourish to revive numerically analytic techniques, suddenly made realizable by the machine that could significantly extend man's computational capabilities. Many of these numerical techniques were associated with the integration of systems of one or more differential equations. The numerical integration algorithms proceed from some initial or boundary condition to compute the stepwise change in a dependent variable as the independent variable(s) moves (move) away from its (their) initial locus. Consequently, whenever the independent variable represents time and the dependent variable measures the status of some physical object (such as its height or its electrical charge), the numerical algorithm may be viewed as a mimic of the physical object in steps of time. Furthermore, by interrelating a number of dependent variables as differential equations with respect to time, the numerical integration techniques may compute the status of each of the several dependent variables each time that the time step (that is, the dependent variable) is advanced.

The numerical integration techniques made possible then a digital computer mimicry of the behavior of systems in time. In the 1940s, the electronic operational amplifier had become sufficiently accurate to permit the analog computer's use as a mimic of systems of differential equations as well. Though in this case the independent variable (time) is best conceptually presented as a continuously flowing variate, the analog computer could also be employed to mime the temporal behavior of a physical system. A recorded trace of the values assumed by specific internal electrical properties of the analog computer serves then as a representation of the behavior of corresponding attributes of the physical objects that are related via the set of differential equations.

From these developments, it was a relatively straightforward, though tedious, generalization to program a digital computer to keep track, via quite general computational algorithms, of the values of large numbers of variates as the computer automatically advanced a preprogrammed clockworks. Thus, networks of queues, such as those extant in production lines or the bureaucratic processing of applications, could be represented in the computer by generating and maintaining a status record of every "customer" and every "server" in the network; such a symbolic model could proceed virtually unattended by programming it to advance customers and to assign servers each time that appropriate conditions are satisfied while the computer's preprogrammed clocking mechanism advances one unit at a time. More generally, it was realized that status descriptors of large numbers of system entities could be maintained and altered as the internal clockworks were advanced. Changes in the descriptors could be accomplished by quite general algorithms, logical formulas that were not necessarily mere recursive applications of some numerical integration procedure.

A new type of model was thus born. Although of a symbolic nature, these models were not merely numerical solutions to time-dependent systems of equations; rather, they employed algorithms of a far more general nature. The algorithms were no longer constrained to provide the time-stepped solution of a few differential equations, but could be employed to simulate the behavior of large numbers of elements in complex, interactive systems. Computerized simulation became a new and viable modeling technique.

Indeed, taking advantage of this methodology is the professional known as the *systems analyst*, or operations research specialist. A *system* may be defined as a collection of components (or elements, or entities) that are interrelated and that interact one with another in a collective effort to achieve some goal. Often the goal of a system will not be clearly recognizable, as is possibly the case for the evolutionary system of organisms on Earth, though in most man-made systems (such as transportation networks and hospitalization systems) one or more goals can be explicitly stated. It is the task of the *systems analyst* not only to define explicitly the goals of the system that he is analyzing, but also to develop a model of the structural elements of the system and of their interactions so as to evaluate the effects of alternative control mechanisms on the attainment of the defined goal. The use of computerized, symbolic models has thus become a primary tool of the systems analyst.

Indeed, Churchman *et al.* (1957) defined a model as "a representation of the system under study, a representation which lends itself to use in predicting the effect on the system's effectiveness of possible changes in the system [p. 155]." Their categorization of models included three types:

(*a*) *iconic models*, those that pictorially or visually represent certain aspects of a system,
(*b*) *analog models*, those that employ one set of properties to represent some other set of properties that the system being studied possesses,
(*c*) *symbolic models*, those that require mathematical or logical operations and can be used to formulate a solution to the problem at hand.

An exemplary iconic model is a globe, which serves to represent the Earth. However, under this categorization, a globe with its surface area colored becomes an analog model, since the colorations on its surface are then taken as properties representative of the aquatic, polar, and land masses of the Earth. (Another exemplary analog model is the use of the measured flow of current in an electronic analog computer as a mimicry of the flow in a hydraulic system.) Symbolic models were apparently to be equated with the formal models of Rosenblueth and Wiener.

Though this categorization recognized the possibility of bifurcating the Rosenblueth–Wiener material model category, the Churchman *et al.* scheme did not explicitly recognize the distinction that was becoming apparent among formal, or symbolic, models. Though symbolic models were subcategorized as analytic or numerical, both types were presumed to be mathematical expressions requiring a solution and were distinguishable one from another only by the methodology applied in their solution. The analytic type of symbolic model was to be solved by accepted mathematical operations on the symbols themselves, whereas the numerical type was solved in particular cases, usually by iterative search techniques.

Monte Carlo procedures were apparently admissible, but were viewed only as providing a means for evaluating certain terms in the symbolic model.

Sayre and Crosson (1963) subsequently defined three categories of models, thereby incorporating the two types delineated by Rosenblueth and Wiener in addition to the new modeling capability of simulation. Their categories are:

(*a*) *replications*, models that display a significant degree of physical similarity between the model and the modeled,

(*b*) *formalizations*, symbolic models in which none of the physical characteristics of the modeled is reproduced in the model itself, and in which the symbols are manipulated by means of a well-formed discipline such as mathematics or logic,

(*c*) *simulations*, symbolic models in which none of the physical characteristics of the modeled is reproduced in the model itself, yet in which the symbols are not manipulated entirely by a well-formed discipline in order to arrive either at a particular numerical value or at an analytic solution.

According to this categorization, Rosenblueth and Wiener's material models become synonymous with the term "replication." The scale model and the orrery become exemplary replications. Other replicas include the *Drosophila* genetic experiment, model train sets, military war exercises, statuette reproductions of the Eiffel or Pisan towers, riverine flow models, counterfeit notes, a relief map, a wind tunnel with model airplane, a pilot plant, and an equestrian statue of Thomas Jefferson. Sayre and Crosson include as replications any mimicry that preserves all three spatial coordinates, though any one or more of these coordinates may be scaled. Thus, a planar map does not qualify as a replica of the portion of the Earth's geography that it represents, because the coordinates for altitude are not physically present on the map. Similarly, a painting or photograph of the Eiffel tower, lacking depth, is not a replica of the tower.

For sake of completeness, physical models in which one or more spatial coordinates have been removed might be referred to as *quasi replications*. The photograph, the cinema, the drawing, and even the planetarium show are exemplary of quasi replicas for they all fail to incorporate physically all three spatial coordinates. Churchman *et al.* presumably did not distinguish between material models on the basis of spatial coordinates. Their iconic and analog models presumably could have included dimensional reductions or exclusions. One should note that coordinate scaling does not necessarily result in quasi replication; rather, the absence of one or more coordinates forbids the model to be a replica and requires that it be a quasi replica. For example, a riverine or estuarine flow model may not be scaled in water depth compatibly with the scaling of the waterway's length or breadth, yet the resulting model is considered as a replication of the waterway; whereas, the planetarium show, despite its deceiving capability to portray relative stellar distances, lacks physically the third dimension and is therefore a quasi replication.

Both the formalization and the simulation are symbolic models and possess none of the physical characteristics of the modeled. The distinction between the two categories rests with the manner in which the model's symbols may be manipulated. The *formalized model* consists of symbols that are manipulated by operations from well-formed disciplines such as mathematics or mathematical logic. For example, a Boolean expression for the status of a switching circuit constitutes a formalized model for the circuit because the symbols (0's and 1's or T's and F's) may be manipulated in accordance with operations of mathematical logic. Similarly, Newton's inverse square, and integrated inverse square, laws become formalized models for representing physical relationships between bodies. Other formalized models include systems of differential equations that become tractable under operations of algebra and calculus, such as a set of differential equations that describes the flow in a water channel or the Lanchesterian laws, which represent the dynamics of a battlefield.

The simulation model, on the other hand, is a symbolic model whose symbols need not always be manipulated by mathematical–logical operations. For example, a model that maintains a dynamic accounting of the passenger load, present location, and desired stops of an office elevator is not readily accomplished with realism by means of a set of mathematical equations or descriptions. Instead, the model requires guidance by certain policies (such as its unidirectional movement unless service demands are absent in the direction of its present flow), which can be described by more flexible algorithms and which can represent the behavior of the elevator even in the presence of randomly arriving demands for its service.

Other exemplary simulation models include: PERT (program evaluation research technique) representations of the network of activities that constitute a complex project; and queueing models, which maintain records of the status of individual customers and servers in a competitive service system. Another example is the *function simulator*, which may be typified by a set of two equations in the five unknowns x, y, u, v, and w.

If one seeks to determine the relationship between x and y, this may be accomplished only by knowledge of the values that have been assigned to u, v, and w. To model the relationship between x and y, representative selections of u, v, and w can be assigned, and the corresponding solutions for x and y then determined. By appropriately searching through assignable values of u, v, and w, one may endeavor to structure or to understand the relationship between x and y.

Thus, a simulation model may be either predictive or investigative. The PERT model would serve to predict the project's completion date and would be comparable to Rosenblueth and Wiener's open-box model. The function simulator, in which the symbols u, v, and w are manipulated heuristically, constitutes an investigative, or closed-box, model. Similarly, the formalized models (directly comparable either to the class of formal models as defined by Rosenblueth and Wiener, or to the symbolic models of Churchman *et al.*) may be categorized according to the intent of their usage: either predictive or investigative.

1.3. A Cross Classification

The reader may question the need for the extensive discussion of model categorizations. The categorizations are of significant value to the organization of a general systems theory, yet to be fully developed [see Mesarović (1964).] In addition, as Sayre and Crosson indicated, the categorizations are of fundamental importance in the exciting activity of efforts to model the human mind. An understanding of the modeling activity appears to be a prerequisite to the successful, artificially intelligent machine.

Other categorizations of models are possible and should also be mentioned. A model is either *dynamic* or *static*, depending upon whether or not any of its features or symbols alter over time. An iconic model, such as the aforementioned statue of Thomas Jefferson, is frequently quite static, whereas other replicas, such as the mobile orrery, may be of the dynamic class. Newton's inverse square law constitutes a static formal model describing the force existent between two or more physical objects, yet its integral provides a time-dependent formal model of the dynamic variety, such as the height of a falling body t time units after its instant of release. Moreover, the simular variety of models may be either dynamic or static, as exemplified, respectively, by the queueing model or the function simulator. Another static simulation model is the symbolic organization chart, which portrays the relationships among elements of a hierarchy.

Dynamic models are further categorized according to the method by which time is represented in the model. The *continuous* dynamic model represents time as a continuous variable, the *discrete* model as a quantized variate. In the dynamic replication, such as the mobile orrery, the passage of time is presumed to be replicated continuously, as it would exist in the modeled planetary system; yet, in the *Drosophila* genetic experiment, recordings of the population's characteristics may need to be made only at the commencement of each generation, so that time is effectively discretized. Similarly, the symbolic models, whether formalizations or simulations, may represent time either as a discrete or as a continuous variable.

The symbolic models of greatest complexity are being implemented more and more frequently on computers. Thus, the dynamic model usually becomes either continuous or discrete in accordance with the type of computer employed—analog or digital. The analog computer may be either mechanical or electrical; in either event, the computer is employed to mime any dynamic system whose behavior may be represented by the same formal expressions that describe the properties of the computer. For example, the electrical capacitor is described by an integral with respect to time and thus measurements of its electrical properties may serve as a dynamic model of any other physical system that could be likewise described by the same integral expression. Consequently, since the trace of the electrical properties of an electrical network can be recorded continuously (subject to constraints imposed by the intrinsic inertia of the recording device), the network may be viewed as a continuous model of the system whose behavior is mimed by the network.

The digital computer is useful in modeling a dynamic system, but its electrical operation is pulsed, thereby making discrete models more appropriate for implementation thereon. Though the same set of differential equations may often be solved dynamically either on an analog or on a digital computer, the digital computer employs the numerically analytic technique of advancing its clockworks forward one unit of time whenever its programmed algorithm has completed the computations necessary for describing the state of the system at its present "clock state." It is precisely because of this pulsed mechanism, however, that the digital computer has become so valuable in the simulation of many systems. To wit, queues that arise because of limited servicing facilities are not always joined or departed in a continuously flowing fashion. (One may view a dam as providing a "queue" for water molecules which are constrained by it, but more usually one would conceive of the water as continuously arriving. In the more usual queueing situation, such as the waiting line of customers at a bank teller's window, one does not conceive of customers as a steady flow of demands for the limited service.) The digital computer has therefore become especially useful in the simulation of these queueing systems, accomplishing the task by updating the symbolically represented status of each element in the system on each occasion that the computer's "clock" is advanced. However, the class of simulation models is not dependent upon its machine implementation.

1.4. Stochasticity and Randomness in Systems

Because of the versatility of the digital computer, modeling has taken significant strides forward since the 1950s. Systems and their dynamic behavior, more general than those representable by time-dependent differential equations, could be mimed in accordance with special-purpose algorithms that could be readily implemented on the new machine. As a matter of fact, the capability to keep track of large numbers of interacting components of a system meant that significantly improved modeling capabilities were available; no longer was the investigator constrained to represent the system's component interactions by formalized representations.

Of course, the accustomed mathematical niceties of the formalized model become lost in a model of the simular variety. Yet the intricacies of a modeled system can be more satisfactorily represented in many cases by a modeling technique that admits more general algorithms than the usual time-dependent differential and integral equations.

Of particular import in this regard is the ability to represent random phenomena in the simulation model. A formalized model of the mixing of two fluids may be readily constructed in terms of differential equations, but the model would prove to be relatively useless in describing completely and accurately the Brownian motion in the solution. A more detailed simulation model of the particulate behavior in the fluid mixture would, if feasible, result in a more meaningful representation of the system, and would likely result in as meaningful a model of the temperature dispersion in the mixture as could be accomplished via the formal set of differential equations. The simular representation would, however, require intricate descriptions of molecular movements in the fluid. Because of their barely finite number, the nature of these movements would need to be summarized, probably statistically.

As a matter of fact, nearly all systems exhibit this feature. The internal relationships may be somewhat modeled by mathematical expressions, but a closer examination of the structure of the system reveals a much more profound and intricate internal behavior. This internal behavior usually comprises such a diversity of contributing effects that it is often best described probabilistically. Thus, for example, the temperature of gas could be described quite roughly by Charles' law in terms of the volume containing it and the pressure exerted upon it, but our more recently acquired knowledge of temperature shows that it is a statistical measure based on the average velocity of the molecules comprising the gas.

Similarly, one may roughly describe the productivity of a work force as a linear function of the number of personnel employed:

$$\text{productivity} = k \cdot (\text{man-hours}).$$

Inevitably, however, one would observe weekly variations in the productivity measure; and an examination of the system would reveal that the productivity cannot be so simply related to the available man-hours. Some employees will be unpredictably ill; others will be constrained in their efforts from time to time because critical machines or other personnel will not perform their assigned tasks perfunctorily and promptly. The causes for machine failures and personnel illnesses are in turn dependent upon a host of possible causes, including for example, the temperature of the working environment and the attitudes of maintenance crews.

Therefore, most analyses of systems result in a statement that insufficient knowledge is available to specify accurately the behavior of all the system's components. One of our best methods for the description of this lack of knowledge is *probability*.

Indeed, the earliest attempts to define the term probability were predicated on the existence of equally likely outcomes, typified by the six numbered faces of a well-tossed die. If n distinguishable outcomes of an experiment, such as a die-tossing (for which $n = 6$), could be listed and if there were no *a priori* reason to suspect that any one of the n outcomes would be more favored as a result, then the probability of any particular one of the outcomes' arising was defined to be $1/n$. Of special importance in this definition of probability is the clause "if there were no *a priori* reason to suspect that any one of the n outcomes would be more favored as a result," since this statement is essentially a substitute expression to the effect that ignorance of the actual experimental conditions leads one to presume that all possible outcomes are equally likely.

A second definition of the probability of a specific outcome is a highly personalized, subjective definition stating that probability is a measure of one's belief that a given event will transpire. Such a definition creates the possibility that different persons will support differing probabilities to describe their measures of belief in the occurrence of some event. Again, in a certain sense, the subjectivist's definition of probability measures his ignorance of the likelihood of observing the event in question.

A third definition of probability, known as the "classical," "empirical," or "frequency" definition, presumes that the "experiment" in question could conceivably be performed an infinity of times. Denoting by $n(A)$ the number of times which a particular event A (such as the appearance of "one", or ace, on the toss of a die) occurs during the first n trials of the experiment (for example, tosses of the die), then the probability of the event A is the limiting value of the ratio $n(A)/n$; that is

$$P(A) \equiv \lim_{n\to\infty} n(A)/n.$$

Since it becomes physically impossible for one to perform an experiment infinitely often in a finite lifetime, the best approximation to $P(A)$ is the value of the ratio after a large number of trials. However, again, one is not *certain* of the true value of $P(A)$, so that probability still remains a measure of uncertainty and, to a degree, of ignorance or lack of knowledge.

Probability and randomness are closely allied terms; indeed, without randomness one would say either that an experimental outcome is certain or that it will occur with probability one. However, if one tosses a die 20 times and observes an ace after each toss, would he accept the approximation

$$n(A)/n = 20/20 = 1$$

as an expression of the probability of an ace on the next toss? Or, if 10 of the first 20 tosses were aces, would one accept betting odds of 1 : 1 on the appearance of an ace on the next toss?

The probability of an event is usually associated with the randomness of the "experiment," or experimental conditions, producing it. If one were to employ the empirical definition of probability, then a coin would be declared fair if the limiting ratio of the number of heads observed in an infinite recording of H's (heads) and T's (tails) were $\frac{1}{2}$. Yet most reasonable persons would reject the notion that the recorded sequence,

$$\overline{HT}\ \overline{HT}\ \overline{HT}\ \overline{HT}\ \overline{HT}\ldots,$$

consisting of alternating symbols, H and T, would constitute a random sequence; neither would such a person presume that the probability of a head is $\frac{1}{2}$ on the next toss if he had just observed a head on the most recent toss of this sequence.

Therefore, the randomness with which an event occurs must be otherwise defined. Mises (1957) suggested that an infinite sequence of symbols (representing, say, the recordings of the outcomes of an infinite series of coin tossings) be considered as random if any preselected infinite subsequence of the symbols would have the same limiting relative frequency; that is, in the coin-tossing experiment, selection of the infinite sub-sequence consisting of the recorded outcomes on the first, third, fifth, seventh, etc., trials should produce the same proportion of heads (H's) as any other infinite sub-sequence whose elements are selected by a means other than an examination of the symbols themselves.

Thus, though one is unable to test a sequence for its randomness just as one is unable to determine the probability of an event, we know a primary characteristic that should exist in order for randomness to be presumed present. This fact will prove of importance in the determination of the rectitude of the term, "random number generator," to be discussed in the next chapter.

1.5. The Uncertainty Principle of Modeling

Of primary import for the moment, however, is the note that probability indeed serves as a measure of our human ignorance of the actual situation and its implications. Thus, it serves a vital role in the description of the internal relationships of the components of systems whose entire behavior is too complicated to be thoroughly comprehended. Thus, the necessity for models of a probabilistic, or *stochastic*, variety.

Deterministic models may be valid for the representation of system behavior on the gross scale (such as the aforementioned productivity of a work force), but if such models do not readily compare with the observed behavior of the actual system, then a more detailed model may be in order. With the increase in detail comes an increase in the likelihood of ignorance of the prevailing state of affairs and, concomitantly, of a *stochastic* model of the system. This uncertainty arises for a number of reasons.

For example, as one delves more and more profoundly into the structure and nature of a system, he finds that the marginal return (i.e., the added gain) decreases with equal portions of additional effort. Thus, the biologist describes the number of unicellular species in a stream in terms of the number of unit volumes he has sampled and examined; yet, the number of additional species located in each succeeding sample decreases in accordance with the familiar exponential law of diminishing marginal returns. Hence, the lack of time (or equivalently, economic considerations) forbids our thorough comprehension of systems and often requires that we express this uncertainty within a model of the system.

Another cogent reason for encountering uncertainty in the modeling of systems lies in the limited capability of our neural receptors and processors. [See Rosenblueth (1970).] Though our individual contact with the world around us will probably eventually be understood mechanically, our ability to discriminate among the minutiae of systems, and their intricate internal interaction, is bounded by the resolution levels of our neural constitution. As indicated earlier, our use of machines to magnify, amplify, or otherwise transform physical phenomena is a blessing indeed in scientific research, but at any given point in time, even our machines are limited in their capabilities. Even with apparently adequate devices, we never seem to be fully certain that the sensations provided us by machine translations of natural phenomena are indeed credible. We are forced to conclude that uncertainty exists in our understanding of the intricacies of systems.

Since uncertainty is best quantified by probabilities, an *Uncertainty Principle of Modeling* can be formulated:

Refinement in modeling eventuates a requirement for stochasticity.

An immediate corollary to this Uncertainty Principle of Modeling would appear to state that the more conscientiously developed model will be more likely stochastic in character.

The Uncertainty Principle of Modeling can also be interpreted from another viewpoint. Whenever a model of a system has been completely structured, it will be desirable to compare the output, or response, of the model with that arising from the modeled system. Often, measurements from the modeled system itself are scarce or difficult to obtain, so that at most one or two such measurements may be available. If, however, the modeled system is intrinsically stochastic, yet has been modeled deterministically, the ability to "prove," or to validate, the completed deterministic model will be greatly jeopardized. Since the simulation model is readily adaptable for the inclusion of random phenomena (as will be shown in the next chapters), these validations are made facile; thus, the use of the *stochastic simulation* model has surged in popularity and effectiveness. Furthermore, since statistical techniques have been well developed for the analysis of random responses of such models, the difficulties of validation of the structured model need not be of burdensome concern.

1.6. Summary

Modeling is one of the more common activities of man. Indeed, man seems especially competent in the evolutionary hierarchy to construct and defend abstract territories called models. In the broad sense, many of man's religions serve as models for the creation of the universe and as models of social and political behavior. In fact, many of his political systems are models for social structure, some being defined or proffered with the same zeal that accompanies a devoutly accepted religious doctrine.

Development and exploitation of the scientific method has led to an increased awareness of the nature of the universe and of man's role therein. Only recently, however, does it seem to be explicitly recognized that the primary goal of the scientist is the development of meaningful *models* of naturally occurring phenomena.

Once this recognition was explicitly formulated, however, the endeavor to categorize the models that man makes was inevitable. In this chapter, an effort has been made to note several such categorization schemes.

Independently of the intended use of a model, it may be categorized as dynamic or static, depending upon whether or not the model or its properties are observed to change with time. Regardless of whether a model be dynamic or static, a cross categorization of models as replications, formalizations, or simulations has been discussed. A further cross categorization of models pertains to the absence or presence of random phenomena: deterministic and stochastic models, respectively.

Possibly a further categorization of models is in order, for a painting (especially of the genre of the present century) may constitute a symbolic model in the eyes of some; the preceding discussion, however, would have led to a categorization of the painting as a replication (or, quasi replication). Perhaps a subjectivist categorization of models should be promoted, so that the viewpoint of the user, or observer, of the model may be taken into account, much as the subjectivist's definition of probability.

Modeling has also become an essential activity for the less scientific disciplines, particularly and appropriately so whenever one contemplates the versatility and accurate mimicry of dynamic simulation models of the stochastic variety. Consequently, it is this particular, though exceedingly general, type of model which shall be the primary topic of this book: the *dynamic*, *stochastic*, *simulation model*.

In the next two chapters, the techniques for incorporating randomness in a simulation model will be presented. Quite general procedures will be derived for the generation of random numbers and random variables; other, more specific techniques shall be developed then for special-purpose, *ad hoc*, modeling requirements, such as correlated variates and time series. The third subsequent chapter will be concerned with the now classical aspects of the Monte Carlo method, and will be followed by a discussion of the methodology of simulation.

Of course, once a simulation model has been completed, its responses must be analyzed and compared in order that one may make inferences regarding the modeled system and its behavior. Since the responses from a stochastic model will be random variables, techniques appropriate to the orderly specification of model iterations and to the analysis of the resulting model responses must be presented; these are incorporated in the five penultimate chapters.

Finally, in the last chapter, a further discussion of modeling in general will be entertained. Some indication will be made of the likely roles that models, particularly those of the simular variety, will play in the control and understanding of our perceptible environment.

Chapter 2

THE STOCHASTIC VARIATE

If a man will begin with certainties, he will end with doubts ; but if he will be content with doubts, he shall end in certainties.

—FRANCIS BACON

2.1. Introductory Remarks

The systems which one does simulate will, in all likelihood, require the use of probabilistic phenomena, since the components of a particular system seldom behave entirely deterministically. Variability in the behavior of, and the interaction among, the diverse components of the actual system will need to be reflected in the simular system. The methods by which such variability and nondeterministic behavior may be described will be discussed in this and the following three chapters.

The fundamental notion of the next two chapters is the concept of a stochastic variate, or *random variable*. A random variable, which we shall almost always denote by an uppercase, Latin character, such as X, is a real number associated with some observable phenomenon. Yet, if the phenomenon were to be observed repeatedly under essentially uniform and identical conditions, the same numerical result would not necessarily obtain.

One may view each observation of the particular phenomenon as an experiment with nondeterministic outcomes, or as a *nondeterministic* experiment, such as:

(*a*) the result of the toss of a pair of coins;
(*b*) the recording of the uppermost face on a tossed die;
(*c*) the determination of the number of living cells in a unit volume of water;

(*d*) the relative proportion of blondes in a Germanic community; and
(*e*) the measurement of the serviceable lifetime of an electronic component or device.

Thus, the outcome of a nondeterministic experiment need not be a random variable at all; for example, a real number does not normally arise as the result of performing an experiment such as (*a*). In these cases, however, we may usually find some technique for mapping the elementary outcomes into real numbers; to wit, in Experiment (*a*) one may define X as the number of heads observed on the tossing of the pair of coins, so that the nondeterministic experiment results in X's being 2, 1, or 0, depending, respectively, upon both coins showing heads, one head and one tail, or both tails. Throughout this book, it will be presumed that any nondeterministic experiment will (through a mapping, if necessary) produce one of a set of real numbers, though the particular outcome will not be predictable beforehand. The resulting mapping is then termed a random variable.

The primary distinction between a random variable and, say, a mathematical function is that the former is presumed to have associated with it a probabilistic description. This description is readily achieved by means of the *cumulative distribution function* $F_X(a)$ for the random variable X; the function gives the probability that the random variable X assumes a value less than or equal to any specific real number a:

$$F_X(a) \equiv P[X \leq a],$$

where P["A"] denotes the probability of the statement "A" being observed as true once the experiment has been conducted.

Therefore, the cumulative distribution function may be plotted for every real number and will necessarily appear as a nondecreasing function of its argument, the function's being bounded below by 0, above by 1. The cumulative distribution function for the random variable X, equal to the number of heads appearing in the tossing of two coins, is given in Fig. 2.1a. Figure 2.1b is a plot of the cumulative distribution function $F_Y(a)$ for the random variate Y, the uppermost face on a tossed die. From these figures, one can see that the probability of observing any particular numerical value of the random variable is given by the size of the jump at that value. For example, the probability of observing $X = 0$ (both coins showing tails) is given by

$$P[X = 0] = F_X(0 + \delta) - F_X(0 - \delta) = \tfrac{1}{4},$$

where δ is an arbitrarily small, positive quantity. Similarly for $X = 1$,

$$P[X = 1] = F_X(1 + \delta) - F_X(1 - \delta) = \tfrac{3}{4} - \tfrac{1}{4} = \tfrac{1}{2},$$

corresponding to the probability that a pair of tossed coins will result in one head and one tail.

As a parenthetical aside, the reader should not fall into the logical trap, known as D'Alembert's fallacy, of presuming that the three probabilities, $P[X = 0]$, $P[X = 1]$, and $P[X = 2]$ are each equal to $\frac{1}{3}$. The trap is

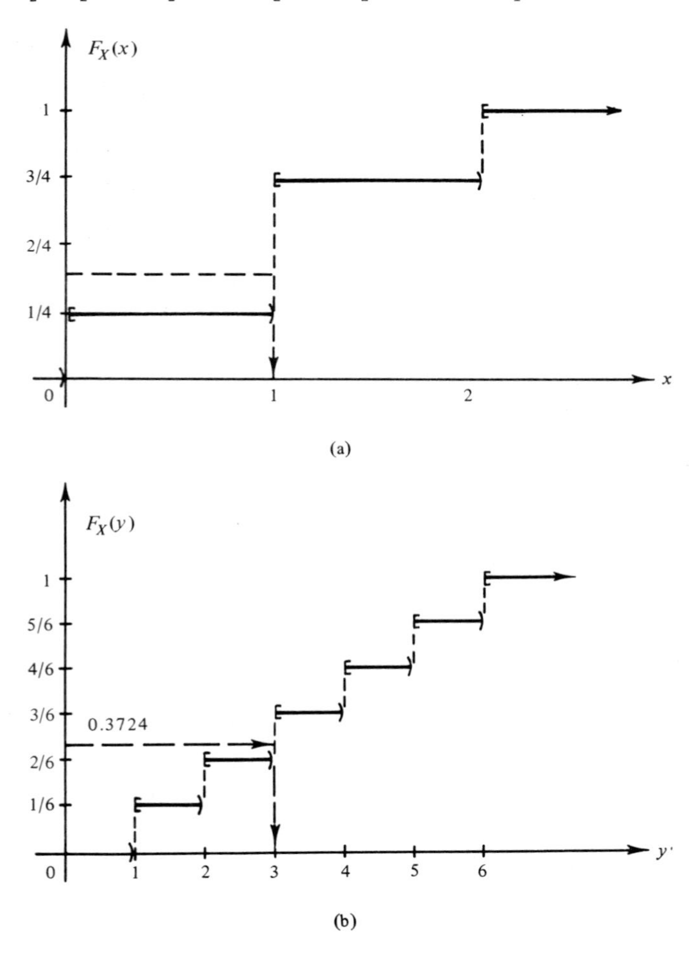

FIG. 2.1. Cumulative distribution functions for coin-tossing (a) and die-tossing (b) experiments.

best avoided by presuming that one is simultaneously tossing a Portuguese escudo and a South African rand, so that $X = 1$ whenever heads appears on either the escudo or the rand, but not on both. That is, two of the four outcomes for the recording (escudo, rand) have one head

$$\begin{array}{cccc} (\mathrm{T}, \mathrm{T}) & (\mathrm{T}, \mathrm{H}) & (\mathrm{H}, \mathrm{T}) & (\mathrm{H}, \mathrm{H}) \\ X = 0, & X = 1, & X = 1, & X = 2. \end{array}$$

The reader will note that the cumulative distribution function for the random variable Y of Experiment (*b*) is plotted so as to imply that the six numbered faces of the (presumably fair) die are equally likely and occur with probability $\frac{1}{6}$ each.

A simulation of either the coin-tossing or the die-tossing game may be readily performed if one has access to a set of random numbers, such as those prepared by the RAND Corporation (1955) or by Clark (1966). An abridged set of random numbers is presented in the Appendix. Such a table of random numbers presumably includes approximately the same proportion of numbers, in a subinterval of length ΔL of the unit interval, as in any other subinterval of the same length. The selection of a random number from this table constitutes a nondeterministic experiment, the number so selected being then a random variable U whose numerical value will lie between 0 and 1. The die-tossing game can then be simulated by the random selection of one of these uniformly distributed variates and by comparing its value with the cumulative distribution function $F_Y(a)$. Since approximately $\frac{1}{6}$ of the random numbers U will lie between 0 and $\frac{1}{6}$, and as many between $\frac{1}{6}$ and $\frac{2}{6}$, one may use the selected number U to give the result Y of the die tossing according to the rule shown in Table 2.1. Thus, if the selected random number were

TABLE 2.1

DIE TOSSING BY COMPUTER

U	Y
$0 \le U < \frac{1}{6}$	1
$\frac{1}{6} \le U < \frac{2}{6}$	2
$\frac{2}{6} \le U < \frac{3}{6}$	3
$\frac{3}{6} \le U < \frac{4}{6}$	4
$\frac{4}{6} \le U < \frac{5}{6}$	5
$\frac{5}{6} \le U < 1$	6

0.3724, the simulated tossing of the die would result in the score of $Y = 3$, as shown in Fig. 2.1b.

In a similar fashion, one can simulate the coin-tossing game. In this case, the selection of $U = 0.3724$ would result in the number of heads being $X = 1$, depicted in Fig. 2.1a. One may note that approximately 50% of all the uniformly distributed random numbers selected from the tables would fall between 0.25000... and 0.74999..., corresponding to the result: one head. The simulation of the coin-tossing game would result in approximately 25% of the trials producing the result $X = 0$ (both tails), and in another 25% of the trials simulating the result $X = 2$ (both heads).

2.2. The Binomial Distribution

In simulating either the coin- or die-tossing game, one may naturally inquire of the relative frequency with which any particular outcome occurs. For example, in the simulation of N coin-tossing games, each simulated tossing corresponds to an independent trial, with the probability $p = \frac{1}{4}$ (of obtaining both heads) remaining the same from trial to trial. There is no reason to believe that the result of any one trial will affect the result of any other trial, assuming that the sequence of random numbers employed are indeed random. Such a sequence of independent trials, each having the same probability p of providing a success S (say, two heads) and the same probability $1 - p$ of resulting in a failure F (i.e., no or one head), is called a sequence of *Bernoulli trials*.

The record of performance of N successive Bernoulli trials could be reported as a sequence of symbols, such as

$$S\,S\,F\,S\,S\,F\,S\,F\,F\,S\,F\,S\,S \cdots F\,S,$$

containing k S's and $N - k$ F's to denote that k of the N trials resulted in the simulation of the coin-tossing game with two heads.

In order to ascertain the probability that the number S_N of successes in N Bernoulli trials is equal to k, one may note that one possible sequence of recorded outcomes leading to exactly k successes is the sequence containing k S's followed by $N - k$ F's:

$$\underbrace{S\,S \cdots S}_{k}\,\underbrace{F\,F \cdots F}_{N-k}.$$

The probability of a particular sequence arising is given by $p^k(1-p)^{N-k}$, since each success occurs with the same probability (p) and each failure with the probability $1 - p$, and since the multiplication rule of probabilities is invokable, owing to the independence of the successive Bernoulli trials.† Now, there are exactly $N!/[k!(N - k)!]$ permutations of N symbols, of which k are S's, $(N - k)$ are F's. Since each of these orderings are equally likely, the addition rule of probability theory may be invoked to give

$$P[S_N = k] = \binom{N}{k} p^k (1 - p)^{N-k} \qquad \text{for any} \quad k = 0, 1, 2, \ldots, N.$$

In particular, if a "success" for the simulation of N coin-tossing games corresponds to the result "two heads," then the probability of obtaining "two heads" exactly k times out of N simulations is

$$P[S_N = k] = \binom{N}{k} \left(\frac{1}{4}\right)^k \left(1 - \frac{1}{4}\right)^{N-k} \qquad \text{for any} \quad k = 0, 1, 2, \ldots, N.$$

Alternatively, if a success is defined as the simulation of the tossing of "an ace" in the die-tossing experiment, the probability of observing exactly k "aces" among the N simular trials is

$$P[S_N = k] = \binom{N}{k} \left(\frac{1}{6}\right)^k \left(\frac{5}{6}\right)^{N-k} \qquad \text{for any} \quad k = 0, 1, 2, \ldots, N.$$

The combinatorial symbol $\binom{N}{k}$ is, in each case, the integer given by

$$\binom{N}{k} = \frac{N!}{[k!(N - k)!]},$$

where $m! = m \cdot (m - 1) \cdot (m - 2) \cdots 2 \cdot 1$ for any integer m.

The variate S_N is also a random variable, called the *binomial random variable* because its probabilities are associated with the binomial expansion $(a + b)^N$, with $a = p$, $b = (1 - p)$. One can indeed tabulate its cumulative distribution function; in fact, Fig. 2.1a represents the cumulative distribution function for a binomial random variable $S_2 = X$, since the tossing of a pair of coins is isomorphic to the tossing of a single coin twice in succession, where, in the latter case, we count the total number of heads (0, 1, or 2) as the number of successes (S_2) observed in two Bernoulli trials ($p = \frac{1}{2}$).

† Two events, A and B, are said to be independent if the probability of their simultaneous occurrence equals the product of their respective probabilities.

Instead of the cumulative distribution function, it is equally usual to plot an associated ***probability distribution function***, which gives the probability that any particular value of a random variable will arise. In the case of the binomial random variable S_N, which may assume any integral value between 0 and N, inclusive, the probability distribution function becomes

$$p_{S_N}(k) = P[S_N = k] = \binom{N}{k} p^k (1-p)^{N-k}, \quad k = 0, 1, \ldots, N. \quad (2.2{:}1)$$

For the random variable, $S_2 = X$, the number of times that two heads are observed to appear in the tossing of a pair of coins, this probability distribution function becomes

$$p_X(k) = \begin{cases} \frac{1}{4}, & k = 0, \\ \frac{1}{2}, & k = 1, \\ \frac{1}{4}, & k = 2. \end{cases}$$

The probability distribution function for the result Y of the tossing of a fair die becomes then

$$p_Y(k) = \tfrac{1}{6}, \qquad k = 1, 2, 3, 4, 5, \quad \text{and} \quad 6.$$

The probability distribution function $p_X(k)$ for an integer-valued random variable X is a real-valued function that satisfies two conditions:

(a) $0 \le p_X(k) \le 1$ for every integer k (positive, negative, or zero),

and

(b) $\sum_{k=-\infty}^{\infty} p_X(k) = 1.$

The reader may readily verify that both conditions obtain for the probability distribution functions for the exemplary random variables, Y and $X = S_2$, as discussed earlier. Indeed, the probability distribution function for the binomial random variable S_N, satisfies both conditions, as $0 \le p_{S_N}(k) \le 1$ for every integer k and

$$\sum_{k=-\infty}^{\infty} p_{S_N}(k) = \sum_{k=0}^{N} p_{S_N}(k) = \sum_{k=0}^{N} \binom{N}{k} p^k (1-p)^{N-k}$$

or

$$\sum_{k=-\infty}^{\infty} p_{S_N}(k) = [p + (1-p)]^N = 1^N = 1,$$

since the binomial expression for $(a+b)^N$ is

$$(a+b)^N = \sum_{k=0}^{N} \binom{N}{k} a^k b^{N-k}.$$

2.3. The Geometric Distribution

Not all integer-valued random variables can assume only a finite number of values. For example, Experiment (c), mentioned on page 18 should result in the observation of an *integral* number of cells in the unit volume, but it may be that the experimenter would not wish to declare an upper bound on the number observable. As a second example, consider the waiting time T until the first success (S) is noted in a sequence of Bernoulli trials. If the first trial ends in success, then the random variable T is set equal to 1; otherwise, a second trial is performed and, if this trial ends in success, T is set equal to 2; otherwise, a third trial is performed, etc. Conceptually, there is no upper bound on the value that the random variable T can assume. The probabilities of its assuming the first few integral values are

$$P[T=1]=p \qquad (S),$$

$$P[T=2]=(1-p)\cdot p \qquad (FS),$$

$$P[T=3]=(1-p)\cdot(1-p)\cdot p=(1-p)^2\cdot p \qquad (FFS),$$

$$P[T=4]=(1-p)\cdot(1-p)\cdot(1-p)\cdot p=(1-p)^3\cdot p \qquad (FFFS).$$

In general, the probability distribution function for the random variable T is

$$P[T=k]=p_T(k)=\begin{cases} p(1-p)^{k-1} & \text{for any} \quad k=1,2,\ldots, \\ 0, & k<1, \end{cases} \tag{2.3:1}$$

where the unsubscripted p denotes the probability of success on any given Bernoulli trial. This distribution is referred to as the *geometric distribution*, since the infinite sum of its terms is the geometric series

$$\sum_{k=1}^{\infty} p\cdot(1-p)^{k-1}=p/[1-(1-p)]=1.$$

As a particular instance, the cumulative distribution function for the waiting time T until the first ace appears in a sequence of tosses of a fair die is presented in Fig. 2.2a. Figure 2.2b presents the cumulative distribution function for S_7, equal here to the number of aces observed in seven tosses of a fair die. In both distributions, then, the underlying parameter p is $\frac{1}{6}$.

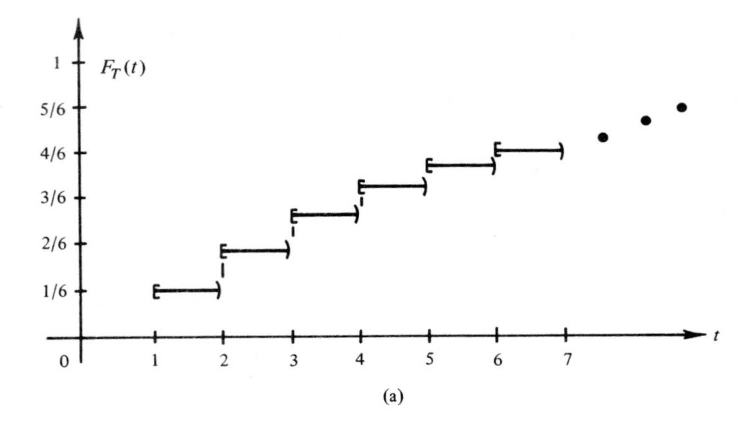

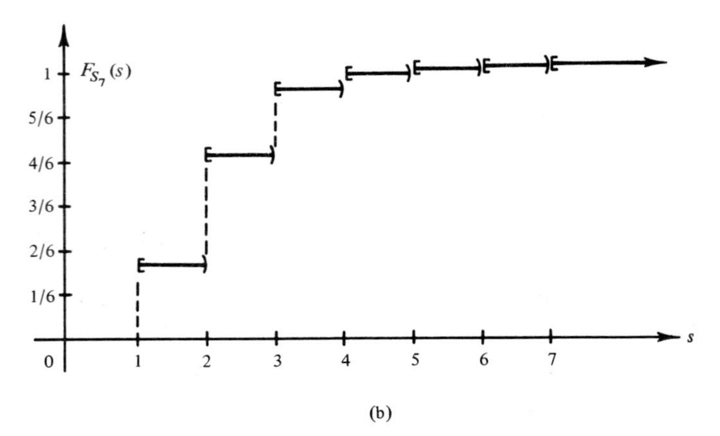

FIG. 2.2. Cumulative distribution functions for geometric (T) and for binomial (S_7) random variables.

One might attempt to simulate the waiting time until the first ace appears on a fair die in either of two ways. In the first, successive random numbers are inspected until the first U of value $\frac{1}{6}$ or less is located. At this point, the number of random numbers so inspected can be recorded as the value of T. Alternatively, one might as well proceed directly to the use of the known cumulative function $F_T(x)$ for the random variable T; i.e., a single, uniformly distributed random number U is located and is employed directly with the sketch of $F_T(x)$, as given in Fig. 2.2a. [If the particular random number U were selected as 0.3724, the reader will note that from $F_T(x)$ the result $T = 3$.]

2.4. Parameters

The reader should note that the binomial and geometric distributions are quite general and require the specification of certain *parameters* before their distribution functions can be explicitly tabulated or plotted. The binomial distribution is indexed by the two parameters: N, the number of Bernoulli trials, and p, the probability of success on any given trial; whereas, the geometric distribution is indexed only by the parameter p. Tables of these distributions are provided for various values of N, p, or both by Eisenhart (1950) and by Williamson and Bretherton (1963), respectively. The discussion will now turn to random variables whose permissible values need not lie in a countable set, such as the set of all positive integers. The reader should note, nonetheless, that the distribution functions for these continuous random variables are also indexed by specific parameters.

EXERCISES

1. Using published tables of uniformly distributed random variates, simulate 100 times the waiting time T until the first occurrence of a pair of heads in the tossing of two fair coins. Record the 100 T-values so computed and note the relative frequency with which $T = 1$, $T = 2$, $T = 3$, etc. Compute aside the theoretical probability distribution function for T and compare it with the relative frequency of simulated T-values.

2. Again using published tables of uniformly distributed random variates, locate a single variate U at random. By comparing the randomly selected variate with the theoretical cumulative distribution function $F_T(x)$ as computed in Exercise 1, ascertain directly a waiting time T. Repeat this procedure 100 times, recording the resultant 100 T-values so as to observe the relative frequency with which $T = 1$, $T = 2$, $T = 3$, etc. Compare the simular results of this exercise with those of the preceeding exercise.

2.5. Continuous Random Variables

In Section 2.1 of this chapter a number of experimental situations that could produce random variables were exemplified. It was noted that a random variable X is associated with a function $F_X(x)$ that gives the probability that the result of the nondeterministic experiment will not exceed the argument x. The random variables thus far discussed have been able to assume either a finite, or at most a countably infinite, number of values. For example, the binomial random variable S_N, which gives the number of observed successes in N Bernoulli trials, may assume any integral value between 0 and N; whereas, the geometric random variable T, which gives the number of Bernoulli trials required until the first success is observed, may assume any positive integral value.

Many phenomena, however, need not be observable in terms of a discrete set of values. For example, the lifetime X of an electronic component would constitute a random variable whose value might be any positive real number: integral, rational, or otherwise. The cumulative distribution function

$$F_X(a) \equiv P[X \leq a]$$

represents the probability that the component will not survive at least a units of time. Since the time of the demise of the component need not be restricted to, say, integral units of time, the cumulative distribution function should appear as a continuous curve, monotonically increasing in terms of its argument, bounded above by unity, and bounded below by zero. Figure 2.3 displays typical cumulative distribution functions for three exemplary *continuous* random variables.

Under a few elementary assumptions, the nature of the cumulative distribution function for the lifetime X of an electronic component may be readily derived. Dividing the initial time interval of length t into n subintervals, each of length Δt such that $t = n\,\Delta t$, one may conceive of the n subintervals as a set of n Bernoulli trials; a success on the kth trial would correspond to the demise of the component in the kth subinterval. If the probability of the first success in any one of the subintervals is presumed constant and given by $p = \lambda(\Delta t)$ for some positive constant λ, then the probability that the component will still be functioning t time units hence is given by the probability that in none of the n subintervals a success (demise) will have occurred. However, this latter probability is that of a Bernoulli variate's being zero, so that

$$P[X > t] = \binom{n}{0} p^0(1 - p)^n = (1 - \lambda\,\Delta t)^n,$$

or

$$P[X > t] = [1 - (\lambda t)/n]^n.$$

Hence, for n sufficiently large, the probability of survival beyond t time

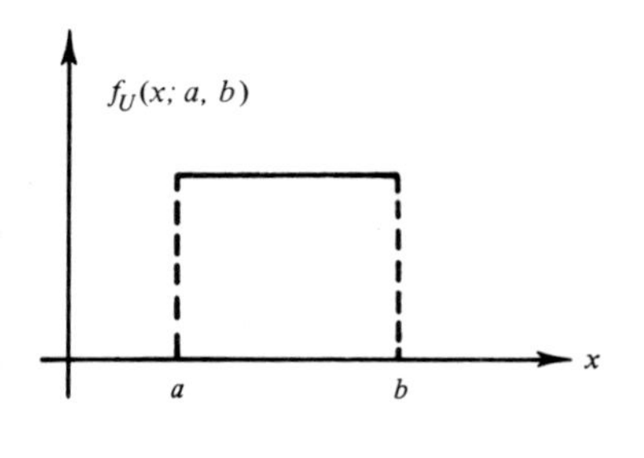

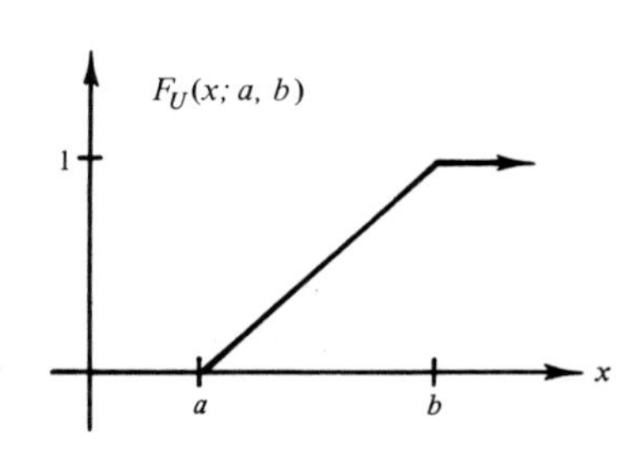

(a)

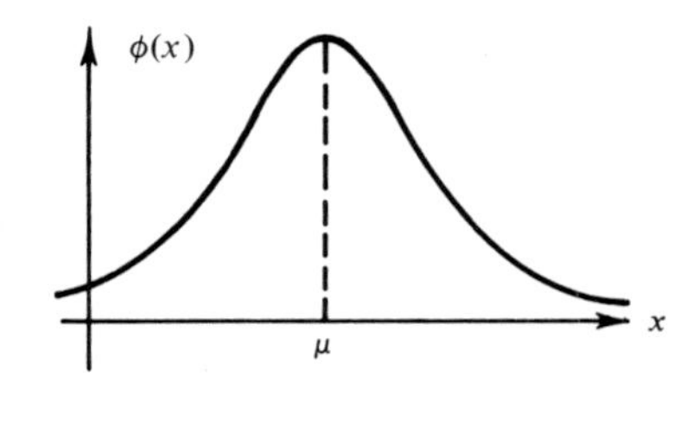

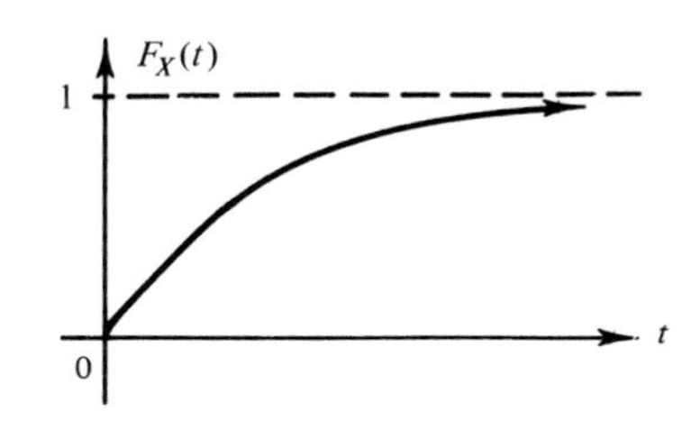

(b)

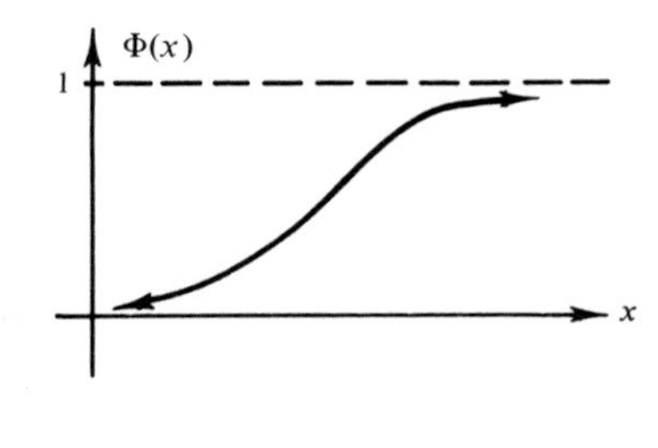

(c)

FIG. 2.3. Probability density and cumulative distribution functions. (a) Rectangular; (b) exponential; (c) Gaussian.

units will be given approximately by

$$P[X > t] \cong e^{-\lambda t};$$

the cumulative distribution function for the random variable X then becomes, when evaluated at t,

$$F_X(t) \equiv P[X \leq t] = 1 - P[X > t] = 1 - e^{-\lambda t} \qquad \text{for} \quad t > 0,$$

as depicted in Fig. 2.3b. Owing to the form of this function, the random variable X, as derived under the above-stated assumptions, is termed an *exponentially distributed random variable with parameter* λ.

Actually, the cumulative distribution function is a family of functions, any member of which is indexed by the assignment of a particular value of the positive parameter λ. More properly, the cumulative distribution function should be denoted

$$F_X(x; \lambda) \equiv P[X \leq x] = \begin{cases} 1 - e^{-\lambda x}, & x > 0, \\ 0, & x < 0 \end{cases}$$

so as to reflect the importance of λ and to emphasize the fact that no accumulation of probability will occur for nonpositive values of the argument x.

Also of particular import is the derivative of a continuous cumulative distribution function, when it exists. The differential quantity

$$F_X(x + \Delta x) - F_X(x)$$

represents the probability that the continuous random variable will be observed in the incremental region between x and $x + \Delta x$. Thus, the differential element $F_X'(x)\, dx$ may be taken to represent approximately this probability. By accepted convention, the function

$$f_X(x) \equiv F_X'(x),$$

denoted by the lowercase letter f corresponding to the uppercase letter F, which is employed to denote the cumulative distribution function, is termed the *probability density function* for the random variable X. Being the derivative of the cumulative distribution function, the probability density function expresses the rate at which $F_X(x)$ increases. Therefore, it is always nonnegative, since $F_X(x)$ is monotonically increasing. Furthermore, antiderivatives (or integrals) of the probability density function provide probability statements regarding the random variable; to wit,

$$P[a < X \leq b] = \int_a^b f_X(t)\, dt = F_X(b) - F_X(a),$$

for any real numbers $a < b$.

One may note the degenerate probability statement

$$P[X = a] = \int_a^a f_X(t)\, dt = F_X(a) - F_X(a) = 0,$$

which is properly interpreted as stating that the probability of observing any particular value of a continuous random variable is zero. This result is in clear contradistinction to corresponding statements regarding discrete random variables, for in the discrete case there do exist values having positive (i.e., nonzero) probabilities of occurrence.

A second note is in order. The probability density function, when plotted in terms of its real argument, is a curve $f_X(x)$ such that:

(*a*) $f_X(x) \geq 0$ for all real x;

and

(*b*) $\int_{-\infty}^{\infty} f_X(x)\,dx = F_X(+\infty) - F_X(-\infty) = 1 - 0 = 1.$

Indeed, if one were to record the results of a large number of repetitions of the underlying, nondeterministic experiment and form these into a frequency histogram, then the histogram should somewhat approximate the curve $f_X(x)$.

If the cumulative distribution function is indexed by one or more parameters, then its derivative, the probability density function, will be similarly indexed and therefore be a *family* of density functions. For example, the exponential family of densities becomes

$$f_X(x;\lambda) = \begin{cases} \lambda e^{-\lambda x}, & x > 0, \\ 0, & x \leq 0, \end{cases}$$

for positive parameter λ.

In addition to the exponential family, several other continuous random variables exist. Particularly noteworthy is the family of rectangularly distributed random variables U having cumulative distribution function

$$F_U(x;a,b) = \begin{cases} 0, & x \leq a, \\ (x-a)/(b-a), & a < x \leq b, \\ 1, & x > b, \end{cases}$$

for particular constants (parameters) $a < b$. Its derivative, the probability density function,

$$f_U(x;a,b) = \begin{cases} 1/(b-a), & a \leq x \leq b, \\ 0, & \text{other} \quad x, \end{cases}$$

is depicted as the rectangular form in Fig. 2.3a. An important special case of this family arises in correspondence to the parametric specifications $a = 0$, $b = 1$, in which case the random variable is said to be *uniformly distributed*:

$$F_U(x; 0, 1) = \begin{cases} 0, & x \le 0, \\ x, & 0 < x \le 1, \\ 1, & x > 1, \end{cases}$$

and

$$f_U(x; 0, 1) = \begin{cases} 1, & 0 \le x \le 1, \\ 0, & \text{other} \quad x. \end{cases}$$

Many authors refer, however, to the rectangularly distributed random variable as the uniform random variable. In any event, the random variable is deemed appropriate to the results of any nondeterministic experiment whose outcomes can be expected to fall "randomly," or uniformly, over some range $b - a$ of finite length. Therefore, the distributional form $F_U(x; a, b)$ is often used when one's knowledge of the distributional behavior of experimental results cannot be based on presumptions other than that the result need be bounded above (b) and below (a).

Other distributional forms will be applicable whenever experimental conditions permit one to prejudge the likely behavior of the measured results, as, for example, were discussed for the exponential distribution. The applicable assumptions leading to other such probability density functions will be presented when appropriate.

EXERCISES

1. From a table of published random numbers, select a sequence of 200 four-digit entries, placing a decimal point before each, thus producing 200 random variables—each between 0 and 1. Build a frequency histogram of 20 cells of equal width, and comment on its comparatively rectangular form over the interval between 0 and 1.

2. Recall the exercises, at the end of Section 2.4 (page 27), regarding the geometric distribution. Compare the frequency distribution of the waiting times T of that exercise with the family of exponential probability density functions, noting that the latter is a continuous approximation to the discrete geometric distribution.

2.6. Transformations of Random Variables

Frequently, one may not wish to deal directly with the measured random variable that arises from a nondeterministic experiment, but may instead transform the response to some other unit of measure. For example, in dealing with a nondeterministic experiment whose response, or random variable, should properly be exponentially distributed, one may wish instead to compute its square root, natural logarithm, or inverse. Clearly, the manner in which these transformed variates are distributed should not all be the same as the original random variable; for, if so, the assumptions leading to the theoretical distributional form would likely imply that one might admit all sorts of transformations on the experiment itself. A need to establish the distributional form for general transformations of random variables then exists.

2.6.1. Bertrand's Paradox

Suppose that, at a French auberge, the innkeeper is suspected of mixing water with the wine, but that the customers do not know the ratio U of water to wine. Most residents of the inn know, or are willing to presume, that the innkeeper is not a total scoundrel and that one could safely assume U to be between $\frac{1}{2}$ and 1; i.e., their carafes would contain no less than 50% wine, yet no more than two-thirds wine. Furthermore, not knowing the degree to which their wine is diluted before serving, a customer might be willing to let ignorance be his guide and to presume therefore that U is a rectangularly distributed random variable with cumulative distribution function

$$F_U(x) = \begin{cases} 0, & x \le \frac{1}{2}, \\ 2(x - \frac{1}{2}), & \frac{1}{2} < x \le 1, \\ 1, & x > 1, \end{cases}$$

so that, for example, the probability that the ratio of water to wine exceeds $\frac{4}{5}$ would be given by

$$P(U > \tfrac{4}{5}) = 1 - F_U(0.8) = 0.4.$$

A second customer, upon arriving at the auberge and hearing of the reputation of the innkeeper, slightly alters his approach to this probabilistic determination. By presuming that a discussion of the ratio V of wine to water is essentially the same as one pertaining to the ratio U of water to wine, he notes that V must be between 1 and 2; i.e., in accordance with the preceding discussion, he presumes that the carafes contain a mixture of at least one, but no more than two, parts of wine per part of water. Furthermore, he presumes no greater knowledge of the likely value of the actual ratio, and assumes that the random variable V is rectangularly distributed between 1 and 2; i.e., the cumulative distribution function for V becomes

$$F_V(x) = \begin{cases} 0, & x \leq 1, \\ (x - 1), & 1 < x \leq 2, \\ 1, & x > 2. \end{cases}$$

From this cumulative distribution function, the probability that the ratio of wine to water is less than $\frac{5}{4}$ is

$$P(V < \tfrac{5}{4}) = F_V(1.25) = 0.25.$$

However, since the two statements, $V < \frac{5}{4}$ and $U > \frac{4}{5}$, connote identical situations with respect to the water–wine ratios, one would expect their probabilities to be equal; i.e., one would *not* expect that

$$P(V < \tfrac{5}{4}) = 0.25 \neq 0.40 = P(U > \tfrac{4}{5}),$$

since $V = 1/U$. Hence, the paradox.

Historically, the paradox was developed to emphasize the futility of presupposing equally likely, or uniformly distributed, values for random variables. The use of ignorance of the experimental situation as a basis for predicting the distribution of random variables arising therefrom could be seen to lead to incompatible results. In another sense, Bertrand's paradox is especially valuable, since it displays the need for care in describing the cumulative distribution function for $V = 1/U$, given that U is rectangularly distributed between $\frac{1}{2}$ and 1.

2.6.2. Equivalent Events

What is sought in transforming one random variable to another is a mapping that will permit complete compatibility in probability statements that should represent the same event, regardless of whether the event is expressed in terms of the new or the original random variable. Now a probability statement concerning a random variable X is a statement of the probability that X will assume a value in some subset $\mathscr{A}_X$ of real numbers. If a new random variable,

$$Y = t(X),$$

is defined according to some transformation t of X, then the event equivalent to $\mathscr{A}_X$ is a subset $\mathscr{B}_Y$ of real numbers such that

$$\mathscr{B}_Y = \{y \mid y = t(x) \quad \text{for} \quad x \in \mathscr{A}_X\}.$$

The events, or subsets, $\mathscr{B}_Y$ and $\mathscr{A}_X$ are then said to be *equivalent events* and would be expected to occur with equal probability.

In particular, the determination of the cumulative distribution function $F_Y(a)$ for the random variable $Y = t(X)$ corresponds to the determination of a subset $\mathscr{A}$ of real numbers such that

$$\mathscr{A} = \{x \mid y = t(x) \leq a\},$$

and therefore depends upon the nature of the transformation t itself. Of particular ease, however, is the determination of the equivalent event $\mathscr{A}$ whenever $Y = t(X)$ is a monotone function.

If $Y = t(X)$ is a monotone increasing function, as exemplified in Fig. 2.4a, then there exists a monotonically increasing inverse function, t^{-1}, such that, if $y = t(x)$, then $x = t^{-1}(y)$, and conversely. Consequently,

$$F_Y(a) \equiv P[Y \leq a] = P[X \leq t^{-1}(a)] \equiv F_X[t^{-1}(a)],$$

that is, the cumulative distribution function for the new random variable Y may be found from the cumulative distribution function for the random variable X by means of

$$F_Y(a) = F_X[t^{-1}(a)] \qquad \text{for any real number} \quad a,$$

whenever $Y = t(X)$ is a monotone, continuously increasing function. The probability density functions become relatable via the expression

$$f_Y(a) = f_X[t^{-1}(a)] \cdot \{dt^{-1}(a)/da\}, \tag{2.6.2:1}$$

obtained from chain-rule differentiation of the right-hand side of the preceding expression.

Similarly, if $Y = t(X)$ is a monotone, continuously decreasing transformation, as depicted in Fig. 2.4b, then there exists a monotonically and continuously decreasing inverse function t^{-1}, such that $y = t(x)$ if and only if $x = t^{-1}(y)$. Furthermore,

$$F_Y(a) \equiv P[Y \leq a] = P[X \geq t^{-1}(a)],$$

or

$$F_Y(a) = 1 - F_X[t^{-1}(a)] \qquad \text{for any real number} \quad a.$$

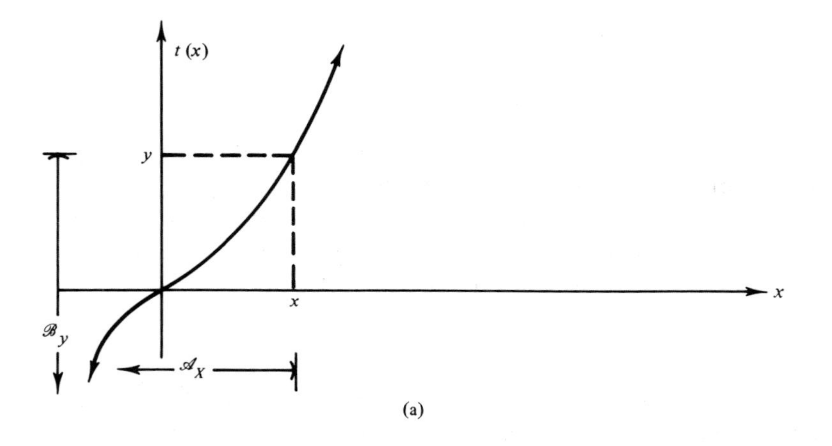

(a)

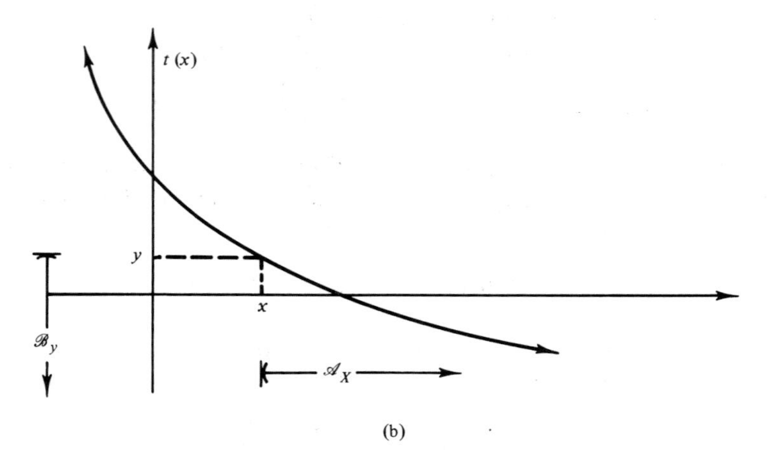

(b)

FIG. 2.4. Equivalent events for transformed random variables. (a) Monotone increasing; (b) monotone decreasing.

In this event, the corresponding probability density functions are related according to

$$f_Y(a) = -f_X[t^{-1}(a)]\{dt^{-1}(a)/da\}; \tag{2.6.2:2}$$

the minus sign is of no concern, since the derivative of $t^{-1}(a)$ will be negative in this case.

In the more general case, one must take care to define explicitly equivalent events, in order to ascertain the cumulative distribution function for the new random variable $Y = t(X)$. Of course, once this function $F_Y(a)$ is specified, probability statements regarding general subsets or events, $\mathscr{B}_Y$, may be made directly therefrom or via integrals of the corresponding probability density function $f_Y(a)$. Throughout this book, however, transformations shall be primarily of the less complicated, monotone type.

2.6.3. The Pareto Family of Distributions

If an exponentially distributed random variable X is transformed by the simple exponential function

$$Y = e^X,$$

then the resulting random variable Y has Pareto cumulative distribution function:

$$F_Y(y; \lambda) = \begin{cases} 1 - y^{-\lambda}, & y \geq 1, \\ 0, & y < 1, \end{cases}$$

where λ is the positive parameter indexing the original exponential distribution. The exponential transformation is a monotonically and continuously increasing function, so that the probability density function for the resulting Pareto variate becomes, in accordance with Eq. (2.6.2:1),

$$f_Y(y; \lambda) = \begin{cases} \lambda/y^{\lambda+1}, & y \geq 1, \\ 0, & y < 1. \end{cases}$$

The Pareto distribution has been used to describe the distribution of incomes in certain populations. It is also frequently cited by engineers because its right-hand extremes decay according to y^{λ}, rather than exponentially, thereby implying that the likelihood of extremely large measurements may not be so readily discounted as in the case of, say, exponentially distributed random variables.

2.6.4. The Rayleigh Family of Distributions

If an exponentially distributed random variable X is transformed according to

$$R = X^{1/2},$$

the resulting family of distributions for R is known as the Rayleigh family. Since $R = X^{1/2}$ is a monotonically and continuously increasing function for $X = 0$, the cumulative distribution function for R becomes

$$F_R(a) = \begin{cases} 1 - \exp(-\lambda a^2), & a > 0, \\ 0, & a \le 0, \end{cases}$$

and the corresponding probability density function becomes

$$f_R(a) = \begin{cases} 2\lambda a \exp(-\lambda a^2), & a > 0, \\ 0, & a \le 0. \end{cases}$$

The importance of the Rayleigh distribution arises in connection with random variables that are planar radial measurements, such as errors from target centers or distances resulting from movements of particles from an original center. The exact conditions under which the Rayleigh distribution may be anticipated shall need attend the discussion of the bivariate normal distributions. Indeed its importance to simulation lies in generating normal random variables.

2.6.5. The Weibull Family of Distributions

Another family of probability distribution functions arises from transformations of exponentially distributed random variables X; for any constant $b > 0$ the transformation

$$W = X^{1/b},$$

a monotone increasing transformation, results in a Weibull random variable whose cumulative distribution function is

$$F_W(w; \lambda, b) = \begin{cases} 1 - \exp(-\lambda w^b), & w > 0, \\ 0, & w \le 0, \end{cases}$$

with corresponding probability density function

$$f_W(w; \lambda, b) = \begin{cases} b\lambda w^{b-1} \exp(-\lambda w^b), & w > 0 \\ 0, & w \le 0. \end{cases}$$

The Weibull distribution is especially applicable whenever the experimental process at hand is yielding a random variable that is essentially the maximum (or minimum) value among a large set of random variables. Thus, if a system, such as an electronic computer, is deemed operational until such time as any one of its components has failed, then the elapsed time S between repairs will likely be a random variable of the Weibull family, since S is essentially defined as the minimum of its components' lifetimes. The applicability of the Weibull distribution to the assessment of systemic reliability is thereby assured.

Of particular note is the fact that the Weibull family is indexed by two positive parameters, λ and b; moreover, whenever $b = 1$, one sees that the exponential family of parameter λ becomes, in reality, a subfamily of the Weibull family of distributions.

EXERCISES

1. Show that, if $Y = t(X)$ is a monotone transformation of the random variable X, then the probability density function of Y is given in terms of that of X by

$$f_Y(y) = f_X[t^{-1}(y)] \cdot | dt^{-1}(y)/dy |$$

regardless of whether t is monotone increasing or is monotone decreasing.

2. Resolve Bertrand's paradox by showing that, if U has rectangular distribution over the interval between $\frac{1}{2}$ and 1, then $V = 1/U$ must have probability density function

$$f_V(a) = \begin{cases} 2/a^2, & 1 < a < 2 \\ 0, & \text{other } a. \end{cases}$$

Plot this function and compare with a rectangular distribution over the interval between 1 and 2.

3. From a published table of random numbers, select a sequence of 100 four-digit random numbers, all between 0.5000 and 0.9999, either by (a) adding 0.5 to each number found to be less than 0.5000 therein; or (b) discarding all numbers found to be less than 0.5000 until 100 numbers have been obtained. From these 100 numbers, build a frequency histogram of 10 cells of equal width over the interval between $\frac{1}{2}$ and 1. Compare it with the presumably rectangular distribution for these variates. Then, for each variate, compute its inverse; for the resulting 100 inverses, construct a frequency histogram of 10 cells of equal width over the interval between 1 and 2. Comment on Bertrand's paradox.

4. Show that the cumulative distribution function for

$$Y = -\ln U,$$

where U is a random variable uniformly distributed over the interval between 0 and 1, is a member of the family of exponential distributions. In this regard, show that the event equivalent to $\mathscr{B}_Y = \{y \mid 0 \le y \le a\}$ is $\mathscr{A}_U = \{u \mid e^{-a} \le u \le 1\}$, for any positive a.

5. Show that the cumulative distribution function for

$$Z = -\ln(1 - U),$$

where U is uniformly distributed between 0 and 1, is the same member of the exponential family of distributions as the density of Y in Exercise 4.

6. Verify the derivation of the Weibull distribution by noting that the event equivalent to $W \le w$ is, in terms of the exponential variate: $X \le w^b$.

7. Show that, for U the uniformly distributed random variable, either

$$W_1 \equiv [-\ln(1 - U)]^{1/b},$$

or

$$W_2 \equiv [-\ln(U)]^{1/b},$$

has the Weibull distribution of parameters $\lambda = 1$ and $b > 0$.

2.7. The CDF Transformation

For the purpose of introducing random variables into a simulation model, one of the most important transformations of random variables is that which transforms a random variable X according to its own cumulative distribution function, $F_X(a)$. That is, for a continuous random variable X, one can discuss the nature of the random variable U defined by

$$U \equiv F_X(X),$$

a monotonically and continuously increasing function due to the nature of the cumulative distribution function. The resulting variate U is restricted to values between 0 and 1, although its distribution between those values may not be so apparent until one computes the cumulative distribution function for U:

$$F_U(a) = P[U \le a] = P[X \le F_X^{-1}(a)], \qquad \text{for} \quad 0 < a < 1,$$

where $F_X^{-1}(a)$ is the inverse cumulative distribution function for the random variable X. That is, $F_X^{-1}(a)$ gives that value x which, when substituted into F_X, produces a. This means then that

$$F_U(a) = P[X \leq F_X^{-1}(a)] = F_X[F_X^{-1}(a)] = a,$$

for any arbitrary argument a between 0 and 1.

Consequently, the random variable, $U = F_X(X)$, *defined as the transformation of* any *arbitrary continuous random variable* X *according to its cumulative distribution function, is a uniformly distributed random variable.*

This result is of special importance in that one can use the inverse of the CDF transformation, itself a monotone increasing function, in order to generate random variables having a particular cumulative distribution function. To accomplish this feat, a source of uniformly distributed random variables U is presupposed; their transformation according to

$$X \equiv F_X^{-1}(U)$$

will result in random variables whose cumulative distribution function is given by $F_X(x)$.

Therefore, in order to generate random variables according to a specific continuous cumulative distribution function $F_X(x)$ one needs only:

(*a*) a method for generating uniformly distributed random variables; and,

(*b*) knowledge of the inverse function $F_X^{-1}(u)$.

Even in those cases for which the inverse of the cumulative distribution function may not be expressible in a convenient analytic form, tabulations of the cumulative distribution function may be employed, much in the manner described earlier for the generation of discrete random variables (see Fig. 2.1). The reader will note then that the use of the CDF transformation, as a method for generating continuous random variables, is a straightforward extension of the procedure employed in the generation of discrete random variables.

Of the essence in either case is the availability of a source of uniformly distributed random variables, such as those published by the RAND Corporation (1955). However, since the use of tabulations of random variables creates slow table searches by the computer (upon which the stochastic simulation model actually depends), the discussion will now turn to methods by which uniformly distributed random variables may be generated mechanically.

EXERCISES

1. Show that if U is a random variable uniformly distributed over the interval between 0 and 1, the random variable $F_X^{-1}(U)$, where F_X^{-1} is the inverse function corresponding to a cumulative distribution function F_X, has as its cumulative distribution function F_X. Note that

$$P[F_X^{-1}(U) \leq a] = P[U \leq F_X(a)]$$

due to the equivalence of these two events.

2. Show that an *exponentially distributed* random variable (of parameter λ) results from the transformation

$$X = (-1/\lambda) \ln(1 - U).$$

3. Show that an alternative method for the generation of an *exponentially distributed* random variable X from an uniformly distributed variate U is via the transformation

$$X = (-1/\lambda) \ln U,$$

by noting that if U is uniformly distributed between 0 and 1, then so is $(1 - U)$.

4. Show that a random variable R, which is *rectangularly distributed* between a and b, may be generated according to the transformation

$$R = a + (b - a)U = a(1 - U) + bU,$$

where U is uniformly distributed between 0 and 1. Alternatively, by noting the equivalence of the distributions of U and $(1 - U)$, show that one can compute instead

$$R = a + (b - a)(1 - U) = aU + b(1 - U)$$

as the random variable rectangularly distributed between a and b.

5. Show that a *Pareto random variable* Y may be obtained from a uniformly distributed random variable U by means of the transformation

$$Y = (1 - U)^{-1/\lambda}$$

or equivalently via

$$Y = U^{-1/\lambda}.$$

6. Show that a *Rayleigh distributed* random variable arises from the transformation

$$R = [(-1/\lambda)\ln(1 - U)]^{1/2}$$

of a uniformly distributed random variable U, noting that R is thence the positive square root of an exponentially distributed random variable. Note also the alternative transformation

$$R = [(-1/\lambda)\ln U]^{1/2}.$$

7. Show that the *Weibull family* of random variables can be generated by alternative specifications of the parameters b and λ in the transformation

$$W = [(-1/\lambda)\ln U]^{1/b}$$

of the uniformly distributed random variable U. Note that, with $b = 2$, the Rayleigh subfamily is generated; whereas, a specification of $b = 1$ produces exponentially distributed random variables.

8. Generate 100 random variables according to

$$S = [-\ln U]^{1/b}$$

by selecting a sequence of 100 uniformly distributed variates U (between 0 and 1) from a published list of random numbers and by specifying: (*a*) $b = \frac{1}{2}$; (*b*) $b = 1$; (*c*) $b = 2$. In each case, prepare a frequency histogram of 10 cells for the 100 variates and note the apparent behavior of this histogram at the ordinate axis.

9. Recalling that the Pareto family of random variables (Y) can be generated from the exponential family of random variates (X) by means of the exponential function,

$$Y = e^X,$$

show that the inverse function,

$$X = \ln Y,$$

can be employed to generate an exponentially distributed random variable from a given Pareto variate (Y).

10. *The Cauchy Distribution.* A random variable V is said to have the Cauchy distribution of parameters a and b if its probability density function is given by

$$f_V(x; a, b) = \frac{b}{\pi[b^2 + (x - a)^2]} \qquad \text{for any real} \quad x.$$

The corresponding cumulative distribution function is

$$F_V(x; a, b) = \pi^{-1}\{(\pi/2) + \operatorname{Arctan}[(x - a)/b]\}.$$

Show then that a Cauchy variate V may be generated from a uniformly distributed random variable U according to

$$V = a + b \cdot \tan[\pi(U - \tfrac{1}{2})],$$

and that the median value of V, given as the solution for x of the equation

$$\tfrac{1}{2} = F_V(x; a, b),$$

is $x \equiv V(\text{median}) = a$.

2.8. Generation of Uniformly Distributed Random Variables

In principle, it is possible to generate random variables from any particular probability distribution provided that either:

(*a*) an adequate tabulation of the cumulative distribution function exists; or,

(*b*) an analytic expression for the inverse of the cumulative distribution function is available.

In the first condition, linear or polynomial interpolations may be required, if the random variate is of the continuous type, in order to avoid generation of random variates from an essentially discrete distribution. Under either contingency, it is presumed that a sequence of uniformly distributed random variables is available.

Recalling the discussion in Chapter 1 of randomness, one observes that the term is defined in terms of the properties of infinite strings of recorded symbols. Thus, if a conceptually infinite string of, say, four-digit, uniformly distributed random variables is to constitute a random string, almost every infinite substring (such as every other, or every third, element of the original string) must possess the same statistical properties as the entire original infinite string. For example, if 10% of the random variates in the original infinite string begin with the digit 9, then a like proportion of the variates in any of the infinite substrings must also begin with 9. More generally, if the infinite string is uniformly distributed, then so must be any qualified infinite substring.

Since randomness is defined in terms of infinite sets, it becomes impossible to ascertain in a practical sense whether any finite string of numbers is truly random. As a working rule, however, one would denominate as random any set of numbers for which there is no reason to believe that the process generating the numbers would not be, if extended infinitely, a random sequence. In addition, one would require that the observed finite record of numbers generated by the process satisfy the particular distributional properties anticipated of the generator.

Therefore, of prime importance are methods by which any observed record of random variates may be compared with the theoretically applicable distribution for these variates. Tests for this purpose will be described in Section 2.9. For the present, however, the discussion will center about various existing techniques for the generation of uniformly distributed random variables, its being presumed that other variates may be generated by inverting the cumulative distribution function, as discussed in the preceding section.

2.8.1. Some Early Sources of Random Numbers

Statisticians were among the first to realize the importance of a source of random numbers; to wit, the determination of the distribution of the important "Student's *t*" statistic was facilitated by Gossett's (1908) experimentation with numbers drawn from a well-shaken stack of numbered chits. A delightful account of this experiment, as well as a history of the subsequent development of sampling experimentation, was prepared by Teichroew (1965).

As these experiments with sampled numbers increased in magnitude (and, concomitantly, in reliability), other and more mechanical sources of random numbers were sought. Kendall and Babington-Smith (1938) relate the relative lack of success in the use of similarly selected numbers from published British telephone directories.

None of these techniques for gathering random numbers was ideally suited for the electronic computer's role in the Monte Carlo computations, which became vitally important in physics during and after World War II, since the computer required a ready source of large numbers of these digits in order to accomplish a single Monte Carlo computation. One approach to the dissolution of this difficulty was to augment the computer with a "black box" which could, when called upon, generate random variates for use by the computer in its programmed computations.

Such a "black box" would need to possess the properties that qualify its variates as random; i.e., no prediction of the nature of any forthcoming variate could be made from the variates generated previously, and no significant deviation from the expected distributional properties could be anticipated. Indeed, the RAND Corporation's tables (1955) of uniformly distributed variates were generated by an electronic "roulette wheel," the details of which are described in the prefatory material accompanying those tables.

Nonetheless, it has not become standard practice to attach an *ad hoc* random number generator to an electronic digital computer, primarily due to economic considerations. Possibly a computing machine that works virtually exclusively on Monte Carlo computations could justify the expenditure associated with building and maintaining such a "black box" as an integral component, though the author is not aware of the existence of such a device now employed in this mode. Perhaps a primary difficulty with such a device lies in defining its maintainability, for the apparatus would need to be declared inoperative whenever it was no longer apparently random in its behavior. Such a definition of the operational quality of a machine is in virtual contradiction to that applicable to any other machine of man's invention. In a pragmatic sense, such a device would require continuous monitoring to ensure its capability to generate random variates.

One feasible alternative is to record in machine-readable format extremely lengthy chains of variates generated from such a device which, incidentally, would be referred to by electrical engineers as a "white noise" generator. The chains could then be subjected to a large number of statistical tests (many of which will be discussed presently, others at appropriate points later in the text) and thereby, if not found wanting, qualified as a source of random numbers. Furthermore, each user of the computing facility would then have common access to the tested, machine-readable chain of digits.

Indeed, in the presence of multiple users' demands, the record could be continually stored on memory devices. Computer operating systems could be programmed to vary the allocation of rapid access memory to the storage of these random numbers in accordance with an algorithm based on the frequency of recent and imminent user requests for numbers from the record. Hence, an observed increase in the number of requests for *ad hoc* simulation languages could be utilized by the operating system to augment either the amount of rapid access storage allocated to random numbers or the frequency with which a table of fixed size is replenished from a more remote storage device. Each program that requests a random number would find itself addressed to a rapid access location, this location then being updated in anticipation of the next request. Presumably, before any particular program would have cycled through the entire rapid-access table of random numbers, the table will have been replenished by the operating system itself.

2.8.2. Generation of Pseudorandom Numbers

The proposed procedure has probably not been implemented for one of the same reasons that the idea of a "black box" appendage has been practically dismissed. Though the use of a pregenerated and pretested chain of random numbers overcomes the undesirable need for the maintenance of a black box (or white noise generator), the requirement for a peripheral device, one dedicated to the storage of the pretested chain, still exists. A more convenient random number generator would be one that could take advantage of the computer's capability to store and use algorithms, one of which might be defined so as to produce, when called, a random number.

However, such a proposal strikes at the very heart of the adjective "random." Any algorithm that could produce these "random" numbers is effectively producing predictable numbers in the sense that, given sufficient information regarding the preceding generated numbers, the next number could be predicted exactly. As a matter of fact, knowledge of the current state of the (necessarily finite-state) machine is adequate to determine the next output of any such algorithm.

Since the digital computer does consist of a finite, though generally quite large, number of states, the use of an algorithm for the generation of random variables implies that eventually the computer must return to a state that had existed at the time of some previous implementation of the algorithm. At this point, repetition of the cycle would begin. Thus, the conceptually infinite sequence of random variates, which could be generated by successively calling upon the algorithm, will contain infinite subsequences that are not random at all; indeed, if the length of the cycle is denoted by C, then any infinite subsequence of the "random" variates U_j having indices of the form

$$j = i + k \cdot C,$$

for any fixed i and for every $k = 1, 2, \ldots,$ would be identical in numerical value.

Therefore, the search for a technique for the generation of random numbers must be compromised by accepting alternative criteria for randomness. Numbers that are generated by means of a stored algorithm are accordingly referred to as *pseudorandom*.

2.8.3. The Mid-Square Technique

One of the earliest techniques for the generation of these pseudorandom numbers is the *mid-square technique* of Neumann (1948). The method proceeds to the generation of a presumably uniformly distributed pseudorandom variate of (2d) digits by squaring the previously generated pseudorandom variate [to produce a number of (4d) digits] and by selecting the middle (2d) digits therefrom as the next number. The algorithmic procedure is readily implemented on a computer by means of integer multiplication and shift operations; if implemented by means of compiler languages, the truncation properties of integral division operations are usually ideally suited to the elimination of the low precision and high precision digits.

A fundamental requirement for the use of the technique was an initial, user-supplied, number of (2d) digits from which the first random number could be generated. This initial number, required in every algorithmic generator, is termed the random number seed, or *seed*. In the mid-square technique, the first random number was taken to be the central (2d) digits of the square of this seed, the second random number being equal to the central (2d) digits of the square of the first random number; in general, the kth random number is generated as the central (2d) digits of the $(k - 1)$st random number following the seed. A decimal point was usually placed before the (2d) digits and the resulting chain of numbers was presumed to be uniformly distributed between zero and one.

A fundamental premise for the use of such a generator was that the seed itself should be selected as randomly as possible. In this way, successive applications of the generator would presumably not employ the same sequence of random numbers. However, the random selection of such a seed posed not only the gnawing philosophical question of *its* randomness, but also the pragmatic problem of the avoidance of seeds that would lead to very short cycles. For example, the selection of zero as the seed would result in the repetitive generation of this same number indefinitely. The elimination of this particular seed would not, however, solve the problem entirely, as the following exercise with the mid-square technique, employing numbers of four digits and seed $U_0 = 0200$, illustrates:

$$
\begin{array}{ll}
\text{Seed:} & 0200 \\
 & \underline{\times 0200} \\
 & \cancel{00}0400\cancel{00} \\
U_1: & \underline{\times 0400} \\
 & \cancel{00}1600\cancel{00} \\
U_2: & \underline{\times 1600} \\
 & \cancel{02}5600\cancel{00} \\
U_3: & \underline{\times 5600} \\
 & \cancel{31}3600\cancel{00} \\
U_4: & \underline{\times 3600} \\
 & \cancel{12}9600\cancel{00} \\
U_5: & \underline{\times 9600} \\
 & \cancel{92}1600\cancel{00} \\
U_6: & \underline{\times 1600} = U_2 \\
 & \cancel{02}5600\cancel{00} \\
U_7: & \times 5600 = U_3.
\end{array}
$$

In general, for $k \geq 2$, $U_{k+4} = U_k$.

As the reader may note from this example, the chain of random numbers generated by means of an algorithmic procedure consists of two segments:

(*a*) an introductory, *aperiodic segment* containing the random numbers generated between the seed and the beginning of the

(*b*) *periodic segment*, consisting of an infinitely repeating cycle of random numbers.

In the exemplary exercise, the aperiodic segment consists of the single number 0400, the periodic segment of the infinite repetition of the cycle (1600, 5600, 3600, 9600) of length (period) four. Depending upon the choice of the seed, the lengths of the aperiodic segment and of the cycles within the periodic segment will vary, as the reader can readily verify by selecting other seeds for the mid-square technique.

Even without the educational crutch provided by such computations, the reader should note that the finite length of the pseudorandom numbers, (2d) digits each, necessitates the eventual cycling of numbers before 10^{2d} pseudorandom numbers are generated. From the preceding discussion of randomness, the chains generated by the mid-square algorithm cannot then qualify as random. Though this fact had been clearly recognized, it was hoped that there would exist many chains of significantly lengthy period (relative to 10^{2d}) to warrant use of the technique. Though in some instances somewhat lengthy aperiodic and periodic segments were found, in general no clearly ascertainable algorithm for defining an admissible set of initiating seeds was evident. Thus, randomly selected seeds would likely lead to undesirable, unsuitable chains.

2.8.4. The Mid-Product Technique

Since no definitive description of the expected length of the aperiodic and cyclic segments of the chains generated from differing seeds seemed to be forthcoming, alternative algorithms were sought for the internal generation of pseudorandom numbers by digital computers. Forsythe (1948) discussed the mid-product technique, in which the kth pseudorandom number is taken as the central (2d) digits of the product of the $(k - 1)$st and $(k - 2)$nd pseudorandom numbers. Therefore, the procedure requires two initial numbers, each of (2d) digits, in order to initiate the generation procedure; the two numbers could be placed, however, in juxtaposition, so that the resulting (4d) digits would again constitute conceptually a single seed.

The mid-product technique might ensure lengthier chains than the mid-square technique, however. The theoretically maximal period of 10^{2d} pseudorandom numbers for the mid-square technique becomes 10^{4d} for the mid-product technique, though in neither case could the maximal period be assured by proper selections of the seed. Indeed, as the preceding numerical examples illustrate, no seed could lead to a period of the maximal length. Furthermore, the lengths of the aperiodic segment arising in the mid-product technique could not be predicted, especially in consonance with the desirability of selecting randomly the seed itself. [The reader may verify that the random selection of the seed (0100, 0600) for the mid-product techniques would lead to a very short aperiodic segment and quite short cyclic portion for the resulting chain.]

Of course, one could restrict from consideration a specified list of undesirable seeds, but the report of Metropolis (1956) reveals the cumbersome nature of such a procedure. In effect, a complete enumeration of the cycles which could arise from the possible 10^{4d} seeds of (4d) juxtaposed decimal digits is required beforehand. Though the mid-product technique augments significantly the number of possible seeds [from 10^{2d} to

10^{4d} for the generation of pseudorandom numbers of (2d) digits], it does not necessarily guarantee a longer period arising from randomly selected seeds.

The possibility of the degeneracy of mid-square and mid-product techniques has probably led to their disfavor. Even a proposal, that selected central digits of products (of three or more of the preceding pseudorandom numbers) be taken as a method for generating pseudorandom numbers, would likely be dismissed because of the unpredictability of both the cycle length and the length of the aperiodic segment in the presence of randomly selected seeds; the dismissal would be warranted (even though the theoretically maximal period of $10^{(2d)k}$ numbers could be arbitrarily increased by augmenting the number k of factors in each product), primarily because of the superior properties of numbers generated by the congruential techniques.

2.8.5. The Multiplicative Congruential, or Power Residue, Technique

Improvement upon the mid-square and mid-product techniques of Neumann and Metropolis was developed in a new technique proposed by Lehmer in 1949 and reported in 1951. Lehmer's technique, known either as the *multiplicative congruential technique*, or as the *power residue method*, selects as the kth pseudorandom number the remainder of the division of the product of a constant integer, a, and the $(k - 1)$st pseudorandom number by some second constant, m. Denoting by U_k the kth variate so generated, one may write the recursive formula:

$$U_k \equiv aU_{k-1} \pmod{m}, \qquad k = 1, 2, \ldots,$$

where the relation "$\equiv x \pmod{m}$" denotes the selection of the remainder from the aforementioned division of x by m.†

Again a seed U_0 is required for the technique and this should be selected as randomly as possible from among an admissible set of seeds, to be defined presently. Thus, the first pseudorandom number to be generated will be

$$U_1 \equiv aU_0 \pmod{m},$$

and the second will be

$$U_2 \equiv aU_1 \equiv a^2U_0 \pmod{m},$$

† Note: $2 \equiv 5 \pmod{3}$ since $5 \div 3$ leaves the integral remainder 2.

as may be shown from number theoretic results relating to the properties of the congruential relationship of integers. More generally, the kth generate becomes

$$U_k \equiv a^k U_0 \pmod{m},$$

from which derives the alternative denomination, *the power residue method*, for the technique.

The technique is ideally suited for implementation on a digital computer, because the selection of the modulus m in accordance with the size of the computer's arithmetic unit implies that the divisions may be implicitly performed by shifting operations. More specifically, one selects $m = p^e$, where p is the number of numerals in the computer's number system and e is the number of such digits in the computer's standard word size. For a binary computer, one selects $m = 2^b$, where b is the number of bits per word; for a decimal computer, one selects $m = 10^d = 2^d 5^d$, where d is the number of digits per word. The results do not appear to be significantly affected by the choice of the modulus m, so that these convenient machine-specific moduli have received widespread acceptance. [See the IBM Corporation (1959).]

As discussed by Hull and Dobell (1962), when using a binary computer, one should appropriately make the choices of U_0 as any odd integer and of a in the form

$$a = 8t \pm 3 \qquad \text{for any integral} \quad t \geq 0,$$

thus providing the maximal attainable period of 2^{b-2} pseudorandom numbers, each between 0 and 2^b. The assignment of the "decimal" point before each number in the resulting chain provides then pseudorandom numbers between 0 and 1; i.e., each succeeding integer of b bits is presumed to represent a b-bit binary fraction.

In the event that a decimal computer is to be employed, the theorems of Hull and Dobell show that one should select the seed U_0 not divisible by either 2 or 5, and should choose the multiplier a as a member of one of the following 32 residue classes, modulo 200:

$$\pm\{3, 11, 13, 19, 21, 27, 29, 37, 53, 59, 61, 67, 69, 77, 83, 91\};$$

that is, one may choose the multiplier a as any integer of the form

$$a = 200t + p,$$

where t is any positive integer, and p is a member of the preceding set of 32 integers. The maximal attainable period becomes in this case: $5 \cdot (10^{d-3}) = (\frac{1}{2})10^{d-2}$, provided that $d \geq 4$. For moduli (m) based on other considerations, maximal periods and the selection of the multipliers and seeds for them are described by Hull and Dobell (1962).

2.8.6. The Additive Congruential Method

Another technique for the generation of pseudorandom numbers also relies on congruences, but forms the kth pseudorandom number in the sequence as the residue from the *addition* of the n most recently generated numbers. For example, with $n = 2$, the recursion formula becomes

$$U_k \equiv (U_{k-1} + U_{k-2}) \pmod{m},$$

where again the modulus m is selected as 2^b (or 10^d) whenever a binary (decimal) machine is employed, and where the two starting values U_{-1} and U_0 are preassigned by the user. (Again, the two values could be considered placed in juxtaposition so as to conceive of the requirement as a *single* seed.)

The theoretically maximal period for the resulting chain is m^n (equal to 2^{nb} for a binary computer of word size b bits, 10nd for a decimal computer of word size d digits), although to the author's knowledge no sequence of such length has been generated by this method. Green *et al.* (1959) report some results of their experimentation with the technique, for a number of values of n, but no definite conclusions regarding the suitability of the resulting numbers or the methods for assuring lengthy chains seem to have been developed, other than the somewhat limited results of Miller and Prentice (1969).

One special case is worthy of note. With $n = 2$, $U_0 = U_{-1} = 1$, the sequence $U_1, U_2, \ldots$ constitutes the elements of the Fibonacci sequence (up to m). Unfortunately, this particular sequence has not proven satisfactory, in that the cycle contains rather unusual and undesired serial properties.

2.8.7. The Mixed Congruential Method

A straightforward generalization of the power residue method employs the recursion

$$U_k = (aU_{k-1} + c) \pmod{m},$$

for $k = 1, 2, \ldots$. The method, known as the *mixed congruential technique*, requires the specification of the additive constant c, the multiplier a, and the seed U_0, all positive integers less than the modulus m (which, again, is chosen for computational convenience as 2^b for a binary computer, 10^d for a decimal computer). In terms of these three initial constants, one may show that the kth pseudorandom variate in the sequence is given by

$$U_k = [a^k U_0 + c(a^k - 1)/(a - 1)] \pmod{m}.$$

The reader should note that the power residue method obtains for the case where $c \equiv 0 \pmod{m}$.

Of primary importance is the fact that the mixed congruential technique is capable of generating a sequence of full period, m; i.e., by proper specification of the constants a and c, each and every one of the nonnegative integers less than m will be generated once before any one of them shall be repeated for the first time. This may be assured by selecting:

(*a*) c relatively prime to m (i.e., so that c and m have no common integral divisors greater than 1);

(*b*) $a \equiv 1 \pmod{p}$ for any p that is a prime factor of m;

(*c*) $a \equiv 1 \pmod{4}$ if 4 be a factor of m,

as shown by Hull and Dobell (1962).

Therefore, if $m = 2^b$, as would be recommended for implementing the mixed congruential technique on a binary computer of word size b bits, one need only specify c as an odd integer and designate a as

$$a \equiv 1 \pmod{4}.$$

Similarly, for a decimal computer of words of d decimal digits, the additive constant c must not be divisible by either 2 or by 5 and the multiplier a must satisfy the congruential relationships:

$$a \equiv 1 \pmod{4} \qquad \text{and} \qquad a \equiv 1 \pmod{5},$$

or, equivalently,

$$a \equiv 1 \pmod{20}.$$

One may note that, for the mixed congruential technique, no restriction need be placed on the seed U_0, since the selection of a and c in accordance with the preceding specifications implies that all integers less than m will appear before the seed reappears. This fact has led to the widespread acceptance of the mixed congruential technique. Indeed, many computational facilities have stored library subroutines, with a and c selected to be compatible with the type and word size of the computer, so that a large pool of users can employ the generator merely by programming a call for the subroutine. Each user need only specify his seed, which is presumed to have been selected randomly from some published source (for example, the RAND Corporation tables).

Of course, additive or multiplicative congruential techniques may also be employed as library subroutines available on demand to computer users. The appropriate seed for these routines can be selected by the user (again, randomly, from some published table) and specified for the first call of the subroutine. Indeed, since some of the early statistical tests of the alternative algorithms revealed a few undesirable serial properties in number sequences from the mixed congruential technique, many software support organizations have been recommending and implementing the power residue method, even though these methods place restrictions on the seed. [See IBM Corporation (1959).]

However, Coveyou (1960) noted that, by properly selecting the constants a and c for the mixed congruential technique, the serial correlation between successively generated numbers could be considerably reduced. Such specifications should then overcome many of the earlier objections to the mixed congruential technique.

It may be of interest to note that, since the mixed congruential technique can be defined so as to generate all the integers between 0 and $(m - 1)$, inclusive, before repeating any one among them, a different choice of a and c (though restricted appropriately) is tantamount to a reshuffling of a deck of m numbered cards; then for any fixed a and c, different choices of the seed U_0 are equivalent to "cutting" this shuffled deck at varying locations.

Consequently, if one's requirements for random numbers is exceedingly great, he would be well advised to "shuffle the deck" occasionally (that is, select new constants a and c: in accordance both with the constraints of Hull and Dobell (1962), so as to obtain the maximal attainable period, and with the suggestions of Coveyou (1960), so as to avoid undesired correlations). Certainly, one will wish to "cut" the shuffled deck by selecting *randomly* his seed value from some published source each time that a program which requires the pseudorandom number generator is submitted for implementation.

One final note is in order regarding the congruential techniques. The mixed congruential technique will eventually generate the number $U = 0$ and thus may create difficulties if certain transformations (such as logarithms) are applied. Possibly the use of $(1 - U)$, instead of U, as the argument of the transform will preclude the difficulty. Or, alternatively, one may wish to employ the power residue method, whose maximal-length cycle, when implemented on a binary computer, contains only odd integers.

Readers might also find instructive certain pseudorandom number generators that are often employed in hybrid computation. Known as *shift register sequences*, these generators may be used to provide pseudorandom number generators of considerable periodic length. (See Golomb, 1967.)

EXERCISES

1. Choosing $m = 32$ (that is, 2^5), show that the multiplicative congruential technique, specified with multiplier $a = 3$, gives the chain:

7, 21, 31, 29, 23, 5, 15, and 13, then 7, *et cetera*,

regardless of which of these eight odd numbers serves as the seed (U_0). Alternatively, show that the remaining odd numbers between 0 and 31 appear in another chain, the seed for which may be any of those remaining eight odd integers.

2. Using the same modulus ($m = 32$) and the same multiplier ($a = 3$) as in Exercise 1, show that even seeds lead to chains having no aperiodic segment, and that the chains shall all be of length 1 (that is, 2^0), 2, or 4 (that is, 2^2): $\{0\}$, $\{16\}$, $\{4, 12\}$, $\{8, 24\}$, $\{20, 28\}$, $\{2, 6, 18, 22\}$, and $\{10, 30, 26, 14\}$.

3. Using the modulus, $m = 32$, show that the selection of an even multiplier ($a = 2r$) cannot provoke odd numbers to appear in the resulting chain (power residue method), regardless of whether the seed is odd or even. From the relationship

$$U_k \equiv a^{k-1} U_0 \pmod{m},$$

conclude that the maximum chain attainable with an even multiplier would be 6, since $U_6 \equiv 32 r^5 U_0 \pmod{m}$.

4. Repeat Exercise 1 with $a = 5$. Note that for odd seeds, the chains arising from odd seeds are

$$\{3, 15, 11, 23, 19, 31, 27, 7, \quad \text{and then} \quad 3, \textit{et cetera}\},$$

or

$$\{5, 25, 29, 17, 21, 9, 13, 1, \quad \text{and then} \quad 5, \textit{et cetera}\}.$$

Compare the respective intersections of these sets of odd numbers with the two of Exercise 1.

5. Convert the chains of Exercises 1 and 4 to binary notation (that is, each number should be represented as a 5-bit binary number). Comment on the common properties of the low-order bits of the numbers in each of the chains.

6. Choosing, for the mixed congruential technique, the modulus $m = 32$, the additive constant $c = 3$, and the multiplicative constant $a = 5$, show that all 32 integers between 0 and 31 appear exactly once before the seed which you select re-appears. Then, changing a to 9, show that the cycle of 32 numbers again appears, yet permuted.

7. As in Exercise 6, for $m = 32$, $c = 3$, yet $a = 3$, show that the resulting chain is either

$$\{1, 6, 21, 2, 9, 30, 29, 26, 17, 22, 5, 18, 25, 14, 13, 10\},$$

or

$$\{3, 12, 7, 24, 11, 4, 15, 16, 19, 28, 23, 8, 27, 20, 31, 0\}$$

depending upon the seed selected.

8. For the mixed congruential method, Coveyou (1960) gives as the approximate correlation (to be defined later) between successively generated pseudorandom numbers:

$$\varrho \cong a^{-1}\{1 - [6c(m - c)/m^2]\},$$

where

$$U_k \equiv (aU_{k-1} + c) \pmod{m}.$$

Show then that the constants a and c that tend to minimize ϱ are: $a \cong m$, $c \cong (m/2)$, though one should respect the conditions for attaining the period of full length (m) as well.

2.9. Testing Random Number Generators

Since the length of the cycles that can be generated by congruential techniques is usually extremely large, and since the length of the cycle can usually be determined beforehand via number theoretic considerations, the congruential techniques have received greater attention and acceptability than the earlier mid-square and mid-product techniques. For example, if a binary computer employs a 30-bit word size, the use of the mixed congruential technique (with its additive constant and its multiplier selected in accordance with the conditions mentioned in Section 2.8) leads to a period of 2^{30}, more than one-thousand million, pseudorandom numbers. Even the appropriately constrained power residue method would, on this computer, produce a sequence of at least 250 million pseudorandom numbers before recycling.

Since many computerized simulation models require fewer pseudorandom numbers than the period of a full cycle, these congruential techniques have received widespread acceptance. This acceptance has been somewhat strengthened as a result of comparisons made of the quality of variates generated by these algorithms with that of variates generated by other means.

These comparisons must be designed to reveal both the randomness and the uniformity of the distribution of the sequences generated by alternative algorithmic schemes. In some cases, it is possible to assure the distributional requirement, at least superficially so. For example, the mixed congruential technique may be initialized so as to produce every nonnegative integer less than the modulus, m; hence, after exactly m pseudorandom numbers have been generated by this procedure, an exactly uniform frequency histogram will, of necessity, result.

However, the usual situation does not call for precisely the same number of pseudorandom variates as the modulus m. Therefore tests of the generative techniques are usually based on sampled subsequences within a cycle. A description of the methodology required to perform these tests will be presented in the following subsections, since the user of a pseudorandom number generator may wish to perform his own tests before blindly accepting and employing the generator.

2.9.1. The Two-Step Testing Procedure

In order to test both the randomness and the distributional properties of a sequence of random or pseudorandom numbers, one must proceed:

(*a*) to *classify* the numbers (or transformations thereof) according to some criterion (or formula); and

(*b*) to *compare* the empirical distribution of these numbers with the theoretical result that would be anticipated if the original variates are indeed random and are uniformly distributed.

Since the randomness of a sequence of numbers implies that periodic effects within the sequence are totally erratic, associative measures relating neighboring variates in the sequence should not reveal interdependencies among successive or near-successive numbers. Therefore, in examining a series for randomness, one may wish to view the sequence of random (or pseudorandom) numbers as a realization of a time series, for which tests of autocorrelation and spectral analyses may be in order.

However, a discussion of these types of tests shall attend a more complete presentation of time series and their analyses. For the present, techniques for testing for randomness will be less complex procedures based on elementary distributional assumptions for the sequence of random (or pseudorandom) numbers.

2.9.2. The Kendall and Babington-Smith Classifications

The first step in testing a sequence is the specification of meaningful classification procedures. The number of possible classification schemes, if finite, is barely so. However, Kendall and Babington-Smith (1938) proposed that the following four classification schemes be employed in order to test for both the randomness and uniformity of distribution of the variates.

(1) *Frequency*. The interval between 0 and 1 is subdivided into k subintervals of equal length, and the proportion of variates, among a selected sequence of n of them, is computed for each subinterval (usually, k is taken as 10 for a decimal computer and 8 or 16 for a binary computer, since this implies that facile testing can be performed on the original integer-valued random numbers without requiring their transformation to decimal form).

(2) *Serial relationship*. By again defining a convenient number k of subintervals, the proportion of variates that are in the ith subinterval *and* are immediately followed by a number in the jth subinterval is computed for $i, j = 1, 2, \ldots, k$.

(3) *Poker relationships.* By representing all numbers in binary form, successive and nonoverlapping sets of, say, 10 bits can be examined to determine the number of zeros appearing in each (in a sense, each 10 bits might be thought of as a "poker hand" of ciphers and units).

(4) *The gap test.* By defining any class $\mathscr{A}$ one can examine the chain of random numbers, determining the number of consecutive variates that remain in the same class—$\mathscr{A}$ or its complement (e.g., $\mathscr{A}$ might be the set of all even integers in the sequence of numbers).

For each of these four classification schemes, a statistical test (to be described presently) needs to be performed. That is, step two of the test procedure compares the results arising from any one of the classification procedures with the results anticipated from theoretical assumptions regarding the randomness and the intended distributional properties of the number sequence.

The observation should be made that the Kendall–Babington-Smith categorizations are not the only classification schemes that could be used. Each of the four classifications is, however, equivalent to a transformation of the random variables U_k appearing in the sequence. For example, the frequency classification (1) transforms each U_k into a class number (say, 1 through 10), each of which occurs with the same theoretical probability (1 in 10) if the variates are presumed to be uniformly distributed; in addition, the mapping of the 10 bits of a random number's binary representation into a "poker hand" results in the performance of 10 Bernoulli trials, if indeed the bits derive from a random sequence of ciphers and units. Conceptually then, one may begin to test any sequence of random or pseudorandom variates by transforming each number, according to any consistent scheme that may come to mind; the transformation, however, should lead to a derivable theoretical distribution for the transformed variates, one against which the actually transformed variates may be compared.

Thus, for example, a set of numbers, each between 0 and 1, may be tested for the uniformity of its distribution by transforming each variate U according to

$$X = -\ln U.$$

If the variates U are uniformly distributed, then the variates X will be exponentially distributed. Thus, the resulting set of X variates could be formed into a frequency histogram which could, in turn, be compared with the appropriate exponential distribution.

2.9.3. The Chi-Squared Test Criterion

The crux of the test procedure is then one or more satisfactory methods for comparing the frequency histogram of the transformed variates with their theoretically applicable probability density function. There exist a number of tests for this type of comparison, the most frequently employed being the Chi-squared test and the Kolmogorov–Smirnov test.

To use the Chi-squared test, one computes from the appropriately derived cumulative distribution function the probabilities p_i, $i = 1, 2, \ldots, k$, associated with k mutually exclusive and exhaustive subintervals of the admissible range of values for the variates; one notes that

$$\sum_{i=1}^{k} p_i = 1.$$

The relative frequencies of a sequence of N (transformed) variates f_i, $i = 1, 2, \ldots, k$ are computed for the corresponding subintervals and the quantity

$$\chi_c^2 = N \cdot \sum_{i=1}^{k} [(f_i - p_i)^2/p_i]$$

is computed, where the subscript "c" indicates "computed." One may observe that the resulting computation is:

(*a*) a value of a random variable χ^2, since different samples or sequences of N original variates would lead to varying χ_c^2 values; and,

(*b*) larger whenever any one (or more) of the differences $(f_i - p_i)$ is unusually large.

Therefore, one would tend to accept a chain of N pseudorandom numbers if the computed value χ_c^2 were not too large, but would reject the generation technique if χ_c^2 were improbably large. Thus, the determination of the improbability of the size of the computed χ_c^2 depends upon the distribution of the χ^2 values, *given that the original N random variates were uniformly distributed.*

The cumulative distribution function for the random variable χ^2 depends upon k, the number of frequency cells into which the variates are placed. There is then an entire family of these variates, and there exists an appropriate cumulative distribution function for each integral value of k ($k \geq 2$). As a matter of fact, these cumulative distribution functions, known as the *Chi-squared distributions of* $(k - 1)$ *degrees of freedom* whenever k frequency cells are employed, have been tabulated, and are readily available in most texts and handbooks of statistics. [See, for example, Lindgren (1968).]

For convenience, a tabulation of the critical Chi-squared values are presented in the Appendix. The row headings $p = 0.95$ and $p = 0.99$ refer to the severity with which the random numbers deviate from their anticipated behavior. For example, with $k = 10$ cells, one would state that the computed χ_c^2 is significantly large whenever it exceeds 16.9, or that it is highly significant if it exceeds 21.7 (9 degrees of freedom, $k - 1$); values of such magnitude (or greater) would only occur with probability 0.05 or 0.01, respectively, even when the random numbers were properly behaved.

2.9.4. The Kolmogorov–Smirnov Test Criterion

A second test criterion is provided by the Kolmogorov–Smirnov statistic. Immediately following completion of the transformation and classification step, one may arrange the N (transformed) variates X in order of ascending magnitude:

$$x_{(1)} \leq x_{(2)} \leq \cdots \leq x_{(N)}.$$

The sample cumulative distribution function, $F^*(x)$, may then be constructed as a step function rising from zero to one in jumps of $1/N$ at each $x_{(i)}$, $i = 1, 2, \ldots, N$. That is,

$$F^*(x) \equiv \begin{cases} 0, & x < x_{(1)}, \\ i/N, & x_{(i)} \leq x < x_{(i+1)}, \quad i = 1, 2, \ldots, (N-1), \\ 1, & x \geq x_{(N)}. \end{cases}$$

The maximum absolute difference D_N between the sample and the theoretical cumulative distribution functions becomes then a random variable itself, because repetitions of its computation will vary among randomly selected sets of N variates. Hence,

$$D_N = \max_x | F^*(x) - F(x) |,$$

if sufficiently large, would not lend credence to the hypothesis that the observed and anticipated cumulative distribution functions are the same; that is, improbably large values of D_N would lead to rejection of the notion that the numbers U qualify as random, uniformly distributed variates.

For the Kolmogorov–Smirnov test, the critically large values of D_N are tabulated in the Appendix. Note: One compares the computed D_N-value with the tabulated critical values corresponding to N, the number of variates selected from the sequence for testing. Further discussion of the Kolmogorov–Smirnov test may be found in Lindgren (1968).

It has become accepted procedure to apply each of the four tests proposed in 1938 by Kendall and Babington-Smith whenever a new random number generation technique is suggested. For example, the RAND Corporation's tables passed all four of these tests: frequency, gap, poker, and serial. However, the Chi-squared test criterion is most usually applied, though it suffers from the possibility of bias in the selection of the k histogram cells prior to the computation of χ_c^2. For this reason, the Kolmogorov–Smirnov test is usually deemed more objective, since no cells are required; nonetheless, many find the Kolmogorov–Smirnov test more computationally cumbersome. Thus, choice between the two becomes a trade-off between these characteristics, though either test criterion is suitable for carrying out the second step of the testing procedure. For a few exemplary computations, one may refer to the monograph by Hammersley and Handscomb (1964).

One may note that the frequency test tends to examine a sequence of random (pseudorandom) numbers for its uniformity of distribution, whereas the remaining three tests of Kendall and Babington-Smith address the *randomness* of the sequence (though all four tests are predicated upon assumptions of the uniformity of distribution of the random numbers, their bits and digits, or both). Therefore, a sequence that passes all four tests is somewhat reassuring. Furthermore, one may observe that the Kendall–Babington-Smith tests are quite flexible, so that many more than four tests can be applied. For example, the frequency test may be applied to the original random numbers or to the variates arising from any direct transformation (such as logarithmic or tangential) thereof; the serial test could just as well be structured to examine the relationships between every second random number (or every kth random number) rather than neighboring variates in the sequence; of course, the poker test may be organized with more than two "suits" (other than zeros and ones), and the gap test may be defined in accordance with any number of classification schemes, $\mathscr{A}$ and $\bar{\mathscr{A}}$; (e.g., $\mathscr{A}$ might be the set of all numbers divisible by 17, or by 20, etc.).

Consequently, one could readily perform a large number n of tests, each on a nonoverlapping portion of a sequence of random numbers. If indeed the numbers are random and uniformly distributed, approximately 5% (1%) of these tests would fail, because this proportion of Chi-squared or Kolmogorov–Smirnov statistics would be improbably large in such an event. Hence, even whenever the random numbers are properly behaved, the number S_n of independent tests that fail among the n tests becomes a binomial random variable with $p = 0.05$ (or 0.01, depending upon the critical values employed). For example, if 20 tests are performed, each using the $\alpha = 0.05$ critical values as tabulated, the probability of at least one test failure is

$$1 - P[S_n = 0] = 1 - (0.95)^{20} \cong 0.637;$$

that is, it would be more likely than not that at least one of 20 tests would fail, even if the numbers were properly behaved. This fact should be borne in mind whenever random number sources are extensively tested, since an increase in the number of tests increases the likelihood that at least one test will fail—even for truly random numbers.

EXERCISES

1. Generate a sequence of 10 pseudorandom numbers U from a power residue generator using $m = 2^{15}$, the multiplier $a = 8t \pm 3$, and the seed U_0 odd. Perform the frequency test by comparing the Kolmogorov–Smirnov statistic D_{10} with the tabulated values in the Appendix. (Presume that the theoretically applicable cumulative distribution function is that of a uniformly distributed random variable.)

2. In Exercise 1, transform first each of the 10 pseudorandom numbers U_i according to $X_i = (-\ln U_i)$, $i = 1, 2, \ldots, 10$. Then perform the frequency test by comparing the Kolmogorov–Smirnov statistic D_{10} with the tabulated values in the Appendix. (*Notes*: 1. The D_{10} value computed here should be exactly the same as that in Exercise 1; one says that the Kolmogorov–Smirnov test is "distribution-free"; 2. The X_i are random variables from the cumulative distribution function $1 - e^{-x}$.)

3. Repeat Exercises 1 and 2 now using a sequence of 10 pseudorandom numbers from the mixed congruential generator specified by $m = 2^{15}$, $a = 4t + 1$, and c odd.

4. Repeat Exercises 1 and 2, using the additive congruential generator

$$U_k \equiv (U_{k-1} + U_{k-2}) \pmod{2^{15}},$$

where seed(s) U_{-1}(odd) and U_{-2}(even) are employed to generate U_1.

5. Generate 100 pseudorandom numbers U from the power residue generator of Exercise 1 and test their uniformity of distribution via a Chi-squared test, defining 10 equiwidth cells over the entire range between 0 and 1. (Note that χ_c^2 is compared with the critical values associated with the Chi-squared variate of 9 degrees of freedom.)

6. Repeat Exercise 5. First use the mixed congruential generator of Exercise 3, then the additive congruential generator of Exercise 4.

7. In Exercise 5, first transform the 100 pseudorandom numbers U_i according to $X_i = (-\ln U_i)$, $i = 1, 2, \ldots, 100$. Then define 10 cells of equal width over the range delimited by $\min(X_i)$ and $\max(X_i)$ and perform the Chi-squared test by comparing the resulting histogram with the exponential density e^{-x}. Is the value of χ_c^2 here necessarily the same as that computed for Exercise 5?

8. Suppose that a pseudorandom number generator has been subjected to four tests, once each for the frequency, serial, poker, and gap classification procedures. Show that the probability of a truly random sequence of numbers failing at least one of these four tests is approximately 0.19 (if critical values corresponding to $\alpha = 0.05$ are employed) or approximately 0.04 (if critical values corresponding to $\alpha = 0.01$ are used).

9. Perform the serial test on sequences of 100 numbers from each of the three generators of Exercises 1, 3, and 4 by defining five cells:

$$0 \leq U < 0.2; \quad 0.2 \leq U < 0.4; \quad 0.4 \leq U < 0.6; \quad 0.6 \leq U < 0.8; \quad 0.8 \leq U \leq 1.0.$$

The Chi-squared test should be performed on the resulting 25 categories in accordance with instructions given by Hammersley and Handscomb (1964).

10. *The Poisson Distribution.* If S_n denotes the number of successes in n Bernoulli trials for which the probability of success is p, then S_n has the binomial distribution. If p is small relative to np $(\equiv \lambda)$, and if λ is, in turn, small relative to n, then one may conveniently approximate the distribution function for S_n by the Poisson distribution:

$$P[S_n = k] \cong e^{-\lambda}\lambda^k/k!, \qquad k = 0, 1, 2, \ldots .$$

In performing n tests on a random number generator, each at the $\alpha = 0.05$

significance level, one may conceive of the tests as constituting n Bernoulli trials with probability p of "success" (the failure to pass a test) equal to 0.05. Supposing that $n = 20$ such tests have been performed on a given random number generator, use the Poisson approximation to show that the probability of the generator's failing none of the 20 tests is approximately $e^{-1} \cong 0.363$, and that the probability of failing exactly one test is also approximately 0.363 even when the random numbers are properly behaved. What is the maximum number of the 20 tests that you would permit to be failed and still accept the random number generator? What is the probability that S_n, the number of failed tests, exceeds 4? exceeds 5? exceeds 6?

11. (A report of actual tests of mixed congruential generators): In a class of 25 students, each was permitted to write his own mixed congruential generators, subject to the following constraints:

(*a*) for the digital computer available, the additive and multiplicative constants were to be selected so as to produce the generator of maximal period; and,

(*b*) the seed value was selected randomly by the instructor and assigned likewise to the student for use in his generator.

Each student was asked to test the uniformity of the distribution of the variates emanating from his generator exactly 100 times, each test to be based on a Kolmogorov-Smirnov criterion appropriate to the particular sample size assigned him. The 5% level of significance was selected, and the following histogram, of the number of students who found exactly t of their 100 test statistics significant, was then prepared by the instructor:

t	0	1	2	3	4	5	6	7	8	9	10	>10
Number of students	2	0	3	2	5	6	1	4	1	0	1	0

Compare this empirical distribution with a binomial distribution of parameters $n = 100$, $p = 0.05$ (or, with its approximation, a Poisson distribution of parameter $\lambda = np = 5$), and comment on:

(1) the number of students among the 25 who are "entitled" to feel that the mixed congruential generators are inadequate for the purpose of providing uniformly distributed random variables; and,

(2) the apparent acceptability of generators of this type.

2.10. Moments of Random Variables

Random variables are defined so that probability statements may be conveniently made regarding the results of a nondeterministic experiment. These statements are formed via computations employing the cumulative distribution function, a nonnegative, monotonically nondecreasing function bounded below by zero and above by one. The manner in which the cumulative distribution steps (discrete random variables) or progresses (continuous random variables) toward unity may often be ascertained from theoretical conditions which are assumed applicable to the experiment at hand.

2.10.1. Definitions and Terminology

In addition to being described by a cumulative distribution function, a continuous random variable X, having cumulative distribution function $F_X(x)$ and probability density function $f_X(x)$, can be essentially described by its moment sequence:

$$\mu_k'(X) \equiv \int_{-\infty}^{\infty} t^k f_X(t)\,dt, \qquad k = 1, 2, \ldots,$$

provided that these integrals exist. If discrete and integer-valued, the random variable's moment sequence becomes

$$\mu_k'(X) \equiv \sum_{a=-\infty}^{\infty} a^k p_X(a), \qquad k = 1, 2, \ldots,$$

where $p_X(a)$ is the appropriate probability distribution function, provided that the series converges absolutely.

The quantity $\mu_k'(X)$ is termed the *kth moment* of the random variable X and will, in general, be functionally dependent upon any parameters that may index the probability density (or probability distribution) function. For example, if X has the exponential distribution of parameter λ, then

$$\mu_k'(X) = 0 + \lambda \int_0^{\infty} t^k e^{-\lambda t}\,dt = \int_0^{\infty} u^k e^{-u}\,du/\lambda^k,$$

or

$$\mu_k'(X) = \Gamma(k+1)/\lambda^k = k!/\lambda^k, \qquad k = 1, 2, \ldots,$$

since the second integral represents Euler's Gamma function of argument $(k+1)$. Similarly, if S_2 is the discrete binomial random variable giving the number of successes in two Bernoulli trials, then

$$\begin{aligned}\mu_k'(S_2) &= (2^k p^2) + 1^k 2p(1-p) + 0^k(1-p)^2 \\ &= p^2[2^k - 2] + 2p, \qquad k = 1, 2, \ldots,\end{aligned}$$

each functionally dependent on the parameter p.

The first few moments of any random variable X are of especial interest. The first moment is known as the *mean* value, or *expected value*, of the random variable and is more usually denoted by

$$\begin{aligned}E(X) \equiv \mu_1'(X) &= \int_{-\infty}^{\infty} x f_X(x)\, dx, \qquad \text{if } X \text{ is continuous,} \\ &= \sum_{k=-\infty}^{\infty} k p_X(k), \qquad \text{if } X \text{ is discrete,}\end{aligned}$$

whenever the integral (infinite sum) is absolutely convergent. The first moment of X is also referred to as its *expectation*, since it provides a measure of the central tendency of the density (distribution) for the random variables.

The *variance* of a random variable X is defined as

$$\begin{aligned}\mathrm{Var}(X) &\equiv \int_{-\infty}^{\infty} [x - E(X)]^2 f_X(x)\, dx, \qquad \text{if } X \text{ is continuous,} \\ &\equiv \sum_{k=-\infty}^{\infty} [k - E(X)]^2 p_X(k), \qquad \text{if } X \text{ is discrete,}\end{aligned}$$

and is often denoted by σ_X^2, provided that the integral (sum) is finite. Its positive root σ_X is termed the *standard deviation* of the random variable. Computationally, the variance can be expressed as a function of the first two moments:

$$\mathrm{Var}(X) = \mu_2'(X) - [\mu_1'(X)]^2,$$

as the reader may verify by expanding the squared quantity in the definiens for the variance.

2.10.2. Chebyshev's Inequality

The mean and the variance play an especially important role in the discussion of random variables. Much of this importance stems from the facts that the mean expresses a measure of the average, or expected, value of the random variable and that the square root of the variance (i.e., the standard deviation) serves as an expression for the dispersion of the random variable. These two results are neatly summarized by an inequality due to the Russian mathematician Chebyshev. For any $k > 0$,

$$P[|X - E(X)| \geq k\sigma_X] \leq (1/k^2),$$

provided of course that σ_X exists. For a proof of this result, the reader is referred to Lindgren (1968). In words, the Chebyshev inequality states that the probability of a random variable X's lying within k standard deviations of its mean is at least as great as $1 - (1/k^2)$, regardless of the underlying distribution applicable to the random variable. Alternatively, the probability that a random variable X shall fall into an interval whose endpoints are given by $E(X) \pm k\sigma_X$ must be at least $1 - (1/k^2)$.

EXERCISES

1. Show that, if X has the exponential distribution indexed by the parameter λ, then its mean and variance are given by

$$E(X) = 1/\lambda$$

and

$$\mathrm{Var}(X) = 1/\lambda^2.$$

Show then that, for $\lambda = 1$, the probability statement

$$P[|X - 1| \leq 2] = 1 - e^{-3}$$

is compatible with Chebyshev's inequality for $k = 2$.

2. Show that the mean value of the binomial random variable, S_n is given by

$$\begin{aligned} E(S_n) &= \sum_{k=0}^{n} kn!p^k(1-p)^{n-k}/[k!(n-k)!] \\ &= np \sum_{j=0}^{n-1} (n-1)!p^j(1-p)^{n-1-j}/[j!(n-1-j)!] \\ &= np[p + (1-p)]^{n-1} = np. \end{aligned}$$

Comment on the result in view of your expectation regarding the number of successes which would be observed in n Bernoulli trials, for which the probability of success is given by p.

3. Show that if X is a continuous and positive-valued random variable having cumulative distribution function $F_X(x)$ such that $F_X(0) = 0$ and such that

$$\lim_{x\to\infty} x(1 - F_X(x)) = 0,$$

then

$$E(X) = \int_0^\infty [1 - F_X(x)]\, dx.$$

Hint: Perform partial integration with $u = 1 - F_X(x)$, $dv = dx$. Verify the result by computing the mean value for the Weibull distribution as

$$E(X) = \Gamma(1 + 1/b)/\lambda^{1/b},$$

an expression given in terms of the two Weibull parameters λ and b.

2.10.3. Theorem of the Unconscious Statistician

If a random variable X is transformed to a new random variable Y according to some transformation, $Y = t(X)$, then it may be desirable to possess knowledge of the moments of the random variable Y. Of course, the moments would be generally available from the probability density function $f_Y(y)$ for Y, where $f_Y(y)$ may be ascertained from $f_X(x)$ in accordance with the procedure outlined in Section 2.6; for example, whenever Y is a continuous random variable,

$$\mu_k'(Y) \equiv \int_{-\infty}^\infty y^k f_Y(y)\, dy, \qquad k = 1, 2, \ldots,$$

whenever this integral exists. However, an alternative procedure exists which precludes the necessity for first determining the probability density function for Y; namely, whenever the definiens exists,

$$\mu_k'(Y) \equiv \int_{-\infty}^\infty [t(x)]^k f_X(x)\, dx, \qquad k = 1, 2, \ldots,$$

a result which might be referred to as the "Theorem of the Unconscious Statistician," since the kth moment of $Y = t(X)$ is computed practically without acknowledging the existence of the probability density function $f_Y(y)$. The proof of the theorem is relatively straightforward and is omitted.

The theorem is especially useful in the computation of moments arising from some of the more elementary transformations. For example, if a random variable X has probability density function $f_X(x)$ and mean value denoted by

$$\mu_X \equiv E(X) = \int_{-\infty}^{\infty} x f_X(x)\, dx,$$

then the random variable, $Y = aX + b$, has mean value

$$\mu_Y \equiv E(Y) = \int_{-\infty}^{\infty} (ax + b) f_X(x)\, dx,$$

by virtue of the Theorem of the Unconscious Statistician; hence,

$$\mu_Y = E(Y) = a \int_{-\infty}^{\infty} x f_X(x)\, dx + b \int_{-\infty}^{\infty} f_X(x)\, dx,$$

or

$$E(Y) = aE(X) + b.$$

(A similar result obtains for discrete random variables.)

Furthermore, the variance of $Y = aX + b$ becomes

$$\sigma_Y^2 \equiv \mathrm{Var}(Y) = \int_{-\infty}^{\infty} [ax + b - aE(X) - b]^2 f_X(x)\, dx$$

or

$$\sigma_Y^2 = a^2 \int_{-\infty}^{\infty} [x - E(X)]^2 f_X(x)\, dx = a^2 \sigma_X^2;$$

i.e.,

$$\mathrm{Var}(Y) = a^2\, \mathrm{Var}(X) \quad \text{whenever} \quad Y = aX + b.$$

Another especially important result derives from the Theorem of the Unconscious Statistician. The square of a random variable X is itself a random variable having its first moment equal to the second moment of X; that is,

$$E(X^2) = \int_{-\infty}^{\infty} x^2 f_X(x)\, dx \equiv \mu_2'(X).$$

From this result, one has the alternative expression for the variance of X:

$$\mathrm{Var}(X) = \mu_2'(X) - [\mu_1'(X)]^2 = E(X^2) - [E(X)]^2.$$

Moreover, since the variance of a random variable is nonnegative, one has

$$E(X^2) \geq [E(X)]^2 \qquad \text{for all random variables} \quad X.$$

2.10.4. The Moment-Generating Function

One particular transformation of a random variable X is especially useful: $Z = e^{tX}$, where t is a real-valued parameter. Of special importance is the first moment of this random variable, given by the Theorem of the Unconscious Statistician as

$$E(e^{tX}) \equiv m_X(t) = \begin{cases} \int_{-\infty}^{\infty} e^{tx} f_X(x)\,dx, & \text{for } X \text{ continuous,} \\ \sum_{k=-\infty}^{\infty} e^{tk} p_X(k), & \text{for } X \text{ discrete,} \end{cases}$$

an integral (sum) which, if finite, is commonly denominated the bilateral Laplace transform of the function $f_X(x)$ [or $p_X(k)$].

A fundamental property of Laplace transforms is their one-to-one correspondence with the transformed functions. Thus, if one knows the Laplace transform of a probability density, he essentially knows the functional form of that probability density. Computations might well be performed (as will be demonstrated in a subsequent section) strictly in terms of moment-generating functions, and the concluding moment-generating function then compared with tabulated Laplace transforms, in order to ascertain the probability density function corresponding to the result, quite analogous to the procedure that adds logarithms in order to obtain a product as the antilogarithm of the resulting sum.

A second useful property of the moment-generating function is implicit in its name. More explicitly, the successive moments of a random variable X may be found by successively differentiating $m_X(t)$ with respect to t and equating therein t to 0. The result may be established by noting that

$$m_X(t) = \int_{-\infty}^{\infty} \left[\sum_{k=0}^{\infty} (tx)^k/k!\right] f_X(x)\,dx,$$

invoking for e^{tx} its Maclaurin series expansion. Under quite general conditions, one can write

$$m_X(t) = \sum_{k=0}^{\infty} \left[t^k \int_{-\infty}^{\infty} x^k f_X(x)\,dx/k!\right]$$

or

$$m_X(t) = \sum_{k=0}^{\infty} [t^k E(X^k)/k!].$$

Hence, again under quite general conditions, the first derivative becomes

$$m_X'(t) = E(X) + \sum_{k=2}^{\infty} [t^{k-1} E(X^k)/(k-1)!],$$

and

$$m_X'(0) = E(X).$$

Successive differentiations lead then to the results

$$E(X^k) = \mu_k'(X) = m_X^{(k)}(0), \qquad k = 1, 2, \ldots,$$

where $m_X^{(k)}(t)$ is the kth derivative of the moment-generating function for the random variable X. Of particular interest is the expression for the variance of X, given by

$$\mathrm{Var}(X) = m_X''(0) - [m_X'(0)]^2.$$

EXERCISES

1. Show that the moment-generating function for the exponentially distributed random variable is given by

$$m_X(t) = 0 + \lambda \int_0^\infty e^{tx} e^{-\lambda x}\, dx = \lambda/(\lambda - t), \qquad (2.10.4{:}1)$$

provided that $(\lambda - t) > 0$. From $m_X(t)$, show that $E(X) = \lambda^{-1}$ and $\mathrm{Var}(X) = \lambda^{-2}$ by computing the necessary derivatives.

2. Show that the moment-generating function for the binomial random variable S_n is given by

$$m_{S_n}(t) = \sum_{k=0}^{n} [e^{tk} n! p^k (1 - p)^{n-k} / k!(n - k)!],$$

or

$$m_{S_n}(t) = [pe^t + (1 - p)]^n.$$

Derive the results that

$$E(S_n) = np$$

and

$$\mathrm{Var}(S_n) = np(1 - p).$$

3. Show that the moment-generating function for the geometrically distributed random variable is

$$m_T(t) = \sum_{k=1}^{\infty} e^{tk} p(1 - p)^{k-1},$$

or

$$m_T(t) = p\left\{\sum_{k=1}^{\infty} [(1 - p)e^t]^k\right\} \Big/ (1 - p),$$

or

$$m_T(t) = pe^t/[1 - (1 - p)e^t].$$

By setting $t = 0$ in the first derivative of $m_T(t)$, show that

$$E(T) = 1/p.$$

Comment on the expected waiting time until the first ace appears in the tossing of a fair die; until the first head appears in the tossing of a fair coin; until for the first time each of two tossed coins falls heads.

4. Show that the moment-generating function for the uniform distribution is

$$m_U(t) = \int_0^1 e^{tx}\,dx = (e^t - 1)/t$$

and, hence, that

$$E(U) = \tfrac{1}{2}$$

and

$$\mathrm{Var}(U) = \tfrac{1}{12}.$$

Compare these results with the mean and variance as obtained directly from their defining integrals.

2.11. Bivariate Distributions

Frequently in performing a nondeterministic experiment, one will wish to record two measurements, X and Y, each of which may constitute a random variable. The two random variables may both be discrete, or may both be continuous, or may be one of each type, though the mixed case shall not be of particular interest in this discussion.

2.11.1. The Joint and Marginal Distributions

The pair of random variables X and Y are described by a *joint cumulative distribution function*,

$$F_{X,Y}(a, b) \equiv P[X \le a \quad \text{and} \quad Y \le b],$$

which provides, for any point (a, b) in the Euclidean plane, the probability that the experimental outcome (X, Y) will be a point in the Euclidean plane that falls below and to the left of the given point (a, b). One may note that this joint cumulative distribution function is monotone nondecreasing in either the x- or y-direction. In fact,

$$\lim_{\substack{x\to-\infty \\ y\to-\infty}} F_{X,Y}(x, y) = 0, \qquad \lim_{\substack{x\to+\infty \\ y\to+\infty}} F_{X,Y}(x, y) = 1,$$

and all values assumed by the joint cumulative distribution function are bounded by these limits.

If X and Y are both discrete random variables, then their joint cumulative distribution function is a set of planar steps, possibly of irregular size and rise, which ascend from zero height in the extreme lower left to unit height above the extreme upper right of the Euclidean plane. The locales and sizes of these steps are given then by a joint probability distribution function

$$p_{X,Y}(x_i, y_j) = P[X = x_i \quad \text{and} \quad Y = y_j],$$

the probability that $X = x_i$ and simultaneously $Y = y_j$. Of course,

$$p_{X,Y}(x_i, y_j) \geq 0 \qquad \text{for all admissible} \quad (x_i, y_j),$$

and

$$\sum_{i=-\infty}^{\infty} \sum_{j=-\infty}^{\infty} p_{X,Y}(x_i, y_j) = 1.$$

Whenever X and Y are both continuous random variables, their joint cumulative distribution function $F_{X,Y}(x, y)$ represents a smooth, rising surface, bounded above by one and below by zero. The rate of increase of the surface is represented by the joint probability density function, expressed at any point (x, y) as

$$f_{X,Y}(x, y) \equiv \partial^2 F_{X,Y}(x, y)/\partial x\, \partial y,$$

a nonnegative function whose double integral over the entire Euclidean plane must be unity:

$$\int_{-\infty}^{\infty} \int_{-\infty}^{\infty} f_{X,Y}(x, y)\, dx\, dy = 1.$$

Since both X and Y are random variables, each possesses its own probabilistic description in the absence of any information regarding the outcome of the other. That is, if no information is made available regarding the outcome Y whenever the experiment is performed, X has its own *marginal probability density (distribution) function*:

$$p_X(x_i) = \sum_{j=-\infty}^{\infty} p_{X,Y}(x_i, y_j), \qquad \text{if } X \text{ and } Y \text{ are discrete,}$$

or

$$f_X(a) = \int_{y=-\infty}^{\infty} f_{X,Y}(a, y)\, dy, \qquad \text{if } X \text{ and } Y \text{ are continuous.}$$

The reader may note that these functions qualify as univariate probability density functions, because each is, being the sum (integral) of a nonnegative function, itself nonnegative, and

$$\sum_{i=-\infty}^{\infty} p_X(x_i) = \sum_{i=-\infty}^{\infty} \sum_{j=-\infty}^{\infty} p_{X,Y}(x_i, y_j) = 1$$

and

$$\int_{-\infty}^{\infty} f_X(a)\, da = \int_{-\infty}^{\infty} \int_{-\infty}^{\infty} f_{X,Y}(x, y)\, dy\, dx = 1.$$

Similarly, Y has its marginal probability density (distribution) function whenever information regarding the outcome of X is absent:

$$p_Y(y_j) = \sum_{i=-\infty}^{\infty} p_{X,Y}(x_i, y_j), \qquad \text{if } X \text{ and } Y \text{ are discrete,}$$

$$f_Y(b) = \int_{-\infty}^{\infty} f_{X,Y}(x, b)\, dx, \qquad \text{if } X \text{ and } Y \text{ are continuous.}$$

Of course, associated with each probability density (distribution) function is a corresponding *marginal cumulative distribution function*: $F_X(x)$ or $F_Y(y)$.

The response, or output, of a stochastic simulation model may not always be conveniently summarized as a univariate quantity and might require a bivariate quantity for its representation. For example, the typical queueing model is constructed in order to measure the idle time of the servers as well as the lost time of the served. In these situations, the bivariate random variable serves as a description of the responses that might be recorded by varying the seeds for the model in successive iterations. In these cases, the marginal probability density (distribution) functions associated with each of the two random variables serves as the description of that random variable whenever no *a priori* knowledge exists regarding the other.

2.11.2. The Conditional Distribution

In many instances, however, one will ask about the nature of the distribution of, say, X given that Y is known. In the aforementioned queueing example, one might well ask of the distribution of servers' idle time given that the customers' average delay Y will be fixed at some amount y^*. Thus, in addition to the need for marginal distribution functions, one defines the *conditional probability density (distribution) function of X* given that $Y = y^*$ by

$$p_{X|y^*}(x_i \mid y^*) = p_{X,Y}(x_i, y^*)/p_Y(y^*) \qquad (X, Y \quad \text{discrete})$$

or

$$f_{X|y^*}(x \mid y^*) = f_{X,Y}(x, y^*)/f_Y(y^*) \qquad (X, Y \quad \text{continuous}).$$

(These functions are defined as zero whenever their denominators are nil.) Analogously, the conditional probability density (distribution) function of Y given that X assumes the value x^* is

$$p_{Y|x^*}(y_j \mid x^*) = p_{X,Y}(x^*, y_j)/p_X(x^*) \qquad (X, Y \quad \text{discrete})$$

or

$$f_{Y|x^*}(y \mid x^*) = f_{X,Y}(x^*, y)/f_X(x^*) \qquad (X, Y \quad \text{continuous}).$$

(Again, these are defined to be zero should their denominators be so.) The reader may readily verify that these functions qualify as probability density (distribution) functions; e.g., the function $f_{X|y^*}(x \mid y^*)$ is the ratio of two nonnegative functions and its integral over all permissible x-values is

$$\int_{-\infty}^{\infty} f_{X|y^*}(x|y^*)\, dx = \int_{-\infty}^{\infty} f_{X,Y}(x, y^*)\, dx/f_Y(y^*) = 1.$$

2.11.3. Independently Distributed Random Variables

In general, one will need to know the conditional probability distribution (density) function in order to prepare probability statements regarding the outcome X, given that the companion variate $Y = y^*$. However, this need not always be the case, for probability statements about X may remain the same regardless of knowledge of the outcome of the second variate. We say that the two random variables are independently distributed (or, simply, *independent*) if and only if their joint cumulative probability distribution function factors, at every point (x, y) in the

Euclidean plane, as the product of the marginal cumulative distributions; that is,

$$F_{X,Y}(x, y) = F_X(x)F_Y(y) \qquad \text{for all} \quad x \quad \text{and} \quad y.$$

The statistical independence of X and Y also implies that the joint probability density (distribution) functions may be expressed as the product of the marginal densities (distributions):

$$f_{X,Y}(x, y) = f_X(x)f_Y(y) \qquad \text{for all} \quad (x, y)$$

and

$$p_{X,Y}(x_i, y_j) = p_X(x_i)p_Y(y_j) \qquad \text{for all} \quad (x_i, y_j).$$

Moreover, the conditional densities become equivalent to the marginal densities:

$$f_{X|y^*}(x \mid y^*) = f_X(x) \qquad \text{for all} \quad x \quad \text{and any given} \quad y^*;$$

and

$$f_{Y|x^*}(y \mid x^*) = f_Y(y) \qquad \text{for all} \quad y \quad \text{and any given} \quad x^*.$$

Similarly, the conditional and marginal distributions become equivalent in the discrete case whenever X and Y are independent random variables. This implies that probability statements about either variate remain unchanged in the presence of knowledge about the other, independent variate.

One of the primary sources of independent random variables is the repetition of a nondeterministic experiment under essentially identical conditions. If the univariate response of the experiment is a random variable X, having a cumulative distribution function $F_X(x)$, then two successive and independently performed repetitions of the experiment will yield the independent random variables X_1 and X_2 whose joint cumulative distribution function will be, for any ordered pair (x_1, x_2) of real numbers:

$$F_{X_1,X_2}(x_1, x_2) = F_X(x_1)F_X(x_2).$$

Such independent random variables would arise from a pair of iterations of a simulation model, all of whose input variables would remain constant save the random number seeds, which should be selected and specified independently for each of the two iterations with the model. Of course, the same seed must be precluded in specifying the two model iterations, so that the pair of independently seeded iterations would yield two independent random variables, X_1 and X_2, each having the same marginal distribution function, $F_X(a)$.

2.11.4. Moments of Bivariate Random Variables

Frequently, it is desirable to transform the bivariate results of a nondeterministic experiment according to some particular transformation, say

$$Z = t(X, Y).$$

Doing so is the equivalent of measuring a single, or univariate, random variable as the result of performing the experiment.

In order to obtain the probability density function for this random variable Z of interest, one may proceed to define a second random variable,

$$W = s(X, Y),$$

as some other transformation of the original random variables (X, Y). The joint probability density function for the bivariate random variables (Z, W), or *random vector*, as it is alternatively termed, may be found from the joint probability density function $f_{X,Y}(x, y)$ for the original bivariate random vector (X, Y). At any point (z, w) in the Euclidian plane:

$$f_{Z,W}(z, w) = f_{X,Y}[x(z, w), y(z, w)] \cdot \left| \frac{\partial(x, y)}{\partial(z, w)} \right|, \qquad (2.11.4{:}1)$$

where $x(z, w)$ and $y(z, w)$ are the inverse transformations, and where

$$\frac{\partial(x, y)}{\partial(z, w)} = \det \begin{bmatrix} \dfrac{\partial x}{\partial z} & \dfrac{\partial x}{\partial w} \\ \dfrac{\partial y}{\partial z} & \dfrac{\partial y}{\partial w} \end{bmatrix} = \left(\frac{\partial x}{\partial z} \cdot \frac{\partial y}{\partial w} \right) - \left(\frac{\partial x}{\partial w} \cdot \frac{\partial y}{\partial z} \right),$$

called the Jacobian of the transformation. The marginal density function of Z follows immediately from the integration:

$$f_Z(z) = \int_{-\infty}^{\infty} f_{Z,W}(z, w)\, dw.$$

Once the marginal probability density function for Z has been determined, it may be employed directly to yield any desired moments of this univariate random variable; for example,

$$E(Z) = \int_{-\infty}^{\infty} z f_Z(z)\, dz.$$

However, if the moments of $Z = t(X, Y)$ are sought, they may be found directly from the joint probability density function for X and Y according to an extension of the Theorem of the Unconscious Statistician. Thus

$$E(Z^k) = \int_{-\infty}^{\infty} \int_{-\infty}^{\infty} [t(x, y)]^k f_{X,Y}(x, y)\, dx\, dy, \qquad k = 1, 2, \ldots,$$

whenever these integrals exist. For example, the mean value of $Z = t(X, Y)$ can be found without need of explicit knowledge of the probability density function of Z:

$$E(Z) = \int_{-\infty}^{\infty} \int_{-\infty}^{\infty} t(x, y) f_{X,Y}(x, y)\, dx\, dy,$$

whenever this double integral exists.

Some particular transformations of bivariate random variables will be of immediate interest. For instance, if $Z = t(X, Y) = X$, then one would anticipate that $E(Z) = E(X)$; indeed,

$$E(Z) = \int_{-\infty}^{\infty} \int_{-\infty}^{\infty} x f_{X,Y}(x, y)\, dy\, dx = \int_{-\infty}^{\infty} x f_X(x)\, dx \equiv E(X).$$

Similarly, if $Z = t(X, Y) = Y$, one may show that $E(Z) = E(Y)$. Another transformation of interest is

$$Z = t(X, Y) = [X - E(X)] \cdot [Y - E(Y)],$$

having expectation

$$E(Z) = \int_{-\infty}^{\infty} \int_{-\infty}^{\infty} \{[x - E(X)] \cdot [y - E(Y)]\} f_{X,Y}(x, y)\, dx\, dy,$$

a quantity known as the *covariance* of X and Y:

$$\text{Cov}(X, Y) \equiv E\{[X - E(X)] \cdot [Y - E(Y)]\}.$$

The reader should note that if X and Y are independent random variables, their covariance is zero, since in this event

$$\text{Cov}(X, Y) = \int_{-\infty}^{\infty} [x - E(X)] f_X(x)\, dx \cdot \int_{-\infty}^{\infty} [y - E(Y)] f_Y(y)\, dy = 0^2 = 0.$$

A closely allied transformation is

$$Z = t(X, Y) = \left[\left(\frac{X - E(X)}{\sigma_X}\right) \cdot \left(\frac{Y - E(Y)}{\sigma_Y}\right)\right],$$

which has expectation given by

$$E(Z) \equiv \text{Corr}(X, Y) = \text{Cov}(X, Y)/(\sigma_X \cdot \sigma_Y) \equiv \varrho_{X,Y}.$$

This quantity is termed the *correlation* (*coefficient*) between X and Y. The correlation coefficient is also zero whenever the random variables X and Y are independently distributed.

The correlation coefficient also has several other interesting properties. First, it is a dimensionless quantity that endeavors to express the degree of relationship between X and Y, even though they are random variables. Second, the correlation coefficient is always bounded above by $+1$, below by -1, as the reader may verify by examining the concise proof of Lindgren (1968). Third, as the subsequent discussion of Lindgren shows, $\varrho_{X,Y}$ is equal to either $+1$ or -1 if and only if the two random variables are linearly related; that is, $|\varrho_{X,Y}| = 1$ if and only if there exist constants a and b such that the experimental result for Y will always be given from that for X according to $Y = aX + b$. Finally, the absolute value of the correlation coefficient is invariant under linear transformations of the form

$$Z = aX + b,$$
$$W = cY + d.$$

[The reader should verify that $\text{Cov}(Z, W) = ac \cdot \text{Cov}(X, Y)$, so that, since $\sigma_Z = |a| \cdot \sigma_X$ and $\sigma_W = |c| \cdot \sigma_Y$, then $\varrho_{Z,W} = \pm\varrho_{X,Y}$.] The primary use of the correlation coefficient, then, is to reveal the *degree of linearity* that exists between the random variables X and Y; this degree of linearity increases as the absolute value of $\varrho_{X,Y}$ approaches unity.

In concluding the present section on bivariate random variables and their transformations, one should note the important moments for the random variable,

$$Z = aX \pm bY;$$

viz.,

$$E(aX \pm bY) = aE(X) \pm bE(Y),$$

and

$$\text{Var}(aX \pm bY) = a^2\,\text{Var}(X) + b^2\,\text{Var}(Y) \pm 2ab\,\text{Cov}(X, Y).$$

The reader should verify these results by completing the exercises which follow.

EXERCISES

1. Show that the Jacobian whose absolute value is required in the expression for the joint probability density function for Z and W, arising from the general linear transformations of the bivariate random vector (X, Y) according to

$$Z = aX + bY,$$
$$W = cX + dY,$$

is the determinant of the inverse of the matrix

$$\begin{bmatrix} a & b \\ c & d \end{bmatrix};$$

in other words, show that

$$\frac{\partial(x, y)}{\partial(z, w)} = \frac{1}{(ad - bc)}.$$

2. Show that the Jacobian, whose absolute value is required in the expression for the joint probability density function for R and θ, a polar random vector which arises from the Euclidean random vector (X, Y) according to

$$X = R \cos \theta$$
$$Y = R \sin \theta,$$

is given by

$$\frac{\partial(x, y)}{\partial(r, \theta)} = r(\cos^2 \theta + \sin^2 \theta) = r.$$

3. Show that the random variable $Z = aX \pm bY$ has mean

$$E(Z) = aE(X) \pm bE(Y)$$

and variance

$$\text{Var}(Z) = a^2 \text{Var}(X) + b^2 \text{Var}(Y) \pm 2ab \text{Cov}(X, Y).$$

If X and Y are independent, show that

$$\text{Var}(Z) = a^2 \text{Var}(X) + b^2 \text{Var}(Y);$$

hence, if X and Y are independently distributed random variables,

$$\mathrm{Var}(X \pm Y) = \mathrm{Var}(X) + \mathrm{Var}(Y).$$

4. Show that, if X and Y are uncorrelated random variables (not necessarily independent), each of mean zero, then their sum and difference, given by

$$Z = (X + Y),$$

and

$$W = (X - Y),$$

are also uncorrelated and of mean zero, provided that $\mathrm{Var}(X) = \mathrm{Var}(Y)$. Note that

$$E(X \pm Y) = E(X) \pm E(Y) = 0$$

and hence

$$\begin{aligned} \mathrm{Cov}(Z, W) &= \mathrm{Cov}(X + Y, X - Y) \\ &= E\{[(X + Y) - E(X + Y)] \cdot [(X - Y) - E(X - Y)]\} \\ &= E\{(X + Y) \cdot (X - Y)\} \\ &= E(X^2) - E(Y^2) \end{aligned}$$

and that, if $E(X) = 0$, $\mathrm{Var}(X) = E(X^2)$.

5. Show that if two random variables are uncorrelated, they need not necessarily be independent. Do so by computing the correlation coefficient and the marginal probability distribution functions for the random variables, X and Y, whose joint probability distribution function is tabulated as follows.

$X \backslash Y$	-1	0	$+1$
-1	$\frac{1}{5}$	0	$\frac{1}{5}$
0	0	$\frac{1}{5}$	0
$+1$	$\frac{1}{5}$	0	$\frac{1}{5}$

entries are $p_{X,Y}(x, y)$

Note that $\varrho_{X,Y} = 0$, yet $p_{X,Y}(-1, -1) = \frac{1}{5} \neq p_X(-1)\, p_Y(-1) = \frac{4}{25}$.

6. Show that if X and Y are independent random variables transformed according to

$$Z = t_1(X) \qquad \text{and} \qquad W = t_2(Y),$$

then Z and W are also independently distributed. Show then that

$$E[t_1(X) \cdot t_2(Y)] = E[t_1(X)] \cdot E[t_2(Y)].$$

7. Therefore, if X and Y are independently distributed, show that the moment-generating function for $S = (X + Y)$ is

$$m_S(t) \equiv E[e^{t(X+Y)}] = E[e^{tX}e^{tY}] = E[Z \cdot W],$$

for $Z = e^{tX}$ and $W = e^{tY}$; thus,

$$m_S(t) = E[Z]E[W] = m_X(t)m_Y(t),$$

so that the moment-generating function of the *sum* of two independently distributed variates equals the *product* of their respective moment-generating functions.

2.12. The Central Limit Theorem

At the core of statistical analysis has resided a quite general theorem, one surmised at first in a very limited context by DeMoivre (1718), then extended to a firmly established result by the Marquis de Laplace (1812) shortly after Gauss (1809) "discovered" the result in the approximation of errors arising in certain astronomical observations. The result is contained in the important *Central Limit Theorem*, which states roughly that, if a random variable arises as the sum of a large number of relatively insignificant contributions, then the probability density function for the random variable X will be of the Gaussian form:

$$\varphi(x; \mu, \sigma) = (\sigma\sqrt{2\pi})^{-1} \exp[-(x - \mu)^2/2\sigma^2] \quad \text{for any real } x, \qquad (2.12{:}1)$$

where μ is real and equals the mean value $E(X)$ for X, and where σ is positive and represents the standard deviation for the random variable; that is,

$$E(X) \equiv \int_{-\infty}^{\infty} x\varphi(x; \mu, \sigma)\, dx = \mu,$$

and

$$\text{Var}(X) \equiv \int_{-\infty}^{\infty} [x - \mu]^2 \cdot \varphi(x; \mu, \sigma)\, dx = \sigma^2.$$

One frequently employs the shorthand notation

$$X \sim N(\mu, \sigma^2)$$

to express the fact that the random variable X has the normal (or Gaussian) distribution of mean μ and of variance σ^2. Especially noteworthy is the observation that the moments μ and σ^2 suffice to describe a distribution function which is presumed to be of the Gaussian family.

2.12.1. The DeMoivre–Laplace Theorem

More specifically, DeMoivre's speculations arose from observations of a binomial random variable S_n that counted the number of successes in n Bernoulli trials, each having probability p of success. The reader may recall that the probability distribution function for S_n is given by Eq. (2.2:1):

$$b_{S_n}(k; n, p) \equiv P[S_n = k] = \begin{cases} \binom{n}{k} p^k (1-p)^{n-k}, & k = 0, 1, \ldots, n \\ 0, & \text{other } k, \end{cases}$$

indexed by the parameters n and p. Feller's compendium (1950, Vol. I, p. 182) on probability theory presents a proof of the DeMoivre–Laplace theorem:

If S_n is a random variable having the binomial distribution $b(x; n, p)$, then asymptotically the distribution of S_n becomes $N(np, np(1-p))$.

One may note that S_n represents the *sum* of n independent random variables X_i each having the distribution $b(x_i; 1, p)$. That is,

$$P[X_i = 0] = p$$

and

$$P[X_i = 1] = (1-p), \qquad i = 1, 2, \ldots, n.$$

Consequently, the random variable S_n having the distribution $b(s; n, p)$, represents the *svm* of n independent random variables X_i each having the distribution $b(x_i; 1, p)$: $S_n = \sum_{i=1}^n X_i$.

2.12.2. Sums of Uniformly Distributed Random Variables

A similar tendency to the normal form can be exhibited from reflections regarding the distribution of sums of independent, continuous random variables. For example, if T_k represents the sum of k independently and uniformly distributed random variables, then

$$f_{T_1}(x) = \begin{cases} 1, & 0 \le x \le 1, \\ 0, & \text{other} \quad x, \end{cases}$$

and (2.12.2:1)

$$f_{T_2}(x) = \begin{cases} x, & 0 \le x \le 1, \\ 2 - x, & 1 \le x \le 2, \\ 0, & \text{other} \quad x. \end{cases}$$

The two densities are plotted in Fig. 2.5. The rectitude of the latter density may be established via moment-generating functions, noting that

$$m_{T_1}(t) = 0 + \int_0^1 e^{tx} \cdot 1 \cdot dx = (e^t - 1)/t$$

is the moment-generating function for the uniformly distributed random variable T_1. Since T_2 is the sum of two such independently distributed random variables, its moment-generating function must be

$$m_{T_2}(t) \equiv E[e^{tT_2}] = E[e^{t(U_1+U_2)}] = E[e^{tU_1}]E[e^{tU_2}],$$

or

$$m_{T_2}(t) = [m_{T_1}(t)]^2 = (e^{2t} - 2e^t + 1)/t^2.$$

A comparison of tabulations of Laplace transforms will reveal that $m_{T_2}(t)$ is the moment-generating function corresponding to the triangular function $f_{T_2}(x)$ of Eq. (2.12.2:1). Owing to the unicity property of Laplace transforms, $f_{T_2}(x)$ must be then the probability density function for T_2, the sum of two independently and uniformly distributed random variables.

Similarly, one can conclude that the probability density function for T_3, the sum of three independently and uniformly distributed random variables, becomes

$$f_{T_3}(x) = \begin{cases} x^2/2, & 0 \le x \le 1, \\ [3 - 4(x - \tfrac{3}{2})^2]/4, & 1 \le x \le 2, \\ (x - 3)^2/2, & 2 \le x \le 3, \\ 0, & \text{other} \quad x. \end{cases}$$

This probability density function, consisting of three smoothly interlocking parabolas, has moment-generating function

$$m_{T_3}(t) \equiv E[e^{tT_3}] = [m_{T_1}(t)]^3 = (e^t - 1)^3/t^3,$$

and is also depicted in Fig. 2.5.

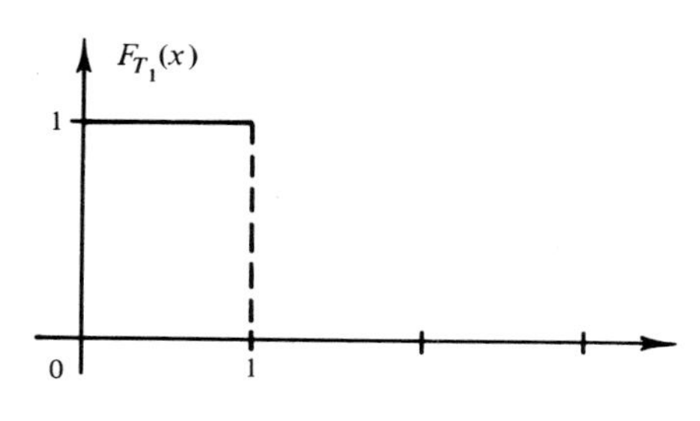

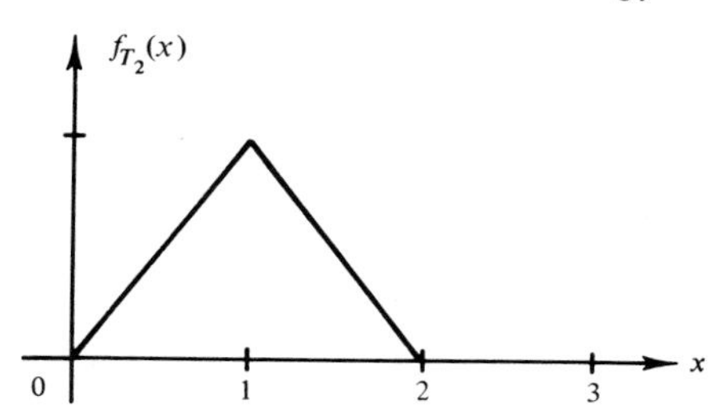

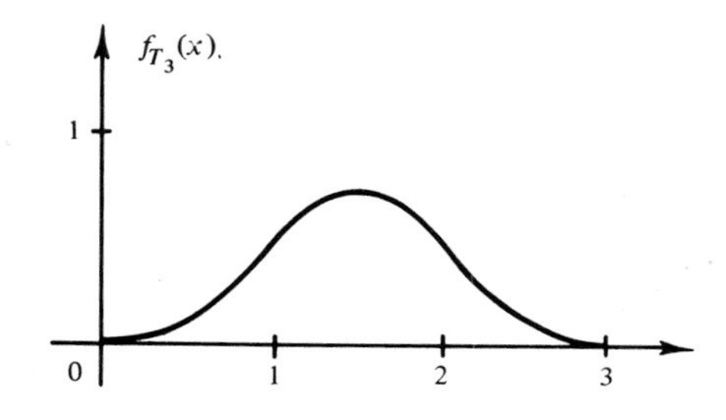

FIG. 2.5. Density functions for sums of k independently and uniformly distributed random variables, $k = 1$, 2, and 3.

Figure 2.5 is intended to display the tendency toward the Gaussian distribution of the sum of an increasing number of independently and uniformly distributed random variables. The expression for the density function of T_n becomes somewhat complicated in the general case, being a set of n smoothly interlocking polynomials, each of degree $(n - 1)$. The expected value of T_n becomes

$$E(T_n) = E(U + T_{n-1}) = nE(U) = n/2.$$

Since the summands are independent, its variance is given by

$$\text{Var}(T_n) = \text{Var}(U + T_{n-1}) = \text{Var}(U) + \text{Var}(T_{n-1}),$$

or

$$\text{Var}(T_n) = n\,\text{Var}(U) = n/12.$$

Consequently, the mean and variance of the approximating Gaussian distribution for T_n may be given by

$$\mu = n/2, \qquad \text{and} \qquad \sigma^2 = n/12.$$

More generally, the Central Limit Theorem applies to sums of independently and *identically* distributed random variables, regardless of their common distribution. Properly stated, the Central Limit Theorem becomes:

If S is the sum of n independently distributed random variables, each possessing the same (marginal) probability density function with finite variance σ^2 and mean μ, then for any real y,

$$\lim_{n\to\infty} P[(S - n\mu)/(\sigma\sqrt{n}) \leq y] = (2\pi)^{-1/2} \int_{-\infty}^{y} \exp(-u^2/2)\, du.$$

Thus, as the number of variates being summed increases, probability statements may be made more accurately by means of the cumulative distribution function associated with the normal distribution $N(0, 1)$. For the proof of this result, the reader is referred to Lindgren (1968); the proof is quite straightforward from the expression for the moment-generating function for the random variable $(S - n\mu)/(\sigma\sqrt{n})$.

The reader should note that the sum S of n independently distributed random variables from any particular probability density (distribution) function will be a random variable (of mean $n\mu$ and variance $n\sigma^2$) about which approximate probability statements can be made (by reference to tabulations of the cumulative Gaussian distribution function). Yet if either condition is relaxed, the resulting distribution of the sum S may or may not be of the Gaussian form. For example, if the contributing random variables are independent, but do not arise from the same probability density, their sum may be a random variable that is not at all Gaussian [see Mihram (1969a)]; conditions on the moments of the contributing random variables may be established that will, nonetheless, require the limiting distributional form of the sum to be Gaussian. [See Feller (1950, Vol. II).]

2.12.3. The Normal (Gaussian) Distribution

The essential result of the Central Limit Theorem is its capability to explain the frequency with which the normal distribution is encountered. For example, many physical measurements are randomly in error because they are the result of the summation of many minute, contributing discrepancies. Indeed, Gauss's acquaintance with measurement errors in the trajectories of heavenly bodies was an exemplary experience of this type.

As will be shown later, many random variables that arise as the response of a large-scale, stochastic simulation model will, in effect, be the sum of a relative large number of contributing effects, so that, in these instances, the resulting random variable of interest may have the Gaussian distribution. [Equally important, there will be cases in which one may anticipate a non-Gaussian distributional form of the simular response; such cases will be illustrated in Chapter 6. See also Mihram (1972a).]

The function

$$\varphi(x; \mu, \sigma) = (2\pi)^{-1/2} e^{-(x-\mu)^2/2\sigma^2}/\sigma$$

qualifies as a probability density, since the square of its integral over the entire real line becomes

$$\begin{aligned}
I^2 &= \int_{-\infty}^{\infty} \exp[-(x-\mu)^2/2\sigma^2]\, dx \int_{-\infty}^{\infty} \exp[-(y-\mu)^2/2\sigma^2]\, dy \cdot (\sigma^2 \cdot 2\pi)^{-1} \\
&= (\pi)^{-1} \int_{-\infty}^{\infty} \int_{-\infty}^{\infty} \exp[-(u^2 + v^2)]\, du\, dv \\
&= (\pi)^{-1} \int_{\theta=0}^{2\pi} \int_{r=0}^{\infty} \exp(-r^2) r\, dr\, d\theta = \int_{t=0}^{\infty} \exp(-t)\, dt = 1.
\end{aligned}$$

Hence, since I is the integral of a positive-valued function,

$$I \equiv \int_{-\infty}^{\infty} \exp[-(x-\mu)^2/2\sigma^2]\, dx \cdot (\sigma\sqrt{2\pi})^{-1} = +1.$$

The family of normally distributed random variables is closed under linear transformations of the form

$$Y = aX + b,$$

since, if $X \sim N(\mu, \sigma^2)$, the random variable Y, being a monotone transformation of X, has the probability density function

$$\begin{aligned}
f_Y(y) &= \varphi((y-b)/a; \mu, \sigma) \cdot |\, a\,|^{-1} \\
&= (|\, a\,|\, \sigma)^{-1}(2\pi)^{-1/2} \exp[-(y - b - a\mu)^2]/2a^2\sigma^2 \\
&= \varphi(y; a\mu + b, \sigma\,|\, a\,|);
\end{aligned}$$

that is,

$$Y \sim N(a\mu + b, a^2\sigma^2).$$

One can confirm that the first two moments of $Y = aX + b$ are given by the quite general results

$$E(Y) = aE(X) + b = a\mu + b \quad \text{and} \quad \text{Var}(Y) = a^2\, \text{Var}(X) = a^2\sigma^2,$$

so that the required parameters for specifying the applicable Gaussian density for Y can be obtained in a straightforward manner.

Of particular interest is the linear transformation

$$Z = (X - \mu)/\sigma,$$

where $X \sim N(\mu, \sigma^2)$. The reader can readily verify that $Z \sim N(0, 1)$, for which the probability density function becomes:

$$\varphi(z; 0, 1) = (2\pi)^{-1/2} \exp(-z^2/2) \qquad \text{for all real} \quad z.$$

The corresponding cumulative distribution function becomes, for any real z,

$$\Phi(z) = \int_{-\infty}^{z} \varphi(t; 0, 1)\, dt = (2\pi)^{-1/2} \int_{-\infty}^{z} \exp(-t^2/2)\, dt,$$

known and tabulated widely as the *error function* of Gauss. An appropriate table of centiles of this distribution is provided in the Appendix (Table A.2). The random variable Z is referred to as the *standardized normal random variable*, and has zero mean and unit variance. Probability statements about any other normally distributed random variable, say $X \sim N(\mu, \sigma^2)$, can be determined from the tabulations of the cumulative distribution function of the standardized variate. For example, for any real numbers c and d,

$$\begin{aligned} P[c \le X \le d] &= P[(c - \mu)/\sigma < Z < (d - \mu)/\sigma] \\ &= \Phi((d - \mu)/\sigma) - \Phi((c - \mu)/\sigma). \end{aligned}$$

Also of note is the fact that the function $\varphi(z; 0, 1)$ is symmetric about zero; i.e., for any real number a,

$$\varphi(a; 0, 1) = \varphi(-a; 0, 1) = (2\pi)^{-1/2} \exp(-a^2/2).$$

As a direct consequence, the cumulative distribution function $\Phi(z)$ for the standardized normal random variate has the property that, for any $a > 0$,

$$1 - \Phi(a) = \Phi(-a).$$

Owing to this relationship, tabulation of the cumulative distribution function $\Phi(z)$ for negative arguments z is frequently omitted (as in Table A.2).

Nonetheless, tabulations of the cumulative distribution function for the standardized normal variate are sufficient for the purpose of deriving probability statements for any normally distributed random variable of known mean μ and variance σ^2.

2.12.4. Methods for Generating Gaussian Variates

Since, as a consequence of the Central Limit Theorem, the normal distribution will frequently be invoked as the distributional form applicable to stochastic elements of a system that is to be simulated, one will need methods by which Gaussian random variates can be generated automatically and efficiently. The use of the CDF transformation is somewhat constrained due to the fact that the Gaussian cumulative distribution function $\Phi(z)$ is represented by the intractable integral

$$\Phi(z) = (2\pi)^{-1/2} \int_{-\infty}^{z} \exp(-t^2/2)\, dt \qquad \text{for all} \quad z.$$

One approach, nonetheless, would be to employ the tabulated cumulative distribution function as though it were the cumulative distribution for a discrete random variable. Uniformly distributed random variables could then be generated and compared with this tabulated cumulative distribution function, much as depicted in Fig. 2.1. Refinements could be made by using linear or polynomial interpolation to arrive at z-values situated between the tabulated values. Of course, the resulting Z-value would then be transformed according to

$$X = \mu + \sigma Z$$

if a Gaussian random variate of mean μ and variance σ^2 is desired. [See Muller (1958).]

Because of the need either for extensive tabulations of $\Phi(z)$ or for the interpolation routines, other means have been sought for the generation of Gaussian distributed random variables. One somewhat acceptable procedure derives from applying the Central Limit Theorem itself, since the sum of n independently and uniformly distributed random variables will provide a random variable S that is approximately normally distributed, with $E(S) = n/2$ and $\text{Var}(S) = n/12$. The n uniformly distributed random variables can be readily obtained as the n successively generated variates from a technique such as the mixed or multiplicative congruential method, as described in Section 2.8. The choice of n then becomes a decision between the increased cost of generating larger numbers of uniformly distributed variates and the corresponding increase in the approximate normality of their sum. For most purposes, the choice of $n = 6$ or $n = 12$ is made because of the corresponding computational simplicity in obtaining standardized normal variates; for $n = 6$, one takes

$$Z = \left[\sum_{i=1}^{6} U_i - 3\right] \cdot \sqrt{2},$$

whereas, for $n = 12$, one computes

$$Z = \left[\sum_{i=1}^{12} U_i - 6\right]$$

as an approximately normally distributed random variable of mean zero and variance unity.

A third, more exact, technique for generating normally distributed random variables will be presented in Section 2.14.5.

EXERCISES

1. Generate 100 approximately normally distributed random variables of mean zero and variance one by employing 100 pseudorandom, uniformly distributed random variables and the tabulated cumulative distribution function $\Phi(z)$ for the standardized Gaussian variate, as given in the Appendix. Use linear interpolation wherever appropriate. Construct a frequency histogram of ten cells for the 100 Gaussian variates and compare with the normal distribution by means of the Chi-squared test.

2. Generate 100 approximately normally distributed random variables of mean zero and variance one by generating 1200 uniformly distributed random variables U_i and computing, for each of the 100 nonoverlapping dozens of numbers:

$$Z = \left[\sum_{i=1}^{12} U_i - 6\right].$$

Compare the 10-cell histogram of Z-values with $\Phi(z)$ via the Chi-squared test criterion.

3. Plot the sample cumulative distribution functions arising in Exercises 1 and 2. By performing the Kolmogorov–Smirnov two-sample test, as described by Lindgren (1968), test whether the two represent the same distribution. Comment on the relative merits of the generation techniques of Exercises 1 and 2, citing computer running, programming, and debugging times as well as the apparent acceptability of the results.

4. From 200 successive, uniformly distributed random (or pseudo-random) numbers, compute 100 sums by adding the elements of the 100 successive pairs of random numbers:

$$T_2 = (U_1 + U_2).$$

Construct a frequency histogram of 10 cells for the 100 T_2-values, and compare this with the theoretically applicable triangular distribution via a Chi-squared test.

5. From 100 successive, uniformly distributed random (or pseudo-random) numbers U_i compute

$$V_i = 2U_i, \qquad i = 1, 2, \ldots, 100.$$

Compare the frequency histogram of these 100 variates V_i with that of the variates T_2 of Exercise 4. Explain why the two random variables V_i and T_2 are differently distributed.

6. Show that the moment-generating function $m_Z(t)$ for the standardized normal random variable Z is

$$\begin{aligned} m_Z(t) &= (2\pi)^{-1/2} \int_{-\infty}^{\infty} \exp(tz) \exp(-z^2/2)\, dz \\ &= \{\exp(t^2/2)\}\, (2\pi)^{-1/2} \int_{-\infty}^{\infty} \exp[-(z - t)^2/2]\, dz = \exp(t^2/2). \end{aligned}$$

Show then that the moment-generating function for $X \sim N(\mu, \sigma^2)$ is

$$m_X(t) = \exp\{\mu t + (\sigma^2 t^2)/2\}.$$

7. Show that if S represents the sum of n independently and identically distributed random variables X_i, then the random variable

$$\bar{X} = \frac{S}{n} = \sum_{i=1}^{n} \frac{X_i}{n}$$

will also have limiting normal distribution, yet of mean μ and variance σ^2/n.

8. Because of its finite word size, a computer is forced to round off the least significant digit in the product of two numbers. If one considers that the decimal point were to occupy the position just to the right of the least significant digit, the round-off error U could be considered as a random variable rectangularly distributed between $-\frac{1}{2}$ and $+\frac{1}{2}$. Show that the probability of an error's exceeding 10 (in absolute value) for the answer to the sum

$$t = \sum_{i=1}^{300} (x_i y_i),$$

may be approximately given by $P[|Z| > 2] = 0.046$, where $Z \sim N(0, 1)$. [Note that the accumulated error is the sum $S = \sum_{i=1}^{300} U_i$ of 300 uniformly and independently distributed random variables, such that $E(S) = 0$, $\text{Var}(S) = 25$.]

2.13. Other Techniques for Generating Random Variables

In this section a number of random variables and their distributional forms will be introduced. In most cases, the theoretical conditions under which nondeterministic experiments are likely to produce each type of random variable will be clearly indicated, just as the previous derivations for the binomial, exponential, geometric, Rayleigh, Weibull, and Gaussian families of random variables have been developed. These developments provide an understanding of the conditions under which these random variates arise and will therefore be of use, in many cases, in suggesting methods by which the random variables can be generated automatically —without resorting to time-consuming searches within tabulations of the appropriate cumulative distribution functions.

One may recall that a random variable X from the Gaussian family can be generated through tabulations of the standardized normal cumulative distribution function; fortunately, in this case, all the members of the Gaussian family are linearly related ($X = \mu + \sigma Z$), so that tabulations of only a single cumulative distribution function are required for the entire family.

In general, however, the parameters indexing a distributional family do not permit rapid and facile transformation from one member of the family to another. In such an event, one is faced either with the awesome choice of preparing separate tabulated cumulative distribution functions for each parametric specification or the alternative of "simulating" the random process (experiment) that, theoretically, should produce the desired random variable. Usually, the second alternative is the more palatable, as was illustrated in the preceding section by the "simulation" of the conditions of the Central Limit Theorem in order to produce approximately Gaussian variates.

2.13.1. The Poisson Family

If the probability of the occurrence of an event is exceedingly small, then its observation should be quite rare. However, if the opportunity for the event to occur is granted sufficiently often, the rare event will be observed from time to time. It is of interest to ascertain the probability of observing the normally rare event once, twice, three times, . . . , etc., in a short observational period. For example, the probability of an air flight's ending in tragedy has become relatively small, yet there are thousands of individual flights daily; an appropriate query is concerned with the likelihood of there being k tragic flights in any given week of the year, $k = 0, 1, 2, \ldots$.

If one presumes that the experiment of interest, such as an air flight, constitutes a Bernoulli trial, for which the probability p of observing the event in question is quite small, then the number S of occurrences (successes) in a large number n of trials of the experiment, will have the binomial distribution

$$P[S = k] = \binom{n}{k} p^k (1 - p)^{n-k}, \qquad k = 0, 1, 2, \ldots, n.$$

Even though n is presumed quite large, the DeMoivre–Laplace form of the Central Limit Theorem might not be deemed applicable because of the moderate size of the product, $\lambda = np$ (e.g., $\frac{1}{10} < \lambda < 10$; in the case of the fatal aircraft flights, λ might represent the average number of accidents among the n flights per week).

Under this condition, the probability of observing zero occurrences of the event among the n trials becomes

$$P[S = 0] = (1 - p)^n = (1 - \lambda/n)^n \cong e^{-\lambda}.$$

Furthermore, the ratio between the successive probabilities is

$$\frac{P[S = k + 1]}{P[S = k]} = \frac{(n - k)p}{(k + 1) \cdot (1 - p)} \approx \frac{\lambda - kp}{(k + 1)},$$

since $(1 - p)$ is approximately 1, or

$$\frac{P[S = (k + 1)]}{P[S = k]} \cong \frac{\lambda}{(k + 1)} \qquad \text{for moderate values of } k.$$

Consequently, the recursion

$$P[S = (k + 1)] \cong \lambda P[S = k]/(k + 1), \qquad k = 0, 1, 2, \ldots$$

can be employed in order to determine the (approximate) probability distribution function:

$$p(k; \lambda) \equiv P[S = k] = \lambda^k e^{-\lambda}/k!, \qquad k = 0, 1, 2, \ldots$$

termed the *Poisson distribution of parameter* λ.

The reader may verify that the infinite sum indeed converges to unity, since

$$\sum_{k=0}^{\infty} (\lambda^k e^{-\lambda}/k!) = e^{-\lambda} \sum_{k=0}^{\infty} (\lambda^k/k!) = e^{-\lambda+\lambda} = 1,$$

the second summation's being the Maclaurin series expansion for e^{λ}. Moreover, the parameter λ represents $E(S)$, the mean value of S, as well as the variance of S, which may be verified via successive differentiations of the moment-generating function:

$$m_S(t) = \exp\{\lambda(e^t - 1)\}, \qquad \text{valid for all real} \quad t.$$

In order to generate Poisson-distributed variates of a given mean (λ), one may employ tabulations of the cumulative distribution function,

$$F_S(x) = \sum_{k=0}^{[[x]]} p(k; \lambda),$$

where $[[y]]$ denotes the greatest integer that is less than or equal to y. A more general technique for generating such variates will attend the discussion of the Poisson process in Chapter 3.

2.13.2. The Pascal Family

In performing a sequence of Bernoulli trials, each of which has probability p of success, the waiting time T_r until the occurrence of the rth success may be of importance (for some fixed r). The reader may recall that T_1, the waiting time until the first success, has the geometric distribution of Eq. (2.3:1):

$$p_{T_1}(k) = p(1-p)^{k-1}, \qquad k = 1, 2, \ldots .$$

If the rth success is to occur exactly on the kth ($k \geq r$) trial, then: (a) the kth trial must end in a success; and, (b) the $(k-1)$ preceding trials must have contained exactly $(r-1)$ successes and $(k-r)$ failures, so that the combined probability of these two events is

$$P[T_r = k] = p\binom{k-1}{r-1}p^{r-1}(1-p)^{k-r}, \qquad k = r, r+1, \ldots.$$

The probability density function for the Pascal random variable T_r becomes then

$$P[T_r = k] \equiv p_{T_r}(k) = \binom{k-1}{r-1}p^r(1-p)^{k-r}, \qquad k = r, r+1, \ldots,$$

also referred to as the *negative binomial distribution* of parameters p and r.

The reader should verify that $p_{T_1}(k)$ is indeed the geometric distribution, as previously discussed, by substituting $r = 1$ in this distributional form. Indeed, T_r is the sum of r geometrically distributed random variables, each having parameter p equal to the probability of success, since the waiting time until the rth success is the sum of the waiting time T until the first success after the $(r-1)$st success and the waiting time T_{r-1} of that $(r-1)$st success:

$$T_r = T_{r-1} + T, \qquad r = 2, 3, \ldots.$$

In generating Pascal random variables, it may prove convenient to sum r independent geometric variates, especially if the parameter r varies among successive generation requirements, yet p remains constant.

2.13.3. The Trinomial Distributions

A straightforward generalization of the binomial random variable, equal to the number of occurrences of one of two dichotomous outcomes among n Bernoulli trials, is a bivariate random vector (X, Y) which gives the number of occurrences of each of two of three possible outcomes (A, B, and C) among n trials. Any given trial may produce one, and only one, of the three mutually exclusive and exhaustive outcomes A, B, and C with respective probabilities:

$$p \equiv P(A), \qquad q \equiv P(B), \qquad (1-p-q) = P(C)$$

the probabilities remaining invariant from trial to trial. Thus, two random variables

X = number of occurrences of A in n successive trials and
Y = number of occurrences of B in the n trials,

can be defined for the experiment (which consists of the n trials). The number of occurrences of event C can be given in terms of X and Y: $(n - X - Y)$.

The joint probability distribution function for X and Y becomes

$$p_{X,Y}(k, l) = n!p^k q^l (1 - p - q)^{n-k-l}/[k!l!(n - k - l)!],$$

for integral k and l such that $0 \leq k \leq n$, $0 \leq l \leq n$, and $0 \leq k + l \leq n$. Generation of the bivariate random variables is probably best accomplished as a two-step procedure, treating first outcomes B and C as a unit outcome (the complement of A) and generating the binomial variate X according to the binomial distribution $b(k; n, p)$. The variate Y, if given that $X = k$, has conditional distribution:

$$p_{Y|k}(l \mid X = k) = (n - k)!\varrho^l (1 - \varrho)^{n-k-l}/[l!(n - k - l)!],$$

for $0 \leq l \leq (n - k)$ and $\varrho = q/(1 - p)$; i.e.,

$$p_{Y|k}(l \mid X = k) = b(l; n - k, \varrho).$$

The resulting pair of variates (X, Y) will have the trinomial distribution of parameters n, p, and q.

2.13.4. The Erlang Family

As has been shown, the exponential family of random variables is often applicable to the description of the distribution of the waiting time until the demise of certain components; for example, with X denoting the (random) lifetime of a component, its probability density function becomes

$$f_X(x; \lambda) = \begin{cases} \lambda e^{-\lambda x}, & x > 0, \\ 0, & \text{other } x, \end{cases}$$

whenever the appropriate conditions obtain. Presuming that failed components are immediately replaced, and that the lifetimes of the successive replacements are independently and exponentially distributed random variables of the same mean $(1/\lambda)$, the total lapsed time until the rth replacement is required becomes a random variable,

$$\tau_r = \sum_{i=1}^{r} X_i,$$

where each X_i has the exponential distribution of mean $(1/\lambda)$ and represents the lifetime of the ith component placed into service.

Proceeding inductively, the moment-generating function for τ_2 becomes

$$m_{\tau_2}(t) \equiv E[\exp(t\tau_2)] = E[\exp(t(X_1 + X_2))] = E[\exp(tX_1)] \cdot E[\exp(tX_2)]$$

since $\tau_2 = (X_1 + X_2)$ and since X_1 and X_2 are independently distributed. Hence

$$m_{\tau_2}(t) = [m_X(t)]^2 = \lambda^2/(\lambda - t)^2, \qquad \text{for} \quad t < \lambda,$$

since X_1 and X_2 have the same exponential distribution. From tabulations of Laplace transforms, one may verify that $m_{\tau_2}(t)$ corresponds to the probability density function

$$f_{\tau_2}(y) = \begin{cases} \lambda^2 y e^{-\lambda y}, & y > 0, \\ 0, & y < 0; \end{cases}$$

owing to the unicity property of these transforms, $f_{\tau_2}(y)$ represents the probability density function for $\tau_2 = X_1 + X_2$.

The moment-generating function for $\tau_3 = (X_1 + X_2 + X_3) = (\tau_2 + X_3)$ then becomes

$$m_{\tau_3}(t) \equiv E[\exp(t\tau_3)] = E[\exp(t\tau_2)]E[\exp(tX_3)] = \lambda^3/(\lambda - t)^3,$$

for $t < \lambda$ and the corresponding probability density function is

$$f_{\tau_3}(y) = \begin{cases} \lambda^3 y^2 e^{-\lambda y}/2!, & y > 0, \\ 0, & y \le 0. \end{cases}$$

More generally, the probability density function for τ_r is

$$f_{\tau_r}(y; \lambda, r) = \begin{cases} \lambda^r y^{r-1} e^{-\lambda y}/(r-1)!, & y > 0, \\ 0, & y \le 0, \end{cases}$$

where the parameters λ (>0) and r (integral, positive) index the family of densities known as the *Erlang distributions*.

The Erlang distributions are also especially applicable in the study of queueing systems and inventory policy studies, where the total time to process r queued customers or r unfilled orders might be represented by a random variable that can be viewed as the cumulative sum of r independently and exponentially distributed random variables of the same mean $(1/\lambda)$.

The mean value of the Erlang variate τ_r can be found either from the probability density function or from differentiating the moment generating function:

$$E[\tau_r] = r/\lambda = rE(X).$$

Similarly,

$$\text{Var}(\tau_r) = \text{Var}[X + \tau_{r-1}] = \text{Var}(X) + \text{Var}(\tau_{r-1}),$$

owing to the presumed independence; or,

$$\text{Var}(\tau_r) = r\,\text{Var}(X) = r/\lambda^2.$$

In order to generate an Erlang variate of parameters r and λ, one essentially "simulates" the component replacement policy, as described above; that is, by generating r independently and uniformly distributed random variables U_i one can obtain the r independently and exponentially distributed variates,

$$X_i = (-1/\lambda) \ln U_i, \qquad i = 1, 2, \ldots, r,$$

according to the parameter λ at hand. The desired Erlang variate then becomes

$$\tau_r = \sum_{i=1}^{r} X_i,$$

or, equivalently,

$$\tau_r = (-1/\lambda) \cdot \left(\sum_{i=1}^{r} \ln U_i \right) = (-1/\lambda) \cdot \ln\left(\prod_{i=1}^{r} U_i \right). \qquad (2.13.4{:}1)$$

Therefore, the generation of Erlang-distributed variates is a straightforward logarithmic transformation of r uniformly distributed random variables. Since $\tau_r = \sum_{i=1}^{r} X_i$, the Central Limit Theorem becomes somewhat applicable if r is sufficiently large. Indeed, with increasing r the Erlang family of distributions approaches more closely the Gaussian distribution of mean r/λ and variance r/λ^2; for $r > 10$, the agreement between the Erlang distribution and the corresponding Gaussian distribution of the same mean and variance becomes quite good for most practical purposes. Unless r is considerably larger, however, one would likely proceed to generate Erlang variates as the sum of r exponentially distributed variates, despite the requisite logarithmic transformation of the uniform variates.

2.13.5. THE CHI-SQUARED FAMILY

The standardized normal variate Z is that member of the Gaussian family of random variables having mean zero and unit variance. Its probability density function is given by

$$\varphi(z; 0, 1) = (2\pi)^{-1/2} \exp(-z^2/2) \qquad \text{for any real} \quad z.$$

Since the mean value of Z is zero, the variance becomes equal to the expectation of Z^2:

$$\mathrm{Var}(Z) = E(Z^2) - [E(Z)]^2 = E(Z^2) = 1.$$

The square of a standardized random variable is of especial interest in the theory of statistics, and is referred to as the Chi-squared random variable (of 1 degree of freedom). Its cumulative distribution function is given by

$$F_{\chi^2}(c) \equiv P[\chi^2 \equiv Z^2 \leq c] = P[-\sqrt{c} < Z < +\sqrt{c}\,]$$

for any $c > 0$; that is,

$$F_{\chi^2}(c) = (2\pi)^{-1/2} \int_{-\sqrt{c}}^{+\sqrt{c}} \exp(-t^2/2)\, dt = (2/\pi)^{1/2} \int_0^{\sqrt{c}} \exp(-t^2/2)\, dt,$$

since the integrand is an even function. Thus,

$$F_{\chi^2}(c) = \begin{cases} \int_0^c (\frac{1}{2})^{1/2} u^{(1/2)-1} e^{-u/2}\, du/\Gamma(\frac{1}{2}), & c > 0 \\ 0, & c < 0 \end{cases}$$

since Euler's Gamma function of argument $\frac{1}{2}$ equals $\sqrt{\pi}$, so that the probability density function for $\chi^2 = Z^2$ becomes

$$f_{\chi^2}(u) = \begin{cases} (\frac{1}{2})^{1/2} u^{(1/2)-1} e^{-u/2}/\Gamma(\frac{1}{2}), & u > 0 \\ 0, & u \leq 0. \end{cases}$$

The corresponding moment-generating function becomes

$$m_{\chi^2}(t) \equiv E[\exp(t\chi^2)] = [(\tfrac{1}{2})/(\tfrac{1}{2} - t)]^{1/2},$$

or

$$m_{\chi^2}(t) = [1/(1 - 2t)]^{1/2} \qquad \text{for} \quad t < \tfrac{1}{2}.$$

From this Laplace transform, the moments can be determined:

$$E[\chi^2] = E[Z^2] = \text{Var}(Z) = 1$$

and

$$\text{Var}[\chi^2] = 2 = \mu_4(Z).$$

Equally important in the theory of statistics, and therefore in the discussion of the analyses of the responses of a stochastic simulation model, is the distribution of the sum of r independently distributed Chi-squared variates:

$$\chi_r^2 \equiv \sum_{i=1}^{r} Z_i^2,$$

where the r Z_i's are independently distributed and standardized normal random variables. The moment-generating function for the sum of the squares of two independent, standardized variates becomes

$$\begin{aligned} m_{\chi_2^2}(t) &\equiv E[\exp(t\chi_2^2)] = E[\exp(t(Z_1^2 + Z_2^2))] \\ &= [m_{\chi^2}(t)]^2 = (\tfrac{1}{2})/(\tfrac{1}{2} - t), \qquad \text{for} \quad t < \tfrac{1}{2}. \end{aligned}$$

However, this moment-generating function corresponds to that of the exponentially distributed random variable of mean 2. [See Eq. (2.10.4:1).]

Therefore, in addition to its role as the descriptor of the distribution of the lifetimes of certain entities, the exponential family of densities is seen to possess a member that represents the sum of the squares of two standardized normal random variables. Indeed, an alternative method for the generation of an (approximately) exponentially distributed variate of mean $1/\lambda$ becomes:

$$X = \left[\left(\sum_{i=1}^{12} U_i - 6\right)^2 + \left(\sum_{i=13}^{24} U_i - 6\right)^2\right] \Big/ 2\lambda,$$

where U_i, $i = 1, 2, \ldots, 24$ are independently and uniformly distributed random variables.

Consequently, the exponential distribution of mean 2 is also known as the Chi-squared distribution of two degrees of freedom, since it may arise as the summation of two squared, independent, and standardized normal random variables. Since the exponential distribution is also the first of the Erlang distributions ($\lambda = \frac{1}{2}$, $r = 1$), the sum of $s = 2r$ independent Chi-squared variates must possess the Erlang distribution of parameters $\lambda = \frac{1}{2}$ and r; that is, the Chi-squared distributions of $s = 2r$ degrees of freedom are also members of the Erlang family.

For any positive integer s, the sum of s independent Chi-squared variates may be shown to possess the probability density function

$$f_{\chi_s^2}(u; \tfrac{1}{2}, s) = \begin{cases} (\tfrac{1}{2})^{s/2} u^{-1+(s/2)} e^{-u/2} / \Gamma(s/2), & u > 0, \\ 0, & u < 0, \end{cases}$$

where $\Gamma(s/2)$ is the Gamma function of Euler. The family of densities indexed by s is called the Chi-squared family of s degrees of freedom. The mean value and variance become

$$E[\chi_s^2] = s \qquad \text{and} \qquad \mathrm{Var}[\chi_s^2] = 2s.$$

In order to generate a Chi-squared variate of s degrees of freedom, one requires s independent, standardized normal variates $Z_1, Z_2, \ldots, Z_s$ each readily obtained by either one of the two approximate methods described in the preceding section or by the more exact method that will be described in Section 2.14.5. The Chi-squared variate then becomes

$$\chi_s^2 \equiv \sum_{i=1}^{s} Z_i^2.$$

(The reader should again note the applicability of the Central Limit Theorem for s sufficiently large. Indeed, as $s \to \infty$, the Chi-squared distribution of s degrees of freedom may be closely approximated by the Gaussian distribution of mean s, variance $2s$. Nonetheless, unless s exceeds, say, 20, more accurate results will probably result from generating Chi-squared random variables according to the preceding formula.) If, however, $s = 2r$ is an even integer, the requisite Chi-squared variate of s degrees of freedom can be obtained from the alternative approach, that for generating an Erlang variate:

$$\chi_{2r}^2 = -2\left(\sum_{i=1}^{r} \ln U_i\right) = -2 \ln\left(\prod_{i=1}^{r} U_i\right),$$

where the U_i are independently and uniformly distributed random variables.

2.13.6. Student's t Family

Another family of probability density functions, which are of prime importance in the theory of statistics, is the collection of densities designated by the pseudonym of W. S. Gossett, an English statistician who published under the alias, "Student." A Student's t variate of k degrees of freedom is a random variable T having the probability density function:

$$f_T(t; k) = \frac{\Gamma[(k+1)/2]}{\Gamma(k/2)\sqrt{\pi k}} \left(1 + \frac{t^2}{k}\right)^{-(k+1)/2} \qquad \text{for all real} \quad t.$$

The Student's t variate of k degrees of freedom arises as the ratio of a standardized normal variate to the positive square root of an independently distributed Chi-squared variate, itself divided by its degrees of freedom (k):

$$T = Z/[\chi_k^2/k]^{1/2}.$$

Therefore, the generation of such a Student's t variate of k degrees of freedom requires the independent generation of a standardized normal random variable Z and of a Chi-squared variate of k degrees of freedom; the two variates are then functionally combined as indicated.

The reader can readily verify the probability density for T by defining a second random variable $S = Z$, by finding the joint probability density function $f_{S,T}(s, t)$, and by integrating this bivariate function over all s-values to produce the (marginal) probability density function for T. Of interest is the fact that

$$\lim_{k\to\infty} f_T(t; k) = \varphi(t; 0, 1) \qquad \text{for any real} \quad t;$$

that is, as the number of degrees of freedom increases without bound, the symmetric family of Student's t densities approaches that of the standardized normal distribution.

Also noteworthy is the fact that the Student's t distribution of a single degree of freedom is a member of the Cauchy family of distributions (with parameters $a = 0$, $b = 1$), so that one may take advantage of the CDF transformation technique when generating Student's t variates of one degree of freedom. (See Exercise 10, following Section 2.7 for the generation of Cauchy variates.) In the general case for k degrees of freedom, however, no such advantage exists; in fact, unlike the Chi-squared and Erlang families, the sum of independently distributed Student's t variates does *not* result in a random variable having density function of the same family.

2.13.7. Snedecor's F Family

Another important family of random variables, whose direct application to simular analyses will be amply illustrated in the forthcoming chapters, is Snedecor's family of F variates. Each member of the family is prescribed by two parameters (or degrees of freedom): m and n, since a Snedecor's F variate of m and n degrees of freedom may be given by the ratio of two independently distributed Chi-squared variates, of m and n degrees of freedom, respectively, each divided by its degrees of freedom:

$$F_{m,n} = \frac{\chi_m^2/m}{\chi_n^2/n}$$

Note that the parameter m is associated with the Chi-squared random variable of the numerator, whereas n is the number of degrees of freedom of the Chi-squared variate appearing in the denominator.

The probability density function for $F_{m,n}$ may be derived as the marginal density associated with the joint probability density function of $F_{m,n}$ and $G \equiv \chi_m^2$, as shown in Lindgren (1968). The resulting probability density function becomes

$$f_F(w; m, n) = \begin{cases} C(m, n)w^{(m/2)-1}\left(\dfrac{n}{m} + w\right)^{-(m+n)/2}, & w > 0, \\ 0, & w < 0, \end{cases}$$

where

$$C(m, n) = (n/m)^{n/2}/\beta(m/2, n/2),$$

where $\beta(c, d) = \Gamma(c) \cdot \Gamma(d)/\Gamma(c + d)$ is Euler's Beta function of positive arguments, c and d.

The cumulative distribution function for the Snedecor's F family is, in general, analytically intractable, but this function is tabulated in most elementary textbooks of statistical theory. This fact also implies that the method for generating random variables from one of the Snedecor family of densities is the generation of independently distributed Chi-squared variates (of the appropriate degrees of freedom) and the computation of the ratios defining the F-variate.

Of special interest is the fact that the Snedecor's F-variate of 1 and k degrees of freedom is equivalent to the *square* of the Student's t-variate of k degrees of freedom. To wit,

$$T^2 = Z^2/[\chi_k^2/k],$$

where Z is the standardized normal variate whose square is therefore a Chi-squared variate of a single degree of freedom; hence,

$$T^2 = [(Z^2/1)]/[\chi_k^2/k] = (\chi_1^2/1)/(\chi_k^2/k)$$

is a Snedecor's F-variate of 1 and k degrees of freedom (the random variables of the numerator and denominator being independently distributed).

2.13.8. The Gamma Distributions

A straightforward generalization of the Erlang and Chi-squared families of random variables is the family of Gamma variates, having, in general, the probability density functions

$$g(x;\, a,\, b) = \begin{cases} x^{b-1}e^{-x/a}/(a^b\Gamma(b)), & x > 0 \\ 0, & x < 0, \end{cases}$$

indexed by the positive parameters a and b, where

$$\Gamma(b) \equiv \int_0^\infty u^{b-1}e^{-u}\, du$$

is the standard Euler Gamma function.

The mean value and variance of a random variable having one of the Gamma distributions are

$$E(X) = ab, \qquad \text{and} \qquad \mathrm{Var}(X) = a^2b,$$

as may be derived from the moment-generating function

$$m_X(t) = (1 - at)^{-b}, \qquad \text{provided that} \quad t < a^{-1}.$$

The cumulative distribution function becomes

$$G(x;\, a,\, b) = \gamma(x;\, a,\, b)/\Gamma(b),$$

where

$$\gamma(x;\, a,\, b) = \int_0^{x/a} u^{b-1}e^{-u}\, du,$$

the incomplete Gamma function, a generally intractable integral unless b is integral. The cumulative distribution function has been extensively tabulated by Pearson (1957) and by Harter (1964). The reader may note that if X has density function $g(x;\, a,\, b)$, then $Y = cX$, for $c > 0$, is a random variable having probability density $g(y;\, ca,\, b)$; thus, the random variable $W = (X/a)$ has density $g(w;\, 1,\, b)$, thereby implying that probability statements of the form

$$P[X \le x] \qquad \text{and} \qquad P[W \le (x/a)]$$

are equivalent. Hence, the cumulative distribution functions $G(x; a, b)$ need be tabulated only for specific values of the shape parameter b.

The generation of the general Gamma variate of parameters a and b requires a tabulation of the incomplete Gamma function ratio:

$$\gamma(w; 1, b)/\Gamma(b).$$

Uniformly distributed variates may then be compared with this table as though a discrete random variable were being generated; linear or polynomial interpolation schemes can then be employed in order to refine W, from which one may obtain

$$X = aW \qquad \text{for the given scale parameter} \quad a.$$

In the event that $b = r$, a positive integer,

$$G(x; a, r) = \int_0^x a^{-r} u^{r-1} e^{-(u/a)} \, du/(r-1)!.$$

Partial integration provides

$$G(x; a, r) = \begin{cases} 1 - e^{-(x/a)}, & r = 1, \\ G(x; a, r-1) - (x/a)^{r-1} e^{-(x/a)}/(r-1)!, & r > 1, \end{cases}$$

so that

$$G(x; a, r) = 1 - \sum_{k=0}^{r-1} \{(x/a)^k e^{-(x/a)}/k!\},$$

directly relatable to the Poisson cumulative distribution function of mean value (x/a). However, in this case, the random variable X is an Erlang variate of parameters r and $a = 1/\lambda$, so that one need not rely on tabulations of $G(x; a, r)$ in order to generate these random variables, but instead can employ the sum of r independently distributed exponential variates, each of mean a. [See Eq. (2.13.4:1).]

Similarly, the Chi-squared family of random variables can be viewed as a subfamily of the Gamma family; in particular, $a = 2$ and $b = s/2$, for s equal to a positive integer, specify the Chi-squared distribution of s degrees of freedom. Therefore, whenever one seeks to generate a Gamma-distributed variate of parameters a and $(s/2)$, he can rely on the previously discussed procedures for generating a Chi-squared variate of s degrees of freedom χ_s^2. The desired variate then becomes

$$X = a \cdot \chi_s^2/2.$$

In the general case, however, one must rely on a tabulation of the appropriate cumulative distribution function, as indexed by the shape parameter b. [See also Phillips (1971).]

2.13.9. The Beta Family

Most of the continuous random variables discussed to this point have density functions that are positive over some infinite extent of real numbers (for example, the Cauchy and Gaussian families over all real numbers; the Weibull, Gamma, and Snedecor families over all positive reals). Virtually the only family of variates confined to a finite range (a, b) has been the rectangularly distributed random variables, of density

$$f_U(r; a, b) = \begin{cases} 1/(b - a), & a \le r \le b \\ 0, & \text{other } r. \end{cases}$$

An important generalization of this family of variates is the Beta family, having density functions

$$f_\beta(x; a, b, c, d) = (x - a)^{c-1}(b - x)^{d-1}/[(b - a)^{c+d}\beta(c, d)],$$

where $a \le x \le b$, and

$$\beta(c, d) = \Gamma(c)\Gamma(d)/\Gamma(c + d) = \int_0^1 u^{c-1}(1 - u)^{d-1}\, du$$

is the standard Beta function of Euler of positive parameters c and d. The parameters a and b serve only to delineate the range of values that the variate can assume; specifically, if β is a random variable having the density $f_\beta(x; a, b, c, d)$, then

$$B = (\beta - a)/(b - a)$$

is a random variable having the density $f_\beta(x; 0, 1, c, d)$ over the unit interval.

Hence, the generation of Beta-distributed random variables can be directly accomplished by transforming variates B derived from employing the cumulative distribution function

$$F_\beta(x; 0, 1, c, d) = \int_0^x u^{c-1}(1 - u)^{d-1}\, du/\beta(c, d),$$

with c and d appropriately specified. Unless the positive parameters c and

d are integers, these cumulative distribution functions become analytically intractable, though they have been extensively tabulated by Pearson (1948) and by Harter (1964). Interpolations within these tables can be used to generate Beta random variables from uniformly distributed variates (via essentially the CDF transformation technique).

Whenever c and d are both positive integers, the cumulative distribution function becomes a polynomial of degree $(c + d - 1)$, so that the CDF transformation technique is especially applicable, though the method requires the solution of the resulting polynomial for a root lying between 0 and 1. In the event that either c or d is unity, considerable simplification of the cumulative distribution function will also result, and the CDF transformation technique shall prove practicable. Of course, if $c = d = 1$, the Beta family of distributions becomes the rectangular (uniform) family, for which facile generation techniques have already been presented.

A general derivation of the Beta family of random variables notes that, if X_1 has the Gamma density $g(x_1; 1, c)$, as given in the Section 2.13.8, and is independently distributed of X_2, which has the Gamma density $g(x_2; 1, d)$, then the random variable

$$B = X_1/(X_1 + X_2)$$

has the probability density function $f_\beta(x; 0, 1, c, d)$. [The reader can verify this result by finding the joint probability density function of B and $C \equiv (X_1 + X_2)$, then finding the (marginal) density for B therefrom.] Consequently, Beta variates can be produced by first generating the independent Gamma variates X_1 and X_2 (with appropriate shape parameter values of c and d, respectively) and then transforming these according to

$$\beta = a + [(b - a)X_1/(X_1 + X_2)].$$

This approach is especially fruitful either whenever c and d are integral, so that X_1 and X_2 are Erlang variates, or whenever c and d are half-integers, so that X_1 and X_2 are proportional to Chi-squared variates; in either of these cases, the necessary Gamma variates can be generated without resorting to tabulations of the cumulative Gamma distribution functions.†

† One may also generate the random variable B having density f_β $(x; 0, 1, c, d)$, as the cth smallest value among $(c + d - 1)$ successively generated, independently and uniformly distributed, random variables, whenever c and d are integral. See also Fox (1963).

The family of Beta distributions is quite versatile in describing the distribution of random variables β which are known to be distributed over a finite range. The parameters a and b delineate this range, whereas the parameters c and d permit a considerable flexibility in shaping the density functions over this range [cf. Hahn and Shapiro (1967)]. The first moments of β are given by

$$E(\beta) = (ad + bc)/(c + d)$$

and

$$\text{Var}(\beta) = (b - a)^2 \cdot cd/[(c + d)^2(c + d + 1)],$$

and might be employed to select the parameters c and d (for given range delimitations, a and b) in a given application. [For further information regarding the generation of Beta variates, see Fox (1963).]

2.13.10. Extreme-Value Distributions

Another important class of random variables arises as the description of extremal measures. For example, the maximum depth of the river Arno at the Ponte Vecchio during any calendar year represents a random variable that is the largest value in the distribution of depths during that year; alternatively, the minimum mark on a final examination given annually to a simulation class represents a random variable that measures an extreme condition. Under appropriate conditions [cf. Hahn and Shapiro (1967)], the probability density function applicable to these extremal random variables X is of the form

$$e(x; a, c) = \{e^{(x-a)/c} \exp -[e^{(x-a)/c}]\}/|c|,$$

for all real x and for real parameters a and c ($\neq 0$). Whenever $c < 0$, the density is that of the maximum random variable; with $c > 0$, the density serves to describe the distribution of minimal variates.

The mean value of the density is given by

$$E(X) \cong a + 0.5772c$$

and its variance is

$$\text{Var}(X) \cong 1.645c^2.$$

One can show that, if X has the density $e(x; a, c)$, then $Z = (X - a)/c$ has the density $e(z; 0, 1)$, so that a method for generating variates, Z, from the density

$$e(z; 0, 1) = e^z \exp -[e^z]$$

will suffice for the more general case by transforming the resulting Z-variate according to

$$X = a + cZ.$$

The cumulative distribution function for the normalized variate Z becomes

$$E(z; 0, 1) = 1 - \exp -[e^z],$$

so that the CDF transformation technique becomes directly applicable to the generation of these variates:

$$Z = \ln[-\ln V],$$

where $V = (1 - U)$ is a uniformly distributed random variable; hence, the generation of a random variable X from the density $e(x; a, c)$ is readily accomplished as

$$X = a + c \ln[-\ln V],$$

where V is uniformly distributed. [The reader may recall that $-\ln U$ is a random variable of mean one having the exponential distribution and may therefore wish to show that the natural logarithm of such an exponentially distributed random variable has the extreme-value density function $e(z; 0, 1)$.]

2.13.11. The Log-Normal Distribution

A positive-valued random variable whose natural logarithm has the Gaussian distribution of mean μ and variance σ^2, is said to possess the log-normal distribution:

$$l(x; \mu, \sigma) = \begin{cases} (x\sigma)^{-1}(2\pi)^{-1/2} \exp[-(\ln x - \mu)^2/2\sigma^2], & x \geq 0 \\ 0, & x \leq 0. \end{cases}$$

The parameter μ is real, whereas σ need be positive. Its mean-value is given by $\exp[\mu + (\sigma^2/2)]$, and its variance by $[\exp(2\mu + \sigma^2)][\exp(\sigma^2) - 1]$. The density may arise as the limiting form applicable to a random variable that is esentially the product of n independently distributed random variables; indeed, a "Central Limit Theorem" for products (rather than sums) of positive-valued random variables would predict the applicability of this family of distributions, as discussed in the monograph of Aitchison and Brown (1957).

There exists then, in addition to the possibility of tabulating the cumulative distribution function L and searching it directly, another method for generating random variates from the log-normal family of densities. The usual approach is the generation of a normally distributed random variable Y of mean μ and variance σ^2, then transformation according to

$$X = e^Y$$

in order to obtain X as the log-normal variate.

2.13.12. The Stacy Family

In 1962, Stacy proposed a generalization of the Gamma family of random variates, the generalized family then being extended by Stacy and Mihram (1965). A random variable S has this extended Stacy distribution if its probability density function is given by

$$f_S(s; a, b, c) = \begin{cases} |c|\, s^{bc-1} \exp[-(s/a)^c]/(a^{bc}\Gamma(b)), & s > 0, \\ 0, & s < 0, \end{cases}$$

where the scale parameter a and the shape parameter b are positive, yet the diminution parameter c may assume any real value. The generality of the distributional forms is depicted in the aforementioned references, its being noted that the extended Stacy family includes as subfamilies the following distributional families:

Exponential	$f_S(s; a, 1, 1)$
Gamma	$f_S(s; a, b, 1)$
Weibull	$f_S(s; a, 1, c),\ c > 0$
Chi-squared	$f_S(s; 2, n/2, 1),\quad n =$ degrees of freedom
Chi $(=\sqrt{\chi_n^2})$	$f_S(s; \sqrt{2}, n/2, 2), n =$ degrees of freedom
Modular normal $(=\lvert Z \rvert)$	$f_S(s; \sqrt{2}, \frac{1}{2}, 2)$
Rayleigh	$f_S(s; c\sqrt{2}, 1, 2),\ c > 0$
Maxwell	$f_S(s; c\sqrt{2}, \frac{3}{2}, 2),\ c > 0$
Erlang	$f_S(s; a, r, 1),\ r$ integral

Of special value are the notes that the random variable $R = kS$ has the density function $f_S(r; ak, b, c)$ and that the random variable $Q = S^p$ has Stacy density $f_S(q; a^p, b, c/p)$; that is, the family of densities remains closed under scale and power transformations. Indeed, the random variable

$$M = (S/a)^c$$

may be termed the normalized Stacy variate, possessing the density $f_S(m; 1, b, 1)$, one of the Gamma densities. Furthermore, the inverse S^{-1} of a Stacy variate is also a Stacy variate.

Therefore, the generation of Stacy variates can be accomplished by the simplest procedure applicable to Gamma variates having the same shape parameter b. For example, if $b = 1$, the cumulative distribution function for the Weibull distributions may be invoked; if $b = r$ (a positive integer), an Erlang variate τ_r of mean r can be generated, and then transformed according to

$$S = a(\tau_r)^{1/c},$$

for the particular scale and diminution parameters a and c at hand; or, if $b = s/2$, an half-integer, a Chi-squared variate χ_s^2, of s degrees of freedom might be first generated, then transformed according to

$$S = a(\chi_s^2/2)^{1/c}.$$

2.13.13. A General Density of Logarithmic Variates

If the natural logarithmic transformation is applied to a Stacy random variable, the resulting random variable is distributed over the entire real axis according to

$$h(m; \alpha, \beta, \gamma) = \{e^{\beta(m-\alpha)/\gamma} \exp -[e^{(m-\alpha)/\gamma}]\}/\{|\gamma| \Gamma(\beta)\},$$

for any real number m and for real parameters α and γ and positive parameter β. Again, $\Gamma(\beta)$ denotes the standard Euler Gamma function. The reader can verify the rectitude of the distribution by deriving the density applicable to the random variable,

$$M = \ln S,$$

where S has the Stacy distribution $f_S(s; a, b, c)$, noting the parametric relationships:

$$\alpha = \ln a, \qquad \beta = b, \qquad \text{and} \qquad \gamma = 1/c.$$

The reader can also verify that the densities $h(m; \alpha, \beta, \gamma)$ are closed under linear transformations. That is, if M has density $h(m; \alpha, \beta, \gamma)$, then $W = pM + q$ has density $h(w; p\alpha + q, \beta, p\gamma)$. Thus, in particular, $Y = (M - \alpha)$ has density $h(y; 0, \beta, \gamma)$ and $Z = (M - \alpha)/\gamma$ becomes the normalized variate of the author's family, having density $h(z; 0, \beta, 1)$, corresponding to the density for the natural logarithm of a Gamma variate of density $g(z; 1, \beta = b)$, as given in Section 2.13.8.

Consequently, the generation of random variables M from the general logarithmic density $h(m; \alpha, \beta, \gamma)$ is readily accomplished via the transformation

$$M = \alpha + \gamma \ln X,$$

where $X \sim g(x; 1, \beta = b)$, the Gamma distribution of unit scale parameter and shape parameter β. Once the Gamma variate has been obtained, the generation of the variates M is then a straightforward task.

The family of densities, $h(m; \alpha, \beta, \gamma)$ can be shown to include the extreme-value densities as a subfamily. The reader can verify directly that

$$e(m; \alpha, \gamma) = h(m; \alpha, 1, \gamma).$$

For further details regarding the family of densities $h(m; \alpha, \beta, \gamma)$, the reader is referred to Mihram (1965), or to Harris (1970).

EXERCISES

1. Generate 100 Erlang variates τ_{10} as the sum of 10 independently and exponentially distributed random variables, each of unit mean. Compare the resulting histogram of 10 cells with the normal distribution of mean 10, variance 10. Comment on the applicability of the Central Limit Theorem to the generation of these Erlang variates instead. Discuss the precautions necessary to avoid negative-valued variates if these Gaussian variates were to be employed in order to approximate the desired Erlang variates.

2. Generate 100 independent Gaussian random variables Y of mean 1 and variance 1. Compute their squares and prepare from these a frequency histogram of 10 cells of equal width. Explain why the resulting histogram is incompatible with the probability density function for the Chi-squared variate of a single degree of freedom. Would a histogram of the squares of the 100 variates $Z = (Y - 1)$ be compatible with this Chi-squared distribution?

3. The reader may recall the earlier discussion of the Chi-squared goodness-of-fit test. If one assumes that k mutually exclusive intervals have been defined to cover the entire range of admissible values for a random variable X, yet in such a way that $100p\%$ of the distribution should lie in each of the k intervals, then the "expected" number, of n independently and randomly selected values from the density, that will fall in each interval is $np = n/k$. In fact, the number S_n of the variates that falls into any particular cell is a binomial random variable, approximately distributed according to the Gaussian density, $\varphi(x; np, np(1-p))$. Therefore, for the ith cell, $(S_n^{(i)} - np)/[np(1-p)]^{1/2} \simeq (S_n^{(i)} - np)/(np)^{1/2}$ is approximately a standardized normal variate. Thus, the random variable

$$(S_n^{(i)} - np)^2/(np)$$

is approximately a Chi-squared variate of one degree of freedom and the sum

$$\chi_c^2 = \sum_{i=1}^{k} (S_n^{(i)} - np)^2/(np)$$

has the (approximate) Chi-squared distribution of $(k - 1)$ degrees of freedom. (The k random variables $S_n^{(i)}$ are not all independent, since $\sum_{i=1}^{k} S_n^{(i)} = n$; hence, the loss of a degree of freedom.) Comment on the appropriateness of the Chi-squared test whenever the p_i differ for the k cells, yet $\sum_{i=1}^{k} p_i = 1$.

4. Exponentially distributed random variables of mean $1/\lambda$ may be generated according to $X_1 = (-1/\lambda) \cdot (\ln U)$, where U is a uniformly distributed random variable or, alternatively, according to

$$X_2 = \left[\left(\sum_{i=1}^{12} U_i - 6\right)^2 + \left(\sum_{i=13}^{24} U_i - 6\right)^2\right] \Big/ 2\lambda.$$

Using a mixed congruential generator for the uniform variates, generate 100 X_1-values and 100 X_2-values of the same theoretical mean. Compare the two histograms and comment on the apparent relative costs of the two techniques.

5. Using moment-generating functions, show that the probability density function for the Chi-squared random variable of s degrees of freedom is as given in the text: $f_{\chi_s^2}(u; \frac{1}{2}, s)$.

6. Comment on the relative merits of the two schemes for generating Chi-squared variates of $s = 2r$ degrees of freedom:

(*a*) χ_{2r}^2 = sum of r exponentially distributed variates, $X_i = -\frac{1}{2} \ln U_i$, $i = 1, \ldots, r$;

(*b*) χ_{2r}^2 = sum of $(2r)$ squared, standardized normal variates Z.

How would you propose to test the alternative schemes?

7. Let Z represent a standardized normal variate, χ_k^2 an independently distributed Chi-squared variable of k degrees of freedom. Show that $T = Z/[\chi_k^2/k]^{1/2}$ has the Student's t distribution described in the text.

8. Show that, as the number k of degrees of freedom becomes unbounded, the Student's t distribution $f_T(t; k)$ approaches the Gaussian curve $\varphi(t; 0, 1)$, for any real argument t.

9. Show that the Beta distribution of parameters $m/2$ and $n/2$ is related to Snedecor's F distribution of m and n degrees of freedom through the transformation:

$$\beta \equiv \frac{mF/n}{[1 + (mF/n)]}.$$

Suggest an alternative method for generating Beta variates from a source of Snedecor's F variates whenever m and n are integral.

10. Select from among 10 standardized Gaussian variates both the smallest and the largest values. Repeat the generation of ten standardized Gaussian variates a total of 100 times, each time recording the largest and the smallest values among the 10 variates. Construct a frequency histogram of the 100 minima and compare with the frequency histogram of the 100 maxima. Compare each with the extreme-value distributions of Section 2.13.10. Are the means of the two histograms significantly different? Are the variances of the two histograms approximately the same?

2.13.14. A Scholium on the Generation of Random Variables

The collection of random variables is difficult to exhaust. The current section has been designed to provide the theoretical bases for many of the more commonly encountered random variables. Often, however, one may have available a frequency histogram of data recorded from the repetitions of a particular experiment, yet insufficient knowledge of the experimental conditions exists, making difficult the prediction of the distributional form that should be theoretically applicable to the measurements (random variables). For example, it may not be clear that the resulting data are arising from the sum of a large number of relatively insignificant effects, so that the applicability of the Central Limit Theorem might remain questionable.

Of course, from the histogram, or the data itself, one may construct a sample cumulative distribution function that may be employed later in a simulation model that requires a random variable of the type in question. Alternatively, one may endeavor to fit curves to the histogram, the curves being appropriately parametrized so that the total area contained beneath them is unity. Systems of such curves have been developed, first by Pearson [cf. Kendall and Stuart (1963, Vol. 1)], and then (another system) Johnson (1949). The interested reader is referred to the text of Hahn and Shapiro (1967) for a discussion of these curve-fitting techniques.

Of historical interest is the fact that the Student's family of t distributions was discovered in just the manner indicated. Gossett (1908) used random numbers to compute a meaningful statistic, or random variable; a frequency histogram of several of these statistics was fitted by curves of the class known today as the Student's t family. [Indeed, Student himself then proved separately that the theoretically applicable densities were indeed the densities $f_T(t; k)$.]

The variety of methods for generating continuous random variables is almost as diverse as the number of density families. Of course, if the cumulative distribution function can be tabulated, its tabulation may be used directly in conjunction with uniformly distributed random variables (U) in order to generate random variables from the desired density. If an analytical and invertible expression for the cumulative distribution function exists (such as is the case for the Cauchy, Pareto, Weibull, Rayleigh, exponential, extreme-value, and rectangular random variables), then the CDF transformation technique is likely to be the most efficient method for generating random variables accordingly distributed.

In many other situations, however, the generation of the desired random variable will be best performed by taking advantage of the theoretical conditions under which the variate arises. One then proceeds to "simulate" these conditions accordingly, such as the approximately normal random variables that arise from the summation of 12 uniformly distributed random variables (an application, or "simulation," of the Central Limit Theorem, though in Section 2.14.5 an alternative and more exact method of generating normally distributed random variables will be presented). Other random variables amenable to this "simulation" approach are the Erlang, Chi-squared, Student's t, and Snedecor's F variates.

Other approaches to the generation of specific random variables would tend to combine the techniques described so far. For example, Gamma variates may need be generated via interpolations within the appropriately tabulated cumulative distribution function; Beta, Stacy, Chi, distended Gamma, and the generalized logarithmic variates of the author may be produced via appropriate transformations of the resulting Gamma variates. Or, as is the case for the logarithmic normal variates, straightforward transformations of the "simulated" Gaussian variables are required.

The monograph of Hammersley and Handscomb (1964) describes a number of other techniques for generating random variables of desired specifications. One of these techniques is of sufficient interest, for present purposes, to be mentioned here: the *rejection technique*, applicable whenever the continuous density function $f(x)$ is finite over a finite range of the real numbers.

Thus, as depicted in Fig. 4.3 (see p. 191), a positive-valued density $f(x)$ is bounded above by a constant c and exists only for arguments x such that $a \leq x \leq b$, for known constants a and b. [The constant c may as well be chosen equal to the maximum value of $f(x)$ over this range; indeed, the rejection method will be made more efficient by this selection of c.] The method proceeds by generating a pair of independently distributed random variables:

$$X_1 = a + U_1(b - a),$$

and

$$X_2 = cU_2,$$

where U_1 and U_2 are independently generated uniform random variables. A comparison of the two quantities is then made: X_2 and $f(X_1)$; if $X_2 > f(X_1)$, then X_1 is rejected, and a new pair (U_1, U_2) of uniform variates is generated, the variates (X_1, X_2) transformed therefrom, and the comparison made again. Whenever $X_2 \leq f(X_1)$ the variate X_1 is taken as a random number from the distribution $f(x)$, as desired.

The rationale for accepting the resulting variates is that of all the variates X_1, which are generated and equal to a specific value x, only that proportion are accepted which, when used as the argument in $f(x)$, provide $f(x) \geq X_2$. Alternatively, only those X_1-values of the coordinates (X_1, X_2) are accepted whenever the coordinate pair lies beneath the curve $f(x)$ describing the density of interest.

EXERCISES

1. Use the rejection technique to generate random variables according to the Beta distribution $f_\beta(x; 0, 1, 3, \pi)$. For what value of c will the method be most efficient?

2. Use the rejection technique to generate random variables according to the Beta distribution $f_\beta(x; -\frac{1}{2}, \frac{1}{2}, 2, 2)$. Compare the efficiency of this method with the CDF transformation technique for generating these Beta variates.

3. Use the rejection technique to generate random variables Y with density:

$$f_Y(y) = \begin{cases} \cos y, & 0 \leq y \leq \pi/2 \\ 0, & \text{other } y. \end{cases}$$

What proportion, approximately, of the pairs generated will be rejected? Compare the efficiency of this procedure with the use of the CDF transformation technique.

2.14. The Bivariate Normal Distribution

2.14.1. Definitions and Terminology

A pair of continuous random variables X_1 and X_2 are jointly distributed according to the bivariate normal distribution if their joint probability density function is given at any point (x_1, x_2) in the Euclidean plane by

$$\begin{aligned} &\varphi(x_1, x_2; \mu_1, \mu_2, \sigma_1, \sigma_2, \varrho) \\ &\quad = C \cdot \exp -\left\{\left[\left(\frac{x_1 - \mu_1}{\sigma_1}\right)^2 - 2\varrho\left(\frac{x_1 - \mu_1}{\sigma_1}\right)\left(\frac{x_2 - \mu_2}{\sigma_2}\right)\right.\right. \\ &\qquad \left.\left. + \left(\frac{x_2 - \mu_2}{\sigma_2}\right)^2\right] \Big/ 2(1 - \varrho^2)\right\}, \end{aligned}$$

where

$$C = [2\pi\sigma_1\sigma_2(1 - \varrho^2)^{1/2}]^{-1},$$

for parameters μ_1 and μ_2 (real), σ_1 and σ_2 (positive) and $-1 \leq \varrho \leq 1$. The function is clearly positive for all (x_1, x_2) and the volume it bounds is one, since

$$\int_{-\infty}^{\infty} \int_{-\infty}^{\infty} \varphi(x_1, x_2; \mu_1, \mu_2, \sigma_1, \sigma_2, \varrho)\, dx_1\, dx_2 = 1,$$

a result that may be readily established upon transforming the variables of integration according to

$$t_1 = \left(\frac{x_1 - \mu_1}{\sigma_1} - \frac{\varrho(x_2 - \mu_2)}{\sigma_2} \right) \Big/ (1 - \varrho^2)^{1/2}$$

and

$$t_2 = \left(\frac{x_2 - \mu_2}{\sigma_2} \right),$$

the Jacobian for which is

$$\frac{\partial(x_1, x_2)}{\partial(t_1, t_2)} = \sigma_1 \sigma_2 (1 - \varrho^2)^{1/2}.$$

The resulting integral then becomes

$$\int_{-\infty}^{\infty} \int_{-\infty}^{\infty} \varphi(t_1, t_2; 0, 0, 1, 1, 0)\, dt_1\, dt_2 = (2\pi)^{-1} \left\{ \int_{-\infty}^{\infty} \exp(-t^2/2)\, dt \right\}^2 = 1.$$

Using essentially the same transformations as given above, one may show that the marginal probability density functions are members of the (univariate) Gaussian families; indeed, one has

$$X_1 \sim N(\mu_1, \sigma_1^2)$$

and

$$X_2 \sim N(\mu_2, \sigma_2^2),$$

from which it follows that the four parameters μ_1, μ_2, σ_1, and σ_2 are directly relatable to the moments

$$E[X_1] = \mu_1, \qquad E[X_2] = \mu_2$$

and

$$\mathrm{Var}[X_1] = \sigma_1^2, \qquad \mathrm{Var}[X_2] = \sigma_2^2.$$

Furthermore, the parameter ϱ is equal to zero if and only if X_1 and X_2 are independently distributed Gaussian random variables. Since one may establish that the correlation coefficient between X_1 and X_2 is given as

$$\varrho_{X_1, X_2} = \varrho$$

it then follows that jointly Gaussian distributed random variables are independent if and only if they are uncorrelated.

One of the more important properties of bivariate normal random variables is that linear functions of each variate will still possess a bivariate normal distribution. For example, if

$$W_1 = aX_1 + b,$$

and

$$W_2 = cX_2 + d,$$

then W_1 and W_2 have joint probability density function

$$\varphi(w_1, w_2; a\mu_1 + b, c\mu_2 + d, |a|\sigma_1, |c|\sigma_2, \pm\varrho),$$

and, hence, $W_1 \sim N(a\mu_1 + b, a^2\sigma_1^2)$ and $W_2 \sim N(c\mu_2 + d, c^2\sigma_2^2)$. Of special interest in this regard are the random variables

$$Y_1 = X_1 - \mu_1,$$

and

$$Y_2 = X_2 - \mu_2,$$

having the joint probability density function

$$\varphi(y_1, y_2; 0, 0, \sigma_1, \sigma_2, \varrho),$$

since it is usually somewhat less encumbering to discuss a density function indexed by only three parameters.

2.14.2. The Variance–Covariance Matrix

Indeed, by employing vector-matrix notation, the joint probability density function for the random vector $\vec{Y} \equiv (Y_1, Y_2)$ becomes

$$\varphi(y_1, y_2; 0, 0, \sigma_1, \sigma_2, \varrho) = [|\mathbf{A}|^{1/2}/2\pi] \cdot \exp -\{[\vec{y} \cdot \mathbf{A} \cdot \vec{y}^T]/2\},$$

where $\vec{y} = (y_1, y_2)$ is any point in the Euclidean plane and where

$$\mathbf{A} = \begin{bmatrix} [\sigma_1^2(1-\varrho^2)]^{-1} & -\varrho/[\sigma_1\sigma_2(1-\varrho^2)] \\ -\varrho/[\sigma_1\sigma_2(1-\varrho^2)] & [\sigma_2^2(1-\varrho^2)]^{-1} \end{bmatrix}$$

is a 2×2 matrix having inverse given by

$$\mathbf{A}^{-1} \equiv \mathbf{\Sigma} = \begin{bmatrix} \sigma_1^2 & \varrho\sigma_1\sigma_2 \\ \varrho\sigma_1\sigma_2 & \sigma_2^2 \end{bmatrix}.$$

Therefore, an alternative notation for the bivariate normal distribution for random variables Y_1 and Y_2 of mean zero is

$$\varphi(y_1, y_2; \boldsymbol{\Sigma}) = [2\pi \mid \boldsymbol{\Sigma} \mid^{1/2}]^{-1} \exp -\{\vec{y} \cdot \boldsymbol{\Sigma}^{-1} \cdot \vec{y}^T/2\}.$$

The matrix $\boldsymbol{\Sigma}$ is referred to as the variance–covariance matrix for the random vector $\vec{Y}$, since its main diagonal's elements are the respective variances of Y_1 and Y_2, and its off-diagonal elements equal their covariance:

$$\mathrm{Cov}(Y_1, Y_2) = \varrho_{Y_1, Y_2} \cdot \sigma_1 \cdot \sigma_2 = \varrho\sigma_1\sigma_2.$$

[One should also note that $\mathrm{Var}(X_i) = \mathrm{Var}(Y_i)$, $i = 1$ and 2, and that $\mathrm{Cov}(Y_1, Y_2) = \mathrm{Cov}(X_1, X_2)$, so that $\boldsymbol{\Sigma}$ is also the variance–covariance matrix for the random vector $\vec{X} = (X_1, X_2) = (\mu_1 + Y_1, \mu_2 + Y_2)$.]

The joint density function for a bivariate Gaussian random vector (X_1, X_2) is, therefore, completely determined by the specification of the mean-value vector

$$E[\vec{X}] \equiv (E(X_1), E(X_2)) = (\mu_1, \mu_2) \equiv \vec{\mu}_X$$

and the variance–covariance matrix

$$\boldsymbol{\Sigma}_X \equiv \begin{bmatrix} \sigma_1^2 & \varrho\sigma_1\sigma_2 \\ \varrho\sigma_1\sigma_2 & \sigma_2^2 \end{bmatrix}.$$

Knowledge of the first and second moments of jointly normally distributed random variables is then sufficient to specify their joint density function. A shorthand notation, indicating that $\vec{X} = (X_1, X_2)$ has the bivariate normal distribution of mean-value vector $\vec{\mu}_X = (\mu_1, \mu_2)$ and of variance–covariance matrix $\boldsymbol{\Sigma}_X$, is given by

$$\vec{X} \sim N_2(\vec{\mu}_X, \boldsymbol{\Sigma}_X).$$

Of course, if the two random variables are independent, then $\varrho = 0$ and $\boldsymbol{\Sigma}_X$ becomes the diagonalized matrix

$$\boldsymbol{\Sigma}_X = \begin{bmatrix} \sigma_1^2 & 0 \\ 0 & \sigma_2^2 \end{bmatrix} \quad \text{for} \quad X_1, X_2 \quad \text{independently distributed.}$$

If, in addition, $\mu_1 = \mu_2 = \mu$ and $\sigma_1^2 = \sigma_2^2 = \sigma^2$, then $\vec{X} \sim N_2(\vec{\mu}, \sigma^2 \mathbf{I}_2)$, where

$$\vec{\mu} = (\mu, \mu)$$

and

$$\mathbf{I}_2 = \begin{bmatrix} 1 & 0 \\ 0 & 1 \end{bmatrix}, \quad \text{the } 2 \times 2 \text{ identity matrix;}$$

in this latter case, we write

$$\vec{X} \sim NID_2(\mu, \sigma^2),$$

indicating that the two random variables X_1 and X_2 are normally and independently distributed, each having the identical Gaussian probability density function, $\varphi(x; \mu, \sigma)$.

2.14.3. Linear Combinations of Jointly Gaussian Variates

Any pair of linear combinations of jointly Gaussian random variables X_1 and X_2 becomes a bivariate random vector having a bivariate normal density function. This is an especially important result regarding jointly Gaussian distributed variates. More specifically, if

$$Z_1 = d_{11}X_1 + d_{12}X_2 + c_1,$$

and

$$Z_2 = d_{21}X_1 + d_{22}X_2 + c_2,$$

then the joint probability density function for the bivariate random vector (Z_1, Z_2) becomes

$$g(z_1, z_2) = \varphi(x_1(z_1, z_2), x_2(z_1, z_2); \mu_1, \mu_2, \sigma_1, \sigma_2, \varrho) \cdot \left| \frac{\partial(x_1, x_2)}{\partial(z_1, z_2)} \right|;$$

substitution for $x_1(z_1, z_2)$ and $x_2(z_1, z_2)$ leads to the result that

$$\vec{Z} \sim N_2(\vec{\mu}_Z, \mathbf{\Sigma}_Z),$$

where

$$\vec{\mu}_Z = (d_{11}\mu_1 + d_{12}\mu_2 + c_1, d_{21}\mu_1 + d_{22}\mu_2 + c_2)$$

and

$$\mathbf{\Sigma}_Z = \begin{bmatrix} \mathrm{Var}(Z_1) & \mathrm{Cov}(Z_1, Z_2) \\ \mathrm{Cov}(Z_2, Z_1) & \mathrm{Var}(Z_2) \end{bmatrix}.$$

The result may be somewhat more readily derived in vector-matrix notation. Noting that

$$\vec{Z} = (\vec{X} \cdot \mathbf{D}^{\mathrm{T}}) + (\vec{c}),$$

where $\vec{c} = (c_1, c_2)$ and where

$$\mathbf{D}^{\mathrm{T}} = \begin{bmatrix} d_{11} & d_{21} \\ d_{12} & d_{22} \end{bmatrix},$$

one can write

$$\vec{V} \equiv \vec{Z} - (\vec{c} + \vec{\mu}_X \cdot \mathbf{D}^{\mathrm{T}}) = \vec{Y} \cdot \mathbf{D}^{\mathrm{T}},$$

where $\vec{Y} = (Y_1, Y_2) \sim N_2(\vec{\Phi}, \mathbf{\Sigma}_X)$. The joint density function for $\vec{V} = \vec{Y} \cdot \mathbf{D}^{\mathrm{T}}$ thus becomes

$$g(v_1, v_2) = |D^{\mathrm{T}}|^{-1}[2\pi\, |\mathbf{\Sigma}_X|^{1/2}]^{-1} \exp - \{\vec{v} \cdot (\mathbf{D}^{\mathrm{T}})^{-1} \cdot \mathbf{\Sigma}_X^{-1} \cdot [(\mathbf{D}^{\mathrm{T}})^{-1}]^{\mathrm{T}} \vec{v}^{\mathrm{T}}/2\},$$

provided that $(\mathbf{D}^{\mathrm{T}})^{-1}$, the inverse matrix of the transpose $\mathbf{D}^{\mathrm{T}}$ of the coefficient matrix $\mathbf{D}$, exists. However, this implies that with

$$\mathbf{S}^{-1} \equiv (\mathbf{D}^{\mathrm{T}})^{-1} \mathbf{\Sigma}_X^{-1} [(\mathbf{D}^{\mathrm{T}})^{-1}]^{\mathrm{T}},$$

such that

$$|\mathbf{S}| = |\mathbf{D}^{\mathrm{T}}| \cdot |\mathbf{\Sigma}_X| \cdot |\mathbf{D}| = |\mathbf{D}|^2 \cdot |\mathbf{\Sigma}_X|, \qquad V \sim N_2(\vec{\Phi}, \mathbf{S}).$$

That is,

$$\vec{V} = \vec{Z} - (\vec{c} + \vec{\mu}_X \cdot \mathbf{D}^{\mathrm{T}})$$

has joint Gaussian distribution of null mean-value vector and of variance–covariance matrix,

$$\mathbf{S} = \{(\mathbf{D}^{\mathrm{T}})^{-1} \mathbf{\Sigma}_X^{-1} [(\mathbf{D}^{\mathrm{T}})^{-1}]^{\mathrm{T}}\}^{-1}.$$

Since the vector $(\vec{c} + \vec{\mu}_X \cdot \mathbf{D}^{\mathrm{T}})$ is merely a vector of constants, it follows that

$$\vec{Z} \sim N_2(\vec{\mu}_Z, \mathbf{\Sigma}_Z),$$

where

$$\vec{\mu}_Z = \vec{c} + \vec{\mu}_X \cdot \mathbf{D}^{\mathrm{T}},$$

and

$$\mathbf{\Sigma}_Z = \mathbf{S} = \{(\mathbf{D}^{\mathrm{T}})^{-1} \mathbf{\Sigma}_X^{-1} [(\mathbf{D}^{\mathrm{T}})^{-1}]^{\mathrm{T}}\}^{-1} = \mathbf{D} \mathbf{\Sigma}_X \mathbf{D}^{\mathrm{T}}.$$

As a corollary to this result, any single linear combination, $Z = aX_1 + bX_2 + c$, of a pair (X_1, X_2) of jointly normally distributed random variables therefore has the (univariate) normal distribution; its mean will be

$$E(Z) = aE(X_1) + bE(X_2) + c = a\mu_1 + b\mu_2 + c,$$

and its variance is given by

$$\begin{aligned} \operatorname{Var}(Z) &= a^2 \operatorname{Var}(X_1) + b^2 \operatorname{Var}(X_2) + 2ab \operatorname{Cov}(X_1, X_2) \\ &= a^2 {\sigma_1}^2 + b^2 {\sigma_2}^2 + (2ab \cdot \varrho \sigma_1 \sigma_2). \end{aligned}$$

It therefore follows that the sum of jointly Gaussian distributed random variables is a univariate random variable whose probability density function is a member of the Gaussian family of densities.

2.14.4. Orthogonal Transformations

If a pair of random variables X_1 and X_2 are independently and normally distributed, then they are jointly normally distributed, and the parameter $\varrho = 0$. In general, a linear transformation of the form

$$\vec{Z} = (\vec{X} \cdot \mathbf{D}^{\mathrm{T}}) + (\vec{c})$$

will not result in random variables Z_1 and Z_2 that are independently distributed. Indeed,

$$\mathrm{Cov}(Z_1, Z_2) = \mathrm{Cov}(d_{11}X_1 + d_{12}X_2, d_{21}X_1 + d_{22}X_2),$$

which becomes

$$\mathrm{Cov}(Z_1, Z_2) = \mathrm{Cov}(d_{11}X_1, d_{21}X_1 + d_{22}X_2) + \mathrm{Cov}(d_{12}X_2, d_{21}X_1 + d_{22}X_2)$$

or

$$\mathrm{Cov}(Z_1, Z_2) = d_{11}d_{21}\sigma_1^2 + (d_{11}d_{22} + d_{12}d_{21}) \cdot \mathrm{Cov}(X_1, X_2) + d_{12}d_{22}\sigma_2^2;$$

due to the independence of X_1 and X_2

$$\mathrm{Cov}(Z_1, Z_2) = d_{11}d_{21}\sigma_1^2 + d_{12}d_{22}\sigma_2^2,$$

which is nonzero unless $d_{11}d_{21}\sigma_1^2 = -d_{12}d_{22}\sigma_2^2$.

The reader may recall the earlier comment that independently distributed random variables quite frequently arise as the result of the repetition of an experiment under identical conditions. Therefore, if the outcome of an experiment can be reasonably expected to be a normally distributed random variable X of mean μ and variance σ^2, then recording of two successive outcomes (X_1, X_2) under such conditions would constitute a random vector

$$\vec{X} \sim NID_2(\mu, \sigma^2).$$

In this event, $\sigma_1^2 = \sigma_2^2 = \sigma^2$, and the random variables X_1 and X_2 are said to be *homoscedastic* (of "equal scatter," or of equal variance). The linear transformation

$$\vec{Z} = (\vec{X} \cdot \mathbf{D}^{\mathrm{T}}) + (\vec{c})$$

then becomes a bivariate normal random vector of independently distributed random variables Z_1 and Z_2 whenever

$$\mathrm{Cov}(Z_1, Z_2) = (d_{11}d_{21} + d_{12}d_{22})\sigma^2 = 0;$$

that is, Z_1 and Z_2 become independent whenever

$$d_{11}d_{21} + d_{12}d_{22} = 0.$$

This condition on the four elements of the transformation matrix $\mathbf{D}^{\mathrm{T}}$ is the same as that which defines $\mathbf{D}^{\mathrm{T}}$ as an orthogonal matrix; i.e., the columns of $\mathbf{D}^{\mathrm{T}}$ (or, equivalently, the rows of $\mathbf{D}$) must be orthogonal vectors—vectors whose scalar product is null.

For example, the transformations

$$Z_1 = X_1 + X_2 \qquad \text{and} \qquad Z_2 = X_1 - X_2,$$

correspond to the vector-matrix equation

$$\vec{Z} = \vec{X} \cdot \mathbf{\Delta}^{\mathrm{T}},$$

where

$$\mathbf{\Delta}^{\mathrm{T}} = \begin{bmatrix} +1 & +1 \\ +1 & -1 \end{bmatrix}$$

is an orthogonal matrix. If the random variables X_1 and X_2 are independently distributed normal random variables of the same variance (σ^2), then so will be Z_1 and Z_2, though $\mathrm{Var}(Z_1) = \mathrm{Var}(Z_2) = 2\sigma^2$. Hence, the sum and the difference of independently distributed and homoscedastic normal random variables are a pair of independently distributed and homoscedastic Gaussian random variables.

More generally, orthogonal linear transformations of independently distributed, homoscedastic, jointly Gaussian variates will preserve the independence of the resulting random variables.

EXERCISES

1. Determine the constant C such that

$$f(x, y) = C \exp\{-x^2 + xy - 3y^2\}$$

is a proper probability density function defined for all points (x, y) in the Euclidean plane. Show that

$$\varrho_{X,Y} = \sqrt{\tfrac{3}{36}} \qquad \text{and that} \qquad \mathrm{Var}(X) = 3 \cdot \mathrm{Var}(Y).$$

2. For the matrix

$$\boldsymbol{\Sigma} = \begin{bmatrix} \sigma_1{}^2 & \varrho\sigma_1\sigma_2 \\ \varrho\sigma_1\sigma_2 & \sigma_2{}^2 \end{bmatrix},$$

show that its determinant is the inverse of the determinant of $\mathbf{A} = \boldsymbol{\Sigma}^{-1}$, as defined in the text.

3. If $\vec{X} \sim N_2(\vec{\Phi}, \boldsymbol{\Sigma}_X)$, and $\vec{Z} = \vec{X} \cdot \mathbf{D}^{\mathrm{T}}$, show directly that

$$\mathrm{Var}(Z_1) = \mathrm{Var}(d_{11}X_1 + d_{12}X_2),$$
$$\mathrm{Var}(Z_2) = \mathrm{Var}(d_{21}X_1 + d_{22}X_2),$$

and

$$\mathrm{Cov}(Z_1, Z_2) = \mathrm{Cov}(d_{11}X_1 + d_{12}X_2, d_{21}X_1 + d_{22}X_2)$$

are the elements of the matrix product

$$\mathbf{S} = \mathbf{D} \cdot \boldsymbol{\Sigma}_X \cdot \mathbf{D}^{\mathrm{T}} = \boldsymbol{\Sigma}_Z.$$

4. If X_1 and X_2 are uncorrelated and homoscedastic random variables, show that any orthogonal transformation $\vec{Y} = \vec{X} \cdot \boldsymbol{\Delta}^{\mathrm{T}}$ produces also uncorrelated random variables, regardless of the distributional form for, and the means of, the random variables X_1 and X_2. Note that

$$\mathrm{Cov}(Y_1, Y_2) = (\delta_{11}\delta_{21} + \delta_{12}\delta_{22})\sigma_X{}^2 + (\delta_{12}\delta_{21} + \delta_{11}\delta_{22})\,\mathrm{Cov}(X_1, X_2).$$

5. If X_1 and X_2 are uncorrelated and homoscedastic random variables, show that any orthonormal transformation $\vec{Y} = \vec{X} \cdot \boldsymbol{\Delta}^{\mathrm{T}}$ produces also random variables which are uncorrelated and homoscedastic. Note that

$$\mathrm{Var}(Y_1) = (\delta_{21}^2 + \delta_{22}^2)\sigma_X{}^2 + 2\delta_{11}\delta_{12}\,\mathrm{Cov}(X_1, X_2),$$

and

$$\mathrm{Var}(Y_2) = (\delta_{21}^2 + \delta_{22}^2)\sigma_X{}^2 + 2\delta_{21}\delta_{22}\,\mathrm{Cov}(X_1, X_2),$$

and that, for an orthonormal matrix $\boldsymbol{\Delta}$,

$$\delta_{11}^2 + \delta_{12}^2 = \delta_{21}^2 + \delta_{22}^2 = 1.$$

6. Verify the result of Exercises 4 and 5 by computing, in each instance, the matrix product

$$\boldsymbol{\Sigma}_Y = \boldsymbol{\Delta} \cdot \boldsymbol{\Sigma}_X \cdot \boldsymbol{\Delta}^{\mathrm{T}} = \sigma_X{}^2\, \boldsymbol{\Delta} \cdot \boldsymbol{\Delta}^{\mathrm{T}}$$

7. If X_1 and X_2 are independently distributed Gaussian random variables of means μ and of common variance σ^2, show that their mean

Y_1 and difference Y_2 are independently distributed, Gaussian random variables, with

$$E(Y_1) = E[(X_1 + X_2)/2] = \mu, \qquad E(Y_2) = E[X_1 - X_2] = 0,$$

and with

$$\text{Var}(Y_1) = \text{Var}[(X_1 + X_2)/2] = 2\sigma^2/4 = \sigma^2/2$$

and

$$\text{Var}(Y_2) = \text{Var}[X_1 - X_2] = 2\sigma^2.$$

Show then that the random variable,

$$S^2 \equiv [X_1 - X_2]^2/2,$$

has expectation σ^2, and that an alternative expression for S^2 is

$$S^2 = [X_1 - (X_1 + X_2)/2]^2 + [X_2 - (X_1 + X_2)/2]^2.$$

Show that all conditions are thus satisfied for the random variable

$$\chi^2 = (S^2/\sigma^2) = Y_2{}^2/(2\sigma^2)$$

to have the Chi-squared distribution of 1 degree of freedom and for the random variable

$$T = \frac{(Y_1/\sigma)}{(S/\sigma)} = \frac{Y_1}{S} = \frac{(X_1 + X_2)/2}{\{(X_1 - X_2)^2/2\}^{1/2}}$$

to have the Student's *t* distribution of 1 degree of freedom.

2.14.5. The Log-and-Trig Method for Generating Gaussian Variates

Let Z_1 and Z_2 be independently and normally distributed random variables, each with mean zero and variance σ^2. Then the joint probability density function for Z_1 and Z_2 beomes

$$\varphi(z_1, z_2) = (2\pi\sigma^2)^{-1} \exp -\{(z_1{}^2 + z_2{}^2)/(2\sigma^2)\}$$

at any point (z_1, z_2) in the Euclidean plane. Thus, the outcome of the underlying experiment produces a random point, or vector (Z_1, Z_2) in the Euclidean space of two dimensions.

Such a random point might be represented as well in polar coordinates (R, θ), constituting transformations of the random variables Z_1 and Z_2 according to

$$Z_1 = R \cos \theta$$

and

$$Z_2 = R \sin \theta.$$

The joint probability density function for the random variables R and θ then becomes, in accordance with Eq. (2.11.4:1):

$$g(r, \theta) = \varphi(r \cos \theta, r \sin \theta) \cdot |\, \partial(z_1, z_2)/\partial(r, \theta)\, |$$

or

$$g(r, \theta; \sigma) = (2\pi)^{-1}\sigma^{-2} r \exp(-r^2/2\sigma^2) \qquad \text{for} \quad r > 0 \quad \text{and} \quad 0 \le \theta \le 2\pi,$$

since

$$\frac{\partial(z_1, z_2)}{\partial(r, \theta)} = \det\begin{bmatrix} \cos\theta & -r \sin\theta \\ \sin\theta & r\cos\theta \end{bmatrix} = r.$$

From this joint probability density function, the marginal density for θ becomes the rectangular distribution, since

$$h_\theta(\theta) = (2\pi)^{-1} \int_{r=0}^{\infty} \sigma^{-2} r \exp(-r^2/2\sigma^2)\, dr = (2\pi)^{-1} \int_{u=0}^{\infty} e^{-u}\, du;$$

i.e.,

$$h_\theta(\theta) = \begin{cases} 1/2\pi, & 0 \le \theta \le 2\pi, \\ 0, & \text{other } \theta. \end{cases}$$

Similarly, the marginal probability density function for R becomes,

$$h_R(r) = \sigma^{-2} r \exp(-r^2/2\sigma^2) \int_{\theta=0}^{2\pi} (2\pi)^{-1}\, d\theta,$$

or

$$h_R(r; \sigma) = \begin{cases} \sigma^{-2} r \exp(-r^2/2\sigma^2), & r > 0 \\ 0, & r \le 0, \end{cases}$$

the Rayleigh distribution of parameter σ.

Noteworthy is the result that the random variables R and θ are independently distributed, for

$$g(r, \theta; \sigma) = h_\theta(\theta) \cdot h_R(r; \sigma) \qquad \text{for all coordinates} \quad (r, \theta).$$

Hence, if (Z_1, Z_2) represents a bivariate random vector of independently and Gaussian distributed random variables of means zero and of common variance σ^2, the random variables R and θ arising from the transformation from Euclidean to polar coordinates, are independently distributed as well. (Such a transformation is often pertinent in weapon accuracy; letting Z_1 and Z_2 denote the coordinates of a hit, then R represents its distance from the center of the target.)

Therefore, by independently generating random variables R and θ having the respective distributions

$$h_R(r; 1) = r \exp(-r^2/2), \qquad r > 0$$

and

$$h_\theta(\theta) = (2\pi)^{-1}, \qquad 0 \le \theta \le 2\pi,$$

one may transform to Euclidean coordinates

$$Z_1 = R \cos \theta$$

and

$$Z_2 = R \sin \theta,$$

obtaining independently distributed, *standardized* Gaussian variates. Since the density function $h_R(r; 1)$ is a member of the family of Weibull densities, one may generate a pair of independent, standardized Gaussian variates from a pair of independent, uniformly distributed random variables U_1 and U_2 according to

$$Z_1 = [-2 \ln(1 - U_1)]^{1/2} \cos(2\pi U_2),$$

and

$$Z_2 = [-2 \ln(1 - U_1)]^{1/2} \sin(2\pi U_2).$$

Of course, since $(1 - U_1)$ has the same distribution as U_1, some computational facility may be derived from the equivalent expressions

$$Z_1 = [-2 \ln U_1]^{1/2} \cos(2\pi U_2),$$

and

$$Z_2 = [-2 \ln U_1]^{1/2} \sin(2\pi U_2).$$

This technique of Muller (1959), termed the *log-and-trig method* for the generation of standardized normal variates, is often preferred to the previously discussed methods for generating Gaussian random variables of mean zero, variance one.

EXERCISES

1. Generate 10 (five pairs) standardized Gaussian variates by the log-and-trig method. Ordering the 10 variates, compare the resulting sample cumulative distribution function with $\Phi(z)$, the cumulative distribution function for the standardized Gaussian random variable, by means of the Kolmogorov–Smirnov test statistic D_{10}.

2. Generate 400 standardized Gaussian random variables from 200 pairs of uniformly distributed random variables by means of the log-and-trig method. Construct a frequency histogram of 10 equiwidth cells over the range -6 to $+6$, noting whether variates fall outside these limits. Compare the frequency histogram with the standardized normal density function by means of a Chi-squared test of 9 degrees of freedom.

3. Generate 400 standardized, approximately Gaussian random variables from 4800 uniformly distributed random variables U_i according to

$$Z_j = \left[-6 + \sum_{i=12j-11}^{12} U_i\right], \qquad j = 1, 2, \ldots, 400.$$

Construct a histogram according to the instructions given in Exercise 2 and compare with the histogram there obtained. Comment on the relative efficiencies of the two procedures for generating standardized normal random variables.

4. Using 400 standardized normal random variables, as generated by the log-and-trig method, transform each of the 200 successive pairs according to

$$Y_1 = (Z_1 + Z_2)/\sqrt{2},$$

and

$$Y_2 = (Z_1 - Z_2)/\sqrt{2}.$$

Noting why Y_1 and Y_2 should be independent and standardized Gaussian random variables, construct a grid of 25 squares of equal area covering the area of the Euclidean plane having coordinates (x_1, x_2) such that $-3 \le x_i \le +3$, $i = 1$ and 2. By separate operations, build a two-dimensional histogram for the 200 (Z_1, Z_2)-vectors over this grid, then for the 200 (Y_1, Y_2)-vectors similarly, noting in each instance any vectors that fall outside the grid. For the bivariate histogram of (Z_1, Z_2), compare the univariate histograms associated with all the Z_2-values that arise in each of the two infinite rectangles:

(1) $-1.8 \le z_1 \le -0.6$ and (2) $0.6 \le z_1 \le 1.8$.

For the bivariate histogram of (Y_1, Y_2), compare the two univariate histograms associated with all Y_2-values whose concomitant variate Y_1 lies in the regions:

$$(1) \quad -1.8 \leq y_1 \leq -0.6 \qquad \text{and} \qquad (2) \quad -0.6 \leq y_2 \leq 0.6.$$

Comment on these results in view of the presumed independence of (Y_1, Y_2), and of (Z_1, Z_2).

5. In consonance with Exercise 4, compute 200 values of the T-statistic

$$T = Y_1/(Y_2^2)^{1/2} = Y_1/|Y_2|.$$

Construct a frequency histogram of the resulting 200 T-values and compare with tabulations of the Student's t distribution of 1 degree of freedom (cf. Appendix).

6. Indicate how you would generate normally distributed random variables of mean a and variance b by means of the log-and-trig method.

7. Noting that $R = (Z_1^2 + Z_2^2)^{1/2}$, where Z_1 and Z_2 are independent and standardized Gaussian variates, has Rayleigh distribution, show that $X = R^2 = Z_1^2 + Z_2^2$ has the exponential distribution of mean 2.

2.15. The Multivariate Normal Distribution

2.15.1. Multivariate Distributions

In many experimental contexts, it is necessary to record the outcome, not as a single random variable, but as a vector of random variables:

$$\vec{X}_n \equiv (X_1, X_2, \ldots, X_n),$$

where each of the n random variables arises as a concomitant in a single performance of the experiment. For example, a stochastic simulation model of a production plant is quite likely to produce a multivariate response, elements of which might properly include the total productivity of the plant, the average daily usage of the plant's equipment, the total man-hour expenditure during the period of interest, the efficiency of any one of the production departments, and the costs attributable to malfunctioning equipment. Probability statements regarding the multivariate outcome of the experiment may then be derived from a joint, n-variate, cumulative distribution function

$$F_X(x_1, x_2, \ldots, x_n) \equiv P[X_1 \leq x_1, X_2 \leq x_n, \ldots, \textit{and } X_n \leq x_n],$$

valid for any point $(x_1, x_2, \ldots, x_n)$ in Euclidean n-space.

The random variables comprising the random vector may be either continuous or discrete, though for the purposes of the present discussion, each of the n variates shall be presumed to be a continuous random variable. The random vector, $\vec{X}_n$, may then be presumed to possess a joint, n-variate, probability density function at any point $(x_1, x_2, \ldots, x_n)$:

$$f_X(x_1, x_2, \ldots, x_n) \equiv \partial^n F_X(x_1, x_2, \ldots, x_n)/\partial x_1\, \partial x_2 \cdots \partial x_n.$$

Analogously to the bivariate case, each random variable of the multivariate response vector possesses its own (marginal) probability density function, obtained by integrating the joint probability density function over all other $(n - 1)$ variables; for example, the marginal probability density function for the random variable X_1 becomes

$$f_{X_1}(a) = \int_{x_2=-\infty}^{\infty} \cdots \int_{x_n=-\infty}^{\infty} f_X(a, x_2, x_3, \ldots, x_n)\, dx_2 \ldots dx_n.$$

Consequently, each random variable has its own mean value,

$$\mu_i \equiv E[X_i] = \int_{-\infty}^{\infty} a f_{X_i}(a)\, da, \qquad i = 1, 2, \ldots, n,$$

and the expected value of the random vector $\vec{X}_n$ is then defined as the vector of expectations

$$E[\vec{X}_n] \equiv (\mu_1, \mu_2, \ldots, \mu_n).$$

Similarly, each pair of random variables has its own (marginal) bivariate probability density function, defined by integrating the joint, n-variate, probability density function over the Euclidean $(n - 2)$-space appropriate to the other $(n - 2)$ random variables. For example, the joint bivariate density for X_{n-1} and X_n would be given at any real numbers (a, b), by

$$f_{X_{n-1},X_n}(a, b) = \int_{x_1=-\infty}^{\infty} \cdots \int_{x_{n-2}=-\infty}^{\infty} f_X(x_1, \ldots, x_{n-2}, a, b)\, dx_1 \ldots dx_{n-2}.$$

Hence, covariances can be found for any pair of random variables among the n elements of the random vector and may be tabulated as the elements of an $(n \times n)$ array $\boldsymbol{\Sigma}$; the element σ_{ij} in the ith row and jth column thereof represents the covariance between X_i and X_j; in general,

$$\sigma_{ij} = \begin{cases} \text{Cov}(X_i, X_j) = \sigma_{ji}, & i \neq j \\ \text{Var}(X_i) \equiv \sigma_i^2, & i = j. \end{cases}$$

As a direct extension of the notion of independence for a pair of random variables, one defines n random variables $X_1, X_2, \ldots, X_n$ as *independently distributed* if and only if

$$f_X(x_1, x_2, \ldots, x_n) = \prod_{i=1}^{n} f_{X_i}(x_i),$$

for every point $(x_1, x_2, \ldots, x_n)$ in Euclidean n-space; i.e., the n random variables are independently distributed if knowledge of the outcomes associated with any subset of them has no effect on probability statements made about any other subset. A typical situation in which n independently distributed random variables may arise is the experiment consisting of n iterations, or *encounters*, with a stochastic simulation model, each encounter producing the same univariate random variable X, and each encounter having been defined by the same input conditions, but initiated by an independently and randomly selected seed for its random number generator. In this case, the n random variables $X_1, X_2, \ldots, X_n$ are not only independently distributed, but also have the same marginal probability density function $f_X(x)$. (More generally, one could obtain n independently distributed random variables by recording the n successive, univariate responses from n independently seeded encounters with the simulation model, the n encounters having been defined by different input specifications. In this case, the marginal probability density functions will not necessarily be the same; namely, the mean values of the successive random variables could probably change as the input specifications are altered from one encounter to the next.)

2.15.2. Transformations of Random Vectors

Just as it has proved of value to discuss the effects that transformations have upon probability statements associated with univariate or bivariate random variables, so it is necessary to describe the joint, n-variate, probability density function associated with transformations of n jointly distributed random variables $\vec{X}_n = (X_1, X_2, \ldots, X_n)$. For example, in the exemplary simulation of a production system, one might wish to ask about the likelihood that the productivity per man-hour will exceed x units, or of the probability that the difference between the value of goods produced and the costs attributable to malfunctioning equipment be positive. Presuming then that n unique transformations of the n original random variables can be defined as

$$\begin{aligned} Y_1 &= t_1(X_1, X_2, \ldots, X_n), \\ Y_2 &= t_2(X_1, X_2, \ldots, X_n), \\ &\vdots \\ Y_n &= t_n(X_1, X_2, \ldots, X_n), \end{aligned}$$

one may form the n-variate random vector

$$\vec{Y}_n = (Y_1, Y_2, \ldots, Y_n).$$

Its joint probability density function may, in general, be found from that of $\vec{X}_n$ according to

$$f_Y(y_1, \ldots, y_n) = f_X[x_1(y_1, \ldots, y_n), \ldots, x_n(y_1, \ldots, y_n)] \cdot |\, \partial(\vec{x})/\partial(\vec{y}) \,|,$$

where $x_k(y_1, \ldots, y_n)$ is the inverse solution, of the n transformations, for the variate x_k, $k = 1, 2, \ldots, n$, and where $\partial(\vec{x})/\partial(\vec{y})$ represents the Jacobian of these inverse transformations:

$$\partial(\vec{x}_n)/\partial(\vec{y}_n) = \det \begin{bmatrix} \partial x_1/\partial y_1 & \partial x_1/\partial y_2 & \cdots & \partial x_1/\partial y_n \\ \partial x_2/\partial y_1 & \partial x_2/\partial y_2 & \cdots & \partial x_2/\partial y_n \\ \vdots & \vdots & & \vdots \\ \partial x_n/\partial y_1 & \partial x_n/\partial y_2 & \cdots & \partial x_n/\partial y_n \end{bmatrix}$$

If the n random variables of the vector $\vec{X}_n$ are transformed by means of n simultaneous linear equations,

$$\vec{Y}_n = \vec{X}_n \cdot \mathbf{B}^{\mathrm{T}},$$

where $\mathbf{B}^{\mathrm{T}} = (b_{ji})$ is an $(n \times n)$ matrix of constant coefficients having inverse matrix $(\mathbf{B}^{\mathrm{T}})^{-1}$ so that

$$\vec{X}_n = \vec{Y}_n \cdot (\mathbf{B}^{\mathrm{T}})^{-1},$$

the joint density function for the random vector $\vec{Y}_n$ is readily obtained, the necessary Jacobian being merely the determinant of the matrix $(\mathbf{B}^{\mathrm{T}})^{-1}$.

For these linear transformations, one should note that the moments are readily obtained from the moments of the original variates. For example, if $\vec{X}_n$ has mean-value vector

$$\vec{\mu}_X \equiv (\mu_1, \mu_2, \ldots, \mu_n),$$

then $\vec{Y}_n = \vec{X}_n \cdot \mathbf{B}^{\mathrm{T}}$ has mean-value vector

$$\vec{\mu}_Y = \vec{\mu}_X \cdot \mathbf{B}^{\mathrm{T}};$$

namely, $Y_i = b_{i1}X_1 + b_{i2}X_2 + \ldots + b_{in}X_n$ has mean

$$E(Y_i) = b_{i1}\mu_1 + b_{i2}\mu_2 + \ldots + b_{in}\mu_n,$$

identically, the ith element of the vector, $\vec{\mu}_X \cdot \mathbf{B}^{\mathrm{T}}$, $i = 1, 2, \ldots, n$. Furthermore, if $\mathbf{\Sigma}_X$ denotes the variance–covariance matrix for the random vector $\vec{X}_n$, then the variance–covariance matrix for $\vec{Y}_n = \vec{X}_n \cdot \mathbf{B}^{\mathrm{T}}$ becomes

$$\mathbf{\Sigma}_Y = \mathbf{B} \cdot \mathbf{\Sigma}_X \cdot \mathbf{B}^{\mathrm{T}}.$$

For example, an element on the diagonal of $\mathbf{\Sigma}_Y$ represents the variance of one of the random variables, say

$$Y_i = \vec{X}_n \cdot \vec{b}_i,$$

where $\vec{b}_i$ is the ith row of the matrix $\mathbf{B}$ (or, the ith column of $\mathbf{B}^{\mathrm{T}}$); that is,

$$\mathrm{Var}(Y_i) = \sum_{j=1}^{n} \sum_{k=1}^{n} b_{ij} b_{ik} \sigma_{jk},$$

or

$$\mathrm{Var}(Y_i) = \vec{b}_i \cdot \mathbf{\Sigma}_X \cdot \vec{b}_i^{\mathrm{T}},$$

the ith diagonal element of $\mathbf{\Sigma}_Y = \mathbf{B} \cdot \mathbf{\Sigma}_X \cdot \mathbf{B}^{\mathrm{T}}$. Similarly, the covariance between Y_i and Y_j becomes

$$\mathrm{Cov}(Y_i, Y_j) = \mathrm{Cov}(\vec{X} \cdot \vec{b}_i, \vec{X} \cdot \vec{b}_j) = \sum_{k=1}^{n} \sum_{l=1}^{n} b_{ik} b_{jl} \sigma_{kl},$$

or

$$\mathrm{Cov}(Y_i, Y_j) = \vec{b}_i \cdot \mathbf{\Sigma}_X \cdot \vec{b}_j^{\mathrm{T}},$$

the element s_{ij} of the matrix $\mathbf{\Sigma}_Y = (s_{ij})$, located in the ith row and jth column.

2.15.3. The Multivariate Gaussian Distribution

The Central Limit Theorem has been acknowledged as applicable to the description of the distribution of a random variable X, which arises as the sum of a large number of independent, and relatively insignificant, contributing random variables. Whenever the outcome of an experiment need be represented by an n-variate vector of random variables, it is frequently the case that the Central Limit Theorem applies to each of the n random variables, so that each has a (marginal) density function from among the Gaussian family.

More generally, the multivariate response $\vec{X}_n$ arising from a nondeterministic experiment may often be shown to possess a joint probability density function known as the *multivariate normal distribution*, an n-variate probability density function specified by the vector of mean values $\vec{\mu}_X$ and the variance–covariance matrix $\mathbf{\Sigma}_X$ for the n random variables; namely, by means of a generalized Central Limit Theorem, one may often anticipate that the joint, n-variate, probability density function shall be of the form

$$\varphi(x_1, x_2, \ldots, x_n) = (2\pi)^{-n/2} \mid \mathbf{\Sigma}_X^{-1} \mid^{1/2} \exp -\{(\vec{x} - \vec{\mu}_X)\mathbf{\Sigma}_X^{-1}(\vec{x} - \vec{\mu}_X)^{\mathrm{T}}/2\},$$

for any point $(x_1, x_2, \ldots, x_n)$ in Euclidean n-space. One may write $\vec{X}_n \sim N_n(\vec{\mu}_X, \mathbf{\Sigma}_X)$ to denote this distributional form.

Of special importance is the fact that any one of the n random variables has (marginal) probability density function which is of the (univariate) Gaussian family. In fact, any pair of the n variates can be shown to be jointly distributed according to the bivariate normal distribution which is specified by the appropriate means, variances, and covariances as extracted directly from $\vec{\mu}_X$ and $\mathbf{\Sigma}_X$.

Even more important for the purposes of this text is the note that a linear transformation of the form $\vec{Y}_n = \vec{X}_n \cdot \mathbf{B}^{\mathrm{T}}$ becomes an n-variate random vector with multivariate normal distribution having mean-value vector

$$\vec{\mu}_Y = \vec{\mu}_X \cdot \mathbf{B}^{\mathrm{T}}$$

and variance–covariance matrix

$$\mathbf{\Sigma}_Y = \mathbf{B} \cdot \mathbf{\Sigma}_X \cdot \mathbf{B}^{\mathrm{T}};$$

i.e.,

$$\vec{Y}_n = \vec{X}_n \cdot \mathbf{B}^{\mathrm{T}} \sim N_n(\vec{\mu}_X \cdot \mathbf{B}, \mathbf{B} \cdot \mathbf{\Sigma}_X \cdot \mathbf{B}^{\mathrm{T}}),$$

whenever $\vec{X}_n \sim N_n(\vec{\mu}_X, \mathbf{\Sigma}_X)$.

A nondeterministic experiment can also be defined by the n recorded variates $X_1, X_2, \ldots, X_n$ of n successive experiments, each of which is conducted under essentially identical conditions and provides an univariate random variable X. The random vector $\vec{X}_n = (X_1, X_2, \ldots, X_n)$ then consists of n independently distributed random variables from the same univariate probability density function and therefore constitutes a ***random sample*** from this density. Whenever the univariate density is the Gaussian probability density function $\varphi(x; \mu, \sigma)$, the joint probability density for the n-variate random sample $\vec{X}_n$ becomes $\varphi(x_1, \ldots, x_n; \vec{\mu}, \sigma^2 \cdot \mathbf{I}_n)$; i.e.,

$$\vec{X}_n \sim N_n(\vec{\mu}, \sigma^2 \cdot \mathbf{I}_n),$$

where $\vec{\mu} = (\mu, \mu, \ldots, \mu)$ and $\mathbf{I}_n$ is the $(n \times n)$ identity matrix. An alternative shorthand notation is $\vec{X}_n \sim NID(\mu, \sigma^2)$.

More generally, if the n successive experiments, each producing a Gaussian random variable, are conducted under differing, yet controlled, conditions, one can reasonably expect that the n resulting independent random variables have Gaussian distributions of different means. If the variances of these n independent Gaussian random variables remain, however, the same, we say that the random variables are *homoscedastic*: $[\mathrm{Var}(X_i) = \sigma^2, i = 1, 2, \ldots, n]$, so that

$$\vec{X}_n \sim N_n(\vec{\mu}_X, \sigma^2 \cdot \mathbf{I}_n),$$

where $\vec{\mu}_X = (\mu_1, \mu_2, \ldots, \mu_n)$ is the vector of the respective means (which need not be equal).

As in the more general case, linear transformations of independent, homoscedastic Gaussian variates will also have the Gaussian distribution; in general, if $\vec{X}_n \sim N_n(\vec{\mu}_X, \sigma^2 \cdot \mathbf{I}_n)$ and $\vec{Y}_n = \vec{X}_n \cdot \mathbf{B}^{\mathrm{T}}$, then

$$\vec{Y}_n \sim N_n(\vec{\mu}_X \cdot \mathbf{B}^{\mathrm{T}}, \sigma^2 \mathbf{B}\mathbf{B}^{\mathrm{T}}),$$

since

$$\mathbf{\Sigma}_Y = \sigma^2 \mathbf{B}\mathbf{I}_n\mathbf{B}^{\mathrm{T}} = \sigma^2 \mathbf{B}\mathbf{B}^{\mathrm{T}}.$$

Therefore, in general, a system of n linear transformations of n independently distributed Gaussian variates will not preserve the independence for the resulting random variables $\vec{Y}_n$.

However, if the n linear transformations are *orthogonal*, then $\mathbf{B} \cdot \mathbf{B}^{\mathrm{T}}$ shall be a diagonalized matrix, all of whose off-diagonal elements are null. However, in this event, $\mathbf{\Sigma}_Y$ will represent the variance–covariance matrix for n independently distributed Gaussian variates, although the resulting variates may be heteroscedastic, since the diagonal elements of $\sigma^2\mathbf{BB}^{\mathrm{T}}$ may well differ. In the specific case of the orthonormal transformation

$$\vec{Y}_n = \vec{X}_n \cdot \boldsymbol{\beta}^{\mathrm{T}}$$

the matrix $\boldsymbol{\beta}$ of coefficients has the property that

$$\boldsymbol{\beta} \cdot \boldsymbol{\beta}^{\mathrm{T}} = \mathbf{I}_n,$$

so that, if $\vec{X}_n \sim N_n(\vec{\mu}_X, \sigma^2 \cdot \mathbf{I}_n)$ and $\vec{Y}_n = \vec{X}_n \cdot \boldsymbol{\beta}^{\mathrm{T}}$ then

$$\vec{Y}_n \sim N_n(\vec{\mu}_X \cdot \boldsymbol{\beta}^{\mathrm{T}}, \sigma^2 \cdot \mathbf{I}_n),$$

so that both independence and homoscedasticity are preserved under the orthonormal transformation.

EXERCISES

1. Let (X_1, X_2, X_3) be independently and identically distributed Gaussian random variables of mean zero and variance σ^2. Show that the transformation to spherical coordinates (R, θ, Φ) according to

$$X_1 = R \cos\theta \sin\Phi,$$

$$X_2 = R \sin\theta \sin\Phi,$$

and

$$X_3 = R \cos\Phi,$$

leads to independently distributed random variables R, θ, and Φ with respective probability density functions

$$f_R(r; \sigma) = r^2 e^{-r^2/2\sigma^2}/[\sigma^3 2^{1/2}\Gamma(\tfrac{3}{2})], \qquad \text{for} \quad r > 0,$$

$$f_\Phi(\varphi) = \tfrac{1}{2}\sin\varphi, \qquad 0 \le \varphi \le \pi,$$

and

$$f_\theta(\theta) = 1/2\pi, \qquad 0 \le \theta \le 2\pi.$$

The density $f_R(r; \sigma)$ is called the Maxwell distribution.

2. Show that a random variable Φ may be generated from a uniformly distributed random variable U according to

$$\Phi = \cos^{-1}(1 - 2U),$$

with Φ having the density $f_\Phi(\varphi)$ of Exercise 1.

3. Let X_1, X_2, and X_3 be independently distributed and homoscedastic Gaussian random variables of mean zero, as in the first exercise. Show that the following transformations are orthogonal, leading therefore to independently distributed, yet heteroscedastic, Gaussian random variables Y_1, Y_2, and Y_3 each of mean zero:

$$Y_1 = (X_1 + X_2 + X_3)/3$$

$$Y_2 = (X_1 - X_3)/2$$

and

$$Y_3 = (X_1 - 2X_2 + X_3)/6.$$

Show that $\text{Var}(Y_1) = \sigma_X{}^2/3$, $\text{Var}(Y_2) = \sigma_X{}^2/2$, $\text{Var}(Y_3) = \sigma_X{}^2/6$, and that therefore

$$Y_1{}^2/(\sigma_X{}^2/3) = (X_1 + X_2 + X_3)^2/(3\sigma_X{}^2),$$

$$Y_2{}^2/(\sigma_X{}^2/2) = (X_1 - X_2)^2/(2\sigma_X{}^2),$$

and

$$Y_3{}^2/(\sigma_X{}^2/6) = (X_1 - 2X_2 + X_3)^2/(6\sigma_X{}^2)$$

are independently distributed Chi-squared variates, each of a single degree of freedom.

4. In Exercise 3, note that the quantity

$$S^2 \equiv \sum_{i=1}^{3} (X_i - \bar{X})^2/(3 - 1)$$

where $\bar{X} \equiv Y_1 = (X_1 + X_2 + X_3)/3$, is the same as

$$S^2 = [Y_2{}^2 + 3 \cdot Y_3{}^2],$$

and that an alternative expression for S^2 is

$$S^2 = \left[\sum_{i=1}^{3} X_i^2 - 3(\bar{X})^2\right]\Big/2.$$

Show that $E(S^2) = \sigma_X{}^2$ and that $2(S^2/\sigma_X{}^2)$ is a Chi-squared variate of 2 degrees of freedom.

5. For the random variables Y_1, Y_2, and Y_3 of Exercise 3, show that

$$Z_1 = Y_1/(\sigma_X/\sqrt{3}) \sim N(0, 1),$$

and therefore that, since $Y_2{}^2/(\sigma_X{}^2/2)$ has the Chi-squared distribution of

1 degree of freedom, the random variable,

$$T = \frac{Z_1}{[Y_2^2/(\sigma_X^2/2)]^{1/2}}$$

or

$$T = \frac{(X_1 + X_2 + X_3)/\sqrt{3}}{[(X_1 - X_3)^2/2]^{1/2}},$$

has the Student's t distribution of 1 degree of freedom.

6. Referring to Exercises 3 and 4, show that

$$T_2 \equiv \bar{X}/(S/\sqrt{3})$$

is a random variable possessing Student's t distribution of 2 degrees of freedom.

7. If X_1, X_2, and X_3 are independent Gaussian random variables, each of mean 0 and of the same variance σ^2, show that

$$\bar{X} = (X_1 + X_2 + X_3)/3,$$

and

$$S^2 = \sum_{i=1}^{3} (X_i - \bar{X})^2/2$$

are independently distributed random variables and that

$$F \equiv 3(\bar{X})^2/S^2$$

is a random variable having Snedecor's F distribution of 1 and 2 degrees of freedom.

8. Generate 100 triplets (X_1, X_2, X_3) of independent, standardized Gaussian random variables by repeatedly using the log-and-trig method. For each triplet, compute

$$\bar{X} = (X_1 + X_2 + X_3)/3,$$

$$S^2 = \sum_{i=1}^{3} (X_i - \bar{X})^2/2,$$

$$T = \bar{X}/(S/\sqrt{3}),$$

and

$$F = 3(\bar{X})^2/S^2.$$

Build a frequency histogram of 10 cells of equal width for each of the four statistics (random variables) and compare with the theoretically applicable distributions by means of Chi-squared tests of 9 degrees of freedom. [The respectively applicable densities are $N(0, \frac{1}{3})$, $\frac{1}{2}\chi_2^2$, Student's *t* of 2 degrees of freedom, and Snedecor's *F* of 1 and 2 degrees of freedom.]

2.15.4. Generation of Correlated Random Variables

In many applications, it will be important to obtain sets of random variables having specified correlations. Theoretically, this should not prove to be difficult, since, from a source of n independent (and therefore uncorrelated) random variables $\vec{X}_n = (X_1, X_2, \ldots, X_n)$ each having variance σ_X^2, the linear transformations

$$\vec{Y}_n = \vec{X}_n \cdot \mathbf{B}^{\mathrm{T}}$$

result in n random variables having variance–covariance matrix

$$\mathbf{\Sigma}_Y = \sigma_X^2 \mathbf{B} \cdot \mathbf{I}_n \cdot \mathbf{B}^{\mathrm{T}} = \sigma_X^2 \cdot (\mathbf{B}\mathbf{B}^{\mathrm{T}}).$$

Therefore, in order to achieve a desired variance–covariance matrix $\mathbf{S} = (s_{ij})$, one may select a square matrix $\mathbf{B}$ such that

$$s_{ij} = \sigma_X^2 \sum_{k=1}^{n} b_{ik} b_{jk}, \qquad \text{for} \quad i, j = 1, 2, \ldots, n;$$

that is, one selects $\mathbf{B} = (b_{ij})$ as a square matrix whose rows have scalar products proportional to the desired covariances (and variances) for the resulting random variables $\vec{Y}_n$.

In general, however, the probability density function for the random vector $\vec{Y}_n$ will remain concealed, unless of course, correlated Gaussian variates are sought. Since a system of linear transformations of jointly Gaussian distributed random variates produces jointly Gaussian variables, one can easily produce an n-variate normal vector from n independent, standardized normal variates: $Z_1, Z_2, \ldots, Z_n$.

Therefore, if one seeks to generate a random vector, $\vec{Y}_n \sim N_n(\vec{\mu}_Y, \mathbf{\Sigma}_Y)$, where $\vec{\mu}_Y$ is a given mean-value vector and $\mathbf{\Sigma}_Y = (\sigma_{ij})$ is a given variance–covariance matrix, he has only to generate the n independent, standardized Gaussian variates Z_n by applying, say, the log-and-trig method, then transforming these according to

$$\vec{Y}_n = (\vec{Z}_n \cdot \mathbf{B}^{\mathrm{T}}) + \vec{\mu}_Y,$$

where **B** is a square matrix chosen such that

$$\mathbf{\Sigma}_Y = \mathbf{B}\mathbf{B}^{\mathrm{T}}.$$

Thus, the rows of **B** are specified such that

$$\sum_{k=1}^{n} b_{ik}^2 = \sigma_{ii} = \mathrm{Var}(Y_i), \qquad i = 1, 2, \ldots, n$$

and

$$\sum_{k=1}^{n} b_{ik} b_{jk} = \sigma_{ij} = \mathrm{Cov}(Y_i, Y_j), \qquad i > j = 1, 2, \ldots, n.$$

These $n(n+1)/2$ conditions on the elements of $B = (b_{ij})$ correspond exactly to the number of elements of a *symmetric* matrix **B** which can be specified; in principle, the matrix **B** is completely determined by these conditions. [See Oplinger (1971).]

As an illustration, one can consider the generation of the highly correlated jointly Gaussian random variables (Y_1, Y_2), having joint probability density function, $N_2(\vec{\mu}_Y, \mathbf{\Sigma}_Y)$, where

$$\vec{\mu}_Y = (3,\ 5)$$

and

$$\mathbf{\Sigma}_Y = \begin{bmatrix} 1 & (1+\sqrt{3})/2 \\ (1+\sqrt{3})/2 & 2 \end{bmatrix}.$$

The requisite three equations become

$$b_{11}^2 + b_{12}^2 = 1$$

$$b_{21}^2 + b_{22}^2 = b_{12}^2 + b_{22}^2 = 2$$

and

$$b_{12} \cdot (b_{11} + b_{22}) = (1 + \sqrt{3})/2,$$

so that

$$\mathbf{B} = \begin{bmatrix} 1/\sqrt{2} & 1/\sqrt{2} \\ 1/\sqrt{2} & \sqrt{3}/\sqrt{2} \end{bmatrix};$$

i.e.,

$$Y_1 = 3 + (Z_1 + Z_2)/\sqrt{2},$$

and

$$Y_2 = 5 + (Z_1 + Z_2 \cdot \sqrt{3})/\sqrt{2}$$

are the desired bivariate Gaussian variates.

General techniques for solving for the elements of $\mathbf{B}$ in terms of those of the desired matrix $\boldsymbol{\Sigma}_X$ have been delineated by Wold (1955) and Scheuer and Stoller (1962).

2.15.5. Other Transformations of the Multivariate Normal

Four transformations of the n standardized random variables, $\vec{Z}_n \sim N_n(\vec{\Phi}, \mathbf{I}_n)$, will be of prime importance in later chapters. These are:

$$\bar{Z} \equiv (Z_1 + Z_2 + \cdots + Z_n)/n,$$

$$S^2 \equiv \sum_{i=1}^{n} (Z_i - \bar{Z})^2/(n-1),$$

$$T = \bar{Z}/(S/\sqrt{n}),$$

and

$$F = (\bar{Z})^2/(S^2/n).$$

The importance of these transformations will become apparent once the discussion of the analysis of simulation experiments is undertaken; for the moment, their consideration will be presented abstractly, in order to provide a foundation for later discussion.

The statistic (random variable) $\bar{Z}$ may be seen to be a linear transformation of the jointly Gaussian distributed random variables, $Z_1, Z_2, \ldots, Z_n$. Therefore, its (marginal) probability density function is the (univariate) Gaussian distribution of mean

$$\mu = E(\bar{Z}) = n^{-1} \sum_{i=1}^{n} E(Z_i) = n^{-1}(n \cdot 0) = 0$$

and of variance

$$\sigma^2 = \mathrm{Var}(\bar{Z}) = n^{-2}\,\mathrm{Var}\left(\sum_{i=1}^{n} Z_i\right) = n^{-2}(n \cdot 1) = 1/n;$$

that is, $\bar{Z} \sim N(0, 1/n)$.

The statistic S^2 may be written in the equivalent form

$$S^2 = (n-1)^{-1}\left[\sum_{i=1}^{n} Z_i^2 - n(\bar{Z})^2\right],$$

so that

$$(n-1)S^2 + n(\bar{Z})^2 = \sum_{i=1}^{n} Z_i^2.$$

The right-hand side of this last equation consists of the sum of the squares of n standardized normal random variables and is therefore a statistic (random variable) having the Chi-squared distribution of n degrees of freedom. On the left-hand side, the quantity

$$n(\bar{Z})^2 = [\bar{Z}/(1/\sqrt{n})]^2$$

is a random variable having the Chi-squared distribution of 1 degree of freedom.

Recalling the additive reproductivity property for independent Chi-squared variates, it would appear that $(n-1)S^2$ should be a candidate for a Chi-squared variate of $(n-1)$ degrees of freedom. Indeed, not only is this the case, but also $(n-1)S^2$ and $n(\bar{Z})^2$ are statistically (though not functionally) independent. These assertions can be established via Cochran's theorem, as shown in Cramér (1946) or Lindgren (1968). Thus, since the Chi-squared distribution of $(n-1)$ degrees of freedom is a member of the Gamma family of distributions, the random variable S^2 has a Gamma distribution, and is distributed independently of $\bar{Z}$.

Hence, the third transformation of interest, given by

$$T = (\bar{Z})/(S/\sqrt{n}),$$

can be written in the alternative form

$$T = \frac{(\bar{Z})/(1/\sqrt{n})}{[(n-1)S^2/(n-1)]^{1/2}},$$

in which T becomes evidently the ratio of a standardized normal random variable to the square root of an independently distributed Chi-squared variate that has been divided by its degrees of freedom. Thus, T has Student's t distribution of $(n-1)$ degrees of freedom.

Finally, the random variable, or statistic

$$F = (\bar{Z})^2/(S^2/n)$$

can be written alternatively as the ratio of two independently distributed Chi-squared variates, each divided by its corresponding degrees of freedom:

$$F = \frac{n(\bar{Z})^2/1}{(n-1)S^2/(n-1)}.$$

The statistic F is therefore distributed according to Snedecor's F density of one and $n-1$ degrees of freedom.

EXERCISE

"Student's" Experiment

Selecting 100 sets, each of four standardized and independently distributed Gaussian random variables, compute for each sample the statistic T. Compare the resulting histogram of the 100 T-values with the Student's t density of 3 degrees of freedom. Then repeat the experiment using 100 samples of size 8 each, comparing the resulting histogram with the Student's t distribution of 7 degrees of freedom, as tabulated in the Appendix.

Chapter 3

TIME SERIES

We do not what we ought,
What we ought not we do,
And lean upon the thought
That Chance will bring us through.

—MATTHEW ARNOLD

3.1. Introduction

In the preceding chapter, the theoretical conditions leading to a number of random variables were examined. In general, each random variable arises as the measured response, or outcome, of a conceptual, nondeterministic experiment, and the nature of the distribution of these responses over the admissible real numbers can be predicted from considerations regarding the existing experimental conditions. More usually, however, it is one of a *family* of random variables that is applicable to the description of the response arising from given experimental conditions; the particular random variable appropriate to a specific nondeterministic experiment has a cumulative distribution function that is indexed by one or more parameters, each of which generally relates to the experimental conditions themselves and which specifies its place in the family.

Even if the random variable X that naturally arises from a nondeterministic experiment is transformed according to some functional relationship $Y = t(X)$ the resulting random variable Y usually has a determinable cumulative distribution function of its own. By taking advantage of these facts regarding the nature of the distributions of random variables (or their transformations), one is able to construct methods for generating random variables in accordance with a large number of families of distribution (density) functions. In effect, in order to produce the desired random variables, one often mimes, or simulates, the experimental conditions leading to them.

Frequently, the outcome or response of a nondeterministic experiment cannot be conveniently summarized by a single real number, so that bivariate or multivariate random vectors need to be considered. Since concomitant measures arising from the same experiment are often somewhat dependent upon, or related to, one another, a joint cumulative distribution function is in order. Studies of the properties of these families of multivariate random variables, especially the bivariate and the multivariate Gaussian families, lead to useful techniques for the generation both of independent, univariate (especially Gaussian), random variables and of correlated, multivariate random vectors.

The discussions of the preceding chapter have therefore revolved about experiments whose outcomes are nondeterministic, yet it is presumed that the conceptual experiment at hand will, when completed, yield one or more random variables. In many contexts, however, the conceptual experiment will provide measures, or observations, serially. For example, the closing price of a particular stock may be recorded serially as days progress; the position of a particle subjected to Brownian motion may be traced as time progresses. Thus, it is frequently important to be able to describe the behavior of such a stochastic process, or *time series*, as it develops in time. Indeed, an understanding of the underlying mechanisms that could generate these unsystematically fluctuating sequences of random variables will be of primary importance in efforts to simulate their behavior in a model.

3.1.1. Definition of Time Series

In observing a time series, the recording $Y(t)$ at t time units from some initial starting position can be considered as a random variable. If the particular experiment were reinitiated repeatedly under essentially identical conditions, then the measurements successively recorded after the passage of t time units would be seen to form a probability density (distribution) function $f_{Y(t)}(a)$, or $p_{Y(t)}(k)$, the nature of which should depend upon the experimental conditions.

Similarly, at any other time s, a random variable $Y(s)$ would be observable and would possess its own density (distribution). Indeed, one might likely anticipate cause-and-effect or other relationships between the random variables, $Y(t)$ and $Y(s)$, so that bivariate probability density or distribution functions would be deemed appropriate to their joint probabilistic descriptions. More generally, for any n time points, $t_1, t_2, \ldots, t_n$, there exist n random variables, $Y(t_1)$, $Y(t_2)$, ..., $Y(t_n)$, about which joint probability statements can be made from computations centered about a multivariate probability density (distribution) function

$$f_Y(y_1, y_2, \ldots, y_n) \qquad \text{or} \qquad p_Y(k_1, k_2, \ldots, k_n).$$

It should be noted that the quantity $Y(t)$ observed at time t is directly comparable to the quantity $Y(s)$ observed at some other time s; i.e., the conceptually infinite collection of random variables $\{Y(t)\}_{t=1}^{\infty}$ consists only of members that tend to measure the same phenomenon, though at different points in time. Thus, the serial record of the positions of a particle in Brownian motion will all be in the same units of measure (say, its distance from some original position, in millimeters), and will not occasionally or regularly represent some other quantity (say, its vectorial velocity). The fact that several variates might be recorded in time, however, is important, yet in these cases, one describes the stochastic process by an infinite sequence of random vectors; e.g.,

$$\{D(t),\ V(t)\}_{t=1}^{\infty}.$$

The multivariate time series will be considered later; for the moment, however, one can view the time series as the infinite collection $\{Y(t)\}_{t=1}^{\infty}$ of univariate random variables. Alternatively, one can choose to view a time series as a set, or *ensemble*, of functions; for example, the record $[D(t),\ t = 1, 2, 3, \ldots]$ of a particle's position constitutes a function of time. By reinitializing the particle (or by starting observation on another particle), another time function, different from the first, can be traced. The conceptually infinite ensemble $[D(t)]$ of these random time functions becomes then an equivalent description of the time series, or stochastic process.

A dynamic, stochastic simulation model can also be viewed as a generator of a time series, an idea to be more fully developed in Chapter 10. The random selection of the required random number seed for the model corresponds to the selection of one of the barely finite random time-functions from the ensemble.

The distribution associated with the response $Y(t)$ at time t will then reflect the experimental conditions at and/or since time zero. Thus, in general, the cumulative distribution function for $Y(t)$ will be somehow associated with that for the random variable $Y(t - 1)$ and more generally, with that for $Y(t - \tau)$. Usually, however, the distribution function for $Y(t)$ differs somewhat from that for $Y(s)$, $s \neq t$.

Especially amenable to analysis, however, are cases in which the cumulative distribution function for the random variable $Y(t)$ is:

(*a*) the same as that for $Y(s)$, for all s,
(*b*) different from that for $Y(s)$, yet with $Y(t)$ and $Y(s)$ independently distributed,
(*c*) dependent upon the results at one, or at most a few, preceding responses; say, $Y(t - 1)$, or $Y(t - 1)$ and $Y(t - 2)$.

3.1.2. Stationarity

If the random variables $Y(t)$ and $Y(s)$ have the same (marginal) cumulative distribution function [Case (*a*)], regardless of the time indices s and t, then the time series is said to be *stationary of order* 1; i.e., regardless of the point t selected in time, the random variable $Y(t)$ will be distributed according to a cumulative distribution function, which will not depend upon t. That is, for any real a,

$$F_{Y(t+\tau)}(a) = F_{Y(t)}(a) = F_Y(a), \quad \text{for all time } t \text{ and for any time shift } \tau.$$

Hence, probability statements about the state of the time series, as measured by $Y(t)$, may be made without regard to the passage of time. Furthermore, all moments of the random variables $\{Y(t)\}$ of the time series would also be independent of time t since

$$\mu_k'[Y(t)] \equiv \int_{-\infty}^{\infty} a^k f_Y(a)\, da,$$

a result independent of time. [Note that the probability density function $f_Y(a)$ is not indexed by t.] Hence, the mean μ_Y and the variance σ_Y^2 of a time series which is first-order stationary will be constants applicable to any one of the random variables $Y(t)$.

The pair of random variables $Y(t)$ and $Y(s)$ will be jointly distributed, yet not necessarily independently so. For example, $Y(s)$ and $Y(t)$ may satisfy conditions for a bivariate central limit theorem such that $Y(s)$ and $Y(t)$ possess the same univariate Gaussian distribution, yet $Y(t)$ and $Y(s)$ are correlated with one another; their degree of correlation would then be expressed parametrically as an element of the variance–covariance matrix for the applicable bivariate Gaussian density function.

A random process, or time series, whose component random variables $Y(t)$ have the same probability density (distribution) function and all of whose pairs $Y(s)$ and $Y(t)$ of random variables are, whenever translated by a time shift τ, jointly distributed according to the same bivariate cumulative distribution function,

$$F_{YY}(a, b) \equiv P[Y(s) \leq a,\ Y(t) \leq b],$$

is said to be *stationary of order* 2; i.e.,

$$F_{Y(s), Y(t)}(a, b) = F_{Y(s+\tau), Y(t+\tau)}(a, b)$$

for all (a, b), s and t, and for any τ.

In the Gaussian case just cited, the first-order distributions were assumed to be stationary, but the associated time series would not be stationary of second order unless the correlation between $Y(s)$ and $Y(t)$ were the same as that existing between the random variables $Y(s + \tau)$ and $Y(t + \tau)$, for any fixed time shift τ. More generally, all joint moments arising from a transformation Tr of the form

$$E[\text{Tr}(Y(s), Y(t))] = \int_{-\infty}^{\infty} \int_{-\infty}^{\infty} \text{Tr}(a, b) f_{YY}(a, b)\, da\, db$$

would be independent of s and t, except possibly via the absolute difference $|s - t|$, whenever the time series is stationary of second order, because the joint probability density function $f_{YY}(a, b)$ will in no other way be dependent upon these time indices; for example, the covariance

$$\text{Cov}(Y(s), Y(t)) \equiv E[(Y(s) - \mu_Y) \cdot (Y(t) - \mu_Y)],$$

and the correlation

$$\begin{aligned} \varrho_{Y(s), Y(t)} &\equiv \text{Cov}(Y(s), Y(t))/[\text{Var}(Y(s)) \cdot \text{Var}(Y(t))]^{1/2} \\ &= \text{Cov}(Y(s), Y(t))/\sigma_Y{}^2 \end{aligned}$$

will depend upon s and t only through the absolute difference $|s - t|$.

Stationarity of order k is defined analogously. If, for any k time indices, $t_1, t_2, \ldots, t_k$, the joint cumulative distribution function for the random variables, $Y(t_1), Y(t_2), \ldots, Y(t_k)$, is the same as that for $Y(t_1 + \tau)$, $Y(t_2 + \tau), \ldots, Y(t_k + \tau)$, for any fixed τ, then the time series $\{Y(t)\}$ is said to be *stationary of order* k. If a time series is stationary for any arbitrary order k, it is said to be *strictly stationary*.

It is usually difficult to assure that a given process is generating a time series that is stationary of all orders. For example, any process having a growth tendency (accumulated sales or cumulative production, for example) cannot be expected to be stationary, since minimally the *mean-value function*

$$E\{Y(t)\} \equiv \mu(t),$$

the trace of the expectations of the random variables $\{Y(t)\}_{t=1}^{\infty}$, will be dependent upon time. However, as will be shown in Chapter 10, many random processes with growth patterns may be transformed to time series whose mean-value function and whose *variance function*,

$$\text{Var}\{Y(t)\} \equiv \sigma_Y^2(t),$$

shall not be dependent upon time; indeed, in many instances, such transformations will leave the first-order distribution functions independent of time.

3.1.3. Wide-Sense Stationarity

A necessary, though not at all sufficient, condition for second-order stationarity is that the covariance between any two random variables $Y(s)$ and $Y(t)$ of the time series $\{Y(t)\}$ be the same as that between $Y(s + \tau)$ and $Y(t + \tau)$, for any time shift τ. Frequently, second-order stationarity cannot be assured because of assumed relationships between $Y(t)$ and $Y(s)$, such as the dependence of their covariance, or correlation, upon both times t and s. An often meaningful relationship for this covariance, however, is its dependence upon the absolute time difference

$$\tau = |t - s|$$

or *time lag* between the observed random variables. In this case, one may often write

$$\text{Cov}[Y(s), Y(t)] = \gamma_{YY}(\tau), \qquad \text{for} \quad \tau = |t - s|,$$

where $\gamma_{YY}(\tau)$ denotes the *autocovariance function of lag* τ, notably independent of the relative position of t and s with respect to the time origin. In terms of correlation coefficients, one can define the *autocorrelation function of lag* τ as

$$\text{Corr}[Y(s), Y(t)] = \varrho_{YY}(\tau), \qquad \tau = |t - s|.$$

Whenever a time series $\{Y(t)\}$ has constant mean-value function and whenever its autocorrelation function $\varrho_{YY}(\tau)$ is dependent only upon time lags, the time series is said to be *weakly stationary*, or *wide-sense stationary*.

Thus, stationarity for a time series imposes some restriction on the behavior of the recorded sequence of observations. For example, a first-order stationary time series would only with very small probability "stray" extremely far from its constant mean-value (function). The study of time series and their behavior is thus greatly facilitated whenever stationarity may be assumed.

3.1.4. White-Noise Processes

Another condition under which considerable progress can be made in analyzing time series is as described in Section 3.1.1 [Case (*b*)]. If the sequence of random variables $Y(t)$, $t = 1, 2, \ldots$, comprises a chain of independently distributed random variates, then, even though these variates have different distributions, the analysis of the time series is greatly facilitated. In many cases, the applicable distributions may be assumed to be of the same family of distribution functions, with the time t representing one of its parameters. Thus a common description for the Brownian motion process $\{D(t)\}$ is that the random variable $D(t)$ have Gaussian distribution of mean zero, variance proportional to t, and is independently distributed of $D(s)$, $s \neq t$. (As time progresses, the expected position of the particle remains its originating position, yet the probability of its remaining within ε units of this position diminishes with time.)

If, however, the time series comprises random variables that are independently distributed according to the *same* cumulative distribution function, the time series is known as *white noise*, or as a *white-noise process*. The reader may recall that a desirable property of a random number generator is that it produce uniform white noise—a sequence of independent random variables from the same (the uniform) probability density function. (If this common density function is a particular Gaussian distribution, then the generator is said to be a *normal white-noise process*.) Since a random sample of size n from an univariate density $f_X(x)$ is defined as a set of n independently distributed random variables from that density, a white-noise process might alternatively be thought of as a random sample of infinite size from a particular density function.

3.1.5. Markov Processes

A third type of time series amenable to analysis is that defined in Section 3.1.1, Case (*c*). Here, probability statements about the state, as measured by $Y(t)$, of the process at time t can be facilitated if knowledge of the state is available for a sufficient number of preceding time points. If the distribution of the random variable $Y(t)$ can be completely specified by knowledge of the value assumed by the time series at time $(t-1)$, then the time series is said to be *Markovian*. Time series having the Markovian property require a probabilistic description that is dependent only upon the most recently acquired information regarding the status of the time series; i.e., given additional information that had arisen still earlier, the probabilistic description of the process would not be altered.

Time series may be categorized in many manners and many of the resulting categories will be quite amenable to analysis. The three types of situations just described are not intended to be more than exemplary, so the interested reader is referred to other sources, such as Doob (1953), Feller (1950), and Jenkins and Watts (1968), for more general information.

EXERCISE

A *normal random process* $\{Z(t)\}$ is one whose n-dimensional probability density function is the n-variate Gaussian density; i.e., any vector of n random variables,

$$\vec{Z}_n \equiv [Z(t_1), Z(t_2), \ldots, Z(t_n)]$$

will have a multivariate Gaussian distribution, $N_n(\vec{\mu}_Z, \mathbf{\Sigma}_Z)$, where

$$\vec{\mu}_Z = (\mu_1, \mu_2, \ldots, \mu_n), \qquad \mu_i \equiv E\{Z(t_i)\}, \quad i = 1, 2, \ldots, n,$$

and where $\mathbf{\Sigma}_Z$ is the variance–covariance matrix. Show that, if the time series $\{Z(t)\}$ is weakly stationary, then it is also strictly stationary by noting the effect of the wide-sense stationarity upon the elements of $\mathbf{\Sigma}_Z$ that serve to parametrize the joint density for $\vec{Z}_n$.

3.2. Some Exemplary Time Series

In this section, several typical random processes, or time series, will be developed and discussed. Each will be examined for stationarity, and its moments, such as its mean-value, variance, autocovariance, and autocorrelation functions, will be presented wherever possible.

The *purely random process*, $\{\varepsilon(t)\}$ is described by an infinite sequence of independently distributed random variables. All its n-dimensional joint probability density (distribution) functions may be factored as the product of the corresponding marginal density (distribution) functions. Since any pair of random variables $\varepsilon(s)$ and $\varepsilon(t)$ from the time series is independent, its autocovariance and autocorrelation functions are, respectively,

$$\operatorname{Cov}\{\varepsilon(s), \varepsilon(t)\} = \begin{cases} 0 & \text{for all} \quad t \neq s, \\ \sigma_\varepsilon^{2}(t) & \text{if} \quad s = t, \end{cases}$$

and

$$\operatorname{Corr}\{\varepsilon(s), \varepsilon(t)\} = \begin{cases} 0 & \text{for all} \quad t \neq s, \\ 1 & \text{for} \quad s = t. \end{cases}$$

Since the variance function, $\sigma_\varepsilon^{2}(t)$, will in general, be time-dependent, the process is neither weakly stationary nor stationary of first order.

If, however, the collection $\{\varepsilon(t)\}$ of independent random variables possesses the same mean and the same variance, even though they be differently distributed, then the resulting purely random process would be weakly stationary; if, in addition, the independent random variables $\{\varepsilon(t)\}$ were presumed to possess the same probability density (distribution) function, then the process would constitute white noise and would be strictly stationary.

Linear time series, or linear processes, are those whose value at time t is a linear combination of random contributors from a white-noise process of mean zero; i.e.,

$$Y(t) = \mu_Y + \sum_{i=0}^{\infty} h_i \varepsilon(t - i)$$

or

$$Y(t) = \mu_Y + h_0 \cdot \varepsilon(t) + h_1 \cdot \varepsilon(t - 1) + h_2 \cdot \varepsilon(t - 2) + \cdots,$$

where $\{\varepsilon(t)\}$ is a white-noise process emanating conceptually from the infinitely remote past. The mean-value function for the time series $\{Y(t)\}$ becomes

$$E\{Y(t)\} = \mu_Y + 0 = \mu_Y, \qquad \text{for all} \quad t,$$

and the autocovariance function is

$$\operatorname{Cov}[Y(t), Y(t + \tau)] = E\left\{\sum_{i=0}^{\infty} h_i \varepsilon(t - i) \cdot \sum_{j=0}^{\infty} h_j \varepsilon(t + \tau - j)\right\};$$

due to the independence and nullity of mean for the time series $\{\varepsilon(t)\}$, expectations of all terms in the doubly infinite sum are zero save those whose indices, $t - i$ and $t + \tau - j$, are equal. Hence,

$$\mathrm{Cov}[Y(t),\ Y(t+\tau)] = \sum_{i=\tau}^{\infty} h_i h_{i-\tau}\ \mathrm{Var}\{\varepsilon(t-i)\}$$

or

$$\gamma_{YY}(\tau) = \sigma_\varepsilon^{\,2} \sum_{j=0}^{\infty} h_j h_{j+\tau},$$

where $\sigma_\varepsilon^{\,2} = \mathrm{Var}\{\varepsilon(t)\}$, so that the (constant) variance function for $\{Y(t)\}$ becomes

$$\mathrm{Var}\{Y(t)\} = \gamma_{YY}(0) = \sigma_\varepsilon^{\,2} \cdot \sum_{j=0}^{\infty} h_j^{\,2}, \qquad \text{for all} \quad t.$$

The autocorrelation function is then

$$\varrho_{YY}(\tau) = \left\{\sum_{j=0}^{\infty} h_j h_{j+\tau}\right\} \Big/ \left\{\sum_{j=0}^{\infty} h_j^{\,2}\right\},$$

so that the linear time series, as defined, is weakly stationary regardless of the distributional form applicable to the underlying white-noise process $\{\varepsilon(t)\}$.

If, in addition, $\{\varepsilon(t)\}$ is normal white-noise, then $\{Y(t)\}$ constitutes a normal process whose autocorrelation function $\varrho_{YY}(\tau)$ is a function of the lag only and whose mean-value function is constant, so that $\{Y(t)\}$ will then be strictly stationary. (It is presumed in all cases that the infinite series, $\sum_{j=0}^{\infty} h_j h_{j+\tau}$, exists for each integral τ.)

The linear time series can be viewed as a moving average in the sense that, as the underlying white-noise process $\{\varepsilon(t)\}$ evolves, the time series $\{Y(t)\}$ becomes a "weighted average" of the past history of the $\{\varepsilon(t)\}$ process. If this averaging process is truncated at some distance k in the past, so that

$$h_i = 0 \qquad \text{for all} \quad i > k,$$

then the linear time series given by random variables of the form

$$X(t) \equiv \mu_X + \sum_{i=0}^{k} h_i \varepsilon(t-i), \qquad \text{where} \quad h_k \neq 0,$$

is termed the *moving average series* of order k.

Again, the mean-value function becomes constant:

$$E\{X(t)\} = \mu_X + 0 = \mu_X,$$

whenever the $\{\varepsilon(t)\}$ time series is white noise of mean zero, and the autocovariance function becomes

$$\operatorname{Cov}[X(t), X(t+\tau)] = \gamma_{XX}(\tau) = \begin{cases} \sigma_\varepsilon^2 \sum_{j=0}^{k-\tau} h_j h_{j+\tau}, & 0 \le \tau \le k \\ 0, & \tau > k. \end{cases}$$

The variance function for the moving average series becomes

$$\operatorname{Var}\{X(t)\} \equiv \gamma_{XX}(0) = \sigma_\varepsilon^2 \sum_{j=0}^{k} h_j^2,$$

so that the autocorrelation function of lag τ is given by

$$\operatorname{Corr}\{X(t), X(t+\tau)\} \equiv \varrho_{XX}(\tau) = \begin{cases} \sum_{j=0}^{k-\tau} h_j h_{j+\tau} / \gamma_{XX}(0), & 0 \le \tau \le k \\ 0, & \tau > k. \end{cases}$$

Hence, the random variables, $X(t)$ and $X(t+\tau)$, of the moving average series are uncorrelated (Indeed, they are independently distributed!) whenever the lag τ exceeds the order k.

Being a special case of the aforementioned linear time series, the moving average series of order k has the same stationarity properties as the linear time series; the moving average series is weakly stationary in any event, yet becomes strictly stationary if the underlying white noise process $\{\varepsilon(t)\}$ is a normal process as well.

An example of a nonstationary time series is the *random walk* $\{W(t)\}$, a time series representing at time t the sum of the initial t terms of a white-noise process $\{\varepsilon(t)\}$:

$$W(t) = \sum_{i=1}^{t} \varepsilon(i).$$

If the white-noise process has mean μ_ε and variance σ_ε^2, then the mean-value function for $\{W(t)\}$ becomes

$$E\{W(t)\} = t \cdot \mu_\varepsilon$$

and the variance function for the random walk is

$$\operatorname{Var}\{W(t)\} = t \cdot \sigma_\varepsilon^2;$$

since both of these quantities depend upon the time index t, the random walk process cannot be stationary in any sense.

The autocovariance function for this nonstationary process can be calculated as

$$\mathrm{Cov}[W(t), W(s)] = \mathrm{Cov}\left[\sum_{i=1}^{t} \varepsilon(i), \sum_{j=1}^{s} \varepsilon(j)\right],$$

or

$$\mathrm{Cov}[W(t), W(s)] = \mathrm{Var}\left[\sum_{i=1}^{m} \varepsilon(i)\right],$$

where $m = \min(s, t)$, since the $\{\varepsilon(t)\}$ time series is composed of independent random variables; i.e.,

$$\mathrm{Cov}[W(t), W(s)] = \sigma_\varepsilon^2 \cdot \min(s, t),$$

so that the autocovariance function for the random walk process is dependent upon both time indices s and t.

If the distribution (density) function of the random variables $\varepsilon(t)$ of the contributing white-noise process is known [this may be denoted by $f_\varepsilon(e)$], then the random walk process qualifies as a Markov process, since given that

$$W(t-1) \equiv \sum_{i=1}^{t-1} \varepsilon(i) = w,$$

the random variable $W(t)$, next in sequence, becomes

$$W(t) \equiv \sum_{i=1}^{t} \varepsilon(i) = w + \varepsilon(t),$$

a random variable having conditional probability distribution (density)

$$f_{W(t)}(x) = f_\varepsilon(x - w) \qquad \text{for all real arguments} \quad x.$$

Similarly, one can show that whenever the distribution (density) of $\varepsilon(t)$ is known, the most recent knowledge of the state of the random walk will suffice to make probability statements about the subsequent behavior of the walk process $\{W(t)\}$. In this sense, $\{W(t)\}$ constitutes a Markov process.

Though the random walk process is not in any sense stationary (save for degenerate cases), its study may be greatly facilitated by noting the stationarity of its associated *incremental process*. For any time series $\{Y(t)\}$, the associated incremental process of lag τ (>0) is the infinite sequence of random variables $\{I_\tau(t)\}$ given by

$$I_\tau(t) = Y(t + \tau) - Y(t) \qquad \text{for each} \quad t = 1, 2, 3, \ldots$$

Thus, each random variable $I_\tau(t)$ in the collection $\{I_\tau(t)\}$ represents the observable change in the $\{Y(t)\}$ process between times t and $(t + \tau)$. In the case of the random walk process, the associated incremental process represents the "progress" which the random walk has made in the most recent τ units of time.

Incrementation of a stochastic process is especially useful as a device for the removal of trends in the distributional properties associated with the initial process. For example, the random walk process $\{W(t)\}$ has mean-value and variance functions linearly dependent on time:

$$E\{W(t+\tau) - W(t)\} = E\left\{\sum_{i=t+1}^{t+\tau} \varepsilon(i)\right\} = \tau \cdot \mu_\varepsilon$$

and

$$\mathrm{Var}\{W(t+\tau) - W(t)\} = \mathrm{Var}\left\{\sum_{i=t+1}^{t+\tau} \varepsilon(i)\right\} = \tau \cdot \sigma_\varepsilon^2,$$

so that the incremental process of lag τ for the random walk process has constant mean-value and variance functions. Indeed, since $\{W(t+\tau) - W(t)\}$ represents the sum of τ identically and independently distributed random variables from $\{\varepsilon(i)\}$, the univariate probability density functions for

$$I_\tau(t) = W(t+\tau) - W(t) = \sum_{i=t+1}^{t+\tau} \varepsilon(i)$$

are identical regardless of the time index t. The *incremental random walk process* is therefore stationary of order 1, a result that frequently arises whenever stochastic processes with linear trends are incremented.

However, the incremental random walk process $\{I_\tau(t)\}$ does not consist of a sequence of independent random variables. One may note that any choice of a pair of time indices, s and t, such that $s < t$ implies either that

(i) $\quad s < (s+\tau) < t < (t+\tau)$,

or

(ii) $\quad s \le t \le (s+\tau) \le (t+\tau)$.

In the first case, $I_\tau(s) \equiv W(s+\tau) - W(s) = \sum_{i=s+1}^{s+\tau} \varepsilon(i)$ and $I_\tau(t) \equiv W(t+\tau) - W(t) = \sum_{i=t+1}^{t+\tau} \varepsilon(i)$ are a pair of random variables, each the sum of independent random variables $\varepsilon(i)$, yet $I_\tau(s)$ and $I_\tau(t)$ possess no common summands. Hence, in Case (i),

$$\mathrm{Cov}[I_\tau(s), I_\tau(t)] = 0;$$

on the other hand, in Case (ii),

$$\text{Cov}[I_\tau(s), I_\tau(t)] = E\left[\sum_{i=s+1}^{s+\tau} \{\varepsilon(i) - \mu_\varepsilon\} \cdot \sum_{j=t+1}^{t+\tau} \{\varepsilon(i) - \mu_\varepsilon\}\right],$$

or

$$\text{Cov}[I_\tau(s), I_\tau(t)] = E\left\{\sum_{i=t+1}^{s+\tau} [\varepsilon(i) - \mu_\varepsilon]^2\right\} + 0,$$

or

$$\text{Cov}[I_\tau(s), I_\tau(t)] = [\tau + (s - t)]\sigma_\varepsilon^2.$$

More generally, for any time indices s and t,

$$\text{Cov}[I_\tau(s), I_\tau(t)] = \begin{cases} 0, & |s - t| > \tau, \\ \sigma_\varepsilon^2[\tau - |s - t|], & |s - t| \leq \tau, \end{cases}$$

as the reader can verify by considering the cases arising whenever $t \leq s$ as well. It therefore follows that the incremental random walk process $\{I_\tau(t)\}$ is weakly stationary, for its autocovariance function is dependent only upon absolute time differences $|t - s|$, and its first-order distribution (density) functions are all identical.

One categorizes a stochastic process $\{W(t)\}$ as a process with *stationary independent increments* if its incremental process $\{I_\tau(t)\} \equiv \{W(t + \tau) - W(t)\}$ consists of identically and independently distributed pairs of random variables whenever the time intervals defining the two random variables do not overlap. The random walk process is, then, exemplary of a process possessing stationary independent increments. Another such process is the Poisson process, to be discussed in Section 3.4.

Another important category of time series is a subclass of linear processes known as the *autoregressive processes* $\{A(t)\}$ of order m:

$$A(t) - \mu_A = a_1 \cdot [A(t-1) - \mu_A] + \cdots + a_m \cdot [A(t-m) - \mu_A] + \varepsilon(t),$$

where $\{\varepsilon(t)\}$ is a white-noise process of mean zero. Thus, the probability distribution (density) function for the random variable $A(t)$ depends upon the values assumed by the process at the most recent m time points in addition to the random contribution arising from $\varepsilon(t)$ at the current time. Thus, the random variable $A(t)$ depends upon the entire remote history of the $\{\varepsilon(t)\}$ process.

In particular, the *first-order autoregressive process* can be written

$$A(t) - \mu_A = a_1[A(t-1) - \mu_A] + \varepsilon(t),$$

or

$$A(t) - \mu_A = \varepsilon(t) + a_1\varepsilon(t-1) + a_1^2[A(t-2) - \mu_A],$$

or more generally,

$$A(t) - \mu_A = \sum_{i=0}^{\infty} a_1{}^i \varepsilon(t - i),$$

in which form the time series becomes more evidently a linear process. The mean-value function for the first-order autoregressive process becomes

$$E\{A(t)\} = \mu_A$$

and the autocorrelation function is given by

$$\text{Corr}\{A(t), A(t + \tau)\} = a_1^{|\tau|}, \qquad \tau = 0, 1, 2, \ldots,$$

provided that $| a_1 | < 1$. Hence, the autocorrelation function decays geometrically (exponentially) in absolute value with increasing lag τ, a frequently useful and anticipated phenomenon.

Also noteworthy is the fact that the first-order autoregressive process is Markovian. Given that $A(t - 1) = \alpha$, it follows that

$$A(t) = a_1 \cdot \alpha + \varepsilon(t),$$

so that probability statements regarding $A(t)$ can be derived from its conditional distribution (density) function:

$$f_{A(t)}(x) = f_\varepsilon(x - a_1 \cdot \alpha), \qquad \text{for all real} \quad x.$$

Indeed, if $\{\varepsilon(t)\}$ is normal white noise of mean zero, then $A(t)$, given that $A(t - 1) = \alpha$, has Gaussian distribution of mean $a_1\alpha$ and variance $\sigma_\varepsilon{}^2$.

For other results pertinent to autoregressive processes, the recent text of Jenkins and Watts (1968) is recommended.

EXERCISES

The Wiener–Lévy Process

The Wiener–Lévy process $\{W(t)\}$ is a time series specified by the following:

(*a*) $W(0) = 0$.
(*b*) For all t, $W(t)$ is normally distributed.
(*c*) The process $\{W(t)\}$ has stationary independent increments.
(*d*) For all t, $E\{W(t)\} = 0$.

From these assumptions show that the following are true.

1. The increments $\{W(t) - W(s)]$ and $[W(s) - W(0)] = W(s)$ have, for $t > s$, bivariate normal distribution with diagonalized variance–covariance matrix.

2. By assuming that for any t, $\text{Var}\{W(t)\} = \sigma^2 \cdot t$, the fact that $[W(t) - W(s)]$ and $W(s)$ are independently distributed implies that

$$E\{W(t)W(s)\} = \begin{cases} E\{W^2(s)\}, & s < t, \\ E\{W^2(t)\}, & s \geq t; \end{cases}$$

i.e.,

$$\text{Cov}\{W(t), W(s)\} = \sigma^2 \cdot \min(s, t); \quad \text{and,}$$

3. The Wiener–Lévy process $\{W(t)\}$ is not stationary.

3.3. Generation of Autocorrelated Random Variables

In the preceding section, a number of time series were developed and their statistical properties delineated. Of special interest is the behavior of the autocorrelation function for such processes, because the autocorrelation provides a partial description of the dependency among the sequence of random variables constituting the time series. In simulation contexts, frequently one requires sources of random numbers having specified autocorrelation functions. For example, simulation models of an economic situation often require the automatic generation of sales (or forecasted sales) at the tth simular time period by means of a moving average process that takes into account the sales generated for the m most recent simular time periods. Alternatively, if "exponential smoothing" is deemed adequate for generating the sales (or demands) in an economic environment, then an autoregressive scheme for generating the sequence of sales would be in order.

In other applications, notably in neutron diffusion processes, simulation models may require sequences of random variables generated in accordance with a random walk process. In other instances, time series may be generated by combining, or mixing, processes, such as mixed autoregressive-moving average processes. The following exercises are intended to be indicative of the usually straightforward procedures for generating autocorrelated sequences of random variables. A brief introduction to the topic is provided in the text of Naylor *et al.* (1966).

EXERCISES

1. The forecasted demand $N(t)$, at the end of the tth time period, for a particular item of stock is a random variable that may be related to the forecasted demand $N(t-1)$ of the preceding time period by means of the recursion

$$N(t) = \begin{cases} a_1 \cdot N(t-1) + (1-a_1)\varepsilon(t), & t > 1, \\ (1-a_1)\varepsilon(t), & t = 0, \end{cases}$$

where $0 < a_1 < 1$ and $\{\varepsilon(t)\}$ is a white-noise process of mean zero and variance σ_ε^2. The random variable $\varepsilon(t)$ can be considered as the random demand arising during the tth time period, so that the forecast computed at the end of any time period is a weighted average of the demand of the tth time period and the forecast of the $(t-1)$st period. Show that $E\{N(t)\} = 0$ for all t and that

$$\text{Var}\{N(t)\} = (1-a_1)\sigma_\varepsilon^2/(1+a_1).$$

Then show that the autocorrelation function of lag 1 is

$$\text{Corr}\{N(t), N(t+1)\} = a_1,$$

and more generally that the autocorrelation function of lag τ (≥ 0) is

$$\text{Corr}\{N(t), N(t+\tau)\} = (a_1)^\tau, \qquad \tau = 0, 1, 2, \ldots .$$

Note then that the $\{N(t)\}$ process is autoregressive of order 1.

2. Presuming now that the forecasted demand $N(t)$ at the end of the tth time period is based on the demands of the immediately preceding k periods according to

$$N(t) = (k+1)^{-1} \sum_{i=0}^{k} \varepsilon(t-i).$$

If the demands $\varepsilon(t)$ are independently and identically distributed random variables of zero mean and variance σ_ε^2, show that

$$E\{N(t)\} = 0 \qquad \text{and} \qquad \text{Var}\{N(t)\} = \sigma_\varepsilon^2/(k+1).$$

Furthermore, show that the autocovariance function becomes

$$\text{Cov}[N(t), N(t+\tau)] = \begin{cases} \sigma_\varepsilon^2[(k-\tau+1)/(k+1)^2], & 0 \leq \tau \leq k \\ 0, & \tau > k, \end{cases}$$

in consonance with the results for a moving average process. Comment on a method for generating a sequence of forecasted demands from a sequence of independent random numbers and compare with the procedure which would be required in the first exercise. In the present exercise, invoke the Central Limit Theorem in order to ascertain the approximate distribution of $N(t)$.

3. Using the log-and-trig method as a source of normal white noise $\{\varepsilon(t)\}$, generate 101 standardized Gaussian random variables from 202 uniformly distributed random variables $U(t)$, $t = 0, 1, 2, \ldots, 201$:

$$\varepsilon(t) = [-2 \ln U(2t)]^{1/2} \cdot \cos[2\pi U(2t+1)], \qquad t = 0, 1, \ldots, 100.$$

Compute the statistic

$$(a) \qquad c(0) = \sum_{t=0}^{100} [\varepsilon(t)]^2/101;$$

then compute

$$(b) \qquad r(1) = \sum_{t=0}^{99} [\varepsilon(t)\varepsilon(t+1)]/\{100c(0)\},$$

$$(c) \qquad r(2) = \sum_{t=0}^{98} [\varepsilon(t)\varepsilon(t+2)]/\{99c(0)\},$$

and

$$(d) \qquad r(3) = \sum_{t=0}^{97} [\varepsilon(t)\varepsilon(t+3)]/\{98c(0)\}$$

as estimates of the autocorrelation function $\varrho(k)$, $k = 0, 1, 2$, and 3.

4. Using the log-and-trig method as a source of normal white-noise $\{\varepsilon(t)\}$, generate a sequence of 101 sales forecasts: $N(t)$, $t = 0, 1, \ldots, 100$, according to the exponential smoothing (i.e., first-order autoregressive) scheme of Exercise 1 with $a_1 = 0.5$. Compute the following statistics from the resulting sequence:

$$(a) \qquad c(0) = \sum_{t=0}^{100} [N(t)]^2/101,$$

$$(b) \qquad c(1) = \sum_{t=0}^{99} [N(t)N(t+1)]/100,$$

$$(c) \qquad c(2) = \sum_{t=0}^{98} [N(t)N(t+2)]/99,$$

and

$$(d) \qquad c(3) = \sum_{t=0}^{97} [N(t)N(t+3)]/98.$$

Comment on the apparent behavior of the sequence of autocovariance *estimates* $c(k)$, $k = 0, 1, 2, \ldots$. Are these compatible with the behavior of the theoretical autocorrelation function?

5. Repeat Exercise 4 with $a_1 = 0.1$.

6. Using the sum of 12 successive uniformly distributed random variables as an approximately Gaussian variate, generate 101 standardized normal variates according to

$$\varepsilon(t) = \left\{-6 + \sum_{j=0}^{11} U(12t + j)\right\}, \qquad t = 0, 1, \ldots, 100,$$

where $U(s)$, $s = 0, 1, \ldots, 1211$ is a sequence of 1212 uniformly distributed random variables generated by means of, say, a mixed congruential technique. Compute $c(0)$, $r(1)$, $r(2)$, and $r(3)$ as defined in Exercise 3. Compare with the results of Exercise 3.

7. As in Exercise 6, generate 101 standardized normal variates, but according to the formula

$$\varepsilon(t) = \left\{-6 + \sum_{j=0}^{11} U(t + j)\right\}, \qquad t = 0, 1, 2, \ldots, 100,$$

where $U(s)$, $s = 0, 1, \ldots, 111$ is a sequence of only 112 uniformly distributed random variables generated as in Exercise 6. Construct a frequency histogram of the 101 $\varepsilon(t)$ variates and test for its normality via the Chi-squared or Kolmogorov–Smirnov statistic. Then compute the statistics $c(0)$, $r(1)$, $r(2)$, and $r(3)$ as defined in Exercise 3, contrasting these with the corresponding results obtained in Exercise 6; in addition, compare these statistics with the theoretical moments anticipated from a moving average series of order 11.

3.4. The Poisson Process

One of the most important and fundamental stochastic processes is the Poisson process $\{N(t)\}$, which counts the number of occurrences of an instantaneous event which have transpired in t time units. The instantaneous events are usually common and happen at random points in time, so that the random variable $N(t)$ is an integer-valued variate giving the cumulative number of occurrences of the event through time t.

A particular realization $[N(t), t \geq 0]$ from the ensemble of time functions constituting the process is then a nondecreasing step function of

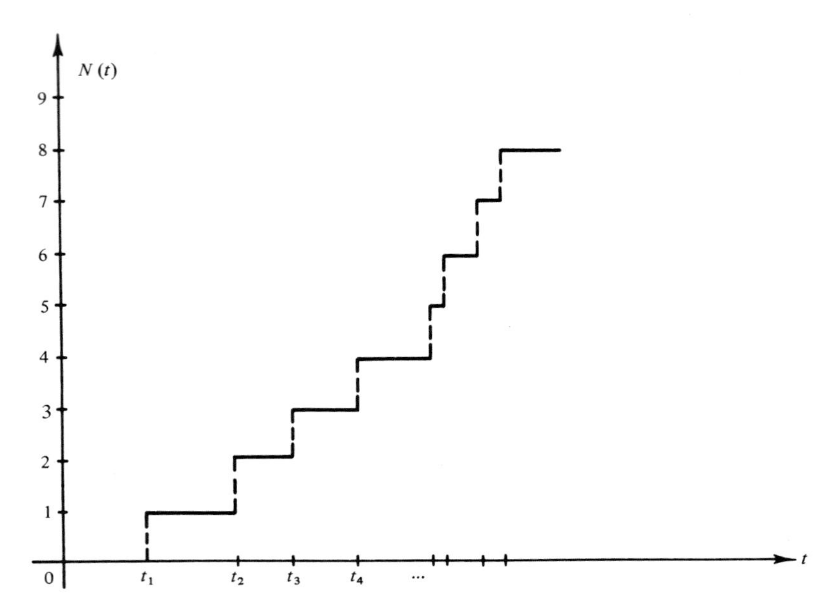

FIG. 3.1. Realization of a Poisson process.

unit jumps, as depicted in Fig. 3.1. The locations t_i of the unit jumps in any realization of the Poisson process are the time points at which the sporadically occurring events arise. These events are usually associated with instantaneous departures from conditions that normally persist; examples are as follows:

(*a*) the demise, or failure, of an electrical component;
(*b*) the sporadic emission of a charged particle in an electronic tube;
(*c*) the commencement of rainfall in a watershed system;
(*d*) the arrival of a merging vehicle into a traffic flow;
(*e*) the occurrence of lightning in a specific volume of atmosphere;
(*f*) the arrival of a customer at a service booth or window.

3.4.1. Axiomatic Specifications

The counting process $\{N(t)\}$ becomes then a *Poisson process* whenever the following assumptions are made regarding it:

1. $N(0) = 0$.
2. For any two instants of time s and t such that $0 \leq s < t$, the increment, or change, in the count is given by

$$N(t) - N(s),$$

a random variable that is stationary in distribution; i.e., for any $h > 0$, the probability distribution function for the incremental random variables $\{N(t + h) - N(s + h)\}$ and $\{N(t) - N(s)\}$ are the same.

3. The process $\{N(t)\}$ has independent increments; i.e., if the time intervals $(s, t]$ and $(\sigma, \tau]$ do not overlap, then the random variables $N(t) - N(s)$ and $N(\tau) - N(\sigma)$ are independently distributed.

4. The incremental random variable $N(t) - N(s)$ has Poisson distribution with parameter (or mean) proportional to $(t - s)$; i.e., for $0 \leq s < t$,

$$P[N(t) - N(s) = k] = e^{-\lambda(t-s)}[\lambda(t - s)]^k/k!, \qquad k = 0, 1, 2, \ldots .$$

Conditions 2 and 3 might be more succinctly presented by noting their equivalence to a condition requiring that $\{N(t)\}$ have stationary independent increments.

The Poisson process can be characterized in a number of ways. One may note that the random variable $N(t)$ has Poisson distribution of mean (λt) since, by Condition 4,

$$P[N(2t) - N(t) = k] = e^{-\lambda t}(\lambda t)^k/k!, \qquad k = 0, 1, 2, \ldots,$$

and, by Condition 2 (with $h = t$ therein), it follows that

$$P[N(t) - N(0) = k] = e^{-\lambda t}(\lambda t)^k/k!, \qquad k = 0, 1, 2, \ldots .$$

Since, via Condition 1, $N(0) = 0$, the probability distribution function of the random variable $N(t)$ is given by the Poisson distribution of mean (λt):

$$P[N(t) = k] = e^{-\lambda t}(\lambda t)^k/k!, \qquad k = 0, 1, 2, \ldots .$$

One can note immediately that the stochastic process $\{N(t)\}$ is not itself stationary, for the distribution of the random variable $N(t)$ depends upon its time index, t; indeed,

$$E\{N(t)\} = \lambda t \qquad \text{and} \qquad \operatorname{Var}\{N(t)\} = \lambda t,$$

so that both the mean value and variance functions for the Poisson process are linear functions of time.

3.4.2. Interarrival Times

The positive parameter λ, which is associated with the process, is termed the *intensity* of the process, and reflects the average rate at which the instantaneous, yet sporadic, events occur during the time interval of length t; i.e.,

$$E\{N(t)/t\} = \lambda.$$

However, the events do not occur regularly in time with an intensity λ, but rather in accordance with an exponential distribution for the *interarrival times*, the ith of which is a random variable T_i defined as the time elapsing between the instant of occurrence of the $(i - 1)$st event and that of the ith event, with $T_0 \equiv 0$.

The interarrival times T_i represent the waiting time between successive events. As random variables, their distribution can be deduced from the four assumed conditions for the Poisson process. For example, the random variable T_1 represents the time until the first event arrives, so that, for any x, the statement "$T_1 < x$" is valid if and only if the statement "$N(x) \geq 1$" can be made. Therefore, the two statements must of necessity represent equivalent events, so that

$$P[T_1 < x] = P[N(x) \geq 1] = 1 - P[N(x) = 0];$$

or

$$P[T_1 < x] = 1 - e^{-\lambda x},$$

since $N(x)$ has the Poisson distribution of parameter (λx). However, by definition, the cumulative distribution function for T_1 is

$$F_{T_1}(x) \equiv P[T_1 < x],$$

so that the distribution of T_1 is the exponential distribution of mean λ^{-1}; i.e., the probability density function for the random variable T_1 is

$$f_{T_1}(t) = dF_{T_1}(t)/dt = \lambda e^{-\lambda t}, \qquad t > 0,$$

having mean

$$E(T_1) \equiv \int_0^\infty t f_{T_1}(t)\, dt = \lambda^{-1}$$

and variance

$$\text{Var}(T_1) = E(T_1^2) - [E(T_1)]^2 = \lambda^{-2}.$$

A similar argument applies to the determination of the distribution of the random variable T_k in general. Indeed, given that $N(s) = (k - 1)$, for some time s, the probability that an additional τ time units will transpire before the kth event is observed is given by

$$P[N(s + \tau) - N(s) = 0] = e^{-\lambda(s+\tau-s)} = e^{-\lambda\tau}, \qquad \tau > 0.$$

Therefore, if the time s were chosen to be the instant of the occurrence of the $(k-1)$st observed event, then the statements "$T_k > \tau$" and "$N(s+\tau) - N(s) = 0$" would be equivalent, so that

$$P[T_k > \tau] = e^{-\lambda\tau}, \qquad \tau > 0.$$

It then follows that the probability density function for T_k, the kth interarrival time in a Poisson process of intensity λ, is the exponential distribution of mean λ^{-1}, for each $k = 1, 2, \ldots$. Thus, the Poisson process can also be characterized by a sequence of independently distributed interarrival times $T_1, T_2, \ldots$, a conceptually infinite random sample from the exponential distribution of mean $(1/\lambda)$. One might indeed pause to reflect on the somewhat logical relationship between λ (the intensity, or time-rate, of the Poisson occurrences) and the mean interarrival time λ^{-1} associated with the Poisson process.

EXERCISES

1. Show that, for a Poisson process of intensity λ, the interarrival times T_1 and T_2 are independently distributed random variables since, if $T_1 = x$, then for any $\tau > 0$,

$$N(\tau + x) - N(x) \qquad \text{and} \qquad N(x) - N(0) = N(x)$$

are independently distributed Poisson random variables, by Condition 3. Since the statement "$T_2 > \tau$ given that $T_1 = x$", is valid if the statement "$N(\tau + x) - N(x) = 0$" can be made, then the conditional distribution of T_2, given that $T_1 = x$, must be the same as its marginal distribution (i.e., must be independent of $T_1 = x$). Indeed, one can note that

$$P[T_2 > \tau \mid T_1 = x] = P[N(\tau + x) - N(x) = 0] = e^{-\lambda\tau}.$$

2. From a similar argument, conclude that the successive interarrival times $T_1, T_2, \ldots$ of a Poisson process are independently and exponentially distributed random variables of mean $(1/\lambda)$.

3.4.3. Waiting Time to the rth Event

The total elapsed time S_r until the occurrence of the rth event (for some fixed $r > 0$) in a Poisson process is the random variable

$$S_r \equiv T_1 + T_2 + \cdots + T_r,$$

where T_i is the ith interarrival time, an exponentially distributed random variable of mean $(1/\lambda)$, $i = 1, 2, \ldots, r$. Hence, S_r represents the sum of r independently and exponentially distributed random variables of common mean, so that S_r has the Erlang density function (see Section 2.13.4):

$$f_{S_r}(s; \lambda, r) = \begin{cases} \lambda^r s^{r-1} e^{-\lambda s}/(r-1)!, & s > 0, \\ 0, & s < 0, \end{cases}$$

a result that can also be established by noting that, apropos of the Poisson process, the statements

$$\text{“}S_r < s\text{''} \qquad \text{and} \qquad \text{“}N(s) \geq r\text{''}$$

are equivalent, since either is valid if and only if the other obtains. Their probabilities are thus equivalent, being

$$P[S_r < s] = P[N(s) \geq r] = \sum_{k=r}^{\infty} \{e^{-\lambda s}(\lambda s)^k/k!\},$$

so that the cumulative distribution function for S_r becomes

$$F_{S_r}(s) = 1 - \sum_{k=0}^{r-1} \{e^{-\lambda s}(\lambda s)^k/k!\} = \int_0^s \lambda^r x^{r-1} e^{-\lambda x}\, dx/(r-1)!,$$

the last result being derivable via partial integrations and induction. The probability density function for S_r then becomes the derivative of $F_{S_r}(s)$, or equivalently, the integrand of the right-most expression for $F_{S_r}(s)$:

$$f_{S_r}(s; \lambda, r) = \begin{cases} \lambda^r s^{r-1} e^{-\lambda s}/(r-1)!, & s > 0, \\ 0, & s < 0. \end{cases}$$

EXERCISES

1. Simulate a Poisson process of unit intensity by generating successively the exponential random variables

$$T_i = -\ln U_i$$

from the uniformly distributed random variable U_i, $i = 1, 2, \ldots$, as the successive interarrival times. Note the value of $N(5)$, the number of occurrences that have been recorded at time $t = 5$, then reinitiate the Poisson process 99 additional times, building a frequency histogram of the 100 integer-valued variates $N(5)$. Using the Chi-squared test, compare this histogram with the Poisson distribution of mean $5 \cdot 1 = 5$.

2. Simulate a Poisson process of unit intensity as in Exercise 1, until such time as the sixth event occurs. Record the random variable

$$S_6 = T_1 + T_2 + \cdots + T_6,$$

and then reinitiate the Poisson process. Repeating this procedure a total of 100 times, construct a frequency histogram of the 100 resulting S_6-values and compare with the Erlang distribution (or, equivalently, with the Chi-squared distribution of 12 degrees of freedom). Compare the proportion of S_6-values that do not exceed 5 with the proportion of the 100 $N(5)$ values (of Exercise 1) that are 6 or greater; do these proportions seem compatible in view of the fact that the statements "$S_6 < 5$" and "$N(5) \geq 6$" are equivalent (in the probability sense)? Will the two compared proportions always be the same if differing random number seeds are employed in Exercises 1 and 2?

3. Note that, in Exercise 1, one cannot determine $N(5)$ until such time as $S_r \geq 5$ for the first time (i.e., until $S_{r-1} < 5$, yet $S_r \geq 5$), at which point $N(5) = (r - 1)$; compute then in each of the 100 instances the "residual" random variable $(S_r - 5)$. Construct a histogram of these residual waiting times and compare these with the exponential distribution of unit mean. Discuss the appropriateness of this comparison by noting that if T is an exponentially distributed random variable of mean λ^{-1}, then

$$P[T > (s + \tau) \mid T > s] = \frac{P[T > (s + \tau), \text{ and } T > s]}{P[T > s]}$$

or

$$P[T > (s + \tau) \mid T > s] = \frac{P[T > (s + \tau)]}{P[T > s]}$$

$$= \frac{e^{-\lambda(s+\tau)}}{e^{-\lambda s}} = e^{-\lambda\tau}$$

a result independent of s. Hence, deduce that T_r and the residual $(S_r - 5)$ are identically distributed random variables.

3.4.4. The Generation of Poisson-Distributed Random Variables

In order to generate Poisson-distributed random variables X of mean μ, one can employ the tabulated CDF:

$$F_X(x; \mu) \equiv \sum_{k=0}^{[[x]]} (e^{-\mu}\mu^k/k!), \qquad \text{for any} \quad x > 0,$$

where $[[t]]$ is the greatest integer t or less. Such a procedure implies the need for tabulations of Poisson distributions, one for each value of μ, the positive parameter and mean of the distribution.

An alternative procedure, one that avoids searches through CDF tabulations, is the mimicry of a Poisson process of unit intensity. One can note that, after μ units of time, this unit Poisson process produces a random variable $N(\mu)$, which has the Poisson distribution of mean $\lambda \cdot \mu = 1 \cdot \mu = \mu$. It therefore follows that the count achieved by a Poisson process of unit intensity at time $t = \mu$ can be taken as a random variable having the desired Poisson distribution.

To achieve this, one need only generate independently and exponentially distributed random variables T_i of unit mean,

$$T_i = -\ln U_i,$$

where the U_i are independently and uniformly distributed, until such time as their cumulative sum

$$S_r \equiv \sum_{i=1}^{r} T_i = -\sum_{i=1}^{r} \ln U_i, \qquad r = 1, 2, \ldots,$$

equals or exceeds μ *for the first time*. Clearly, at this point

$$N(\mu) = (r - 1),$$

since $S_{r-1} < \mu$ and $S_r \geq \mu$. However, as noted above, $N(\mu)$ is a random variable X possessing the desired Poisson distribution of mean μ.

One can note that, since $S_r = -\sum_{i=1}^{r} \ln U_i$, an equivalent expression becomes $S_r = -\ln\{\prod_{i=1}^{r} U_i\}$. The algorithm for determining the Poisson-distributed random variable consists then of the successive generation of uniformly and independently distributed random variables U_i until such time as

$$-\ln\left\{\prod_{i=1}^{r-1} U_i\right\} \equiv S_{r-1} < \mu \leq S_r \equiv -\ln\left\{\prod_{i=1}^{r} U_i\right\},$$

at which point one takes $X = (r - 1)$ as the Poisson variate. The inequalities can also be written as

$$\left\{\prod_{i=1}^{r} U_i\right\} \leq e^{-\mu} < \left\{\prod_{i=1}^{r-1} U_i\right\},$$

so that a more expeditious algorithm for generating Poisson-distributed random variables of a given mean (μ) is as follows:

1. Compute $e^{-\mu}$, for the given parameter μ.
2. Successively generate uniformly distributed random variables, computing their "cumulative" product $\prod_{i=1}^{k} U_i$, $k = 1, 2, \ldots$.
3. Terminate the generation of uniformly distributed random variables as soon as, for the first time, the product $\prod_{i=1}^{r} U_i \leq e^{-\mu}$.
4. At this point, set $X = (r - 1)$, a Poisson variate of the desired mean.

EXERCISES

1. Generate 100 Poisson random variables of mean 2.5 by the algorithm just discussed. Compare the histogram of these 100 variates with the theoretical Poisson distribution of mean 2.5 by means of a Chi-squared statistic.

2. Denoting by R the (random) number of uniformly distributed random variables that must be generated in order to produce a Poisson variate X by means of the algorithm just discussed, conclude that, in any instance,

$$R = (X + 1),$$

so that the mean number of required uniformly distributed random variables is $(\mu + 1)$, with standard deviation $\mu^{1/2}$. Confirm this result by repeating several times Exercise 1, noting the mean number of required uniformly distributed random variables and their standard deviation.

3. Comment on the use of this algorithmic generator to test the uniformity and independence of the distribution of the random numbers emanating from, say, a mixed congruential generator. Would not Chi-squared statistics, of the type suggested in Exercise 1, constitute this test?

3.4.5. Application of Poisson Processes

From the conditions (or axioms) of the Poisson process, it follows that the probability of an event's occurring in any time interval of sufficiently small width h is approximately proportional to h. One may note that, at any time t, the process is engaged in one of its exponentially distributed interarrival times, say T, so that

$$P[T < (t+h) \mid T \geq t] = P[t \leq T < (t+h)]/P[T \geq t],$$

or

$$P[T < (t+h) \mid T \geq t] = \{[1 - e^{-\lambda(t+h)}] - [1 - e^{-\lambda t}]\}/e^{-\lambda t},$$

or

$$P[T < (t+h) \mid T \geq t] = 1 - e^{-\lambda h},$$

which becomes, from Maclaurin's series expansion of $e^{-\lambda h}$:

$$P[T < (t+h) \mid T \geq t] = (\lambda h) + o(h),$$

where

$$\lim_{h \to 0} \frac{o(h)}{h} = 0.$$

Furthermore, since the process has stationary independent increments, one might view the Poisson process as the observation of a sequence of successive subintervals, each of width h, until such time as an interval contains an event. At that time, a new interarrival time is initiated and the procedure of observing another sequence of subintervals of width h is begun again. The reader may recall the earlier derivation of the exponential distribution, in which this procedure of recording the results of Bernoulli trials, each providing a success (event) with probability $p = \lambda h$, led to the exponentially-distributed waiting time T (see Section 2.5).

Indeed, that preceding derivation was motivated by the desire to describe the distribution of the lifetime T of an electronic component whose probability of failure (the event of interest in the present context) was assumed to be proportional to the length h of the subinterval and to be independent of the age of the component (i.e., $p = \lambda h$, for any subinterval). Under these assumptions, the Poisson process becomes a descriptive recording for the successive failures of a sequence of essentially identical components in which any failed component is immediately repaired or replaced. The successive lifetimes $T_1, T_2, \ldots$ become independently and exponentially distributed random variables of the same mean, and the count $N(t)$ of the number of replacements made in time t represents a realization from a Poisson process.

In a certain sense, the Poisson process describes events that occur "uniformly" in time, because, regardless of the length of time that the process has been observed, the probability of an event's occurring in the next small time interval of width h remains the same. Hence, in reliability contexts, the assumption of the applicability of the Poisson process is frequently warranted whenever the components in a renewable system are presumed not to age with use [see Cox (1962)].

The Poisson process also proves especially useful in other areas as well. For example, the occurrence of "shot noise," electrical disturbances that arise with independently and exponentially distributed interarrival times, is a random process that can frequently satisfy the axioms of the Poisson process. Hence, its value to communications and electrical engineers is paramount; closely akin to these applications is the *random telegraph signal* [see Papoulis (1965)].

In addition, the Poisson process has been shown to be applicable to transportation studies, especially to the spatial distribution of one-way traffic along a conceptually infinite roadway whenever the automobiles on it maintain randomly assigned velocities, as discussed by Breiman (1963).

The arrival of orders for goods from a stock or depot can also be represented by a Poisson process over time. Hence, the study of inventory policies is often expedited by assuming the applicability of the Poisson process to describe the arrival pattern of orders over time. Quite similarly, the arrival of customers for limited service or sales facilities can be patterned after a Poisson process described by the intensity λ, or average rate, of the customer arrivals. The consideration of the role of the Poisson process in waiting lines, will be discussed in Section 4.7.

In summary, the Poisson process is central to the description of many random phenomena that are observable in natural and/or man-made systems. It may often be invoked in order to overcome an encounter with the Uncertainty Principle of Modeling, for the Poisson process is ideally suited to the description of repeatedly, but randomly, occurring phenomena whose probabilities of occurrence are distributed "uniformly" in time or space (i.e., whose probabilities of occurrence do not depend on the past history or status of the process and whose random arrivals appear at an effectively constant rate, as denoted by the intensity parameter λ).

3.5. The Spectrum

The study of time series or stochastic processes is especially important to the systems analyst because it provides a mechanism by which the analyst can structure a framework for time-dependent, probabilistic phenomena as they are observed in an intricate system. For example, the Poisson process can be invoked to describe randomly recurring phenomena whenever the probability of subsequent occurrences does not depend upon the past history of occurrences and whenever the inter-arrival times of the successive occurrences are exponentially distributed random variables; such conditions are frequently observed in the arrival patterns of customers in a queuing system or in the arrival mechanism of orders into an inventory system.

Of primary importance in the description of the behavior of a sequence $\{X(t)\}$ of random variables (i.e., of a time series) is the autocovariance function

$$R(s, t) \equiv \text{Cov}\{X(s), X(t)\},$$

which, if weakly (or strictly) stationary, becomes a function of the time difference, $\tau = (t - s)$, only:

$$R(\tau) = \text{Cov}\{X(t - \tau), X(t)\} = \text{Cov}\{X(t), X(t - \tau)\} = R(-\tau).$$

The autocovariance function for stationary processes provides then a measure of the dependency between random variables in the time series and is usually expected to wane (i.e., to decrease in absolute magnitude) as the time lag τ between the random variables increases.

The variance function of a stationary random process becomes then the constant $R(0)$, so that the correlation between any pair of random variables $X(t)$ and $X(t + \tau)$ in the time series becomes also a function of the time lag τ:

$$\varrho_{XX}(\tau) \equiv R(\tau)/R(0), \qquad \tau = 0, 1, 2, \ldots,$$

called the autocorrelation function of lag τ, which usually tends to diminish in absolute value with increasing time lag.

Another measure of dependency among the terms in a time series is the *spectrum*, given by the Fourier transform of the autocovariance function:

$$I(\omega) \equiv \pi^{-1}\left\{R(0) + 2 \cdot \sum_{k=1}^{\infty} R(k) \cos k\omega\right\}, \qquad 0 < \omega < \pi.$$

The spectrum reveals the relative strength of any intrinsic periodicities in the series' behavior, as measured by the relative magnitude of $I(\omega)$ for any frequency ω between 0 and π.

3.5.1. Components of a Time Series

A time series can be viewed as receiving contributions from several sources, usually categorized as

(*a*) *trends*—time-dependent, deterministic contributions, such as an inherent linear or exponential growth or decay;
(*b*) *periodic*—time-dependent, deterministic, and fluctuating contributions in addition to a trend, such as a set of sinusoidal components; and
(*c*) *random*—nondeterministic, fluctuating contributions in addition to trend and periodic components.

In Section 3.2, it had been indicated that time series having linear trends (such as the Poisson and Wiener–Lévy processes) in their statistical descriptions may often have these trends eliminated by considering instead their associated incremental processes. Indeed, in Chapter 10, it will be indicated how the nature of this trend effect may be estimated and how, via successive incrementation procedures, trends can be somewhat removed.

For the present, however, it will be presumed that no trend effects exist; more specifically, the ensuing discussion will assume that the time series $\{Z(t)\}$ of interest is stationary (weakly), and that the spectrum will exist for all frequencies ω between 0 and π.

3.5.2. Some Exemplary Spectra

The white-noise process $\{\varepsilon(t)\}$ has been defined as a sequence of independently and identically distributed random variables, so that the autocovariance function becomes

$$R(\tau) = \begin{cases} \sigma_\varepsilon^2, & \tau = 0, \\ 0, & \tau \neq 0. \end{cases}$$

Thus, the spectrum of the white-noise process becomes

$$I(\omega) = R(0)/\pi = \sigma_\varepsilon^2/\pi, \qquad 0 < \omega < \pi,$$

a constant function for all frequencies ω. Indeed, if a time series consists of a sequence of independently and identically distributed random variables, no particular periodicities would be expected to arise with frequency greater than any other. By analogy with the spectrum for the frequency components constituting white light, such a time series (i.e., one having a constant spectrum) is said to be a *white-noise process*.

It may be noted in passing that the relationship between autocovariance functions and their Fourier transforms (spectra) is unique, just as there exists a one-to-one correspondence between probability density functions and their Laplace transforms (moment-generating functions). In a sense then, the degree to which the autocovariance function characterizes a time series is not amplified by its spectral representation. The two approaches are of value in their own right, however, since the autocovariance function describes the relationship among the random variables of the time series in terms of time lags (i.e., in the "time domain"), whereas the spectrum emphasises the latent periodic effects among the random variables (in the "frequency domain").

As an example in this regard, the aforementioned white-noise process might be assumed to consist of a sequence of independently Gaussian distributed random variables of mean $\frac{1}{2}$ and variance $\sigma_\varepsilon^2 = \frac{1}{12}$. The corresponding autocovariance function would be null (save for $\tau = 0$, for which $R(0) = \sigma_\varepsilon^2 = \frac{1}{12}$), and the spectrum would be

$$I(\omega) = 1/12\pi, \qquad 0 < \omega < \pi.$$

If, instead, the time series were constituted by a sequence of independently and *uniformly* distributed random variables, the autocorrelation and spectral functions would not change, though any realization $[X(t), t = 1, 2, \ldots]$ from this latter process would differ remarkably from any recording of the normal white-noise process [e.g., negative values for $X(t)$ could never be observed for the uniform case, but would frequently be anticipated in the case of the normal white-noise process].

Consequently, the summary information contained in the autocovariance function (or, equivalently, in the spectrum) is not a complete representation of the character of a time series. Other examples can be constructed to illustrate this point, notably the similitude of the autocovariance function for the Wiener–Lévy process to that for the random telegraph signal, as illustrated by Jenkins and Watts (1968).

Furthermore, though two autocovariance functions may not be identical, their relative behavior can often mask the nature of the underlying mechanism for the random process. The reader may recall that the autocovariance function for the moving average series of order k is given by

$$R(\tau) = \begin{cases} \sigma_\varepsilon^2[1 - |\tau|/(k+1)] & \text{for } |\tau| = 0, 1, \ldots, k, \\ 0 & \text{for } |\tau| \geq (k+1), \end{cases}$$

whenever the moving average assigns equal weights to the $(k + 1)$ most recent contributing white-noise terms. On the other hand, the autocovariance function for the first-order autoregressive series also decays with increasing lag:

$$R(\tau) = a_1^{|\tau|}, \qquad \tau = 0, 1, 2, 3, \ldots .$$

Though the first-order autoregressive series has autocovariance function that is theoretically nonzero for all positive lags, it is often practicably difficult to discern between this form and that associated with a series which is essentially a moving average scheme. Yet the difference may be quite profound; for the first-order autoregressive scheme is a Markov process of order one, whereas the moving average series is not.

Similarly, the spectra of such processes do not always shed a great deal of light on the character of the time series. The spectrum of the general linear process may be shown to be given by

$$I(\omega) = | H(\omega) |^2 \sigma_\varepsilon^2,$$

where

$$H(\omega) \equiv \sum_{k=0}^{\infty} h_k e^{-i\omega k} = \sum_{k=0}^{\infty} h_k[\cos k\omega - i \sin k\omega]$$

is the Fourier transform of the series of impulse responses h_k and where $i = (-1)^{1/2}$, the imaginary particle. Hence, for the first-order autoregressive series, $h_k = a_1^k$, $k = 0, 1, 2, \ldots$, so that

$$I(\omega) = \sigma_\varepsilon^2/[1 + a_1^2 - 2a_1 \cos \omega], \qquad 0 < \omega < \pi;$$

in the case of the moving average process of order k [with weights equal to $1/(k + 1)$], the spectrum becomes

$$I(\omega) = \sigma_\varepsilon^2[1 - 2 \cos(k + 1)\omega]/[(k + 1)^2\{1 - 2 \cos \omega\}],$$

for $0 < \omega < \pi$, a somewhat similarly behaving function.

3.6. Summary

The spectrum, like the autocovariance function, may be of limited utility to the categorization of time series, but may be of immense assistance in revealing latent or disguised periodicities in the time series' behavior. Subsequently, in Chapter 10, a discussion will be undertaken of methods by which recorded time series data can be employed to estimate the spectrum and thereby to reveal hidden periodicities. The reader might be directed to the report of spectral analyses, performed by Fishman and Kiviat (1967) on time series data arising from the responses (or state descriptions) of several simulated queue discipline alternatives.

In the next chapter, techniques known collectively as the Monte Carlo method will be investigated. Then in Chapter 5, attention will be directed away from direct discussions of random variables and time series; though the exposition will concern the aspects and stages of an orderly modeling effort, the reader should be aware of the many probabilistic mechanisms (i.e., random variables arising from theoretical considerations such as the Central Limit Theorem or time series developing from, say, underlying considerations such as those leading to the Poisson process). It will be imperative in the analysis of systems, especially when that analysis is directed toward the construction of a meaningful model of the system, that the Uncertainty Principle of Modeling be appropriately acknowledged by the inclusion of *pertinent* and *appropriate* stochastic phenomena.

Chapter 4

THE MONTE CARLO METHOD

Patience—and shuffle the cards!

—M. CERVANTES

The ease with which random variables can be generated for subsequent use in a (usually computerized) symbolic model has been amply illustrated in the preceding two chapters. Numerous techniques have been described for providing random or pseudorandom numbers, which may be construed as independently and uniformly distributed variates. Methods for transforming these random numbers into essentially independent random variables of specific, desired distributions have also been delineated. Furthermore, in the event that correlated pairs, sets, or sequences of random variables are required by a simulation model, specific linear combinations of generated random variables have been shown to be often appropriate, and are thus readily employed by means of computer program algorithms.

The reader may note that, in addition to using straightforward functional transformations, one may also be able to mime certain random processes in order to generate random variables from a given distribution. For example, geometrically distributed random variables are known to arise as the waiting time T (in terms of the number of trials), until the first success occurs in a sequence of Bernoulli trials, success occurring with probability p on each trial; one method for generating such random variables, then, is to imitate the performance of the Bernoulli trials by generating independently and uniformly distributed random variables U until such time as $U \leq p$, at which point T is set equal to the number of variates (U's) so examined.

The imitation of some random process leads then to a random variable whose distributional properties may be well-known in advance, because the random process itself may have been studied analytically and the distribution of random variables generated therefrom may be determined rigorously. The procedure of generating (approximately) normally distributed variates by the addition of, say, 12 independently and uniformly distributed random variables, thereby miming the Central Limit Theorem, is another case in point.

From the generation of random variables of known distributional properties by means of imitations of certain random processes, it is a short step to the imitation of random processes whose associated random variables are of unknown distributional properties. In these cases, the unknown distributional properties can be ascertained (or at least estimated) by iterated imitations of the random process in question. A discussion of these techniques, referred to as the *Monte Carlo Method*, will constitute the primary concern of this chapter. In a later section, a discussion of the elements of queues will reveal an exemplary class of complex systems whose simulation often is best accomplished by application of the Monte Carlo method.

4.1. Experiments with Random Numbers

In their concise monograph, Hammersley and Handscomb (1964) define the Monte Carlo method as "that branch of experimental mathematics which is concerned with experiments on random numbers [p. 2]." Thus, sequences of random (or, presumably, pseudorandom) numbers are assumed to be available and are subjected to transformations, investigations, and comparisons in order to ascertain the statistical (distributional) properties resulting from such experimentation.

An experiment on random numbers then becomes a generator of random variables, since the experimental measurements can be only transformations of the original random numbers. However, the cumulative distribution function of the random variable arising from a Monte Carlo experiment may not be of a known functional form. If the functional form $F(x)$ *is* known, then r repetitions of the experimentation on successive segments of the sequence of random numbers will result in r

random variables, mutually independently distributed according to $F(x)$ [i.e., the r repetitions of the experiment will provide a *random sample* from $F(x)$]. Thus, statistical comparisons of the distribution of the r sample values with the theoretically applicable cumulative distribution function can, if found compatible, serve only one of two purposes:

(*a*) Verify that the transformations of the uniformly distributed random numbers are indeed being correctly applied.

(*b*) Presuming that the experiment's transformations are being properly applied, establish that the original source of the experimental numbers is a generator of randomly and uniformly distributed random variables.

Consequently, experiments with random numbers whose experimental outcomes are random variables of *known* distributional forms are of little value, other than in the verification of the source of the numbers as a random number generator. For example, if the log-and-trig method is invoked in order to generate p successive pairs of independently distributed Gaussian random variables (of mean zero and known variance σ^2) by transforming successive, mutually exclusive pairs of uniformly distributed random numbers, then the absolute magnitude of the mean

$$\bar{X} = (2p)^{-1} \sum_{i=1}^{2p} X_i,$$

can be employed to ascertain whether the random numbers are indeed a source of independently and uniformly distributed random variables. The reader may recall that $\bar{X}$, being a linear combination of independently (and therefore jointly) distributed Gaussian variates, has itself a (univariate) Gaussian distribution of mean zero and of variance $\sigma^2/(2p)$; hence, for any real number $c > 0$,

$$P[|\bar{X}| \geq c] = P[|Z| \geq (c/(\sigma/\sqrt{2p}))]$$

or

$$P[|\bar{X}| \geq c] = 2\Phi[(-c/(\sigma/\sqrt{2p}))],$$

where Z is the standardized Gaussian variate and Φ is its cumulative distribution function. One would then assume something to be amiss whenever $|\bar{X}|$ was improbably large [relative to σ, the underlying (and presumably known), standard deviation].

Therefore, another test of a random number generator can be performed by computing the sample mean of the $2p$ Gaussian variates generated by transforming $2p$ random numbers and by assigning a critical constant $C > 0$ such that the probability

$$P[|\bar{X}|/(\sigma/\sqrt{2p}) \geq C] = 2\Phi(-C)$$

is so small as to be considered unlikely to have arisen by chance alone. [Usually, C is chosen as 1.96 so that $2\Phi(-1.96) = 0.05$ or as $C = 2.58$ so that $2\Phi(-2.58) = 0.01$, it being then presumed that events occurring so infrequently as 1 time in 20 (or 1 time in 100) will be construed so unlikely as to place in jeopardy any hypothesis that the source of random numbers is well behaved.] Of course, unless the source of random numbers were suspect, such a Monte Carlo experimentation would not need to be undertaken, since the distributional properties of $\bar{X}$ are quite well known whenever the underlying parameter σ is precisely specified.

It was, however, a closely allied problem that provoked Gossett under the pseudonym "Student" (1908) to perform a sampling, or Monte Carlo, experiment. Whenever the underlying parameter σ is not known for a Gaussian random variable, one can "estimate" this parameter by the "root mean square" of the $n(=2p)$ data $X_1, X_2, \ldots, X_n$:

$$S = \left[\sum_{i=1}^{n} (X_i - \bar{X})^2/(n-1)\right]^{1/2}.$$

However, if the statistic

$$T_n \equiv \bar{X}/(S/\sqrt{2p})$$

is employed instead of $\bar{X}/(\sigma/\sqrt{2p})$, then the applicable distribution function could no longer be assumed to be of Gaussian form, for T_n is no longer a simple linear combination of the data (random sample).

In one of the earliest recorded applications of a sampling experiment, "Student" employed hundreds of samples of size 4 (and size 8), drawing each sample from a well-shaken collection of numbered chits, and then computing for each sample the statistics T_4 (and T_8, respectively). By constructing a frequency histogram for each of the two random variables, he was able to fit systems of curves to them until a satisfactory agreement was attained. The theoretical proof of the applicability of these frequency curves, known as the family of Student t distributions, then followed.

Of course, now that this pioneering experimental effort has been brought to a successful end, any repetition of Gossett's experiment becomes merely another test of the underlying random numbers, for the random variates T (provided that the X_i are normally and independently distributed, homoscedastic, Gaussian random variables) should produce a histogram compatible with an appropriate member of Student's t family. In general, however, if random variates can be repetitively generated from a sequence of random numbers by means of a fixed transformation process, then the experimentation shall result in empirical information regarding the distributional properties of the resulting random variables. Such numerical experiments, an elementary application of the Monte Carlo Method, are conducted extensively today, especially whenever the desired distributional properties are otherwise analytically intractable.

EXERCISES

1. Generate 1200 uniformly distributed random numbers and convert by dozens to 100 Gaussian (approximately) random variables by means of the transformation

$$Z_i = \left[\sum_{j=12i-11}^{12i} U_j - 6\right], \qquad i = 1, 2, \ldots, 100.$$

Note that here $\sigma = \sigma_0 = 1$, the known standard deviation of the Z_i. Compute

$$\bar{Z} = (100)^{-1} \sum_{i=1}^{100} Z_i$$

and test the random number generator by comparing $|\bar{Z}|$ with the extremely improbable values arising from a normal distribution of mean zero, variance $\frac{1}{100}$. Supposing that your sample were to lead to an improbably large value of $|\bar{Z}|$, comment on how you might attribute its cause either to the transformation procedure or to the generator itself.

2. As in Exercise 1, generate 100 approximately Gaussian distributed variates, Z_i, $i = 1, 2, \ldots, 100$ of mean zero and of unit variance. Also compute the statistic

$$T = \bar{Z}/(S/10),$$

where

$$\bar{Z} = (100^{-1}) \sum_{i=1}^{100} Z_i$$

and

$$S = \left\{(99)^{-1}\left[\sum_{i=1}^{100} (Z_i - \bar{Z})^2\right]\right\}^{1/2}.$$

Compare the value of $|T|$ with the improbably large values of Student's

t distribution of 99 degrees of freedom. Comment on the possible reasons for obtaining an improbably large value of $|T|$.

3. Repeat both Exercises 1 and 2, but generate the 100 Gaussian variates by the log-and-trig method. Comment as in Exercises 1 and 2 and compare results obtained by both of the transformational techniques for generating Gaussian variates.

4.2. Buffon's Needle Problem

One of the earliest known examples of the use of probability games for the purpose of ascertaining physical parameters is the solution of a problem posed by Georges Louis Leclerc, Comte de Buffon (1733): If a needle of length l units is tossed randomly upon a floor composed of parallel planks of equal width d ($>l$) units, what is the probability that the needle, once it comes to rest, will cross (or touch) a crack (of conceptually zero width) separating the planks of the floor?

One analytical approach to the problem is to assume that the resting position of the needle can be uniquely described by a pair of random variables:

$Y =$ the distance from the needle's midpoint to the nearest line, L;

and

$\theta =$ the acute angle between the line L and the needle (the needle conceptually extended if necessary).

By assuming that Y and θ are independently distributed random variables, that Y has rectangular distribution over the region between 0 and $d/2$, and that θ has rectangular distribution, yet over the region between 0 and $\pi/2$, then in accordance with Fig. 4.1, a toss of the needle corresponds to the specification of a set of coordinates (θ, Y) within the rectangle of length $\pi/2$, height $d/2$.

One may note that only those coordinates falling beneath or upon the curve

$$Y = (l/2) \cdot \sin \theta$$

correspond to needle tossings that cross or touch one of the cracks separating the planks of the floor. Since the coordinates (θ, Y) are uniformly distributed over the rectangular region of area $\pi d/4$, the probability p of the needle's crossing or touching a crack is given by the proportion

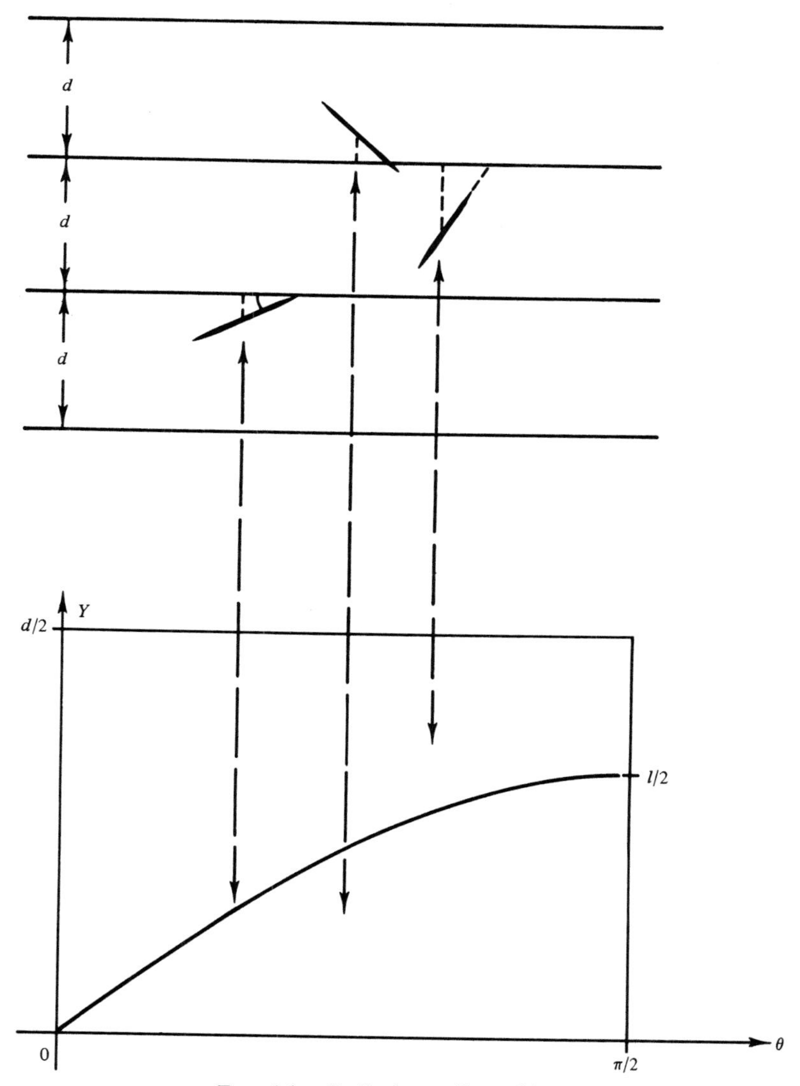

FIG. 4.1. Buffon's needle problem.

of this area beneath the sine curve, as depicted in Fig. 4.1;

$$p = \int_{\theta=0}^{\pi/2} (l/2) \sin \theta \, d\theta / (\pi d/4) = 2l/\pi d.$$

Le Comte de Buffon had suggested that, by tossing as randomly as possible a needle n times upon a floor, one could compute f, the relative frequency of occurrences of the needle's contacting a line. (Note that $l < d$, so that at most one line will be crossed whenever the needle comes to rest.) For n sufficiently large, one could anticipate that

$$f \cong p = 2l/\pi d.$$

Alternatively, since l and d are fixed lengths, le Comte de Buffon noticed that one could *estimate* the constant π by means of the expression

$$\tilde{\pi} = (2l)/(f \cdot d),$$

once the ratio f of crossings to tossings had been experimentally determined.

EXERCISE

By generating 100 pairs of coordinates (θ, Y) uniformly distributed over the rectangle of length $\pi/2$ and height $d/2$, simulate Buffon's needle problem and compute your estimate of π.

4.3. Numerical Integration Techniques

Suppose that $f(x)$ is a bounded, positive, function over an interval, (a, b), and that the value of the integral

$$I = \int_a^b f(x)\,dx$$

is sought. We examine three approaches to the evaluation of I.

4.3.1. Simple Numerical Technique

Let the interval (a, b) be divided into N equal subintervals of width Δ so that $(b - a) = N\Delta$. Letting $x_i = a + i\Delta$, we may evaluate $f(x_i)$, $i = 1, 2, \ldots, N$. The product $f(x_i) \cdot \Delta$ is the area of a rectangle of height $f(x_i)$, width Δ, so that

$$I^* = \sum_{i=1}^{N} f(x_i) \cdot \Delta$$

is a numerical approximation for I. Since $\Delta = (b - a)/N$,

$$I^* = \left\{\sum_{i=1}^{N} f(x_i)/N\right\} \cdot (b - a)$$

is an alternative expression for the approximation. Note that this is the equivalent of finding the "average" value of $f(x)$ and multiplying by $(b - a)$, the length of the interval, as depicted in Fig. 4.2.

Note also that there is *no* statistical variation in the answer: Given Δ, a, b, and $f(x)$, each of several persons computing the approximation would arrive at the same value I^* within limits of round-off errors.

One may note that other numerically analytic techniques, such as the trapezoidal and Simpson's rules, also employ weighted averages of the $f(x_i)$. [The reader is referred to Kunz (1957) for these techniques.]

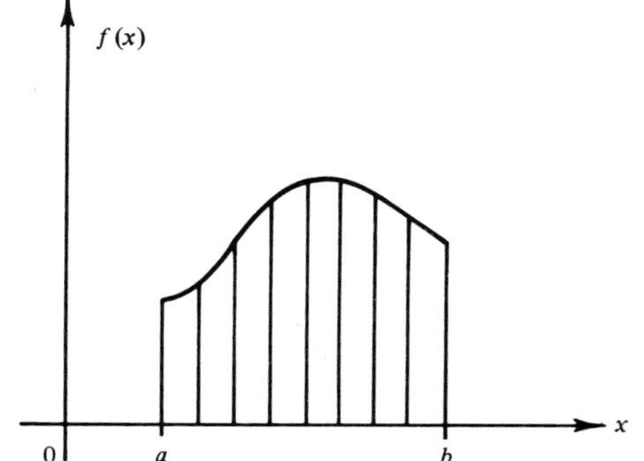

FIG. 4.2. Simple numerical integration.

4.3.2. THE CRUDE MONTE CARLO METHOD

Generate N uniformly distributed variates $U_1, U_2, \ldots, U_N$, and transform to N variates uniformly distributed over (a, b), the interval of integration, to provide $X_i = a + (b - a)U_i$, $i = 1, 2, \ldots, N$. Then, evaluate $f(X_i)$, $i = 1, 2, \ldots, N$, and compute

$$\tilde{I} = \left\{ \sum_{i=1}^{N} f(X_i)/N \right\} \cdot (b - a)$$

as an "estimate" of I. Each of k persons, proceeding independently to select N random numbers, would arrive at his own individual estimate of I. This is another way of stating that $\tilde{I}$ is a random variable, being proportional to the sum of N independent random variables, $Y_i = f(X_i)$, $i = 1, 2, \ldots, N$. Since $E(Y_i) = \int_{x=a}^{b} f(x)\, dx/(b - a)$, we have $E(Y_i) = I/(b - a)$, so that

$$E\left\{ \sum_{i=1}^{N} f(X_i)/N \right\} = I/(b - a),$$

and hence

$$E\{\tilde{I}\} = \{I/(b-a)\} \cdot (b-a) = I.$$

One says, then, that $\tilde{I}$ is an *unbiased statistical estimate* of I, since its mean value is the quantity I whose estimate is sought. Note also that, since

$$\operatorname{Var}\left\{\sum_{i=1}^{N} Y_i/N\right\} = \{\operatorname{Var}(Y_i)\}/N,$$

due to the statistical independence of the Y_i, then

$$\operatorname{Var}[\tilde{I}] = (b-a)^2\{\operatorname{Var}(Y_i)\}/N.$$

The quantity $\{\operatorname{Var}(Y_i)\}$ is equal to $E(Y_i^2) - \{E(Y_i)\}^2$, so that

$$\operatorname{Var}(Y_i) = \int_{x=a}^{b} f^2(x)\,dx/(b-a) - \{I^2/(b-a)^2\}$$

hence,

$$\operatorname{Var}[\tilde{I}] = \left\{(b-a) \cdot \int_{x=a}^{b} f^2(x)\,dx - I^2\right\}\Big/N.$$

One sees then that the precision of the estimate $\tilde{I}$ as measured by $\{\operatorname{Var}[\tilde{I}]\}^{-1/2}$, increases with $\sqrt{N}$.

4.3.3. The Hit-or-Miss Monte Carlo Method

Since $f(x)$ is a bounded, positive, function over (a, b), one can find some positive constant c such that

$$f(x) \leq c \qquad \text{for all} \quad x \quad \text{between} \quad a \quad \text{and} \quad b$$

One may proceed by generating N *pairs* of uniformly distributed random numbers: (U_1, U_1'), (U_2, U_2'), ..., (U_N, U_N'), and then computing both

$$X_i = a + U_i \cdot (b-a), \qquad \text{and} \qquad f(X_i), \qquad i = 1, 2, \ldots, N.$$

One scores a "hit" if the computed $f(X_i) \geq cU_i'$, a "miss" if $f(X_i) < cU_i'$ $i = 1, 2, \ldots, N$, as depicted in Fig. 4.3.

The ratio of the number N_{H} of hits to the number N of trials estimates the proportion of the total area $c \cdot (b-a)$ of the rectangle, given by the points $\{(x, y): x \in (a, b), y \in (0, c)\}$, which lies beneath the curve $y = f(x)$.

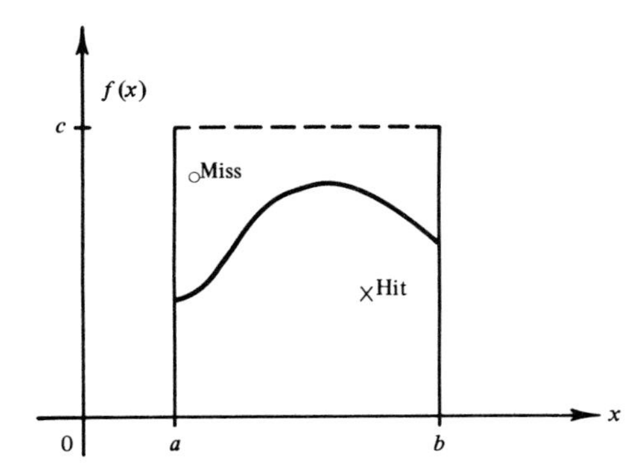

FIG. 4.3. Monte Carlo integration.

An estimate of I is then

$$\hat{I} = (N_H/N) \cdot (b - a) \cdot c.$$

Since each of the N trials constitutes a Bernoulli trial with probability, $p = I/[(b - a) \cdot c]$, of a hit, then

$$E[\hat{I}] = [(Np)/N] \cdot (b - a) \cdot c = pc(b - a) = I$$

and

$$\text{Var}[\hat{I}] = [Np(1 - p)/N^2] \cdot (b - a)^2c^2 = p(1 - p) \cdot (b - a)^2c^2/N$$

or

$$\text{Var}[\hat{I}] = I \cdot [c(b - a) - I]/N.$$

Again, one should note the result, typical of Monte Carlo computations, that the precision of the Monte Carlo estimate, as measured by the inverse of its standard deviation, increases with $\sqrt{N}$.

One notes that the mean value of the estimate $\tilde{I}$ arising from the crude Monte Carlo method is the desired integral I, as is the mean value of $\hat{I}$, the estimate from the hit-or-miss technique. Any comparison between the two then should be based upon the precision (variances) as well as the computational difficulties.

Note that each method requires that $f(x)$ be evaluable at the selected $X_i = a + U_i \cdot (b - a)$, $i = 1, 2, \ldots, N$. However, each trial of the hit-or-miss technique requires the generation of twice as many pseudorandom numbers. Thus, computational simplicity favors the crude method.

In addition, the comparison of the variances, in general, proves more favorable to the crude method. For,

$$\begin{aligned}
\operatorname{Var}[\hat{I}] - \operatorname{Var}[\tilde{I}] \\
&= [I \cdot (b-a)c/N - I^2/N] - \left[(b-a)\int_{x=a}^{b} f^2(x)\,dx/N\right] + I^2/N \\
&= [(b-a)c/N] \cdot \left[I - \int_a^b f^2(x)\,dx/c\right] \\
&= [(b-a)c/N] \cdot \left[\int_a^b f(x)\,dx - c^{-1}\int_a^b f^2(x)\,dx\right] \\
&= [(b-a)c/N] \cdot \left\{\int_a^b f(x)[1 - c^{-1} \cdot f(x)]\,dx\right\} \geq 0,
\end{aligned}$$

since the constant multiplier and the integrand are both positive. [Note that $1 - c^{-1} \cdot f(x) \geq 0$ since $f(x) \leq c$.]

Therefore, $\operatorname{Var}[\hat{I}] \geq \operatorname{Var}[\tilde{I}]$. One would conclude then that the crude Monte Carlo technique is somewhat superior to the hit-or-miss method. Whichever technique is selected [one should note that other techniques are available; cf. Hammersley and Handscomb (1964)], however, an *estimate* of the variance of the estimated integral may be readily computed:

$$s^2(\tilde{I}) = (b-a)^2 \cdot \left\{\sum_{i=1}^{N} f^2(x_i) - N^{-1} \cdot \left(\sum_{i=1}^{N} f(x_i)\right)^2\right\} \Big/ N \cdot (N-1)$$

and

$$s^2(\hat{I}) = \{[\hat{I} \cdot (b-a)c] - \hat{I}^2\}/N.$$

Thus, not only can each estimate be computed, but also a measure of its accuracy is available as well.

One may note that Buffon's needle problem, as discussed in the preceding section, is essentially the Monte Carlo evaluation of the area contained beneath a quarter-period of a sine wave. More generally, the reader should note that the aforementioned *rejection technique* for the generation of random variates is essentially a hit-or-miss Monte Carlo procedure, one that would be associated with finding the area beneath a given probability density function $f_X(x)$; since, in this event, the area is known in advance to be unity, the Monte Carlo evaluation may be turned into a procedure for generating random variables from the probability density function $f_X(x)$ (see Section 2.13.14.)

4.4. The Law of Large Numbers

The Monte Carlo method is founded upon one of the more fundamental principles of probability: The "law of large numbers," which states that, if $X_1, X_2, \ldots, X_N$ are independent random variables from a probability density (distribution) function $f_X(x)$, having mean μ and variance σ^2 $(<\infty)$, then, for N sufficiently large, their arithmetic mean,

$$\bar{X}_N \equiv N^{-1} \sum_{i=1}^{N} X_i$$

will, in all probability, approach μ as a limiting value.

The proof is quite straightforward, depending directly on Chebyshev's inequality. Noting that $\bar{X}_N$ is a random variable having mean μ and variance σ^2/N, so that, for any arbitrary $\varepsilon > 0$,

$$P[|\bar{X}_N - \mu| > \varepsilon] \leq \sigma^2/(N \cdot \varepsilon^2),$$

due to the inequality of Chebyshev. However, this implies that, as $N \to \infty$, the probability that X_N will not deviate from μ by any preassigned quantity ε can be made as near unity as desired.

Therefore, if one takes a sufficient number of observations on X, their arithmetic mean will converge (in this probabilistic sense) to the (usually unknown) parameter μ. For example, in the crude Monte Carlo method, one is seeking to estimate the average heights of $f(x)$ over the range $a \leq x \leq b$ by using the arithmetic mean of the random variables

$$Y_i = f(X_i), \qquad i = 1, 2, \ldots, N;$$

i.e.,

$$\bar{Y}_N = N^{-1} \sum_{i=1}^{N} f(X_i)$$

should approach the average height of $f(x)$ as N becomes sufficiently large, and

$$\tilde{I} \equiv \bar{Y}_N \cdot (b - a)$$

should therefore approach the desired integral I, as N becomes infinitely large, in the probability sense of limit.

Of note in this regard is the fact that, since the variance of $\bar{X}_N$ is given by σ^2/N, the precision of the Monte Carlo estimate of μ_X is directly proportional to $\sqrt{N}$, whenever the precision is measured in terms of the inverse standard deviation of the estimate. Hence, a twofold improvement in precision requires a fourfold increase in the amount of Monte Carlo sampling.

4.5. Random Walks

Since the Monte Carlo method can be used to provide random variables that estimate the values of integrals, it is not surprising that the method is also applicable to the estimation of the solutions of certain differential equations. Essentially, the Monte Carlo method is applicable to such estimations whenever a probabilistic problem, or game, produces a random variable whose expectation is functionally related to the solution sought in the original, nonprobabilistic problem; for example, the estimate f of the probability of a crack-crossing for Buffon's needle is proportional to a Bernoulli random variable

$$f = S_n/n,$$

so that

$$E(f) = np/n = p = (2l)/(\pi d).$$

The application of the Monte Carlo method to the estimation of the solution for a differential equation is generally best accomplished by finding a probabilistic game or problem, whose solution involves a difference equation that is directly analogous to the differential equation whose solution is sought. Since the solutions of sufficiently similar difference and differential equations are essentially the same, one can estimate the solution of the probabilistic problem (i.e., the solution of its difference equation) by playing, or imitating, the probabilistic game a large number of times and recording the outcomes of the successive plays of the game.

A typical class of probabilistic games that leads to suitable difference equations is the collection of *random walk* problems. One-dimensional random walks are often referred to as the classical ruin problem, in which the random walk is the recorded sequence of the holdings of a gambler as his repetitive playing of a game of chance leads either to his eventual ruin or, more rarely, to that of the casino.

A particular two-dimensional random walk is especially illustrative of the Monte Carlo method's applicability to the solution of a particular partial differential equation; namely, Laplace's equation in two dimensions:

$$\frac{\partial^2 u(x, y)}{\partial x^2} + \frac{\partial^2 u(x, y)}{\partial y^2} = 0.$$

The solution $u(x, y)$ is a functional description of the state of a two-dimensional object at any point (x, y) within some closed boundary curve Γ given that for any point R on Γ, the value $u(R)$ is known; i.e., there exists a known function f such that

$$u(R) = f(R) \qquad \text{for any point} \quad R \quad \text{of} \quad \Gamma.$$

[An exemplary physical situation defines $u(x, y)$ to be the temperature at points (x, y) on a lamina, the temperatures at the boundary points being fixed or given.]

The difference equation that corresponds to Laplace's equation is

$$u(x + h, y) + u(x - h, y) - 2u(x, y) + u(x, y + h) + u(x, y - h) - 2u(x, y) = 0$$

or

$$u(x, y) = \tfrac{1}{4}[u(x + h, y) + u(x - h, y) + u(x, y + h) + u(x, y - h)].$$

Thus, by establishing a mesh, or grid, of squares of side h over a planar region, one notes that the difference equation expresses $u(x, y)$ as the arithmetic mean of its value at the four nearest grid points, as depicted in Fig. 4.4. Of course, if (x, y) is one of the boundary points R, then the difference equation is replaced by

$$u(x, y) = f(R) \qquad \text{for} \quad (x, y) = R \quad \text{on} \quad \Gamma.$$

This difference equation also arises in connection with the solution of a stochastic problem known as the "drunkard's walk." One must try to envision a medieval city, surrounded by an irregular moat and comprising a regular grid of streets. From a local pub, situated at (x, y), one of the street corners of the city, a drunk walks from corner to corner, choosing any one of the four available directions with equal probability, until he falls into the moat at some point R.

The ensuing commotion leads to the arrest of the drunkard who is the next morning fined an amount $f(R)$ dependent upon the social and political status of those disturbed by the drunkard's predawn swim. In con-

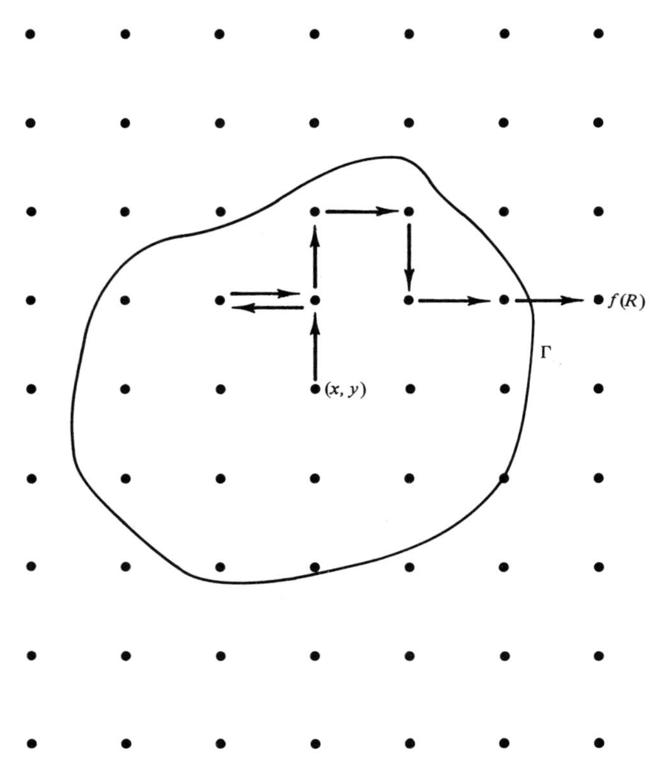

FIG. 4.4. The drunkard's walk.

sonance with the "theory of revolving door justice," the drunkard is then released, only to appear again at (x, y) the next evening for another walk to the moat and its concomitant fine.

The probability of beginning a walk at intersection (x, y) and ending it at R_i, the ith point of intersection of the moat and the streets of the city, can be denoted by $p(R_i \mid x, y)$ and is therefore given by

$$p(R_i \mid x, y) = \tfrac{1}{4}\{p(R_i \mid x + h, y) + p(R_i \mid x - h, y) + p(R_i \mid x, y + h) + p(R_i \mid x, y - h)\},$$

for any intersection (x, y) interior to Γ (the moat), since the probability of being at point (x, y) and moving to each of the neighboring points is $\frac{1}{4}$; i.e., the probability of eventually arriving at R_i, having started at (x, y), must be $\frac{1}{4}$ the sum of the probabilities of arriving eventually at R_i, having passed through a neighboring point. Of course, if an intersection (x, y)

is at the moat itself, then $(x, y) = R_i$, for some index i, in which case

$$p(R_i \mid x, y) = p(R_i \mid R_i) = 1.$$

The drunkard's walk thus defines a random variable $Y(x, y)$, equal to the fine $f(R_i)$ assessed for falling into the moat at point R_i, $i = 1, 2, \ldots, s$ (s being the number of boundary points). The probability distribution function for $Y(x, y)$ becomes $p_Y(R_i) \equiv p(R_i \mid x, y)$, $i = 1, 2, \ldots, s$, and is dependent upon the starting location (x, y).

The expected fine can then be determined as

$$\begin{aligned} E[Y(x, y)] &= \sum_{i=1}^{s} f(R_i) \cdot p(R_i \mid x, y) \\ &= \tfrac{1}{4} \sum_{i=1}^{s} f(R_i)\{p(R_i \mid x + h, y) + p(R_i \mid x - h, y) \\ &\qquad + p(R_i \mid x, y + h) + p(R_i \mid x, y - h)\}, \end{aligned}$$

or

$$\begin{aligned} E[Y(x, y)] = \tfrac{1}{4}\{&E[Y(x + h, y)] + E[Y(x - h, y)] + E[Y(x, y + h)] \\ &+ E[Y(x, y - h)]\}. \end{aligned}$$

Denoting by $u(x, y)$ the expected fine $E[Y(x, y)]$ for a drunkard's walk beginning at (x, y), one has the difference equation

$$u(x, y) = \tfrac{1}{4}\{u(x + h, y) + u(x - h, y) + u(x, y + h) + u(x, y - h)\},$$

with boundary condition,

$$u(x, y) = f(R_i) \cdot 1 = f(R_i) \qquad \text{for} \quad (x, y) = R_i \quad \text{at} \quad \Gamma.$$

Consequently, the difference equation associated with the drunkard's expected fine is identically that corresponding to Laplace's equation. One can, of course, estimate this expected fine by averaging the fines encountered from n independent walks, each commencing at the point (x, y) in question. Denoting by $y_k(x, y)$ the fine imposed for the kth such random walk, the quantity

$$y^*(x, y) = n^{-1} \sum_{k=1}^{n} y_k(x, y)$$

serves as an estimate of the expected fine $u(x, y)$ and, hence, of the solution $u(x, y)$ of Laplace's equation.

4.6. Accuracy of Monte Carlo Estimates

In general, the use of Monte Carlo estimates as approximate solutions to differential equations depends upon the ability to locate a probabilistic situation whose solution can be described in terms of the difference equation corresponding to the given differential equation. Thus, by miming the particular probabilistic game, an estimate of its solution may be had. The estimate can thereby serve as a approximate solution of the differential equation. For other differential equations amenable to this type of solution, the reader is referred to the excellent monograph of Hammersley and Handscomb (1964) and to Feller (1950, Vol. I) for material relating to random walks.

More general still is the applicability of the Monte Carlo method to the estimation of solutions of problems that can be classified by their affinity to linear operators. The reader is referred to the papers of Bauer (1958) and Curtiss (1953) for discussions of the utility of the method in finding eigenvalues, inverting matrices and other applications. Further material on the application of the Monte Carlo method can be found in Shreider (1964).

The accuracy of the Monte Carlo method, as opposed to that of techniques available from numerical analysis, is of course a matter of concern. The result of a Monte Carlo estimate is a random variable whose mean value is often equal to the solution being sought. However, the approximate solution arising from applying a technique of numerical analysis is not a random variable, the resulting solution being approximate only in the sense that its grid size must be necessarily positive and that its exceedingly large number of requisite computations may lead to rounding errors.

Reportedly the error in applying a numerically analytic technique to approximate the solution to a particular problem is of the order of $h^{-1/d}$, where h is the grid size and d is the number of dimensions (independent variables) involved. As was indicated in the discussion of the application of the Monte Carlo method to integration, the "error" in such applications, if expressed in terms of the standard deviation of the resulting estimate, is of the order of $h^{-1/2}$, regardless of the dimensionality of the problem [see Kac (1949)].[†] Hence, for a given mesh size, consideration might be given to the use of the Monte Carlo method whenever multidimensional

[†] Referenced by Chorafas, D. N. (1965). *Systems and Simulation*, p. 152. Academic Press, New York.

($d > 2$) integral and differential equations are to be solved. In this regard, one may take advantage of techniques, such as *antithetic variates*, discussed by Hammersley and Handscomb (1964) for the purpose of reducing the "errors" involved in the application of the Monte Carlo technique.

EXERCISES

1. A PERT (Program Evaluation Review Technique) network is a diagrammatic description of the elementary tasks constituting an overall project. The component tasks may be considered as possessing a head event, corresponding to their start or commencement, and a tail event, corresponding to their completion. Each task is presumed to require a positive amount of time for its completion and each is not permitted to begin until all prerequisite tasks preliminary to it have been completed. The following PERT network consists of six fundamental tasks, as indicated by the six arcs in the graph:

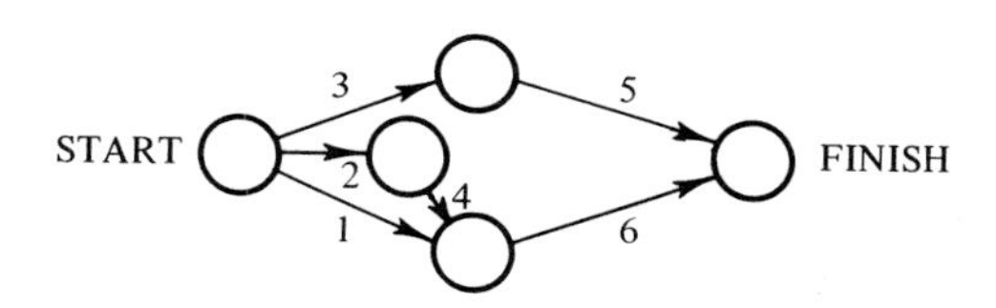

Presuming that task i requires exactly as many hours to complete as are given in the column headed MEAN (T_i) in Table 6.1, $i = 1, 2, \ldots, 6$, show that the overall project cannot be completed in less than 20.5 time units. Note that the project completion time may be computed as:

$$T \equiv \max\{T_3 + T_5, T_6 + \max[T_1, T_2 + T_4]\}.$$

2. Repeat the computations for the preceding exercise, but select the times for the completion of task i as positive-valued random variables from the density $f_i(t)$, as given in Eqs. (6.3.2:1).

3. Repeat Exercise 2 a total of 10 times, computing for the 10 project completion times their arithmetic mean. Compare this mean with the result of Exercise 1 and comment on the reasonableness of any observed discrepancy in view of the computational formula for T, as given in Exercise 1.

4.7. Queues

Queues, or waiting lines, form whenever there exists competition for service or attention that can be extended simultaneously only to a limited degree. A queuing system is generally considered to be composed of three physical elements:

(a) *customers*—those objects, animate or inanimate, which enter the queuing system seeking its services;
(b) *servers*—those objects, again either animate or inanimate, which provide the system's services to arriving customers;
(c) *the facility*—the physical location of the queuing system.

The list of conceptual queuing systems is indeed immense and any effort to enumerate these would need to include ticket windows, license bureaus, telephone systems, medical systems, mail-order institutions, dams and reservoirs, power distribution facilities, transportation networks, air and naval ports, and machinery maintenance facilities. In any queuing system, the limitations on service may create congestion, or the queue, either whenever customer arrivals are, though regular, too frequent or whenever the demands for service arrive quite sporadically, though infrequently. In either event, occasional or perpetual queues may be formed for the limited service.

Queuing systems require for their study a number of specifications, primary among which are the system's:

(1) *arrival mechanism*—the description of the pattern of customer arrivals at the facility;
(2) *service mechanism*—the description of the capacity, availability, and method for dispatching the service at the facility;
(3) *queue discipline*—the procedure by which customers are selected from any extant queue for service.

4.7.1. Arrival Patterns

The arrival mechanism, or arrival pattern, for a queuing system is best viewed in terms of its *interarrival* times; i.e., in terms of the times that transpire between successive customer arrivals. These interarrival times might be completely deterministic or periodic, as is often the case with automated production lines, or may require probabilistic descriptions, as is more usually the case whenever the customers of the system are animate. Cox and Smith (1961) have endeavored to categorize these arrival patterns in their monograph on queues.

Regular, punctual arrival mechanisms are those in which the customers arrive at equally spaced intervals, such as is the case for items being processed along a conveyor belt. At the other extreme are arrivals that are termed "completely random," meaning that the arrival mechanism is a Poisson process of a given intensity, or arrival rate, denoted by λ; hence, if the arrivals are completely random, the interarrival times are independently and exponentially distributed random variables of mean $1/\lambda$; and, the probability of arrivals in any time interval is completely independent of the past record of arrivals, and their number is distributed according to Poisson's law. The completely random arrival pattern for a queueing system is also referred to as a Poisson (traffic) input, as discussed by Riordan (1962) and Karlin (1966).

An immediate generalization of the Poisson traffic input is the use of independent interarrival times from some distribution function, defined for a positive-valued random variable only, other than the exponential distribution. This type of arrival pattern is referred to as a general independent (GI) arrival mechanism, or as a recurrent input (of which the Poisson input process is a special case). Most theoretical developments regarding queuing systems depend upon arrival mechanisms that are stochastic processes with independent increments; i.e., the interarrival times are independently and identically distributed, positive-valued, random variables.

One other special case is worthy of mention; namely, the arrival process by which customers are scheduled to arrive at regular times $n \cdot \tau$ but actually arrive at $n \cdot \tau + \varepsilon_n$, $n = 1, 2, \ldots$, where the ε_n are then presumed to be independently and identically distributed random variables; such an arrival pattern, typical of a doctor's office or an executive's waiting room, might be referred to as regular, unpunctual arrivals.

Other arrival mechanisms are conceivable and indeed observed. The preceding patterns have all been presumed to be stationary, in the sense that the distribution of interarrival times remains the same as time evolves. Certainly, nonstationary arrival patterns are conceivable, though inevitably these increase the theoretical difficulties regarding the study of the queue. In addition, the preceding discussion of arrival patterns has implicitly presumed that, at each arrival instant, exactly one customer arrives. Such a presumption is necessarily restrictive, as contemplation regarding the essentially simultaneous arrival of several cars in a railway yard should reveal. One standard mechanism for handling the multiple and simultaneous arrival pattern is to presume that interarrival times may be described by one of the aforementioned methods, yet the number of customers at any given arrival instant constitutes a separate, integer-valued, random variable.

4.7.2. The Service Mechanism

Once a customer arrives at the facility and requests service, he conceptually has joined the queue at the facility. In many instances, he may find upon arrival a server unoccupied and may therefore be immediately released from the queue without incurring any delay. The act of service depends, however, on a description of the service mechanism as it exists at the facility.

A primary feature of the service is its capacity c denoting the maximum number of customers that (who) may be served simultaneously. In most queuing systems, complete availability of all servers is presumed as time progresses, though in many practical situations, both the availability of, and the capacity for, service may vary in order to accomodate work schedules, server breakdowns, or scheduled maintenance activities. The capacity c is, in most queuing studies, presumed to be a stationary constant, and its availability is usually assumed to be the complete availability of the entire service capacity throughout the period of interest.

Of the essence in describing the service mechanism is the distribution of service times. The *service time* for a customer is the amount of time transpiring between the release of the customer from the head of the queue until the instant that his service is completed. In general, one presumes that these service times are positive-valued random variables, though completely deterministic service times exist in many situations (e.g., in production lines). Hence, the classification of service times follows that of the interarrival times: constant (regular), exponentially distributed, general (independently, identically) distributed, and nonstationary service times. To further complicate the issue, however, it is possible to have service times dependent on such factors as queue length or time of day, but more tractable results seem to follow from the stationary cases. Indeed, most analytic results have depended upon either the Poisson arrival mechanism or the exponentially distributed service time, primarily because probabilities associated with these arrival and service mechanisms are independent of the past history of the queuing system.

4.7.3. The Queue Discipline

The third essential feature of a queuing system is the manner in which queued customers are called forward for service. The *queue discipline* describes the method by which priorities are assigned to customers awaiting service, and it is usually easier to discuss these priority schemes first in the context of a queuing system of capacity 1, then for multiple capacity systems.

The usual queue discipline is first-come, first-served (FCFS), implying that as each customer arrives, his arrival time (or some monotone function thereof) is recorded as an attribute, or characteristic, describing the customer. Among all queued customers requiring service, that customer having the earliest arrival time is selected once the server becomes free. An exemplary FCFS queuing system might be a delicatessen, which gives each customer a serially recorded number upon his arrival; this numbered ticket then serves to determine priorities in the queue.

An obverse of this priority system is the last-come, first-served (LCFS) queue discipline. Again, arriving customers are assigned an attribute as a monotone function of their arrival times, yet the most recent arrival receives service first, as is often the case for the unloading of a crowded passenger, or filled cargo, vehicle.

Other queue disciplines may be maintained by the assignment to customers of attributes other than those based on arrival time. Typical priorities of this category include military or social rank; or, urgency of need of service, as would exist for critically ill patients at a hospital; or, predicted service times, such as is often implemented by computer queuing systems. The latter queue discipline is of special interest, since, as Leibowitz (1968) shows in one or two elementary examples, the mean waiting time of customers may often be significantly reduced if the server(s) can take advantage of knowledge regarding the customers' requirements for attention. The concept of the special server to handle the small purchase customer at a supermarket endeavors to acknowledge this fact.

The virtual antithesis of queue discipline for a single-server queuing system is the complete absence of a priority system, in which customer selection by the freed server is a completely random choice from the queue. This frequently encountered queue "discipline" is often implemented by business agencies, especially during holiday seasons.

If priorities among waiting customers depend upon some attribute other than their arrival times, then it may be desirable to permit the preempting of service if a higher priority customer arrives while a customer of lower priority is being served. This typically occurs in medical systems, but other preemptive systems may need to be investigated, as particular situations warrant. One such example is the preempting of runway rights by an aircraft whose fuel supply has become critically low while awaiting his regular turn in an otherwise FCFS queue discipline.

Whenever a queuing system has multiple capacity, a number of queue disciplines become available. The usual situation is that any server, once freed, selects the customer of highest priority among those queued, but frequently this is not possible due to the layout of the facility. For example, at many banks, customers select one teller's queue, itself usually of FCFS queue discipline, though frequently customers in these circumstances are free to shift from queue to queue. In other situations, the customers may be somehow classified, as frequently happens in multiply lined cafeterias, so that servers attend only to customers in their "class." Still other queuing systems require that server assignments be conducted on a strict rotation basis, so that queued customers may remain so even though some servers are idle.

4.7.4. Classification of Queues

Further refinements of, and policies for, queue management are possible. For example, at some facilities, such as parking lots, no queue is permitted at all; when all servers are occupied, customers are dismissed. The dismissed customers may or may not have a subsequent effect on the arrival rate; for example, the telephone caller who receives a "busy" signal is likely to try to complete the call again in the near future. Another feature of queuing systems might be *balking* by customers who either refuse to wait initially or who tire of waiting subsequently (the second alternative is often referred to as *defection*).

Of course, queuing systems may consist of a sequence of service booths, so that queues exist in tandem, the released customers from one queue becoming the arrivals for the next.

In most studies of queues, these refinements are absent unless clearly stated otherwise. Kendall (1951) proposed the following categorization for queues, a three-place mnemonic $A/B/c$ which denotes:

c = capacity of the queuing system

A = the interarrival time distribution

B = the service time distribution,

where A or B may be assigned the symbols:

M = exponential distribution;
G or GI = general independent;
E_k = Erlangian distribution of parameter k ($E_1 \equiv M$);

or

D = deterministic, or scheduled, pattern of constant interarrival or service times.

One may note that this classification scheme does not reflect the queue discipline, a feature that must be specified separately depending upon the queuing situation at hand. In the absence of such a specification, however, the queue discipline is usually the FCFS, though the queue discipline does not always affect every measure of the system's performance.

4.7.5. Performance Measures for Queuing Systems

A queuing system's performance may be reviewed from two primary vantage points: that of the servers and that of the served, or queued. To the former, the distribution of idle times is an important measure, as is the probability that a server will be idle at any given instant in time. For the latter, the waiting time (defined as the sum of his queuing and service times) or its probability distribution are important, as is the number of customers at the facility at any given point in time.

Most queuing systems' performances are measured in the *steady state*; i.e., their probabilistic descriptions are presented after sufficient initial time has passed that the probabilities are essentially independent of time. Thus, attention is focused primarily on the nontransient behavior of the queuing system.

Another measure of performance is the *traffic intensity* ϱ defined as the ratio of the mean service time to the mean interarrival time. In the case of the $M/M/1$ queue, where the exponentially distributed interarrival times have mean $1/a$ and the service times have mean $1/b$, the traffic intensity becomes

$$\varrho = (1/b)/(1/a) = a/b,$$

the ratio of the arrival rate to the service rate. In this case, if $\varrho \geq 1$, the mean queue length has infinite expectation in the steady state. Such a queuing system, for which the arrival rate exceeds the service rate, is said to be saturated so that indeed, infinite queues could be anticipated. (One might note parenthetically that this result, often typical of queue performance, is frequently independent of the queue discipline.)

If, for the $M/M/1$ queuing system, the traffic intensity ϱ is less than unity, then the expected queue length is the mean of a geometric distribution with parameter $p = (1 - \varrho)$; i.e., the steady-state probability of their being exactly n customers in the system is given by

$$p_n = (1 - \varrho)\varrho^n, \qquad n = 0, 1, 2, \ldots \quad (\varrho < 1),$$

so that the expected number of customers is

$$\sum_{n=0}^{\infty} np_n = \varrho/(1 - \varrho)$$

and the variance of the number of customers in the system at steady-state is $\varrho/(1 - \varrho)^2$. Thus, the nearer ϱ is to unity, the lengthier the queue to be anticipated.

This last result is quite standard in queuing theory. If the mean arrival rate exceeds the mean service rate, then service deteriorates and excessively lengthy queues develop; whenever the traffic intensity is nonsaturating (i.e., $\varrho < 1$), then the queues are generally manageable, though their expected lengths will increase with ϱ.

4.7.6. Summary

The detailed analysis of queuing systems is not a major purpose of this exposition. It is of primary importance that the simulation analyst be aware of the elements of queues, for queuing systems, especially networks of queues in tandem, become virtually intractable. Simulation of these probabilistic queuing systems becomes then a primary method by which comparisons of alternative arrival mechanisms, alternative service mechanisms, and differing queue disciplines may be made. It is essential, however, that the systems analyst be aware of the fundamentals of queuing systems, for queues, as indicated earlier, arise in a number of modeling contexts; the measures of a queuing system's performance and the likely effect of alternative policy formulations are of great importance to the systems analyst. For further details, the reader is referred to the review paper of Bhat (1969) or to the earlier paper by Saaty (1957).

4.8. Monte Carlo versus Simulation

As defined earlier in this chapter, the Monte Carlo method is a branch of experimental mathematics that is concerned with experiments on random numbers. However, since the quite general class of dynamic, stochastic simulation models will require the generation and subsequent "experimentation" with random numbers and their transformations, it is probably wise to try to distinguish such models from the Monte Carlo method.

As has been suggested in this chapter on Monte Carlo methods, the Monte Carlo approach is especially valuable if an equation, which has arisen in an entirely nonprobabilistic context, has a solution that is related to the expected value of a random variable arising from some probabilistic game. In this context, then, the use of the term "Monte Carlo method" might be properly restricted to situations in which statistical estimates are used to give solutions to formalized models [as these are defined by Sayre and Crosson (1963)] that should theoretically be amenable to solution by means of operations defined by mathematics or numerical analysis.

Nonetheless, such a restriction might not incorporate, as a Monte Carlo method, the determinations of statistical distributions by numerical experiments, such as those leading Gossett to his family of Student's t distributions. Indeed, many authors prefer to separate these "sampling experiments" of the statistician from the Monte Carlo method, but the contagious affectation of the latter term makes the distinction practically impossible to enforce.

The present author would, however, prefer that the distinction be made between these sampling experiments, and the use of experiments with numbers in order to estimate the solutions of deterministic equations. Furthermore, he would generally tend to distinguish between the Monte Carlo method and the simulation model on the basis of their respective exclusion or inclusion of time as an explicit, required variable. Though simulation models need not be dynamic (cf. Section 1.3), a more restricted definition of this class of models (as employed in the next chapter) presumes that an explicit accounting for the passage of time is an integral part of the simulation model. On the other hand, the Monte Carlo method requires at best an implicit accounting for time; even if one of the independent variables in the differential or integral equation would be interpreted as time, the Monte Carlo method, when applied to its solution, seldom *requires* an explicit mechanism for advancing time or for keeping track of its passage.

The suggested distinctions that could possibly be made among the terms: sampling experiments, the Monte Carlo method, and stochastic simulation modeling, will not likely be universally adopted. The famous hill in Monaco seems to have been accepted, for better or for worse, as an adjective to describe all three of these related activities.

Chapter 5

MODELING

> *"Le hasard sait trouver ceux qui savent s'en servir."*
>
> —R. Rolland

5.1. Stages of Model Development

As indicated in Chapter 1, the Uncertainty Principle of Modeling implies that any conscientious mimicry, or model, of a system of interdependent and interacting elements will probably require the inclusion of random phenomena within the model's structure. Hence, simulation models often become generalized Monte Carlo procedures that serve as mimics of the random processes being modeled and whose associated random variables are usually of unknown distributional form. The preceding study of the Monte Carlo approach showed that randomness can often be put to good use in order to "solve" otherwise difficult or intractable problems. In addition, the study reveals the ease with which difficulties, which might be presumed to arise due to the need to generate and analyze random variables, may be readily dispelled.

Modeling is the act of mimicry, yet the conscientious model builder will probably find that his model needs to mime the random effects present in the system he is modeling. In the context of this book, a *system* may be defined as a collection of interdependent and interactive elements that act together to accomplish a given task. Most systems of interest develop their characteristic behavior over time, so that any proposed model of this behavior needs to be of the dynamic (as opposed to the static) class. Furthermore, the Uncertainty Principle of Modeling will usually impose upon the modeler the need to construct a model of the stochastic variety. Since a stochastic, dynamic, and symbolic model does not always lend itself to construction as a formalized model (except as a formal time series of usually limited complexity), the remainder of this chapter will be concerned with the fundamental aspects of constructing the dynamic, stochastic, and simular variety of model. In the current section, emphasis will be placed on an overview of the essential stages in the development of a successful simulation model.

The decision to attempt modeling should never be made until a clear and concise statement of the requirement for the model has been formulated. One must answer as explicitly as possible questions of the following nature:

(1) For whom is the model to be constructed?

(2) What kinds of questions is the model expected to answer?

(3) If the model had already been built, what kinds of comparisons would be made with the actual system in order to determine the validity of the model?

(4) Furthermore, if the model were already constructed, and were satisfactorily validated, what experimental plans for manipulating its parameters and input conditions would provide the best answers to still outstanding questions regarding the actual (i.e., the simulated) system's behavior?

The construction of a model that does not provide considerable information about the behavior of the system of interest for the system's controllers becomes a waste of time (and money) to all concerned; if a model is constructed so that essential bits of information are either omitted, inaccessible, or improperly defined, then the success of a modeling effort is probably doomed before it is properly begun.

The modeling effort can properly commence once it is clear what analyses must be performed, and once the model is actually completed. The kinds of organised analyses that may be employed, once a model is completed, will be discussed fully in later chapters. For the present, however, it is assumed that a clear understanding exists of the questions that will be directed to a completed model.

The initial stage of model development becomes then a study of the system: its component parts, their interactive behavior, their interrelationships, and the aspects of the system requiring a probabilistic description. This stage of the development is termed the *systems analysis*, a term that is equally well applied to the now reasonably well-organized science of studying a system in order to determine and isolate its salient features, interactions, and behavior mechanisms. The techniques required of the systems analyst are presented in the next section, but it may be mentioned in advance that the resulting model of a system will probably reflect the educational background of the analyst; for example, it is not likely that a neurophysiologist and a sociologist view the same system in the same manner. However, general principles and procedures for efficiently accomplishing the systems analysis stage will be outlined in detail in the next section.

The systems analysis stage of model development is centered about organized, analytical procedures for studying a system before its formulation as a simulation model. Once systems analysis has been completed, the investigator is in a position to begin the *system synthesis*, which is concerned with organizing the diverse results, as ascertained in the systems analysis stage, into a unified, logical structure. Flow charts, data array structures, and programming instructions constitute the essence of the system synthesis stage, especially if (as is most likely the case) the model is implemented on an electronic computer. Another element, then, of the system synthesis stage in model development must be concerned with selecting the hardware (computer) and software (programming language) that are most amenable to the implementation of the model. A short description of the features of special-purpose *ad hoc* simulation languages will constitute an integral portion of the discussion of the second modeling stage.

The system synthesis stage ends whenever the computer program, which describes the system and which therefore represents the symbolic model, is completed. (Without great loss of generality, one may presume that the model is to be implemented on an electronic digital computer; other implementations, such as the use of analog devices, require quite similar developmental stages.) The third stage of model development may then commence: the *verification stage*. Though the developed program will have been constructed as cautiously and logically as possible, the modeler will necessarily wish to experiment with the resulting model in order to verify that its behavior is compatible with that intended in its programmed structure. Elementary aspects of model verification include the correction of syntax in the model's component statements, as well as the other activities, normally called *debugging*; yet, the more fundamental questions which must be answered in the verification stage relate to whether the model will produce the behavior anticipated of it.

In this regard, one shall wish to experiment with the model (or, possibly, with component modules thereof). A simulation model may be viewed as an input–output device which produces a *response* (i.e., output) corresponding to a given set of *environmental conditions* (input). More specifically, the environmental conditions for a simulation model comprise the specification of all those parameters, initial conditions, and governing policies that are applicable to the mimicry of the actual system. Their specification and the following observation of the simulation model's behavior by means of its responses constitutes an *encounter* with the model.

Thus, the verification stage of model development is concerned with the specification and observation of model encounters so as to investigate the compatibility of their response with outputs that would be anticipated if indeed the model's structure had been programmed as intended. Once this compatibility has been assured, there remains the question of the model's ability to mime reality. This fourth, or *validation*, stage of model development is undertaken, when possible, in order to establish the degree of comparability between the model and the system it is miming. Techniques applicable to the validation of a simulation model are introduced in Section 5.6.

The fifth and final stage of model development is the most exciting and challenging: the use of the verified and validated model in order to make inferences regarding the behavior and response of the modeled system. Some confusion appears to exist regarding an appropriate designation for this stage of model development, primarily because many have chosen to describe this aspect of the modeling activity by the term "systems analysis." Indeed, as was indicated earlier, the initial stage of model development (which this author prefers to designate as "systems analysis") should not be undertaken without stipulating definitely preconceived notions as to the types of analyses that will be expected of the completed model and as to the types of queries it will be expected to investigate. Thus, both the initial and final stages of model development are properly within the domain of the systems analyst.

Nonetheless, the distinction between the activities comprising these stages will be maintained here by referring to the final stage of model development as *model analysis*, or as the *analysis stage*. During this stage, the analyst is engrossed in experimentation with encounters of the model so as to ascertain either: (*a*) the static effects of the model's responses, or (*b*) the dynamic effects of the model's behavior. The introduction of procedures for efficiently conducting this experimentation is the topic of Section 5.7.

The essential goal of the analysis stage of model development is the organized, experimental use of the model in order to make inferences efficiently about the modeled system itself. The degree of success with which this is accomplished is therefore dependent upon the conscientious manner in which the modeler proceeds in an orderly fashion through the five stages of model development:

1. Systems Analysis
2. System Synthesis
3. Model Verification
4. Model Validation
5. Model Analysis

Of course, if any one of the stages prove to be lacking in some regard, it is often necessary to backstep to an earlier stage and repeat the necessary stages again until satisfactory results are obtained. For example, if the model validation stage reveals a shortcoming in the model's structure, then the analyst must return to the system synthesis stage (and possibly even the systems analysis stage), thereby restructuring the model; he must then reverify and validate the restructured model before proceeding to the ultimate analysis stage. Such cycling among the stages is necessary until such time as all desired analyses have been completed with a verified and validated model.

5.2. Elements of Systems Analysis

In the first chapter were several schemes introduced for categorizing models. Rosenblueth and Wiener's (1945) dichotomous categorization describes a model as being either *material* or *formal*, depending upon whether the model is a physical or symbolic representation of the modeled. Sayre and Crosson (1963) subsequently categorized models as *replications*, *formalisations*, and *simulations*; subject to certain constraints, replicas correspond in their scheme to the material models of Rosenblueth and Wiener, whereas formalizations and simulations are symbolic models. Sayre and Crosson differentiate between the formalization and the simulation by defining the former as a class of symbolic models whose symbols are capable of manipulation by a well-formed discipline, such as mathematics or mathematical logic.

There exist several cross-classification schemes applicable either to the Rosenblueth–Wiener categories or to the Sayre–Crosson classification. For example, a model may be either *dynamic* or *static*, whether it be of the material or formal variety. The distinction is made in accordance with whether the model has properties that alter meaningfully and distinguishably with the passage of time; i.e., the dynamic model clearly recognizes time as a variable of interest, the static model does not acknowledge time at all. A second cross-classification procedure notes that models are either *stochastic* or *deterministic* depending upon whether or not the model attempts to mime the random behavior of the modeled.

For the most part, this section will discuss the analytical efforts required in order to develop dynamic simulation models of the stochastic variety. In many ways, this particular model category is the most general of the model types. Namely, the dynamic, stochastic simulation model, though a symbolic representation, acknowledges the passage of time and represents reality by recognizing explicitly the Uncertainty Principle of Modeling. The remaining types of symbolic models are lacking in one of these two regards, so that the dynamic, stochastic, simular class of models would appear to be the most general category. It is for the purpose of generality that the forthcoming discussion is restricted to this class of models.

5.2.1. System Boundary and Environment

Thus, of primary concern will be the description of the activities required of the *systems analyst* as he prepares to undertake a modeling activity. This breed of analyst is devoted to the study and analysis of systems, a system being defined as a collection of interdependent and interactive elements that act together to accomplish a specific task.

It is a fundamental premise of system methodology that the elements of a system can be isolated. However, in many systems, the goal (or task) of the system may not be clearly evident (e.g., any self-organizing system, such as is presumably the case for biological evolution), in which case it is partly the duty of the systems analyst (or operations researcher, or systems engineer) to define this goal.

The distinction between a systems analyst and a scientist (who indeed is engaged in analyzing and studying naturally occurring phenomena in the search for explicative models of their behavior) or between a systems analyst and an engineer (who is engaged in the use of wisdom, the application of knowledge for the improvement of systems and systemic relationships) is essentially one of outlook and of scope. All three individuals—the scientist, the engineer, the systems analyst—are engaged in efforts to ferret out the important characteristics describing a system and to describe relationships among them. Usually, however, the engineer (and especially the scientist) has sought expressions or logical developments that would tend to relate two, three, or at most several physical properties, though clearly there are exceptions (e.g., Darwin's theory of evolution). The systems analyst seeks to develop models that will encompass as many variables, as many relationships, as many relevant elements as are necessary for an adequate representation of a system. The techniques by which such a conceptually large body of relevant information can be coalesced into meaningful models constitute the curriculum of the systems analyst.

A fundamental concept in the analysis of a system is the specification of a (closed) *boundary* for the system. The system's interactions occur within the boundary, giving the system its characteristic behavior. This boundary is defined so that its interior elements and interactions do not have a relevant effect on components outside the boundary. Though it will be conceded that no system can be entirely bounded, at some point the systems analyst must apply judgment, as objectively and impartially as possible, in order to isolate those elements and interactions that do not significantly affect elements outside the boundary. A conceptual boundary can then be placed about the isolated elements and interactions.

One should note that elements outside the boundary can influence those within it, however. In deciding whether to include the second of two elements within a system boundary, presuming that the first is clearly within and that the first would definitely affect the second, it is important to inquire whether effects felt by the second will themselves produce subsequent effects either on the interior element or on any other elements within the boundary. If so, the system boundary should include the second element. (In the terminology of the engineer, all pertinent *feedback mechanisms*, or feedback loops, should be contained within the system's boundary.) If, however, the first (interior) element is only quite relatively indirectly affecting the second element, then, even though the second may occasionally significantly affect elements within the boundary, it need not be incorporated within the boundary.

The collection of all elements of this second type are included in the environment, or *system environment*, which contains all those influences outside the boundary which may affect the system's internal behavior, yet which do not closely *interact* with internal influences and elements. Thus, the concept of a closed boundary does not preclude external occurrences; it merely implies that outside influences do not provide the intrinsic behavior pattern of the system [cf. Forrester (1969)].

Though the simulation model of a system need not be dynamic, no great loss of generality results from considering it so. Indeed, many authors explicitly recognize that the inclusion of time as a variable in a simulation model is essential. Kiviat (1969) defines simulation as "a technique used for reproducing the dynamic behavior of a system as it operates over time." Gordon (1969) defines system simulation as "the technique of solving problems by following changes over time of a dynamic model of a system." Though recognizing that such definitions are somewhat more restrictive than necessary, this text will assume that a dynamic simulation model will probably arise as the analyst's description of the system at hand and, for the moment, will assume that mechanisms exist for miming the passage of time. The dynamic system that is being simulated will be frequently referred to as the "simuland," in order to distinguish the modeled system from the model, or simulation. Furthermore, the adjective *simular* will relate to the simulation model, as opposed to the *simuland*; e.g., "simular time" shall refer to the status of the symbol representing time in the simulation model and a "simular property" is one of the model, not of the modeled.

5.2.2. Components of the Simulation Model

A system simulation is then composed of a large number of simular symbols, or *state variables*, which collectively serve to describe the state of the system at any point in simular time. As will be discussed momentarily, certain of these state variables will designate properties of the elements that are contained within the system boundary; others may either be functions thereof (such as the average size of all the elements within the system boundary at any point in simular time) or be endemic to the overall system simulation (such as the status of the simular time itself). In general, it is the goal of the systems analyst to establish a one-to-one correspondence between state variables and the meaningful properties of the simuland. That this goal is essentially unattainable (cf. the Uncertainty Principle of Modeling) should not deter the objective analyst from conscientiously pursuing it. According to Beer (1965), a system simulation constitutes a transformation, or mapping, of the elements of the simuland into an admissible set of symbols or state variables. In addition, however, the system simulation must act to alter its symbols as simular time evolves in a manner quite like the changes that would be observed for the simuland as real time develops; in the terms of the mathematician, the building of a simulation model is a *homomorphism* between the elements of the simuland and the state variables of the simulation model. Ideally, the homomorphism should be an *isomorphism*, or one-to-one correspondence, between the two sets of elements as they evolve dynamically. Because of considerations akin to the Uncertainty Principle of Modeling, we inevitably must be satisfied with homomorphic, rather than isomorphic, models.

Since the simulation model seeks to follow the changes occurring in its state variables as simular time evolves, it is imperative that the mechanism by which these changes are implemented be understood. There are two methods for this implementation, each leading to a separate category of simulation models:

(*a*) *Discrete-change models*—only at discrete instants of time may any one or more of the state variables alter. (Such models are ideally suited for implementation on a digital computer.)

(*b*) *Continuous-change models*—the state variables are permitted to change continuously. [Usually, the continuous-change model would be implemented on an analog computer, though accuracy requirements might lead to its implementation on a digital computer, in which case the resulting program is referred to as a *digital-analog simulation*, as discussed by Brennan and Linebarger (1964).]

Thus, the discrete-change and continuous-change models differ primarily in their respective methods for representing the particular state variable: simular time. The type of computer on which a simulation model is implemented may not reveal whether conceptually the model is discrete-change or continuous-change, however.

Since much more modeling flexibility and resulting accuracy is generally available with the discrete-change model, the remainder of this chapter will presume time to be represented discretely rather than continuously; i.e., one may consider simular time as being quantized, capable of being advanced only in unit "ticks" or integral multiples thereof.

The discrete-change system simulation thus comprises a collection of symbols that are capable of alteration as simular time is advanced. The resulting model is organized so as to be composed of two kinds of structures: *static* and *temporal*. The static structures of a dynamic, discrete-change simulation model are those symbolic representations of the elements comprising the simuland. These static structures consist of

(*a*) the *entities*—the symbolic representations of the objects, or elements, of the system,

(*b*) their *attributes*—the recorded characteristics of the entities,
(*c*) their *relationships*—the connections existing among the entities.

Thus, entities represent those components that change in the simuland, the attributes signify by how much and to what states the entities have changed, and the entity relationships connote affiliations existing among the entities at any point in simular time.

The temporal structures of a simulation model define the internal behavior existing within the system boundary. These structures, known as *activities*, define the relevant manners in which the state variables alter with time. Now, the state variables of a simulation model comprise (*a*) arrays of entity attributes, (*b*) relationships among entities, and (*c*) system properties. Thus, the activities of a simulation model must be designed so as to

(1) alter any of the attributes of one or more entities,
(2) change the relationships among entities,
(3) transform system properties, such as augmenting or decreasing the number of entities within the system.

Activities constitute the simulation model's representation of the internal interactions existing within the simuland. They are therefore always viewed as transpiring entirely within the system boundary. Each activity is generally composed of three components:

(*a*) a starting action,
(*b*) a duration of time, during which the activity occurs,
(*c*) a completion action.

However, the duration of an activity need not be positive; certain activities, such as the status of a queue or some other relationship among entities, might change instantaneously. In these cases, the activity may be composed simply of a single action, implemented merely by the alteration of the applicable state variables at the appropriate point in simular time.

The essence of the temporal structures of a simulation model are the actions, or *events* (as they are usually denominated), which describe the commencement or completion of simular activities. Activities are begun by an *initial event* and are ended by a *terminal event*. Nonetheless, other events may exist in a simulation model, because any development that requires the alteration of one or more state variables will also constitute an event. For example, internal activities of duration zero will be composed of only a single event; moreover, effects from the system environment may periodically impinge upon elements within the system boundary, though no specific activity of positive duration within the boundary will be associated with their occurrence.

Events, then, are conceived of as occurring instantaneously and of being capable of altering state variables. Each event can be viewed as an algorithm that examines the current states of the simular state variables and then, if conditions be satisfactory, endeavors to alter pertinent state variables as required. Events may therefore be viewed as consisting of: (*a*) *a test section*, which ascertains those (if any) state variables that should undergo alteration; and, (*b*) *an action section*, which accomplishes the necessary and permissible changes.

Usually, one of the characteristics of an event is its determination of the need for other events to transpire. For example, if customer service is completed at a particular instant in simular time, not only does the customer's relationship with the server and with other entities change, but also a determination of the ensuing activity of both the customer and the server is in order. In this way, other activities may be ordered to commence either at the current simular time (such as the immediate initiation of service for the highest priority customer present in the queue) or at some determinable future simular instant (such as the arrival of the served customer at the next processing station, for example).

Events can also be categorized by their originating mechanisms. *Endogenous events* are those which are caused, or created (in the manners just described), by some previous event. *Exogenous events* are those which are initiated from the simular environment and may cause subsequent endogenous events to occur. Either endogenous or exogenous events may generate the need to "schedule" other endogenous events; however, exogenous events are not scheduled by either an endogenous or any other exogenous event of the simulation model.

For example, exogenous events may be required in order to introduce into the simular system certain entities that will interact with the usual entities of the system. The entering entities could be referred to as *temporary*, since they enter the simular system subsequent to its initiation and will often depart the simulation prior to its conclusion. Entities whose attributes are defined from the start of the simulation and which can be expected to remain within the conceptual system boundary are referred to as *permanent*.

Events, then, constitute the basic mechanism by which the state variables of a simulation model alter with the passage of simular time. In fact, if all events of a simulation model are endogenous, the simulation is called *closed*; with one or more exogenous events, the simulation model is categorized as *open*. Though these categories, introduced by Gordon (1969), do not appear to be directly comparable to the categorization of Rosenblueth and Wiener (1945) for their formal models, one should note that experimentation with a closed model would differ somewhat from investigations with an open model, since the effects of exogenous events must be taken into account in the latter case.

The need for, and the use of, events in discrete-change simulation models should also serve to clarify the distinction that Sayre and Crosson (1963) make between formalizations and simulations. In the simulation model, the events may be quite complicated algorithms that require not only numerical computations from algebraic expressions, but also decision rules based on logical relationships among the simular state variables of the system. In addition, the event algorithm may need to generate and/or employ random numbers. This complexity is not usually amenable to straightforward solution by the usual procedures of mathematics or mathematical logic; hence the need for a separate category of symbolic models, termed "simulation."

5.2.3. Summary

The systems analyst who undertakes the modeling of some particular system is, in a pragmatic sense, endeavoring to isolate

(*a*) a necessary and adequate set of state variables to represent the system,
(*b*) the essential events that transpire in the interactions among the elements of the system.

By specifying these arrays (tables of attributes and relationships) and program structures (event routines), the analyst will have completed the systems analysis stage of model development.

The degree of success associated with the resultant model will depend a great deal upon the analyst's ability to ascertain the relevant components of the simuland and the important manners in which these components interact. More abstractly, if a system consists of a set S of elementary states which undergoes changes in time due to a set F of transformations of S into itself, the system itself should be representable as

$$\{f: S \to S \mid f \in F\},$$

i.e., as the set of all mappings f which transform the elementary states of the system into other elementary states.

The systems analyst is likely to find that he cannot begin to cope with the enormity of the set S; every element of the system could probably be described by a barely finite number of characteristics and could probably be divided into subelements in essentially as many ways. He is then forced to select the apparently pertinent features of the system and describe these by a more limited set of symbols; the collective set of states for these symbols may be denoted by Σ. The instantaneous values assumed by the entire collection of symbols at any point in simular time is referred to collectively as the *simular state.* Wisdom and sound judgment play no minor role in this selection, though ignorance of (or lack of adequate information about) the components of the simuland will inevitably force the analyst to keep in mind the Uncertainty Principle of Modeling.

Even with the set Σ of symbol states adequately defined, and the corresponding arrays of attributes and relationships delineated, the analyst must have, implicitly or explicitly, selected some set Φ of transformations which will constitute the events (or event algorithms, or event routines) of the simulation model. Any element φ of Φ is then a mapping of Σ into itself, so that the simulation model becomes the set of mappings:

$$\{\varphi: \Sigma \to \Sigma \mid \varphi \in \Phi\}.$$

Presumably, the set Φ of transformations will need arise from the study and/or observation of the simuland and its interactions; hence, the set Φ should be determined so as to duplicate F, the set of actual transformations extant in the simuland, as closely as possible.

Both Σ and Φ need be determined concomitantly, the ideal being the establishment of one-to-one correspondence between Σ and S and between Φ and F such that if φ and f are corresponding elements of Φ and F, respectively, then $\varphi(\sigma)$ will correspond to $f(s)$ for any states σ and s that are in one-to-one correspondence. The likelihood of the establishment of such an isomorphism is, however, extremely unlikely in view of the Uncertainty Principle of Modeling.

The incorporation of random features in a simulation model must not, of course, be undertaken capriciously. Inclusion of stochasticity for its own sake shall not necessarily imply that a proper transformation set F of events has been isolated. Events which induce randomness into a simular system should be structured on the basis of known probability laws, such as the Central Limit Theorem, Bernoulli's theorem, or the theory of extreme values. Only detailed observation of, and reflection concerning, the stochastic aspects of a system can lead to adequate probabilistic descriptions of the simuland.

Much of the systems analyst's efforts shall need be tempered not only by the incompleteness of his understanding of the simuland and all its components and interactions, but also by the unavailability of data to support certain modeling structures. For example, an event routine might conceptually require a probabilistic description, but there may be no applicable probability law discernably appropriate from analytical observations of the simuland; in such an event, limited experimentation with the appropriate portions of the modeled system may be necessary in order to construct a frequency histogram as a representation of the unknown probability law. Should such experimentation be infeasible, alternative probabilistic structures, other event routines, and/or additional state variables may need to be defined.

In conclusion, it is the task of the systems analyst to define the pertinent entities (and their attributes and relationships) and the essential events (and their probabilistic description, if necessary) which will constitute the completed model of a system. The task may need to be tempered by the experience, judgment, and wisdom of the analyst, or even by the harsh realities of the lack of adequate data and information regarding the elements of the simuland and their interactions. Nonetheless, the goal is to structure entities and events that are as nearly isomorphic as possible to the elements and transformations which exist in the simuland.

EXERCISES

For each of the following exercises, indicate:

1. whether your model would be static or dynamic,
2. whether your model would be formal or simular,
3. whether your model would be deterministic or stochastic,
4. your concept of the system boundary appropriate to this model,
5. whether you would require any exogenous events in the model,
6. the state variables which you would employ,
7. the entities, and the number thereof, which you would require,
8. the attributes which you would use to define each entity type,
9. the events for which you would define event routines,
10. the activities of the system, denoting each by its initial event and its terminal event,
11. whether you would employ any temporary entities.

1. A municipal transportation authority owns KB buses which operate over a fixed route. Buses depart singly from a depot each day at intervals of 15 min., moving through the same sequence of KS stops (the depot is included in KS). The expected, or scheduled, transit time is τ minutes between successive stops, though these times are independently distributed random variables from a rectangular distribution over the range $(\tau - \varphi, \tau + \varphi)$, where φ is a positive constant less than τ.

At the end of H hours of any given day $(1 \leq H \leq 24)$, no further buses are dispatched until the start of the next day, though any dispatched bus continues its appointed rounds until such time as it returns to the depot. It is desired to simulate 30 continuous days of the transportation cycle in order to ascertain whether service is adequate.

At each stop (including the depot), the time required for the buses to discharge and receive passengers is in accordance with a single independent exponential distribution of mean λ (minutes). At any given stop (save the depot), only a single bus may discharge passengers, the buses' being effectively queued at the stop on a first in–first out (FIFO) basis. (Buses arriving simultaneously at a stop are given preferential treatment dependent upon their dispatch times from the depot; i.e., the bus dispatched earlier receives priority.) Upon discharging passengers at the depot, buses are placed in a FIFO queue there until the next requirement for a bus arises.

Presume that you are to model this system in order to determine

(*a*) the distribution of the buses' *waiting time* at each individual stop in the network,
(*b*) the day-by-day total of time spent by buses in the depot,
(*c*) the distribution of the times required for buses to make the circuit.

2. Telephone calls arrive at a switchboard during a busy hour (60 min) with a Poisson process such that the mean interarrival time for calls is λ (seconds). The switchboard is capable, at any point in time, of making up to NC connections, any customer being rejected (given a "busy" signal) should his demand for switchboard service arrive when all connections are occupied. However, a customer (caller) so rejected does not reenter the Poisson stream of incoming calls, but then repeatedly dials each 60 sec thereafter until his call is connected. Once a call is connected, it remains so for a random time selected from a Weibull distribution having mean $\mu = 4$ (seconds) and variance 25 (scale parameter $a \cong 3.54$, shape parameter $b \cong 0.81$). When completed, the disconnected switch is returned to the idle state. Incoming calls always use the free idle switch having the lowest index number, the switches being indexed by the integers 1–NC, inclusive. It is desired to measure the switchboard's effectiveness as a function of λ and NC, its being presumed that decreasing λ implies increasing NC in order to maintain the same effectiveness. Measures of this effectiveness have been selected as:

(*a*) proportion of the hour during which the ith switch (connection) is idle, $i = 1, 2, \ldots, NC$;
(*b*) the number of calls rejected in each of the 12 5-min intervals constituting the busy hour,
(*c*) the number of switches occupied at the start of each minute during the hour.

5.3. Elements of System Synthesis

The systems analysis stage of model development concludes with a reasonably complete description of the system under study. Its essential components, or entities, will have been delineated, and the attributes necessary to their description listed. Important relationships among the system's internal entities will have been isolated and a note taken of the necessity for including temporary entities in a model of the system. If temporary entities are required in order to simulate the system realistically, then their requisite attributes and possible relationships with all other entities will need to have been determined.

Furthermore, the systems analysis stage will have outlined the events, endogenous and exogenous, necessary to describe changes in the attributes and relationships of the entities and/or other state variables that may be required to describe a mimicry of the system. Thus, the systems analysis stage is directed toward the description of the system's events and the necessary state variables required to represent the system.

The set Σ of all possible combinations of values of the state variables contains, then, as elements possible state representations of the system at any point in time. Each state representation is a vector of values for the state variables, the vector representing a conceptually possible state of the system at any point in time. As events occur within the system, the state representations change accordingly, so that the essential idea of a dynamic model is to mime the sequence of events likely to take place in the actual system. In essence, the dynamic model is intended to produce a *state history*, a time-indexed record of state representations that will provide a description of a possible evolution and development of the modeled system.

Consequently, from studying the system itself, the systems analyst should have a reasonably well-defined system boundary in mind. The mechanisms by which exogenous events and temporary entities penetrate this boundary from the system environment should also be reasonably well established. Since the activities required of the analyst during the system synthesis stage may alter somewhat the definition of the system boundary, it is sometimes difficult to determine at what point in model development the system synthesis stage is actually entered.

5.3.1. The Event Graph

One of the transitional activities between the systems analysis and system synthesis is the preparation of an *event graph*. The event graph for a system is a collection of directed arcs and nodes that indicate cause-and-effect relationships among the events that are necessary to describe the dynamic behavior of the system. The nodes of the event graph represent the events, both endogenous and exogenous, that have been isolated for this purpose. These nodes may be conveniently labeled by any mnemonic scheme convenient to the analyst. In the following discussion, a number will be assigned to each event.

The directed arcs of an event graph indicate the relationships among the events. As the reader may recall, the occurrence of an event will probably alter one or more state variables of the system; but, in so doing, the event shall likely permit or preclude the possibility of other events' transpiring. In accordance with the notation of Evans *et al.* (1967), these relationships may be categorized and denoted as:

(*a*) (1) ——→ 2 : Event 1 will generate (cause) Event 2.

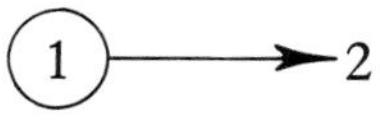

(*b*) (5) - - -→ 3 : Event 5 may generate Event 3.

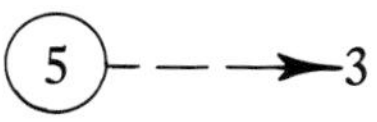

(*c*) (4) ——+——→ 2 : Event 4 will cancel Event 2.

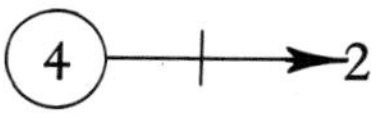

(*d*) (2) - + -→ 5 : Event 2 may cancel Event 5.

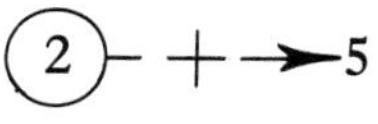

Of course, in the event graph, any event may influence the occurrence of several other events; for example, the graph

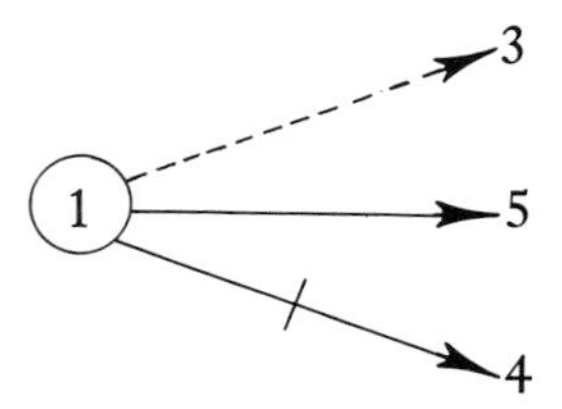

indicates that the occurrence of Event 1 will definitely generate Event 5 (though perhaps at a later time), will cancel Event 4, and might generate Event 3 (dependent upon the presence of other system conditions). The resulting graph of nodes (events) and connecting arcs may possess several *source nodes* (those which are not caused, cancelled, or generated by any other events); if so, any such event must either be the event which would initiate the system's activity or be an exogenous event. Thus, an inspection for relevant source nodes in the event graph serves as a guide to the rectitude of the systems analysis stage of model development.

Furthermore, if it is presumed that every event may (or will) generate or cancel at least one other event, then one sees that, from any source node, there must be at least one path (i.e., sequence of alternating arcs and nodes) leading to a *sink node* (which is representative either of the simulation terminus event or of an event that removes temporary entities from the simular system). Indeed, under this presumption, one should be able to trace at least one path leading from any given node to some sink node of the event graph. In the language of the graph theorist, the event graph consists of *connected components* with respect to the sink nodes of the graph. If the analyst's event graph violates this condition, then he will need to reinvestigate the system until such time as the system boundary becomes clearly defined and the event graph does consist of connected components [cf. Berge (1962)].

5.3.2. The Continuous-Change Model Reconsidered

The structuring of an event graph is also especially useful in delineating the types of state variables and entity attributes that will necessarily be incorporated within the model. For example, if Event X generates or cancels Event Y conditionally, then the system state representation at the time of the occurrence of Event X will likely determine whether conditions are proper for (the scheduling of) Event Y's occurrence; hence, among the system state-variables must be symbols whose values can indicate whether Event X will generate (or cancel) Event Y. In this manner, one may also determine whether his list of entities, their attributes, and their permissible relationships is adequate for the modeling task remaining in the system synthesis stage.

In addition, the systems analyst should find himself in a position to evaluate whether the system can be represented as a continuous-change or discrete-change simulation model. Although the preceding presentation has been oriented toward the analysis of a system in order to isolate those events that occur instantaneously in time and that cause changes in one or more state variables, the systems analyst may find that the system he has investigated is composed of elements that appear to flow smoothly through space and time. If so, one might well reconsider implementing the model in a continuous-change formulation. Such a formulation requires, however, that the events of the system be definable as sets of relatively well behaved though interrelated expressions involving time as an independent variable. Though not necessary, the resulting expressions are often representable on an electronic analog computer; or, alternatively, may be implemented by means of digital-analog simulation techniques such as those described by Brennan and Linebarger (1964).

One should note that the digital-analog simulation technique implicitly advances and accounts for the passage of simular time in time increments, since usually any formulation of a system's behavior by this method involves differential equations—ordinary or partial. The procedures of numerical analysis suitable to the integration (i.e., solution) of these differential equations on a digital computer are accomplished by advancing the independent variable, time, in fixed incremental steps.

The decision between a continuous-change and a discrete-change formulation may, however, be founded upon several characteristics of the event-graph representation of the system. One of the primary difficulties with the implementation of a model by means of a continuous-change philosophy is the representation of conditional transfers among events; for example, if Event 1 may generate Event 5, say, but does so if the system state-representation is appropriate for the transition, then a continuous-change formulation would need to incorporate decision rules based on logical comparisons of the truth or falsity of hypothesized conditions of the model each time Event 1 occurs. Such decision rules, which are essentially discretized choice mechanisms, are fundamentally discontinuous phenomena. Furthermore, the need to examine whether certain thresholds have been attained by one or more state variables is a procedure more naturally implemented on a digital computer as a part of a discrete-change formulation; for example, a model of a highly interactive ecological environment will probably need to include thresholds prescribing population growth characteristics and predator–prey relationships.

In addition to the need to represent intricate, conditional transfer mechanisms in a model, the systems analyst may also encounter the need to acknowledge the Uncertainty Principle of Modeling; i.e., he may need to incorporate random phenomena as integral components of the model. For example, if the systems analysis stage indicates that the system has within itself an imbedded Poisson process or, say, a probabilistic queueing structure, then a discrete-change formulation may be especially useful, since the times between successive events now become random variables that are not easily incorporated in the equations and expressions normally accompanying a continuous-change formulation. (Indeed, most queueing systems are concerned with tracing activities of "customers" and "servers," represented ideally as entities in a discrete-change model formulation.) However, recent developments in hybrid computation (cf. Hampton, 1965) imply that the incorporation of probabilistic effects in continuous-change models is becoming less difficult to accomplish, so that the need for randomness may no longer imply the necessity for a discrete-change formulation.

The systems analyst may wish to formulate his model of a system as a continuous-change simulation if studies of the event graph reveal that both simular time and the state variables of the system change values in an essentially continuous fashion. (In such a case, the event graph should appear somewhat superficial.) For example, a model of the mixing of dissolved and suspended particles in a reservoir might be adequately prepared as a continuous-change model; but, if the model is to concern itself adequately with the instantaneous, and often catastrophic, effects that influence the growth of populations of aquatic species within the reservoir, then a discrete-change model will more likely be in order. In addition, the analyst who proceeds to a continuous-change formulation should be forewarned of the difficulties of stability and scaling that are necessary considerations in analog computer implementations.

5.3.3. The Simular Clockworks

For the present, then, the discussion will presume that the systems analyst has decided to proceed with a discrete-change formulation of the model at hand. Furthermore, since the digital computer is especially adept at symbol manipulation, the discussion will presume that the model is to be implemented on one. Under these conditions, the discrete-change simulation model becomes a (software) computer program including an internal timing mechanism and capable of producing a state history of the transitions and events affecting the state variables as the internal time mechanism, or clockworks, is advanced.

An integral part of a computerized simulation model, then, is a subprogram, or routine, that maintains the *simular clock* and ensures that events are sequenced in their proper order as simular time advances. The simular time is maintained as the status of a system state-variable, usually referred to simply as the *clock* and usually initiated at simular time zero.

Historically, two or three methods have been developed for the "automatic" advancing of the simular clock. All of these are implemented in close association with a larger subroutine, referred to as a simulation control program, or *executive routine*. This executive routine is a master to all other routines that comprise a discrete-change simulation model and it serves to determine the proper sequence in which the routines are required and called.

Since the executive routine maintains control over all other subprograms and has access to information regarding all the state variables describing the simular system, then it can employ information from either in order to advance its clock. One such mechanism is to have the executive routine examine each of the event routines at a particular point in simular time, so as to determine which routine to implement at that instant of simular time. If a particular event routine is to be employed, control is released to it in order that appropriate state variables be changed in accordance with the event routine's algorithm; in addition, the event routine will ascertain which other events are to transpire in the simular system and at what simular time they are to occur. In this way, the event routine "notifies" the executive routine of the simular times at which event routines will need be selected. When control is returned by the event routine to the executive routine, the latter cycles through the remaining routines in order to determine which one is required at the current simular time. As soon as this cycling through the event routines is completed, the executive routine advances the simular clock one unit of time and proceeds again through the event-selection cycle.

An alternative approach to this time-stepping algorithm is one in which the executive routine cycles through the entities of the simular system, rather than the events. Each entity possesses, then, an attribute called its *entity clock*, which maintains a record of the event number of, and the simular time for, the next most imminent event routine that is to apply to that entity. If the executive routine finds that an entity's clock equals its own simular time, then the required event routine is called by the executive; again, the event routine could perform changes in one or more simular state variables, including a determination of appropriate entity clock values. Control is then returned to the executive routine, which continues to cycle through all entities until all entity clocks exceed the current simular time. At this point, the executive routine advances simular time one unit and begins to recycle through the entities again.

Both of the preceding techniques for advancing and maintaining simular time can be achieved by moving simular time forward one unit at a time; in this case, the procedure is referred to as *time-stepping* simulation and is implemented by search procedures which scan either the entity clocks or the event routines in order to sequence the activation of the event routines, and hence, the simulation itself.

A distinct improvement in the time-stepping mechanism can be seen by incorporating a list of predetermined instructions as a fundamental array in the simulation model. This event list, or *calendar*, becomes a time-ordered sequence of events which, at any given point in simular time, represents the currently known future behavior of the simular system. The calendar is composed of *event notices*, each of which may be considered a vector with entries describing the "who, what, when, where, and why" of the scheduled event; i.e., each event notice informs the executive routine of the simular time of, the entities involved in, and the pertinent conditions surrounding the occurrence of some simular event. The executive routine merely selects the uppermost event notice from the calendar, records the event notice's time as the current simular time, and releases control to the event routine indicated by the notice. The event routine then proceeds to alter state variables in accordance with its algorithm; since the event may itself generate (or cancel) other endogenous events, the calendar is adjusted accordingly by the addition (or deletion) of scheduled event notices. The event routine then returns control to the executive routine, no advance in simular time having been made. However, the executive routine, in selecting the next most imminent event from the calendar, will note the event's scheduled time, and since no other event is scheduled to occur before this time, the executive routine safely advances simular time to the time of this next most imminent event.

This *event-stepping* algorithm has become much more popular than the earlier time-stepping algorithm, not only because it avoids rather needless searching through event routines or entity clocks but also because it permits simular time to advance in accordance with the current simular state; if no system changes are required for, say, 10 units of simular time, then the simular clock is automatically advanced by this (variable) amount.

5.3.4. The Executive Routine

Regardless of the mechanism by which the simular clock is implemented, the modeler must, of course, select a basic time unit common to all internal simular events and activities. If the simular clock is maintained as an integer-valued state variable (i.e., as a fixed-point variable), then this selection of a time unit is tantamount to the assignment of a minimal "time-step" size and must therefore preclude a smaller time interval in which any single activity could begin and end. Therefore, the system synthesis stage of model development should be concerned with the determination of the appropriate time units and, if necessary, the establishment of a basic time step size.

The system synthesis stage must also decide upon the type of executive routine to be employed: time-stepped or event-stepped. Usually, the latter is the more efficient mechanism and is selected, in which case a calendar must be organized [cf. Conway *et al.* (1959)]. The calendar must be of adequate dimensions in order to incorporate all the information required by each of the event routines.

Included in these considerations regarding the executive routine must be a mechanism for specifying or generating an initial event which may be depended upon to spawn succeeding endogenous events and thus initiate the simular activity. As the simular events transpire, there are likely to be simular instants at which two or more events are simultaneously scheduled to occur. For example, if two or more simular entities are scheduled to begin use of a simular facility (or server) at the same simular time, provision must be made to ensure that proper priorities are involved so that the executive routine (or event routine itself, for that matter) can make a proper selection between the competing entities. (A typical example of this is the competition for the single runway by arriving and departing aircraft at a busy airport, where priority would normally be given to arriving aircraft unless some congestion threshold were exceeded on the ground.) Therefore, for simultaneously scheduled events, the executive routine must be programed to sequence the event routines in the appropriate order, the algorithm for this purpose possibly having to rely upon simular conditions at the time of the scheduled simultaneity.

5.3.5. The Event Routines

Within the framework of the executive routine and its clockworks, the modeler must define completely the event routines which give the simulation model its dynamic character. It must be remembered that each event routine is designed primarily to alter one or more simular state variables if simular conditions will permit the event to occur. Thus, the first segment of an event routine is a *test section* which performs the necessary computations and logical checks in order to establish whether the event should indeed take place. If the test is passed, then the event routine proceeds to its primary task of altering state variables in accordance with the general algorithm defined by this event; this segment of the event routine might be referred to as the *action section*. The event's occurrence may well have produced the need to schedule other event routines on the calendar; thus, a third segment of the event routine will be composed of a *scheduling section*, one or more decision algorithms which select the simular times for which, and the simular entities to which, subsequent events must attend. These events are then properly annotated as event notices and placed on the simular calendar. Then, the event routine may fulfill its final obligation: A *report section* should record the event (if any, for the test section may have been failed) for subsequent analysis. Control is then returned to the executive routine.

As an exemplary event routine, one might reconsider the airport runway simulation, for which a typical event may well be an event routine called SEIZE (the runway). Presumably at the time of the calling of this event routine, a high-priority aircraft is prepared to "seize" the runway for its own use. The test section of SEIZE would then determine if simular conditions are appropriate for the seizure (such as the runway's general availability for use and its specific availability for use by the aircraft requesting it, the latter condition possibly determined by weather conditions and runway length). If all conditions are satisfactory according to the test section of the event routine, then the action section would alter the status of the runway and would change the status of the aircraft seizing it. The scheduling section would then ascertain the length of time which the runway (and aircraft) would remain in their newly assigned relationship, and, possibly based on the generation of a random variable, would schedule another event routine (called perhaps RELEASE) at the appropriately determined instant. Before returning control to the executive routine, however, the report section of the event routine SEIZE would probably file a report of the activity that has been begun.

5.3.6. Report Generation Routine

There remains one other essential structural element of a discrete-change simulation model: its record of what has transpired in the simular encounter. As was just suggested, the event routines may themselves be used to report the actions that were initiated each time that they were called. One unified approach to this reporting is to assign a fixed format of information to be used by every event routine. One might view the procedure as a "newspaper headline," a vector of elements indicating:

WHEN:	the simular time of the event's occurrence
WHAT:	which event
WHO:	which simular entity(-ies) was (were) involved
WHERE:	at which simular location
WHY:	the degree to which simular state variables were altered.

The executive routine can then be engaged to dispose of each report as the modeler desires. Indeed, with such standardized report formats, one has a great deal of flexibility. For example, the executive routine may be programmed to:

(*a*) ignore or destroy a particular report if the analyst does not specify a requirement for its use;

(*b*) perform immediate computations for eventual statistical summaries of the simular behavior;

(*c*) place the report on a peripheral device for subsequent data analysis and/or report generation; or

(*d*) store the report in a journal file that can be periodically (in simular time) reviewed, used in computations, or ejected onto peripheral devices for subsequent analysis.

Alternatives (*c*) and (*d*) have an especially important feature which should be acknowledged. If every event's report is recorded, then stored on a peripheral device (such as magnetic tape), then a complete, time-sequenced history of the simular encounter will be available at the end of the simulation run. For large-scale models, whose individual encounters may require extensive, costly computer time, the taped recording of the state history of an encounter might prove of value, especially whenever separate requests for data analysis arise subsequent to the model's running. In these instances, one may select "from the shelf" an appropriate state history and subject this recording to *ad hoc* sort–merge–compute routines in order to obtain the special data analysis, thereby avoiding the need for a complete repetition of the model.

5.3.7. The Cornerstone of Successful Modeling: Data

The system synthesis stage of model development requires primarily that the necessary state variables, entities, attributes, and relationships be delineated and that the requisite event routines be programmed in consonance with an executive routine and report generation routine. Of course, the structure of a simulation model may be considerably improved by taking advantage of any noticeable similarities (such as, for example, the "newspaper headline" reporting) common to all event routines. Similarly, the entities included in the static structure of a simulation model can be naturally associated in sets, in the sense that each entity's attributes may specify the same properties as the corresponding attributes of all other entities in the same set; evidently one could arrange such sets of entities as two-dimensional arrays, the elements of the ith row being associated with the ith entity of the set, and the elements of the jth column pertaining to the jth attribute of the entities of the set.

Such structural simplifications may be readily incorporated in discrete-change simulation models that are implemented on the electronic digital computer. Other simplifications, or programming "tricks," need not be delved into here, yet a paragraph or so on data requirements for simulation models is in order.

The reader may recall that, prior to engaging in the first stage of model development (the systems analysis stage), one should have clearly in mind the types of analyses expected of the completed simulation model. In particular, any controllable operating conditions of the modeled system should be represented by flexible input parameters (environmental conditions) if indeed it is anticipated that

(*a*) in the validation stage, the verified model will be used to test its adequacy as a mimic of the modeled system, or

(*b*) in the model analysis stage, the verified and validated model will be iterated in order to make inferences regarding the possibility of alternative specifications of the operating conditions for the modeled system.

Hence, the system synthesis stage must also explicitly recognize the requirements of the subsequent stages of model development by ensuring that the analyst will have adequate flexibility in specifying environmental conditions for the model in a manner directly relatable to meaningful operating conditions for the modeled system itself. Policy specifications (such as alternative queue disciplines or diverse stock reordering policies) should be parametrized if possible, and the algorithms of the event routines thus developed to acknowledge the intended differences in these policy specifications.

The activities of the system synthesis stage must also acknowledge that, once they are completed, the verification stage will commence. Since the event routines may need to generate random phenomena, their structures may possibly be improved if each event routine has access to its own random number chain, in which case, the input data structure of the model must be aligned so as to accommodate multiple seeds. Indeed, separate seeds may be desired for the allocation of randomly assigned initial conditions for certain attributes and state variables; another chain of random numbers might prove useful for the generation of exogenous events or the introduction of temporary entities. Whenever structured multiple seeds could prove useful in the verification stage, they should be included in the input data structures that are defined for the model in the system synthesis stage; of course, the corresponding event routines, which introduce and promulgate the stochasticity of the simular system, shall need to be programmed so as to employ random variables generated from the appropriate chain.

Definite care must be taken during the system synthesis stage to ensure that the randomness introduced into a stochastic simulation model is appropriate and in accord with the stochasticity encountered in the modeled system itself. Thus, uniformly distributed random variables should not necessarily be introduced into a stochastic simulation model, just because an observable phenomenon of the system is "obviously random" in character. [The reader may wish to refer to Bertrand's paradox (p. 33) for an illustration of this point.] Neither should the pattern of arrivals in a queueing model be presumed to be a Poisson process, simply because the arrival pattern at the actual service facility is obviously "random."

Of special interest in this regard is the author's experience with a particular difficulty in data representation [cf. Mihram (1970).] A rather large-scale simulation model of a transportation network required hourly weather representations at each of several nodes in the transportation network. An immense amount of worldwide weather information, of the explicit type required by the simulation model, was available, having been dutifully recorded each hour for dozens of years and conveniently stored in machine-readable form. Since the vehicles in the transportation network could be impeded by weather conditions (especially at the traffic nodes), it was deemed especially important to impart meaningful weather patterns into the simulation model.

However, a proposal that these recorded meteorological conditions be statistically summarized, then employed as statistical distributions to generate simular weather conditions at each node, was discarded because:

(*a*) random selections of weather observations from such empirical distributions might reproduce appropriate stationary statistics, but accurate autocorrelations of the simular weather at each node would be more difficult to assure; and,

(*b*) even if the autocorrelations in weather conditions were effectively represented, random selections of weather variates would not likely reflect extant cross-correlations between (and among) nodes in proximity to one another.

These difficulties were overcome by employing the sequence of weather data actually recorded from a randomly selected season. Indeed, subsequent encounters with the model were initiated at randomly selected time locations in the sequence of data and the subsequent weather data were read into the simulation model as required (hourly, for each 30- or 40-day encounter). Such mechanisms for using actual data from the modeled system should be used whenever possible and reasonable, since costs associated with their collection and preparation are often quite minimal.

The moral of the present discussion is that the validity of a model will depend upon its adequacy as a representation of the modeled system, and that the adequacy of an otherwise secure foundation (the structural properties of a well-analyzed, well-synthesized model) might be proven faulty if the resulting model depends upon inadequate data. During the system synthesis stage, then, it is especially imperative that note be taken of the availability of meaningful data for supporting the structure of a simulation model. If certain data, as would be required by a particular model structure, do not exist or could not be collected except at prohibitively expensive cost, then the systems analyst may need to undertake another approach to the system synthesis stage. Indeed, if certain related data are currently collected, yet are not usable for the purposes of a meaningful simulation model, then investigations into the feasibilty of alternative data collection and/or presentation schemes should be undertaken.

5.3.8. Summary

The system synthesis stage of model development concludes with the detailed, programmed specification of the model, including

(*a*) the necessary sets of entities and their arrays of attributes,
(*b*) the requisite state variables for describing the model's state history and evolution,
(*c*) the event routines,
(*d*) the input data specifications, including both the requisite environmental conditions, the necessary random seed(s), and the desired flexibility of report generation capabilities.

Both the systems analysis and the system synthesis stages rest firmly on the cornerstone of adequate and meaningful data, preferably data taken directly from the modeled system itself in order to corroborate the model structure being developed in the system synthesis stage.

The system synthesis stage is especially important in model development, since it must not only conclude with a rather firm definition of the system's boundary, but also incorporate those features within the structure which will be useful in the ensuing verification, validation, and model analysis stages. Thus, the second stage of model development is that about which much of the success of the modeling effort revolves.

Fortunately, much of the tedious work of the system synthesis stage can be overcome by employing one of the available, *ad hoc* simulation languages. Although each of these languages imposes upon the systems analyst its own "world-view," thereby restricting somewhat modeling flexibility, the fact that these special-purpose languages perform powerful operations (such as sort–merge operations, statistical analyses, report generation, executive routine duties, and random number generation), with a concomitant reduction of programmer effort, means that the systems analyst is left considerably free to concern himself with his major task: the adequate representation of the modeled system as a symbolic, stochastic, and dynamic simulation model. The discussion will turn now to these *ad hoc* simulation programming languages.

EXERCISES

For each model described in the exercises of Section 5.2, indicate:

1. the basic time-step size you would use,
2. your event graph,
3. whether a time-stepped or an event-stepped clockworks would be preferred,
4. the types of event notices you would use,
5. the contents of the test section (if any) of each event routine,
6. the contents of the action section (if any) of each event routine,
7. the contents of the scheduling section (if required) of each event routine,
8. the contents of the report section (if required) of each event routine,
9. how you would prepare reports for the report generator,
10. how you might acquire any data required by the model,
11. how you would enter each stochastic element into the model.

Note that answers to Questions 4–9 may depend upon one's selection of an *ad hoc* simulation language.

5.4. *Ad Hoc* Simulation Programming Languages

As should be reasonably clear from the preceding discussions of the systems analysis and system synthesis stages of model development, a number of common features exist in the modeling of virtually all systems. For example, in miming a dynamic, stochastic system, one must have methods for tracing simular behavior in time and for acknowledging the Uncertainty Principle of Modeling. Since such features are commonly required by the systems analyst, whether he be modeling a communication network, a transportation facility, a queueing situation, a bureaucratic information channel, or a critical path structure, it is hardly surprising to learn that a number of special-purpose simulation languages have been developed.

Once the systems analysis stage has been reasonably well completed, and the event graph for the system has been examined, the analyst would be well-advised to investigate the applicability of one of these *ad hoc*, simulation languages to his particular modeling effort. The alternative is to begin the programming, in the system synthesis stage, either by developing a machine-language or assembly-language (i.e., machine-oriented language) program or by implementing the discrete-change model via a compiler (i.e., problem-oriented, general-purpose) language.

These alternatives imply the need to construct for oneself virtually all those features that are common to most simulation efforts. The present section will indicate the scope of these elements of commonality, although no specific comparisons of currently available *ad hoc* simulation languages will be presented. For such comparisons, the reader is referred to the papers of Krasnow and Merikallio (1964), Tocher (1965), Teichroew and Lubin (1966), and Kiviat (1967). Other discussions of particular languages are sprinkled throughout the literature of the computing, operations research, statistics, and management science fields; for specific references, one can consult the bibliography in Teichroew and Lubin (1966) or the extensive bibliography of Martin (1968).

5.4.1. Desirable Features of a Simulation Programming Language

Regardless of the nature of the dynamic system being modeled on a digital computer, the systems analyst seeks a programming language that will facilitate the system synthesis stage of his modeling activity by

(*a*) providing a convenient means for initializing the status of the model,
(*b*) permitting one to introduce changes in both the static and temporal structures of the model as simular time evolves,
(*c*) providing simple methods by which different reports and statistical summaries of the model's behavior can be specified, and
(*d*) allowing the analyst considerable flexibility in the specification of experimental designs for the ensuing validation and model analysis stages.

Each of the special-purpose simulation languages attempts to accommodate these four desires as features within its structure. Some of the *ad hoc* languages are more flexible than others in one or more of these aspects, but in general, each language endeavors to accommodate, to some degree, all four. For example, the initialization of a stochastic simulation model must not only attend specification of the initial values of all state variables (including the initial attributes for, and relationships among, entities) and introduction of at least one initial event, but also should permit the user to ensure that the model has advanced sufficiently far from its transient state before an accumulation of statistics regarding the encounter commences. In not all cases need a simulation model be advanced to the "steady state" by a procedure that reinitializes simular time once the model's behavior has become somewhat characteristic. For example, one might intentionally attempt to determine the duration of the "transient state," or one might indeed be interested in studying the model's behavior only in this portion of its state history. Nonetheless, a simulation programming language may possess features permitting one to reinitialize state variables, including the simular clock, after an initial phase of an encounter's state history.

The second feature of simulation languages deals with its executive routine. Each language has its own intrinsic clock (invariably accessible to any event routine of the model), possesses subroutines that maintain ordered lists (such as queues of priority-labeled elements, arrays of entity attributes, and elements of sets of entities), contains its own mechanism for sequencing both its own routines and the user-prepared event routines (often by maintaining a calendar, or ordered-event list). The executive routine is generally implemented as a *simulation algorithm*, the major features of which are outlined in the flow chart of Fig. 5.1. Included in the algorithm, of course, is the language's mechanism for processing simultaneously occurring events and it is incumbent upon the user of any simulation language to ascertain how these embedded priorities can be manipulated to meet his own needs. The executive routine of a simulation language usually imposes a somewhat rigorous structure upon the resulting model, since it requires that the systems analyst (or his programmers) develop separate routines for each of the events of the simular system. This imposition is often a favor instead, for the resulting simulation model becomes a collection of associated routines and subprograms. This modular structure is frequently valuable in implementing design changes if, say, the validation stage of model development finds the adequacy of the model somehow lacking; any subsequent reprogramming of the model is often facilitated in the presence of an organized, modular structure.

The simulation model is of dubious value if pertinent information about any one of its encounters' state histories cannot be retrieved and presented to the analyst in a meaningful manner. Thus, a simulation language invariably incorporates efficient data processing routines (such as sort–merge routines and statistical computation subprograms) which the executive routine may invoke in accordance with input parameter specifications as defined by the analyst. Indeed, most simulation programming languages include standard report generation formats and

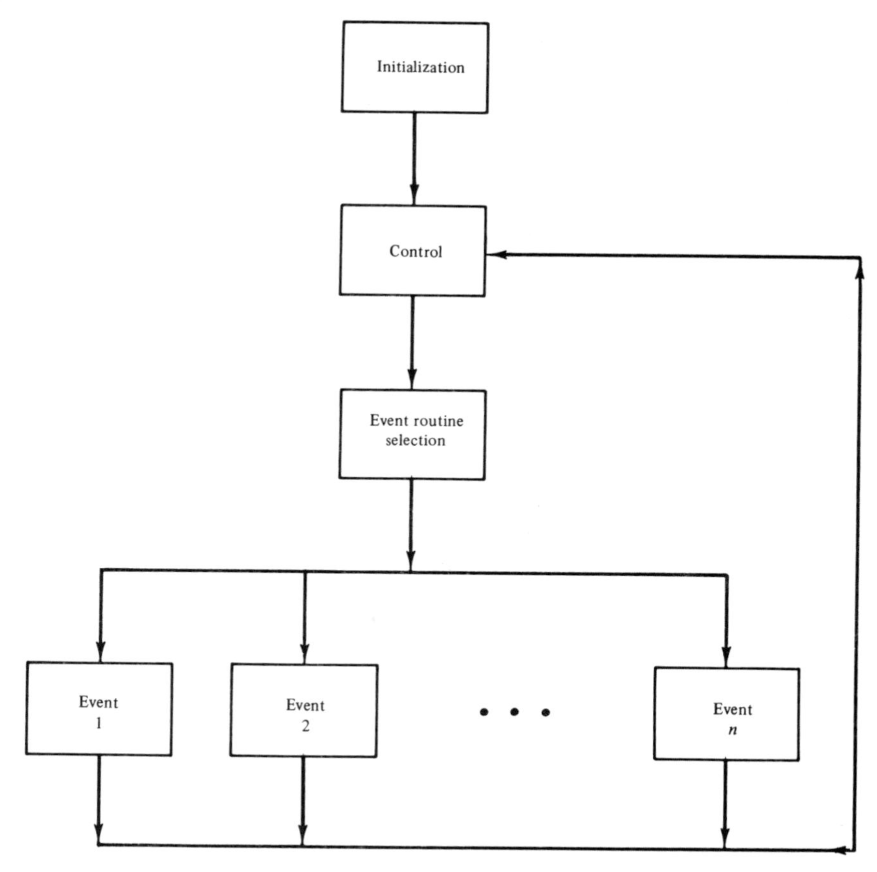

FIG. 5.1. The simulation algorithm.

provide input specification parameters that permit analysts to request (or suppress) their use.

A fourth feature normally provided within a simulation programming language is a convenient scheme by which experimentation with the completed model can be undertaken. Methods by which "stacks" of encounters can be specified in accordance with organized experimental plans (or designs) are becoming more common to simulation programming languages. Although experimentation with a model is usually associated with the validation and model analysis stages of its development, the capability to define successive model encounters (with or without seed repetitions) may be an essential component of the verification stage as well.

5.4.2. Additional Features

In addition to providing the four fundamental aspects of dynamic model synthesization, *ad hoc* simulation programming languages usually provide rapid access to efficient mathematical and computational routines. Indeed, many simulation programming languages are compiler-based, or are derivatives of problem-oriented compiler languages such as FORTRAN or ALGOL, in which cases one frequently has access to the entire library of prepared mathematical functions (logarithmic, trigonometric, hyperbolic, etc.). Included among the routines of virtually all simulation programming languages are subprograms that will accumulate standard statistical information, such as frequency histograms, means, standard errors, parametric estimates, autocorrelations, and spectral estimates.

Another important feature of most simulation programming languages is an inherent debugging feature. Since the simulation algorithm of the executive routine sequences the event routines as they are required by simular developments, one may monitor, or trace, the succession of computing and/or simulation activities in which the executive routine engages. Such a monitoring capability implies that debugging may proceed at a "macro" level; i.e., at or above the level of source language statements. A simulation program, which is prepared in a machine-oriented language or in a compiler language such as FORTRAN or ALGOL, may require extensive debugging and the analyst will probably need to be concerned both with errors that may have arisen due to a faulty executive routine and with those due to logical faults in modeling. Most simulation programming languages have been sufficiently used and tested so that the concerns of debugging usually rest only with the second type of error: the logical fault in modeling. Again, this emphasizes the value of the *ad hoc* simulation language: It frees the analyst's mind from all details except those concerned with construction of an adequate and effective model.

5.4.3. The "World View" of a Language

The convenience supplied to the analyst by an *ad hoc* simulation language is not without its compensating costs. Primary among these is the fact that, because the simulation language incorporates certain fixed structures and "housekeeping functions," such as the simular clockworks and the decision priorities for handling simultaneously occurring events, the simulation programming language must necessarily restrict one's view of the modeling synthesis. One must take care to define and recall entities and attributes in accordance with the particular simulation language's idiosyncracies. Some simulation programming languages do not admit temporary entities; others do not permit sets to be defined, reduced, or augmented as simular time evolves. (Even elementary queues cannot be formed in some language structures.) Some simulation programming languages are "richer" in their intrinsic capability to generate random variables or correlated time series, while others require that all inherent procedures for the generation of random variables employ analyst-supplied histograms.

These difficulties should seem rather small in comparison with the many advantages offered by the *ad hoc* simulation programming languages. Indeed, they usually are. Moreover, as the languages themselves evolve, their annoying features seem to disappear.

Nonetheless, the analyst who selects an *ad hoc* simulation language must anticipate that each language is the embodiment of a modeling viewpoint, usually quite characteristic of the language's developers. Of necessity, this viewpoint is imposed, to a degree, on every user of the language and has been referred to by Krasnow and Merikallio (1964) as the language's "world view."

The "world view" of some languages is that any system is representable as sets of interdependent differential and integral equations. Several languages of this type have been developed for continuous-change simulation models to be implemented on hybrid computers [cf. Brennan and Linebarger (1964) and Gordon (1969)]. Other languages require that the system be viewed as feedback mechanisms capable of description in terms of difference equations. These languages are often severely restricted both in their set-forming capabilities and in the number of representable entities and state variables. Still other simulation programming languages are oriented toward the modeling of queues. Some languages require that one emphasize the active elements of a system (such as might be typified by unsatisfied customers), whereas others tend to stress the more passive components (such as fixed installations awaiting requests for service).

The choice among *ad hoc* simulation programming languages rests necessarily then with the systems analyst who, in completing the event graph associated with the system, will be in a good position to evaluate that "world view" among the diverse languages which most nearly corresponds to his needs, as implied by the event graph. Therefore, the selection of the programming language should be one of the initial steps of the system synthesis stage of model development. Of course, this decision must also take into account costs attributable to procuring and implementing any of the languages, including implicit costs of education as the analyst–programmers adapt to the world view of the selected language. In this regard, some of the special-purpose simulation programming languages are supported directly (sometimes, indirectly) by computer manufacturers; other languages are reasonably well documented through texts printed by some of the better-known scientific publishing houses.

In summary, there is seldom a dynamic simulation modeling effort that would not be enhanced by the use of an *ad hoc* simulation language. The choice among these is inevitably situation-dependent and may require an instructive (and usually beneficial) exercise in cost-effectiveness.

EXERCISES

1. Describe the PERT network of Exercise 2, Section 4.6, in terms of the following modeling concepts: (*a*) system boundary, (*b*) system environment, (*c*) entities, (*d*) attributes, (*e*) events, (*f*) activities, (*g*) event routines. Write a program, using a simulation clock and an executive routine, that will simulate dynamically the described project from start to finish and that will determine the project completion time T via the simulation rather than by direct computation.

2. A computer facility of six machines is maintained by a service agent who visits each of the six machines exactly once each week in the same cyclic order. Presuming that the service time for the ith machine is a positive-valued random variable T_i (in hours) having probability density $f_i(t)$, as given by Eqs. (6.3.2:1), write a simulation program (including a simular clockworks and an executive routine), that will mime the activities of the service agent for 100 consecutive weeks. For each week, compute

$$H = \sum_{i=1}^{6} T_i$$

and from a histogram of the 100 H-values. Also compute the proportion of weeks in which the agent requires overtime pay (i.e., for which $H > 40$). Discuss for this model each of the seven modeling concepts, (*a*)–(*g*), as defined in Exercise 1.

5.5. Model Verification

5.5.1. The Simulation as an Input–Output Device

Once the analyzed system has been synthesized into a computer program, the analyst will want to establish the credibility of the model before embarking upon a plan of experimentation with it. There are two phases in this operation, each of which constitutes a fundamental stage of model development:

(*a*) *the verification stage*—concerned with establishing the rectitude of the programmed, logical structure of the model,
(*b*) *the validation stage*—concerned with establishing the verisimilitude of the responses from the verified model and comparable observations taken on the modeled system.

The current section will be concerned with techniques for accomplishing efficiently the model verification stage of model development.

Though the response of a simulation model may be univariate or multivariate, each encounter with the model corresponds to a transformation of the environmental (i.e., input) conditions to responses (output) conditions. In many instances, one will be aware of elementary relationships that should exist between these input–output pairs, even if the model is especially complicated, so that a proper verification of the model should include the specification of encounters to test these known relationships. If any one of these tests should fail, an intensive search of the program's structure should be undertaken to correct its faults.

It should be noted that the existence of known input–output relationships for the verification stage is *not* predicated on the behavior anticipated of the model if it were to mime accurately specific relationships which have been recorded for the actual system. Rather, the verification stage is based upon comparisons of the model's observed behavior and that anticipated if indeed the programming structure had been accomplished as desired.

For example, consider a deterministic model of a system. One may be aware of certain environmental specifications (e.g., particular policy specifications, such as a queue discipline) which, under specific input conditions, should yield a known result at the end of T simular time units. If these conditions are specified for the deterministic model, yet the model's response is not in accordance with its anticipated behavior, one must infer that the model's structure remains intrinsically in error. Of course, such verification of the individual event routines of the deterministic model can be used to isolate any such logical faults.

Other verification tests for deterministic models are available even when specific input–output relationships are not known. For example, if it may be safely presumed that the deterministic model's response should be a monotonic function of some input parameter, then a verification test based on defining successive model encounters at successively larger values of that input parameter should result in a monotone sequence of responses. Similarly, if the deterministic model's responses should be some unknown, yet continuous, function of a specific input parameter, then a verification test may be accomplished by observing the model's responses at successively nearer values of that input condition.

5.5.2. The Random Number Seed

In the presence of stochasticity, however, it may not appear evident that such comparisons could be validly undertaken. Stochastic features are incorporated in a simulation model in order to enhance its ability to mime systems that inevitably appear random to some degree in nature. This randomness is spawned in the typical simulation model by the incorporation of a *random number seed* as an integral part of the input specifications of a model. In many instances, one will wish to define several of these seeds, each seed to be used in generating the randomness of certain aspects of the model; e.g., one seed may be used to generate a chain of random variates that will be used only to provide the random service times in a queueing model, whereas a second seed might generate a sequence of random weather conditions, which affect the arrival pattern of customers for service. More generally, one might define a random number seed for each stochastic event routine, as well as a separate seed for initiating, at a random point, each prerecorded set of exogenous events; other seeds might be used to initiate certain model parameters. Conceptually, it is often valuable to consider these separate seeds in juxtaposition (much as they might appear on a computer input data card) and therefore as a single seed, or single input parameter s.

In either case, the (juxtaposed) random number seed is a necessary feature of the input specification, since it provides the randomness necessary for the adequate representation of the stochastic aspects of the modeled system. Since the randomness which ensues is directly related to the seed(s), the seeds themselves should be selected as randomly as possible; i.e.,

> *Any seed specification should in general result from its random selection from among the set of all admissible seeds.*

As a direct consequence of this principle, the response Y which derives from a simulation model becomes a random variable, because it is a transformation not only of the environmental specifications, but also of the randomly selected seed(s); i.e.,

$$Y = Y(x_1, x_2, \ldots, x_p; s)$$

where $x_1, x_2, \ldots, x_p$ represent the environmental conditions other than the seed s. If, in specifying a set of n encounters with the model, the seeds are randomly selected and, furthermore, are selected so as to preclude the possibility of repetitions among the selections, then we say that the n encounters are *independently seeded.*

There are two immediate consequences of this principle of randomly selecting the seed(s). First is the fact that analyses of model responses, such as those required in the verification phase, will necessarily deal with recordings of random variables. This fact will not prove especially disturbing, since the specification of n independently seeded simulation encounters will result in a set of n independent random variables Y_1, $Y_2, \ldots, Y_n$. That is, a *random sample* of n responses will exist, the analysis of which may be accomplished by existing statistical methodology. The discussion of many of these statistical techniques will constitute the subject matter of the next five chapters; many of these analytical procedures are applicable to the verification of a stochastic simulation model.

A second immediate consequence of using the principle of randomly selecting the seed(s) exists. In the verification phase of model development, it is especially important that any known input–output relationships be tested by correspondingly defining the environmental specifications and observing the resulting responses for their compatibility with the anticipated responses. Thus, whenever randomness is to become an integral part of a model, it is imperative that, whenever possible, the resulting random variables be compared with those which would be expected to arise if the model were known to be properly programmed. If, therefore, for a particular set of environmental conditions $(x_1, x_2, \ldots, x_p)$, the nature of the distributional form of the simulation model's responses (output random variables) is known, then the comparison of the frequency histogram, of the responses that arise from *n independently seeded encounters*, with this known distributional form should also serve as a verification tool. Such comparisons may be effected by the Chi-squared test criterion, or by the Kolmogorov–Smirnov test, as introduced in Chapter 2 for testing random number generators (see p. 61).

The principle of randomly selecting the juxtaposed seed for stochastic models is primarily applicable to the experimental stages of model development; that is, the validation and analysis stages in which the intrinsic randomness of the simuland must be mimed fully in the stochastic model. For the purposes of model verification, during which one seeks only to establish the rectitude of the stochastic model's programmed logic, the principle may be suspended in several ways.

For example, one may know of particular input–output conditions that should obtain if indeed stochasticity were suppressed entirely. A queueing model whose environmental specifications and whose interarrival and service times can be set at constant values may permit such input–output relationships to be specified. Indeed, practically any stochastic model can be subjected to deterministic verification tests in a similar manner, provided that sufficient foresight regarding the need for such verifications was present during the system synthesis stage. For example, one may wish to program the model so that each call for a random variable will provide a given constant (i.e., the provision of degenerate random variables is a desirable feature of most stochastic models and, consequently, of most *ad hoc* simulation programming languages).

Once such deterministic verifications have been accomplished by suppressing all randomness in the stochastic model, other verification tests may be performed by partially suppressing stochasticity. Namely, one may know that certain stochastic relationships should obtain as a result of a properly programmed model that uses degenerate random variables in all but, say, one of the seeded random number chains of the model. For example, one might specify a complicated queueing network model in terms of input conditions which could correspond to the $M/M/1$ queue, as discussed in Section 4.7.5; observation of the queue length distribution should then be in accordance with the theoretically applicable geometric distribution, as could be verified by means of nonparametric statistical procedures such as the Chi-squared or Kolmogorov–Smirnov tests.

The model verification procedure may therefore be terminated more satisfactorily if, whenever multiple seeds have been defined, one is assured that the responses are indeed random variables as a direct consequence of the effect of the seeds. Therefore, an acceptable procedure during the verification stage is the temporary suspension of the principle by fixing in turn all but one of the seeds. The fixed seeds then become the equivalent of constant input parameters, so that a set of k encounters of the model, each differing in its specification only by the random selection of a value for the one remaining seed, will produce k random variables. If this randomness does not appear in these responses, or if the resulting random variables do not possess the distributional properties anticipated of them (if indeed the model structure were programmed as was intended), then a more thorough examination of the model's logical structure is in order. One may proceed to the same kind of verification tasks by selecting exactly one of the remaining seeds as the unique, random seed and by comparing model responses arising from its randomly selected specifications. In this way, module-by-module verification of a model can be obtained, even in the presence of stochasticity.

The verification stage of model development is therefore concerned with determining whether or not the model is properly programmed. To repeat, the verification stage is *not* composed of comparisons of the model's responses with known measurements, or recordings, of the modeled system. In the language of the computer programmer, model verification is simply debugging, although the special nature of dynamic simulation models, particularly those of the stochastic variety, requires comparisons somewhat different from those usually associated with debugging.

In summary, if the correctly programmed model structure were designed to produce a given response as a direct consequence of specific, fixed input conditions (including suppressed stochasticity, where appropriate), then the verification stage should incorporate a test of this known input–output relationship. Particularly useful in these cases are programmed mechanisms by which the randomness of a stochastic model can be completely extinguished; a typical example is a programmed switch that would normally select, say, interarrival times from an exponential distribution, but that would (for the verification stage) generate an interarrival time equal to, say, the mean interarrival time regardless of the value of the random number seed (or its successor values). However, if input–output relationships are known for the model in its stochastic "form" (i.e., the distributional nature of the random variable, defined by the responses to independently seeded encounters of the model, is known), then appropriate statistical tests should also be invoked in the verification stage. This can be accomplished even with multiply seeded models by selecting randomly values for only one set of seeds, maintaining all other seeds constant, in order to ensure that each separate source of random variation is being properly instilled in the responses from the model. Thus, verification of a model's programmed structure may require manipulations of its seeds in "unnatural" ways so that known input–output relationships can be examined and verified. [See Mihram (1972b).]

EXERCISES

1. Referring to Exercise 1, Section 5.2, describe precisely the theoretical distribution of the buses' circuit times if: $KB = 1$, $KS = 5$, $\tau = 3$, $\varphi = 0$, $H = 20$, $\lambda = 1$. Indicate how information could be obtained from a particular encounter with the model, when defined according to these environmental specifications, so as to compare the results with the theoretical distribution by means of a Kolmogorov–Smirnov test.

2. Referring to Exercise 2, Section 5.2, describe the switchboard as an $M/M/1$ queue whenever: $NC = 1$; the Weibull shape parameter is set equal to 1; and, the rejected calls are merely held in a first-in–first-out (FIFO) queue. From the preceding discussion of such queues, suggest a verification test for the model based on the theoretical distribution of the equilibrium queue lengths for a $M/M/1$ queue.

5.6. Model Validation

5.6.1. Comparisons with the Real World

The fourth stage of model development is its *validation*. Once the verification stage has been satisfactorily completed, one will have established only that the model is emitting the responses that would have been anticipated if indeed the programmed structure of the model was accomplishing those tasks that were designed into it. Another proper inquiry relates to the manner in which the model is miming reality; i.e., the degree of comparability between the model's responses and recorded results of the modeled system.

A complete discussion of the techniques applicable to the validation stage would require an extensive presentation at this point. Fortunately, one may make the observation that both the independently seeded stochastic simulation model and the modeled system itself will produce random variables; thus, techniques suitable to the comparison of random variables (namely, statistical procedures) are in order for comparing model responses with observations recorded from the modeled system. The following five chapters contain a quite representative sampling of techniques suitable to model validation.

Of particular importance at this point, however, is the observation that knowledge of the behavior of the modeled system will presumably have been recorded under known system conditions. The conscientiously developed model should therefore include environmental specifications, or input parameters, whose assigned values would designate model encounters that would be directly comparable to these known system conditions; the resultant responses of the model should therefore be in consonance with the recorded observations of the modeled system.

This then is the essential idea of the model validation stage and it should be clearly distinguished from the activities of the verification stage. Model validation is concerned with the comparison of the model's responses with those of the modeled system, whenever the conditions producing each are essentially the same; model verification, on the other hand, is directed toward establishing whether or not the logical structure of the model is compatible with its programmers' intentions.

5.6.2. A Principium of Seeding

The validation stage also acknowledges the fact that the responses of a stochastic simulation model are direct consequences of the random number seeds which, among successive encounters, should be selected randomly and independently of one another. Of course, if the successive encounters with the model are defined by the same environmental conditions $(x_1, x_2, \ldots, x_p)$, then the successively selected seeds must not only be independently and randomly selected, but also preclude the possibility of repetition. Even if the environmental conditions should be altered between successive encounters, it is good practice to make certain that these encounters are *independently seeded*; i.e., each successive seed should be randomly determined by selecting its value from some published source of random numbers [such as those of the RAND Corporation (1955)] *and* the successive seeds should be inspected to forbid repetitions in the set.

One is therefore led to formulate a *Principium of Seeding* for stochastic simulation models:

In any sequence of n encounters with a verified stochastic simulation model, the n successive random number seeds shall be selected randomly and independently, though repetitions shall be forbidden.[†]

The seeds which may be employed in specifying a model encounter are, of necessity, restricted to a specific set of admissible seeds. For

[†] The Principium of Seeding is not intended to preclude the use of antithetic variates (cf. Hammersley and Handscomb, 1964), although the applicability of this Monte Carlo technique to the analysis of general, large-scale, dynamic, stochastic simulation models remains conjectural. (See also page 495.)

example, if a pseudorandom number generator of the power residue type is being used in the model, then it is likely that the seed will be odd (on binary computers); even if the mixed congruential generator is engaged, there will usually exist a largest positive integer that the computer will accept as a seed value. If nonalgorithmic procedures are being used for the chains of random numbers required in a simulation model (such as the use of a pretested sequence of digits from a white-noise generator), then the seed may be thought of as an index location that indicates a (randomly selected) starting position in this sequence; in this event, the set of admissible seeds corresponds to the set of allowable index values for the prerecorded random number chain.

Therefore, a set of n *independently seeded encounters* can be defined as the specification of the environmental conditions for, and the recording of the resulting responses from, n iterations of a stochastic simulation model, with the required n seeds having been selected

(*a*) *randomly* from among the admissible set of seeds, preferably by selecting random entries from a published list of random digits,
(*b*) *independently*, save for the preclusion of any possibility of repetition among the n seeds so selected.

Comparisons of the model's responses with recorded observations of the modeled system's behavior will depend upon independently seeded encounters with the model. To repeat, it is the seed that initiates the characteristically random behavior of the modeled system into the model. Since the "random initial conditions" would never be known precisely for any observation made on the actual system, neither should the "random conditions" (as are necessarily reflected by the seed) for the model be subjected to bias or to the possibility of bias.

Conceptually, validation is to be accomplished by concomitant experimentation with the model and the modeled. Environmental conditions for the model and operating conditions for the modeled system should be compatibly altered, and comparisons made of the respective responses (outputs). Indeed, a number of experimental designs, useful for such comparisons, will be presented in the following chapters. However, in many situations, experimentation with the modeled system is prohibitively costly (e.g., the operation of a large-scale production line under alternative policies and workshop configurations). In these cases, model validation will be necessarily curtailed and applicable only to the comparison of the model's responses with observations arising from economically feasible experimentation with the actual system. In other instances, experimentation with the modeled system will be physically impossible (e.g., the system may still be on the drawing-boards); in these cases, model validation cannot, strictly speaking, be accomplished, although the modeler should be aware of the consequences that will probably follow from a haphazardly constructed and invalidated simulation model.

Validation techniques for deterministic models become somewhat limited in number. The presumption that a conscientiously conducted systems analysis will lead to the need for probabilistic representations in any model of the system has prompted the discussions of this book to center about the stochastic model. Indeed, the appropriateness of comparisons of the response arising from a deterministic model with a set of corresponding observations made on a complex, stochastic simuland, would appear somewhat conjectural.

Nonetheless, one could (presuming the modeled system to be available for repeated experimentation under some given operating condition) record n observations $z_1, z_2, \ldots, z_n$ for the simuland, constructing therefrom, say, an approximate 90% nonparametric confidence interval for its mean value by means of Chebyshev's inequality (see Sections 6.2.2 and 2.10.2). If the recorded response from the deterministic model fails to fall into such an interval, a complementary degree of confidence in the model would be in order.

Other nonparametric validation tests are available whenever both the model and the modeled are stochastic and observable under comparable environmental (operating) conditions. Presuming that m observations can be made on the simuland, and that any arbitrary number n of corresponding model encounters can be defined, one may distinguish between two cases.

Case I ($m = 1$). If the stochastic model is indeed a valid representation of the system, then the n model responses (from independently seeded encounters) and the single observation on the system should constitute a random sample of size $(n + 1)$ from the same cumulative distribution function. If these data are rearranged in numerical order, one would tend to suspect the model's validity, should the single observation on the simuland lie at either extreme. Under the assumption that all $(n + 1)!/(n!1!) = (n + 1)$ orderings of the sample into the two groups are equally likely (as would be the case if indeed all values were drawn from a common cumulative distribution function), the probability of the simuland's observation being at one extreme or the other becomes $2/(n + 1)$. For proper n, this probability becomes sufficiently small so as to cast doubt on the conjecture that the $(n + 1)$ recordings are from the same distribution; i.e., one would likely conclude that the stochastic model is not yet a valid representation of the system at hand.

CASE II $(m > 1)$. Whenever the simuland itself can be observed repeatedly under some specific operating condition, one can apply a number of nonparametric tests for model validation. The *runs test* (see, e.g., Lindgren, 1968) may be applied by ordering all $(m + n)$ recordings and by counting the resulting number U of runs. (A *run* is an uninterrupted sequence of ordered observations of the same type, model or simuland.) Using tabulations of the theoretical distribution of U, one can determine whether the observed number of runs is statistically unusual (i.e., highly improbable); if so, the model is probably not a valid representation of the simuland at this environmental condition. Another nonparametric test applicable in this case is the two-sample Kolmogorov–Smirnov test; this test compares the two-sample cumulative distribution functions, each formed by ordering the values of each sample separately, in much the same way that the one-sample Kolmogorov–Smirnov test was used to test random number generators, although the appropriate critical values are different from those tabulated in the Appendix, beginning on page 506 (see Lindgren, 1968).

5.7. Model Analysis

As was indicated in the first section of this chapter, the fifth and final stage of model development is concerned with the performance of experiments with the verified, validated model. This experimentation is conducted in order to address precisely those questions which prompted construction of the model initially. Typically, the experimentation is undertaken in order to determine

(*a*) the static effects of the model's responses, or
(*b*) the dynamic effects of the model's behavior.

In the case of a dynamic, stochastic simulation model, the responses recorded at the end of, say, T time units, becomes a random variable,

$$Y(T) = Y(x_1, \ldots, x_p; s),$$

where again $x_1, x_2, \ldots, x_p$ represent the variable input conditions (environmental specifications) other than the (single, juxtaposed, if necessary) random number seed s. The static effects of a stochastic simulation model are thus studied by analyzing responses $Y(T)$ that arise from experimental plans or designs which systematically alter the environmental conditions $(x_1, x_2, \ldots, x_p)$ among successive, yet independently seeded, model encounters of T simular time units.

There are two general categories of analyses concerned with the static effects of a stochastic simulation model:

(1) those undertaken in order to determine the relative importance of alternative policies, dissimilar environmental conditions, or differing parametric specifications as they affect the model's response at some point T in simular time, and

(2) those developed in order to determine that particular combination of policies, environmental conditions, and parametric specifications which will provide, according to some preassigned criterion, the optimal response at the end of T simular time units.

The discussion of these two analytical aspects of the study of model static effects constitutes the major portion of the following four chapters. As for the analysis of dynamic effects, appropriate procedures are presented in Chapter 10. In this regard, however, it should be noted that the dynamic, stochastic simulation model produces a random variable $Y(t)$ at the end of t simular time units, for any $t = 1, 2, 3, \ldots$. Typically, $Y(t)$ is some univariate (or multivariate) quantity that measures the status of the model as simular time advances. However, it is instructive to note that there exists a sequence, or conceptually infinite collection $\{Y(t)\}_{t=1}^{\infty}$ of these random variables, so that the dynamic, stochastic simulation model can also be viewed as a generator of a stochastic process, inferences about which are properly conducted by statistical procedures applicable to the analysis of *time series*.

5.8. A Scholium on Modeling

In this chapter numerous principles for the efficient modeling of dynamic systems have been presented. The presentation has emphasized the stochastic, dynamic simulation model, since this category of models would appear to be more general than combinations containing any among the adjectival descriptors: deterministic, static, or formalized. Special emphasis has also been placed upon the discrete-change formulation of such models, although repeated reference has been made to the possibilities of using continuous-change formulations wherever such formulations would prove appropriate.

The symbolic model is a conceptual transformation of the modeled. This chapter has been devoted to the delineation of the five stages of orderly model development. These are:

1. The Systems Analysis Stage
2. The System Synthesis Stage
3. The Verification Stage
4. The Validation Stage
5. The Model Analysis Stage.

It has been a presumption of this discussion that the intention of the systems analyst, once the model's goals and intended analyses have been precisely formulated, is to achieve, as nearly as possible, a one-to-one correspondence between the behavior of his model's symbols as simular time advances and the behavior of the modeled system as real time evolves. That this is rarely achieved has not been, and should not be, a deterrent to the systems analyst. The Uncertainty Principle of Modeling would appear to complicate further the endeavor, since a randomly behaving model could hardly be expected to mime precisely the random fluctuations which exist in the system under study. However, one may ask *statistically* whether a model is behaving characteristically as the modeled system would behave, so that the analyst who might prefer to establish the desired isomorphism will, nonetheless, probably be able to produce an adequate model of the system.

The initial stage of model development, then, is the analysis of the system in order to isolate its salient features and interactions. As greater detail regarding the system's intrinsic behavior is sought, the more likely a probabilistic description (i.e., a stochastic model) shall be required. The systems analyst must endeavor in this initial stage to segregate the state variables required to represent the status of the system at any point in time and must locate the mechanisms by which these state variables are altered dynamically.

The systems analysis stage is concerned with the definition of state variables, entities, attributes, relationships, and events; the stage merges into the system synthesis stage with the construction of the event graph, which serves as a partial check for the completeness of the systems analysis stage. The state graph also serves to determine the model formulation methodology: continuous-change or discrete-change. In consonance with this choice is the selection of an *ad hoc* simulation programming language.

Note that less frequently is size of the model a constraint in discrete-change formulations than in continuous-change formulations when implemented by means of the special-purpose languages.

The system synthesis stage develops both the static structure and the temporal structure of the dynamic simulation model. State variables, including simular entities and their attributes and relationships are delineated more explicitly; a basic time unit is selected and event routines are defined in accordance therewith and in a manner facilitating the dynamic representation of the modeled system. The Uncertainty Principle of Modeling is acknowledged by incorporating stochastic features in the model's structure: randomly specified initial conditions for the model's state variables, randomly arriving exogenous events and/or temporary entities, or intrinsic elements in the endogenous event routines. These random features are imposed in the model's structure in consonance with the findings of the initial stage (systems analysis) of model development. Indeed, throughout system synthesis, continuous emphasis should be placed on the availability of meaningful data, in the requisite form, for the simulation model.

The system synthesis stage must not be conducted in isolation of the intentions of the stages to follow. Once the model has been constructed, it must be verified and validated before it can be used to make acceptable inferences regarding the modeled system's behavior. The system synthesis stage should incorporate features in the model's structure that will facilitate these verifications, validations, and subsequent inferences.

The verification stage is concerned with debugging of the programmed model. Noting that the simulation model is essentially an input–output device, the systems analyst should try to determine whether there exist input–output relationships which must appear, if the model has been properly programmed. Deterministic input–output relationships would probably not exist for a stochastic simulation model, but fortunately the computerized stochastic model can, with a bit of ingenuity, be forced to play the role of a deterministic model. By suppressing all random number generation (a feat most easily accomplished by temporarily defining all random variables as degenerate, so that each provides a specific, unique value with probability one), the stochastic model (or any of its modules or event routines) can be temporarily converted to a deterministic model with the property of completely reproducible responses arising from identical environmental specifications. [See Mihram (1972b).]

One can also reduce stochasticity by defining multiple random-number seeds. (Juxtaposed, these seeds may be considered as a single-integer seed for the subsequent, experimental stages of model development.) By having each of the three stochastic features of a model depend upon its own seed, one can verify whether randomness of the proper description is being instilled into the model's behavior. For this purpose, each appropriate event routine could be defined so as to depend upon its own random number seed and its stochastic effects observed against the background of suppressed stochasticity for other parts of the model.

In the verification stage, advantage should be taken of the probability laws of statistics, which can often be invoked to predict the nature of the distribution for the random variables that are the responses (outputs) of the stochastic simulation model. More will be said of this in Chapter 6, but the analyst should be prepared to ensure that the random variables being emitted by a stochastic model are in consonance with those that would be anticipated from a properly programmed model. If incompatibilities are discovered, the analyst must return to the first two stages of model development in order to effect the necessary changes in the programmed logic or to ensure the adequacy of data.

The verification stage of a model is therefore concerned with determining the soundness of its programmed logic, independent of any consideration as to the veracity of the simular responses and any observations recorded on the behavior of the modeled system. Of course, the two activities cannot be clearly divorced one from another, since the conscientious analyst will be concerned with the adequacy of his model, in comparison with the modeled system. However, the model's structural components should be verified before experimentation with the model is undertaken.

The validation stage of model development seeks to determine whether the simulation model's behavior is consistent with that of the modeled system. Since the modeled system is presumably stochastic in character, yet shall not likely possess random behavior that can be suppressed, it will be necessary to forbid (in specifying all model encounters which are defined subsequent to the verification stage) experimentation with the random number seeds. Indeed, one may consider a multiply seeded model as requiring only a single seed, composed by the juxtaposition of the multiple seeds.

Once a model has been presumed to be verified, there is no longer need to conduct experiments with the juxtaposed seed *per se*.[†] Moreover, since the model's stochastic behavior is spawned from this seed, each

[†] The possibility exists of constraining some of a model's randomization by holding constant among several model encounters some, but not all, of the components of the juxtaposed seed. (See page 401.) Variance components (random block effects) then become part of the simular response (see Mendenhall, 1968, for example).

encounter of the verified model should be defined by the *random* selection of a specification value for the seed, preferably from a published table of random numbers, such as those of the RAND Corporation (1955). In addition, since the simular experimentation of the validation stage and of the ensuing model analysis stage will possibly include model encounters (iterations) defined by the same environmental (input) conditions, then the seeds required for successive encounters should be independently and randomly selected, yet with the proviso that repetitions of seed values be forbidden.

One could conceivably quarrel with the necessity of precluding repetitive seeds in encounters that are otherwise defined by differing environmental specifications (input data), but absolutely no harm will arise as a result of religiously adhering to this Principium of Seeding. Indeed, much realism may be gained by ensuring that each encounter with the model is initiated with a proper random impetus.

In the validation and analysis stages of model development, the analyst will be observing model responses that are random variables. The corresponding observations of the modeled system will also be random variables, so that their comparison with the model responses will need to be statistical in character. Consequently, the mechanism for defining orderly and meaningful experimental designs for the required comparisons of these experimental stages of model development is necessary. Such designs will be discussed in the next five chapters.

In the event that the validation tests of a model's adequacy are failed, the analyst will probably need to return to either the initial stage (systems analysis) or its sequel (system synthesis) in order to alleviate inadequacies such as

(*a*) accuracy or precision of the simular responses,
(*b*) level of detail in model structure,
(*c*) assumptions regarding the system boundary and/or effects arising from the system environment,
(*d*) static or temporal structures of the model,
(*e*) the data supporting the model's structure.

A further verification of the (presumably altered) model's structure will then be in order; validation tests can then be again organized in accordance with the Principium of Seeding. (See Fig. 5.2.)

The extent to which a stochastic simulation model must be verified and validated is always a moot question. The more, the better. Of course, many systems will not be available for the experimentation required by the validation stage of their model's development. If this be the case, then the analyst will proceed to the fifth and final stage (model analysis) on a somewhat less secure footing. In any event, the analyst should note that a model that passes a validation test has passed a somewhat empty test. If the model fails such a test, it becomes highly suspect; whereas, if it passes, no strong statement regarding the confidence in the model can be formulated.

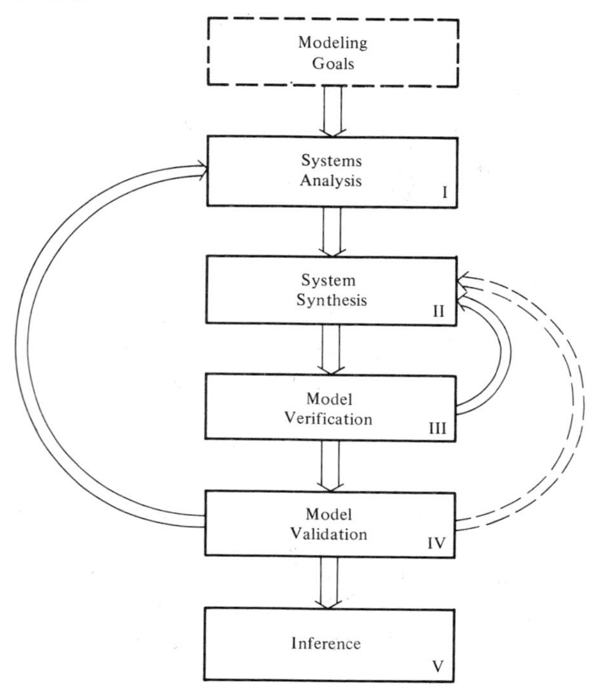

FIG. 5.2. Cycling among model development stages.

However, simular experimentation, the topic of the next five chapters, will produce results and recommendations that are much more readily accepted whenever confidence in the model has been established by adequate verification and validation techniques. Since the fundamental purpose for construction of a model is increased understanding of the modeled system so as to implement wisely changes therein, the acceptability of analyses made with independently seeded encounters of the system's model is of paramount concern to the analyst. A model that has been well-defined (Stage 1), well-structured (Stage 2), verified (Stage 3), and validated (Stage 4) will more likely produce acceptable responses than one that has not.

Chapter 6

FUNDAMENTALS OF SIMULAR EXPERIMENTATION

Those oft are stratagems which errors seem.

—ALEXANDER POPE

6.1. Introduction

Once a simulation model has been verified, it can then be used to provide inferences about the modeled system. Regardless of the type of dynamic model at hand, this simular analysis is usually conducted to determine

(*a*) *static effects*: the state of the model *at the end of* a stipulated period T of simular time; or,

(*b*) *dynamic effects*: the changes occurring in the model *during* a stipulated period of simular time.

Static effects are represented by the state of one or more of the attributes of the model after T units of simular time have passed, whereas the dynamic effects are represented by the record of the changes that an attribute (or set of attributes) undergoes during the T units of simular time. Thus, a single attribute might be employed to evaluate both effects, although in the first case only the "final state" of the attribute is of interest; if the analyst wishes to study the dynamic effects of his model, then he seeks information not only about the final state of an attribute but also about the transitions that led to that observed final state.

Whether a dynamic simulation model is deterministic or stochastic, one inevitably proceeds to the analyses (of static *or* dynamic effects) by iterating the model a number of times, usually by performing each iteration under differing environmental conditions (as reflected in alternative input data specifications). Comparison of successive model responses (outputs) reveals the effect of altered environmental conditions.

This process of altering the simular environmental conditions and of the comparing of the successive simular responses may be termed simulation experimentation, or *simular experimentation.* In the manner of any scientific investigation, simular experimentation involves both the organization of the succession of model iterations and the analysis of the resultant responses: i.e., both the *design* and the *analysis* of the experiment. The efficacy of the second must rely heavily on the cleverness of the first, so that experimental designs that are capable of providing information useful to the subsequent analysis are of prime importance.

One should note that the invocation of the principles of simular experimentation need not attend the validated model. Indeed, during the verification and validation stages of model development, experimental designs and analyses are appropriate in order to assure that the model is programmed in consonance with its intended effects (verification) and in order to be sure that the model's behavior simulates realistically the system being modeled (validation). However, throughout most of this and the following two chapters, the emphasis will be placed upon experimentation with the completed simulation model. Whenever and wherever appropriate, passing reference will be made to the applicability of the given design or technique to the verification or validation stages.

For the study of either static or dynamic effects, a well-developed methodology currently exists. Dynamic effects are amenable to analysis by means of such mathematical tools as Fourier decompositions and, in the case of the dynamics of a stochastic simulation model, the correlogram. Topics related to the design and analysis of simulation experiments for the purpose of determining the dynamic effects of altered environmental conditions shall be outlined four chapters hence.

As for the static effects, the analyst is usually asking about either the relative importance of suspected contributors to the system's response or the environmental conditions (i.e., input conditions) that will provide the optimal system response. In both of these enquiries, it is fruitful to view the response (or output), $Y(T)$, of the simulation model at simular time T as a function of p pertinent environmental conditions (or input parameters):

$$Y(T) = Y(x_1, x_2, \ldots, x_p; s),$$

where s is the random-number seed, required of the stochastic model.

For example, some particular specification of these simular environmental conditions, say

$$(x_{1,0}, x_{2,0}, \ldots, x_{p,0}),$$

would correspond to the standard, or basic, operating conditions for the system being simulated, so that the resultant simular response, say $Y_0(T)$, should correspond reasonably well to the anticipated or known system response under these conditions.

Of course, if the simulated system itself is stochastic in nature, then the simular response (whether it arises from a deterministic or stochastic model) at the standard environmental conditions can hardly be expected to reproduce a particular response observed under these conditions in the system being simulated. This point seems often overlooked in the validation stage of model development; indeed, there seems to exist a basic misunderstanding whenever the builder of a deterministic model considers his model validated if and when he finds it possible to locate a particular specification of the model's input parameters which will reasonably contort the simular system's behavior so that its response is in accordance with some (probably stochastic) observation made on the simulated system.

This very point is probably one of the greatest justifications for the construction of a stochastic, as opposed to a deterministic, simulation model, because most of the systems subjected to analysis for the purpose of modeling do indeed comprise stochastic elements (or, at least, elements the ignorance of whose behavior is best described probabilistically).

If one is dealing with a stochastic simulation model, then its response $Y(T)$ at simular time T may be rightfully viewed as a random variable for each permissible specification of environmental conditions. For example, at the simulated system's standard operating conditions

$$(x_{1,0}, x_{2,0}, \ldots, x_{p,0}),$$

the set of all possible responses (one of which conceptually arises from each different assignment of the seed s) might be formed into a frequency histogram, so that it becomes meaningful to presume the existence of a probability density function $f_Y(y)$ and, of course, a cumulative distribution function $F_Y(y)$ for the simular response. Similarly, probability density and cumulative distribution functions are appropriately conceived for the simular response at any other specification of the environmental conditions.

The attempt to analyze the stochastic simular response becomes then a statistical problem, so that both the design and analysis of the simular experimentation comes to be based on statistical considerations. One might well anticipate difficulties in locating the optimal simular response in the presence of stochasticity, but, as will be demonstrated in Chapter 9, most of these difficulties can be overcome if the search is directed toward the location of environmental conditions under which the optimal *expected* simular response is attained.

A primary assumption in each of the next four chapters, nonetheless, will be that the simular response $Y(T)$ is a function of the environmental conditions producing it. These environmental conditions can be categorized as either qualitative or quantitative.

As an example of the qualitative variety, one might well consider a binary policy, each form of which might constrain differently the activities and entities of the simular system: for example, the absence or presence of a redundant component in the system. Another example of such qualitative conditions would include the choice of one of several sets of exogenous events, as might be more particularly exemplified by reels of recorded weather data, any one of which could be used to impart stochastic weather effects in the simulation model. Qualitative variates might be assigned numerical values in order to distinguish between their alternative states; e.g., $0 =$ winter, $1 =$ spring, $2 =$ summer, and $3 =$ autumn. Such a numerical assignment would appear essentially superficial; yet, the variate would be qualitative.

On the other hand, a quantitative simular environmental condition may be deemed an input parameter whose numerical value is not so strikingly superficial. Indeed, we shall usually mean to imply by the term, *quantitative simular environmental condition*, an input parameter whose values are conceptually extant throughout some continuous range of real numbers. As an example, the flow rate of a simulated liquid pump is suggested.

The next three chapters will be concerned with the design and the analysis of experiments with simulation models whose environmental conditions are either qualitative or quantitative. Since the approach will presume that the simular response $Y(T)$ is stochastic, the reader will find that the techniques recommended are, as suggested by the subtitle, statistical. After the presentation in this chapter of some of the more fundamental applications of the statistical methodology to the analysis of simular responses, a statistical technique appropriate to the determination of the relative importance of alternative input specifications will be presented: the analysis of variance (ANOVA).

Under appropriate conditions, ANOVA is particularly applicable to such a determination, especially so when alternative enviromental conditions are of the qualitative type and are each representable by a finite number of possible states (values). This discussion will constitute the subject matter of Chapter 7.

It will be shown that the technique of ANOVA is also appropriate to the analysis of simular responses whenever the environmental conditions are of the quantitative type; the treatment of this application of ANOVA will also be presented in Chapter 8.

Determination of the effects of altering a continuously variable environmental condition are best stated in terms of the *rate* at which the simular response $Y(T)$ varies with respect to incremental or unit changes in the value of the input variable. In the language of the economist, such a rate might be referred to as a *marginal simular response* with respect to the particular environmental condition. Determination of the relative significance of various environmental conditions can then be meaningfully translated to the isolation of the largest marginal simular response; if, on the other hand, one would like to ascertain the optimal simular response, then these marginal simular responses should provide some insight into the directions in which one might proceed so as to locate the optimal response more readily.

In the next four chapters, then, the emphasis will be placed on the design and the analysis of simular experiments for the purpose of determining the structure of the model's static effects. This and the next two chapters will be devoted to the delineation of a few of the more important statistical techniques, including the analysis of variance and the technique of multiple regression, whereas Chapter 9 will illustrate procedures by which information relating to the marginal simular response may be gainfully employed in locating the optimal simular response. As stated earlier, Chapter 10 will outline the extant statistical methodology appropriate to the analysis of the dynamic effects present in a simulation model.

6.2. Statistical Analyses

As indicated in the preceding section, the response $Y(T)$ of a stochastic simulation model at the end of T simular time units is a random variable that is functionally dependent upon the environmental conditions for which a particular encounter is defined:

$$Y(T) = Y(x_1, x_2, \ldots, x_p; s),$$

where $x_1, x_2, \ldots, x_p$ are the p environmental conditions other than the random number seed s. The functional form for $Y(T)$ is, of course, unknown; however, by experimentation with the model, one may hope to ascertain some of its properties and possibly obtain an approximating polynomial form for the function.

Once a stochastic simulation model has been structured and verified, the model must undergo a validation stage, during which comparisons of the simular responses are made with those of the simulated system. For this purpose, it is likely that observations on the simulated system either have been, or may readily be, acquired: at least for some particular specification of the environmental conditions, those corresponding to, say,

$$\vec{x}_0 \equiv (x_{1,0}, x_{2,0}, \ldots, x_{p,0})$$

as input conditions for the simulation model. (The conditions $\vec{x}_0$ can be regarded as representative of the system's *standard operating conditions.*)

For the present, it will be assumed that both the simular response $Y(T)$ and the corresponding system response are univariate quantities of the same units and the same relative magnitudes; i.e., each does not require a vector of quantities for its description, and the simular response is indeed a quantity directly comparable to the simulated system's response. In a subsequent section the more general case of a vector of simular responses will be discussed.

The question may well be raised as to the distinction made between the random number seed s and the remaining input parameters defining a stochastic simulation model and to which the collective, *environmental conditions*, will be applied. As was stated in Section 6.1, the environmental conditions can be categorized as either qualitative or quantitative, with the latter category conceptually containing input parameters whose permissible values lie in some continuum of real numbers. Therefore, since the random number seed s may be conceptually defined as a real number between 0 and 1, it should constitute a continuous input parameter, another quantitative environmental condition.

However, the simular response function,

$$Y(T) = Y(x_1, x_2, \ldots, x_p; s)$$

will generally be presumed to be a continuous function of each of its quantitative environmental conditions; i.e., if x_k is a continuous input parameter, then the difference between the two simular responses, given by

$$Y(x_1, \ldots, x_{k-1}, x_k + \Delta x_k, x_{k+1}, \ldots, x_p; s) - Y(x_1, \ldots, x_p; s)$$

can be made arbitrarily close to zero by choosing Δx_k successively nearer zero. The random number seed, however, would not be expected to produce such a behavior in the response of a sufficiently intricate simulation model, and is thus separated from the set of input parameters called the environmental conditions.

If the simulation model is iterated N times at the same environmental conditions $\vec{x}_0$ save for independently selected and differing random number seeds $s_1, s_2, \ldots, s_N$, then the jth encounter will produce a simular response $Y_j(T)$. The set of N responses, which one can denote as the vector

$$Y_1, Y_2, \ldots, Y_N \qquad \text{or} \qquad y_1, y_2, \ldots, y_N$$

constitutes a *random sample of size* N from the probability density function $f_Y(y)$ of simular responses at $\vec{x}_0$.

The intrinsic nature of the density function $f_Y(y)$ for simular responses $Y(T)$ arising from environmental specifications $(x_{1,0}, \ldots, x_{p,0})$ may or may not be known to the analyst. As will be shown in Section 6.3, the very nature of the simulation model itself may provide a clue as to the form that might be anticipated for this probability density function.

6.2.1. Estimating the Mean Simular Response

Nonetheless, for the purpose at hand, discussion will center about the moments of the distribution; specifically, the mean simular response at $\vec{x}_0$,

$$E[Y(T)] = \int_{-\infty}^{\infty} y f_Y(y)\, dy \equiv \mu$$

and the variance of the simular responses at $\vec{x}_0$,

$$\mathrm{Var}[Y(T)] = \int_{-\infty}^{\infty} (y - \mu)^2 f_Y(y)\, dy \equiv \sigma^2$$

will be of primary concern. Knowledge of these distributional properties can assist in the specification of a particular density function for the simular responses; indeed if the structure of the simulation model implies that a simular response will probably be normally distributed, knowledge of μ and σ^2, as the reader may recall, is sufficient to specify the probability density function completely.

In order to estimate these unknown distributional properties (μ and σ^2), one may proceed to employ the random sample $(y_1, y_2, \ldots, y_N)$ of simular responses. An estimate of the true mean μ can be obtained by computing the arithmetic average of the independent simular responses; namely,

$$\tilde{\mu} = N^{-1} \sum_{j=1}^{N} y_j \equiv \bar{y}.$$

Of course, this estimate cannot be expected to equal the actual, yet unknown, mean μ, as it would vary depending upon the particular set of N simular responses used in its computation. Consequently, $\tilde{\mu}$ is used to denote that the quantity $\bar{y}$ is an *estimate* of μ.

In fact, the estimator $\tilde{\mu}$ is itself a random variable, as it is a particular transformation of the N random variables $Y_1, Y_2, \ldots, Y_N$. Any transformation of the elements of a random sample of size N (i.e., of N independent observations taken randomly from the same probability density function) is called a *statistic* and becomes then a random variable itself. Indeed, for every possible sample of size N, one may proceed to compute the statistic according to the particular transformation rule at hand, say,

$$\tilde{\mu} = N^{-1} \sum_{j=1}^{N} y_j,$$

so that a frequency histogram of the statistics could then be constructed.

If one has reason to believe that the simular response $Y(T)$ has the normal (Gaussian) distribution, then since the random sample $Y_1, Y_2, \ldots, Y_N$ is composed of N independent and normal random variables, the joint probability density function for the random sample is the multivariate normal distribution having a mean-value vector (with every element being μ therein) and with variance–covariance matrix $\sigma^2 \mathbf{I}_N$, where $\mathbf{I}_N$ is the $N \times N$ identity matrix. Hence, the statistic $\tilde{\mu}$, being a linear combination of the variates of a multivariate normal distribution, will have as its probability density function the (univariate) normal distribution, as the reader may recall from the previous discussion regarding the multivariate normal distribution (cf. Section 2.15).

Furthermore, the reader may recall the earlier discussion of the Central Limit Theorem, as it was employed to justify the generation of normally distributed random variables by forming the sum of several independently distributed random variables. The quantity,

$$N \cdot \tilde{\mu} = (Y_1 + Y_2 + \cdots + Y_N),$$

is also such a sum and may therefore be expected to have a probability density function which may be approximated by the normal distribution, regardless of the distributional form of the Y_j. Moreover, since any multiple of a normally distributed random variable is also normally distributed, the statistic

$$\tilde{\mu} = N^{-1}(N \cdot \tilde{\mu}) = N^{-1} \sum_{j=1}^{N} y_j$$

can be expected to have an approximately normal distribution, provided that the Y_j arise as independent observations from a common probability density function, $f_Y(y)$.

This approximate (or exact) normal distribution for $\tilde{\mu}$ is described of course by its own mean and variance. The mean of the distribution of $\tilde{\mu}$ is μ itself; i.e., it is the same as the mean of the distribution from which the simular responses arise, whether that underlying distribution be itself Gaussian in form or not. This result may be readily verified by considering the expectation of the statistic $\tilde{\mu}$:

$$E(\tilde{\mu}) = E\left(N^{-1} \cdot \sum_{j=1}^{N} Y_j\right) = N^{-1} \cdot \sum_{j=1}^{N} E(Y_j),$$

since the expectation operator is linear; hence,

$$E(\tilde{\mu}) = N^{-1} \sum_{j=1}^{N} (\mu) = N^{-1}(N\mu) = \mu.$$

The property just exhibited is a realization of a more general phenomenon, worthy of a short aside. Whenever a statistic has expectation equal to a particular parameter or distributional property, then the statistic is said to be an *unbiased estimate* of that parameter (or property). As will be noted frequently in subsequent sections, unbiased estimators will be particularly useful, since an unbiased statistic for a parameter or distributional property is one which, if employed many times successively, would have an average value equal to the parameter or property of interest.

The variance of the approximate (or exact) normal distribution for $\tilde{\mu}$ becomes then

$$\text{Var}(\tilde{\mu}) = E\{[\tilde{\mu} - E(\tilde{\mu})]^2\} = E[(\tilde{\mu} - \mu)^2],$$

since, as we have seen, $E(\tilde{\mu}) = \mu$. Alternatively,

$$\text{Var}(\tilde{\mu}) = \text{Var}\left(N^{-1} \cdot \sum_{j=1}^{N} Y_j\right) = N^{-2}\, \text{Var}\left(\sum_{j=1}^{N} Y_j\right),$$

so that,

$$\text{Var}(\tilde{\mu}) = N^{-2}\Big[\sum_{j=1}^{N} \text{Var}(Y_j)\Big] + 2N^{-2}\Big[\sum_{j=1}^{N} \sum_{k>j} \text{Cov}(Y_j, Y_k)\Big],$$

or

$$\text{Var}(\tilde{\mu}) = N^{-2} \sum_{j=1}^{N} (\sigma^2) = \sigma^2/N,$$

since the Y_j are presumed to be independently distributed and, therefore, uncorrelated random variables.

The approximating normal distribution for $\tilde{\mu}$, the arithmetic mean of N simular responses arising from independently seeded encounters which are defined at the otherwise identical environmental conditions $(\vec{x}_0)$, has mean μ and variance (σ^2/N). Because of the inverse relationship between the variance of $\tilde{\mu}$ and the sample size N, the arithmetic average of a sample is deemed to be an increasingly better estimate of the underlying and unknown mean μ. Statisticians refer to an estimate as *consistent* whenever the probability of its deviating from the unknown parameter value by more than any arbitrary quantity tends to zero as the sample size N increases. Whenever an estimator is unbiased for a parameter and has variance that diminishes with increasing sample size, it is necessarily a consistent estimator. The importance of consistency is worthy of note, for the consistent estimator tends to "improve" as an increasingly large sample is recorded. As will be demonstrated in Chapter 10, certain spectral estimators are not consistent, and since unbiasedness and consistency are valuable properties of statistical estimates, alternative estimators will need be derived in those instances.

6.2.2. Confidence Intervals for the Mean of the Simular Responses

For the present, however, one may note that the estimator $\tilde{\mu}$ is approximately normally distributed, so that the random variable

$$Z \equiv (\tilde{\mu} - \mu)/(\sigma/\sqrt{N})$$

is the standardized normal variate; i.e., Z has the normal distribution of mean zero and variance one. Hence, it becomes possible to locate points $z_{\alpha/2}$ and $-z_{\alpha/2}$, such that, for any α between 0 and 1,

$$P[-z_{\alpha/2} < Z < z_{\alpha/2}] = (1 - \alpha)$$

from tabulations of the normal distribution. For example, with $\alpha = 0.10$,

$z_{\alpha/2}$ is that value such that 95% [$=(1-\alpha/2)\times 100\%$] of the standardized normal distribution appears to its left; namely, $z_{0.05}\cong 1.645$.

A compatible probability statement can then be made for the random variable $\tilde{\mu}$. Since $\tilde{\mu}=\mu+(\sigma\cdot Z/\sqrt{N})$, one may write

$$P[\tilde{\mu}-(\sigma\cdot z_{\alpha/2}/\sqrt{N})<\mu<\tilde{\mu}+(\sigma\cdot z_{\alpha/2}/\sqrt{N})]=(1-\alpha),$$

a statement that should properly be read:

Of all the intervals bounded by the quantities, $\tilde{\mu}\pm(\sigma\cdot z_{\alpha/2}/\sqrt{N})$, and as computed from the statistic $\tilde{\mu}$, $(1-\alpha)\cdot 100\%$ of these will include the unknown mean μ. The interval of values, bounded on the left by $\tilde{\mu}-(\sigma\cdot z_{\alpha/2}/\sqrt{N})$ and on the right by $\tilde{\mu}+(\sigma\cdot z_{\alpha/2}/\sqrt{N})$, is termed a $100\cdot(1-\alpha)\%$ confidence interval for the parameter (distributional property) μ. In a certain sense then, one would have approximately 90% confidence of enclosing the mean of the simular responses $Y(T)$ by computing the interval bounded by the points

$$\tilde{\mu}\pm(1.645\cdot\sigma/\sqrt{N}).$$

6.2.3. Estimate of the Variance of the Simular Response

Such a confidence interval cannot be computed in the absence of the value of σ, the standard deviation of the simular responses $Y(T)$ arising at the environmental conditions specified by $\vec{x}_0$. Since it is quite unlikely that the analyst will know beforehand the standard deviation of the simular responses, it will be necessary to define an estimator for this distributional property as well.

An unbiased estimate for the variance σ^2 of *any* probability density function $f_Y(y)$ can be computed by transforming a random sample, $(y_1, y_2, \ldots, y_N)$ from the density, to obtain the statistic

$$S^2=(N-1)^{-1}\sum_{j=1}^{N}(y_j-\bar{y})^2,$$

where $\bar{y}=N^{-1}\sum_{j=1}^{N}y_j$. Again, as for any statistic, the quantity S^2 is a random variable possessing its own probability density and cumulative distribution functions and therefore having its own distributional properties and moments. Regardless of the distributional form for the simular responses, $Y(T)$, and regardless of the distributional form which would result for the histogram of S^2 statistics, the mean value of S^2 is given by

$$E(S^2)=(N-1)^{-1}E\left[\sum_{j=1}^{N}(Y_j-\bar{Y})^2\right]=(N-1)^{-1}E\left[\left(\sum_{j=1}^{N}Y_j^2\right)-N(\bar{Y})^2\right],$$

since the quantity representing the sum of the squared deviations of a set of real numbers from their arithmetic average may be written in the (computationally superior) form on the right-hand side of

$$\sum_{j=1}^{N} (Y_j - \bar{Y})^2 = \sum_{j=1}^{N} (Y_j^2 - 2\bar{Y}Y_j + (\bar{Y})^2) = \sum_{j=1}^{N} (Y_j^2) - N(\bar{Y})^2.$$

Hence, due to the linearity property of the expectation operator,

$$E(S^2) = (N-1)^{-1}[N \cdot (\sigma^2 + \mu^2) - N\{(\sigma^2/N) + \mu^2\}]$$

or

$$E(S^2) = (N-1)^{-1}[(N-1)\sigma^2] = \sigma^2.$$

Thus, an unbiased estimate for the variance σ^2 of the distribution for the simular response $Y(T)$ can be computed directly from a random sample of responses $(Y_1, Y_2, \ldots, Y_N)$, the jth of which arises from a simulation encounter defined by the same input conditions, excepting the independent selection of its random number seed $s_j, j = 1, 2, \ldots, N$.

The distributional form of the statistic S^2 is not, in general, known. Yet if the simular responses from which it is computed may be deemed to be a random sample from a normal distribution, then $(N-1)S^2/\sigma^2$ has the Chi-squared distribution of $(N-1)$ degrees of freedom. Furthermore, under this same assumption, the random variables S^2 and $\tilde{\mu}$ are independent; that is their joint probability density function is everywhere equal to the product of their (marginal) probability density functions (cf. Section 2.15.5).

The proof of both of these two results can be found in Kendall and Stuart (1963, Vol. I); their importance may be illustrated by a return to the discussion of a confidence interval for the unknown mean μ whenever the variance σ^2 is likewise unknown.

6.2.4. Application of Student's *t* Statistic

The reader may recall that the random variable,

$$T = Z/(\chi^2/k)^{1/2},$$

where Z is a standardized normal variate distributed independently of χ^2, a Chi-squared variate of k degrees of freedom, has the Student's t distribution of k degrees of freedom. Therefore, whenever a simular response $Y(T)$ may be presumed to be normally distributed with an unknown mean μ and an unknown variance σ^2, and whenever these parameters are estimated, respectively, by $\tilde{\mu}$ and S^2, as defined earlier, then the random variable

$$T = \frac{(\tilde{\mu} - \mu)/(\sigma/\sqrt{N})}{\{(N-1)S^2/[\sigma^2(N-1)]\}^{1/2}} = \frac{\tilde{\mu} - \mu}{(S/\sqrt{N})},$$

where S is the positive square root of the statistic S^2, has the Student's t distribution with $(N - 1)$ degrees of freedom (cf. Section 2.13.6).

Consequently, probability statements of the form

$$P[-t_{k,\alpha/2} < T < t_{k,\alpha/2}] = (1 - \alpha),$$

where α is between 0 and 1 and where $t_{k,\beta}$ is that value to the left of which $100(1 - \beta)\%$ of the Student's t distribution of k degrees of freedom lies, may be made. In particular, upon computing $\tilde{\mu}$ and S^2 from N normally distributed simular responses, one may make the statement

$$(1 - \alpha) = P[\tilde{\mu} - (S \cdot t_{N-1,\alpha/2}/\sqrt{N} < \mu < \tilde{\mu} + (S \cdot t_{N-1,\alpha/2}/\sqrt{N})],$$

which is most properly read as follows:

Of all the intervals whose endpoints are computed as

$$\tilde{\mu} \pm (S \cdot t_{N-1,\alpha/2}/\sqrt{N}),$$

approximately $100 \cdot (1 - \alpha)\%$ of these will include the unknown parameter μ of the normal distribution from which a random sample of size N was selected and from which the statistics $\tilde{\mu}$ and S^2 were computed.

In consonance with the preceding discussion of such a probability statement, made when the underlying variance σ^2 of the simular responses was presumed known, the resultant interval between $\tilde{\mu} - (S \cdot t_{N-1,\alpha/2}/\sqrt{N})$ and $\tilde{\mu} + (S \cdot t_{N-1,\alpha/2}/\sqrt{N})$ is also termed a $100 \cdot (1 - \alpha)\%$ *confidence interval* for μ. The present interval is appropriate, however, when the underlying variance σ^2 of simular responses is, as shall likely be the case, unknown.

6.2.5. Confidence Interval for the Variance of the Simular Response

The reader might properly inquire of the availability of a method for defining a confidence interval for σ^2; indeed, since $(N - 1)S^2/\sigma^2$ frequently has the Chi-squared distribution of $(N - 1)$ degrees of freedom, an appropriate confidence interval is defined by the probability statement:

$$(1 - \alpha) = P[(N - 1)S^2/\chi^2_{N-1,(\alpha/2)} < \sigma^2 < (N - 1)S^2/\chi^2_{N-1,1-(\alpha/2)}],$$

where $\chi^2_{N-1,\beta}$ is a real number such that $100(1 - \beta)\%$ of the χ^2 variates of $(N - 1)$ degrees of freedom are to its left.

This confidence interval will be *exact* (that is, the probability statement will be correct) if the simular responses $Y_1, Y_2, \ldots, Y_N$ are normally and independently distributed, as would be the case for a set of N independently seeded encounters with a simulation model whose responses could be expected to have the Gaussian distributional form. For other distributional forms, the confidence interval is merely approximate.

6.2.6. Summary

Regardless of the distributional form appropriate to the simular response $Y(T)$, the first two moments of this random variable can be estimated unbiasedly by

$$\tilde{\mu} = \bar{y} = N^{-1}\left[\sum_{j=1}^{N} y_j\right]$$

and

$$\tilde{\sigma}^2 = S^2 = (N - 1)^{-1} \cdot \left[\left(\sum_{j=1}^{N} y_j^2\right) - N(\bar{y})^2\right],$$

whenever the responses $(y_1, y_2, \ldots, y_N)$ constitute a random sample from the probability density function $f_Y(y)$ of simular responses arising from the same environmental specification $\vec{x}_0 = (x_{1,0}, x_{2,0}, \ldots, x_{p,0})$. Whenever the underlying distributional form of the responses is the normal distribution, the statistics $\tilde{\mu}$ and $\tilde{\sigma}^2$ have known probability density functions; namely, $\tilde{\mu}$ is itself normally distributed with its own mean of μ and with variance given by σ^2/N, whereas the quantity $(N - 1) \cdot \tilde{\sigma}^2/\sigma^2$ has the Chi-squared distribution of $(N - 1)$ degrees of freedom. Furthermore, $\tilde{\mu}$ and $\tilde{\sigma}^2$ are, under the normality condition, statistically independent random variables. Even though the simular responses may have a probability density function not of the Gaussian form, then by means of the Central Limit Theorem, the estimator $\tilde{\mu}$ has an approximate normal distribution of mean μ and variance σ^2/N.

These facts have been used to determine confidence intervals for the unknown mean μ of simular responses deriving from the same (usually the standard) environmental conditions, $\vec{x}_0$. The interval is constructed with the aid of tabulations of the cumulative normal distribution if the underlying σ^2 should be known, yet with the aid of tabulations of the cumulative Student's t distribution of $(N - 1)$ degrees of freedom if, as is more likely, the underlying variance σ^2 of simular responses need be estimated by $S^2 = \tilde{\sigma}^2$.

Fundamental to these discussions has been the assumption that the N simular responses can be regarded as statistically independent. In order to ensure that this assumption is as valid as possible, the random number seed that initiates any one of the N encounters should be selected independently of the seeds selected for all previous encounters. It is for this reason that this author recommends the use of a standard set of random digits (e.g., those of the RAND Corporation, 1955). A starting location, or initial seed s_1 should be selected as randomly as possible from such a table, and the subsequent chain of digits, as they appear sequentially in the table, employed as the seeds $s_2, s_3, \ldots, s_N$ for the second, third, $\ldots$, Nth encounters, as required by the simular experimentation.

However, since two simular encounters, each specified by the same environmental conditions $\vec{x}_0$ and the same random number seed s will provide identical simular responses whenever the model is iterated on the same class of computing machines, one must maintain a record of the seeds previously employed in the experimentation and must forbid duplication of these seeds. As a general rule, it is wise to maintain such a list and to forbid repetition of any seed, even if other environmental conditions are to be altered between successive model encounters, a tactic which has been delineated as a Principium of Seeding in Chapter 5.

Before proceeding to a discussion of the distributional forms likely to arise, in general, for simular responses, one might note that, for the present, the discussion has presumed that a single quantity or state variable $Y(T)$ will serve satisfactorily to summarize the state of the simulation at simular time T. Of course, in many modeling contexts, a single response variable is not adequate, so that pairs or n-tuples of random variables may be required to define the simular response. This generalization of the simular response as a bivariate or multivariate random vector will be the subject of a subsequent section on the analysis of simulation experiments. First, however, the discussion will move to the limiting forms attainable in simular responses.

EXERCISES

1. Using the mixed congruential method in order to generate uniformly distributed random variables, prepare a random sample of size

10 from a normal distribution of mean 4 and variance 25 by generating 120 uniformly distributed variates, then by summing successive and non-overlapping dozens, adjusting to obtain the proper mean and variance for each of the 10 normal variates. Compute for the sample of size 10 both $\bar{Y}$ and S^2, as discussed in Section 6.2. By continuing to generate these variates, prepare a total of 100 such samples, each of size 10, and for each compute its arithmetic average $\bar{Y}$ and its estimate S^2 of σ^2 (=25). Form a frequency histogram for the 100 $\bar{Y}$'s, and for the 100 S^2's similarly; in each case, perform a Chi-squared test in order to compare the histogram with the corresponding normal and Chi-squared distributions; namely, the normal distribution of mean 4 and of variance 2.5 (=25/10) and the Chi-squared distribution of 9 (=10 − 1) degrees of freedom. Test also by the Kolmogorov–Smirnov criterion, using tables of the Appendix.

2. Repeat Exercise 1, excepting the generation of the normally distributed random variables, which should instead be accomplished by the log-and-trig method as described in Section 2.14.5. Based on the Chi-squared or Kolmogorov–Smirnov tests which you perform, comment on the preference of the two methods for the generation of normally distributed random variables, in view of the fact that one of the two methods is approximate only.

3. For each of the 100 random samples of Exercise 1 (or 2) compute the 95% confidence interval for the mean under the presumption that σ^2 (=25) is known. What proportion of these intervals *do* contain the true mean, μ (=4)?

4. For each of the 100 random samples of Exercise 2 (or 1) compute the 95% confidence interval for the mean, now presuming that the statistic S^2, as computed for the sample, must be used to estimate σ^2. What proportion of these intervals do contain the true mean, μ (=4)? (Is this proportion unusually high or low?) Your answer should be based on 100 Bernoulli trials with $p = 0.05$, or approximately therefore on the Poisson distribution with parameter $\lambda = 100 \cdot p = 5$.

5. For each of the 100 random samples of Exercise 2 (or 1), compute the Student's t statistic,

$$T = (\bar{y} - 4)/(S/\sqrt{10}),$$

and form a frequency histogram of the 100 resulting variates. Compare this frequency histogram with the Student's t distribution of 9 degrees of freedom by means of a Chi-squared test, then via the Kolmogorov–Smirnov statistic.

6. Independently of the random variates generated in Exercises 1 and 2, compute 100 values of a Chi-squared variate with 9 degrees of freedom by summing directly the squares of nine standardized normal random variates. Placing these 100 variates in one-to-one correspondence with the 100 means of normal random variates, as computed in Exercise 1 (or 2), compute the 100 random variables:

$$\tau = (\bar{y} - 4)/(25\chi^2/90)^{1/2}.$$

Prove that τ has the same distribution as T of the preceding exercise. Do the sample cumulative distributions of the 100 τ-values and of the 100 T-values differ significantly? [Refer to the Kolmogorov–Smirnov two-sample test, as described in Lindgren (1968).]

6.3. Distributional Forms Attainable by Simular Responses

As indicated in the preceding two sections, the stochastic simulation model yields as its output (*response*) one or more random variables, so that a set of successive iterations of (or *encounters* with) the model provides a random sample whose analysis is best performed by extant statistical methodology. Although many fundamental properties of random variables and their probability density functions can be adequately estimated by means of sample moments, many essential aspects of these distributions require specialized statistical procedures.

Of primary importance in the analysis of simular responses is knowledge of the nature of the distributional form of the random variables emitted by the stochastic simulation model. In this section, two of the fundamental limiting laws of probability theory are presented; the applicability of each is illustrated by an exemplary simulation. Of considerable interest is the note that increasing model complexity may not serve to promote the applicability of the Central Limit Theorem, and the corresponding normal (Gaussian) probability density function, but rather may serve to guarantee the appropriateness of the asymmetric Weibull distribution. A subsequent section will indicate some of the differences in statistical techniques required in the analyses of simular responses having these differing distributional forms.

If we denote by Y the simular response of a stochastic simulation model, we see that this response is a function of the environmental conditions as stipulated by a set of values for the input data:

$$Y = Y(x_1, x_2, \ldots, x_p; s),$$

where the x_i are input variables other than the seed, s. For a given vector of environmental conditions $(x_1, x_2, \ldots, x_p)$, the set of all responses Y constitutes then a random variable, there being conceivably as many responses Y as there are seeds s. For this random variable, then, there exist: (a) the standard probability density function (p.d.f.) $f_Y(y)$ defined for all real y such that the probability of a simular response falling between y and $(y + dy)$ is, for dy sufficiently small, $f_Y(y)\,dy$, and (b) the usual cumulative distribution function (c.d.f.) $F_Y(y) = \int_{-\infty}^{y} f_Y(t)\,dt$, defined such that $F_Y(-\infty) = 0$ and $F_Y(+\infty) = 1$. The probability that a simular response is less than or equal to any real number (say, a) is given by $F_Y(a)$.

Moreover, if n encounters of a simulation model are defined by maintaining the same environmental conditions, except for an independently selected random number seed (repetitions forbidden) for each encounter, the resulting n simular responses $(y_1, y_2, \ldots, y_n)$, constitute a statistical *random sample* from the p.d.f. $f_Y(y)$. Of course, the analysis of the random sample depends on the nature of the probability density function $f_Y(y)$ as will be shown in Section 6.4.

6.3.1. Limiting Forms for the Distribution of Simular Responses

The response Y of a dynamic, symbolic, and stochastic simulation model is then a random variable whose p.d.f. is dependent upon the nature and structure of the simulation itself. Conceptually, one may view any particular response Y as the result of the intrinsic interactions which have been incorporated as the activities and events within the model structure. For a model sufficiently complicated, the simular response effectively represents the limiting effect of a large number of these inherent interactions, so that the random variable Y often has a p.d.f. which is characterized by one of the "limit laws" of probability theory.

The reader is referred to the first and second volumes of Feller (1965) for a discussion of the Central Limit Theorem of probability theory. Historically, a series of central limit theorems has developed, but the essence of each is that a random variable Y which is the limit of the *sum* of a large number of relatively unimportant random effects has p.d.f.

given by Gauss's distribution:

$$\phi(y; \mu, \sigma^2) = (2\pi)^{-1/2}\sigma^{-1/2} \exp -\{(y-\mu)^2/2\sigma^2\}, \qquad (6.3.1:1)$$

where the parameters μ and σ are real and positive, respectively, and represent the mean and the standard deviation (which is the positive square root of the variance) of Y, respectively:

$$E(Y) \equiv \int_{-\infty}^{\infty} y\phi(y; \mu, \sigma^2)\, dy = \mu,$$

and

$$\mathrm{Var}(Y) \equiv \int_{-\infty}^{\infty} (y-\mu)^2\phi(y; \mu, \sigma^2)\, dy = \sigma^2.$$

More specifically, the usual form of the Central Limit Theorem states that if a random variable Y is the sum of a large number of independent random variables from a common p.d.f., then the p.d.f. for Y may be closely approximated by Eq. (6.3.1:1). The condition of independence of the many contributing random variables is not necessary, as the interested reader will see from the article by Diananda (1953). Neither is the condition necessary that the many contributing random variates arise from a common p.d.f., although in its absence extreme care must be taken with respect to the relative magnitudes of the variances of the contributing variates. [See the cautionary note of the author (Mihram, 1969a) for an example in which the limiting distribution of an infinite number of independently and Gamma distributed random variables is not at all that of the Gaussian form.]

Nonetheless, for a sufficiently complex stochastic simulation model, one might well anticipate that the simular response Y will be a Gaussian random variable. Nonetheless, one should note that the requisite complexity should be of a nature which implies that the simular response is the *additive sum* of a large number of contributing effects arising due to the model's structure.

In 1928, Fisher and Tippett published a well known (but apparently little used) result regarding the distributional forms applicable to random variables that are limits of the maximum or minimum value in a random sample. The interested reader is referred to the excellent compendium published subsequently by Gumbel (1958) on this subject. Of the three limiting forms, two are quite Gaussian in appearance, though somewhat skewed; the third form, however, known more widely as the Weibull distribution (Weibull, 1951), is quite asymmetrical in many of its forms.

A Weibull random variable W has the probability density function

$$g_W(\omega; a, b) = \begin{cases} b\omega^{b-1} \exp[-(\omega/a)^b]/a^b, & \omega > 0, \\ 0, & \omega < 0, \end{cases} \quad (6.3.1:2)$$

where the positive parameters a and b are referred to as the scale and shape parameters for the density. The parameters are not directly related to the mean and variance of the random variable; rather, they are related through the relationships:

$$E(W) = a\Gamma(1 + b^{-1})$$

and

$$\mathrm{Var}(W) = a^2\{\Gamma(1 + 2b^{-1}) - \Gamma^2(1 + b^{-1})\},$$

where

$$\Gamma(v) = \int_0^\infty x^{v-1}e^{-x}\,dx$$

is the standard Euler Gamma function for positive argument v.

The Weibull random variable can be characterized, as was shown by Fisher and Tippett (1928), as the limit of the minimum value among a random sample of ever-increasing size drawn from a p.d.f. of positive-valued random variates. Thus, the minimum value among n independent, positive-valued, random variables from the same p.d.f. has a density which can be approximated by the Weibull distribution.

In particular, if the lifetime T of a Christmas-tree light has the exponential distribution of mean a:

$$h_T(t; a) = \begin{cases} a^{-1}e^{-t/a}, & t > 0, \\ 0, & t < 0, \end{cases} \quad (6.3.1:3)$$

then the time W between successive replacements of lights in a series system of n such lights is the random variable

$$W = \min(T_1, T_2, \ldots, T_n).$$

In this instance, the random variable W can be shown to possess the exponential distribution, but with mean a/n; that is $h_T(w; a/n)$. The reader may verify that $h_T(w; a) = g_W(w; a, 1)$ by comparison of Eqs. (6.3.1:2) and (6.3.1:3), so that the distributional form of W, the system lifetime, becomes indeed the Weibull distribution $g_W(w; a/n, 1)$.

More generally, if a series system is composed of n independent components, the ith of which has lifetime represented by a random variable

T_i having the p.d.f. $h_T(t_i; a_i)$, $i = 1, 2, \ldots, n$, then

$$W = \min(T_1, T_2, \ldots, T_n)$$

is a random variable having the p.d.f. $g_W(w; \alpha/n, 1)$, where α is the harmonic mean of the a_i; namely,

$$\alpha = n[a_1^{-1} + a_2^{-1} + \cdots + a_n^{-1}]^{-1}.$$

Little is known about the applicability of the Weibull distribution, however, whenever the elementary (contributing) random variables are not independent; nor is much known about the utility of the Weibull distribution whenever the elementary random variables are independent but do not arise from a common probability density function. There is indeed much speculation, somewhat justified [see, e.g., Gumbel (1958)], that the distributional form (6.3.1:2) obtains for the minimum value among a large set of positive-valued random variables.

Therefore, if a simular response Y is, for all practical purposes, the smallest value among a set of random quantities, or is the result of a minimization procedure or selection (such as may arise in PERT studies), one may well anticipate that the response will be a Weibull random variate.

6.3.2. Examples of Simular Responses

In the remainder of this section, two stochastic simulation models will be discussed, each related to the reliability and maintainability of computer systems. Each model will require the availability of positive-valued random variables from the following six probability density functions:

$$\begin{aligned}
f_1(t) &= 2^{-1}e^{-t/2};\\
f_2(t) &= (4\pi)^{-1/2}e^{-(t-9)^2/4};\\
f_3(t) &= t^{5-1}e^{-t/2}/(2^5.4!);\\
f_4(t) &= 6^9t^{9-1}e^{-6t}/8!;\\
f_5(t) &= (2\pi)^{-1/2}e^{-(t-6)^2/2};\\
f_6(t) &= t^{10-1}e^{-t}/9!.
\end{aligned} \qquad (6.3.2{:}1)$$

Each of these densities is defined only for positive arguments ($t > 0$) with units in hours, though the densities $f_2(t)$ and $f_6(t)$ are Gaussian density functions, which theoretically also permit nonpositive arguments.

However, each of these two densities contain an insignificant proportion of nonpositive values (<0.0001); in any event, the two simulation models, to be presently described, excluded any nonpositive sampled values, thereby truncating these normal distributions at the origin.

One should note in passing the remaining four probability densities, each of which is one from the Gamma family of densities. The density $f_1(t)$ is of the exponential subfamily [cf. Eq. (6.3.1:3)], whereas the density $f_3(t)$ is of the Chi-squared subfamily. The means and the variances of the six random variables are provided in Table 6.1.

TABLE 6.1

MOMENTS OF SIMULAR RANDOM VARIATES

i	Family	Mean (T_i)	Variance (T_i)
1	Exponential	2	$2^2 = 4$
2	Normal	9	2
3	Chi-squared	$(2)(5) = 10$	$(2^2 \times 5) = 20$
4	Gamma	$(6^{-1} \times 9) = 1.5$	$(6^{-2} \times 9) = 0.25$
5	Normal	6	1
6	Gamma	10	10

The discussion will now turn to the description of two simulation models that will use these distribution functions.

6.3.3. THE SIMULATION OF THE CUSTOMER ENGINEERS' WORKWEEK

A team of customer engineers rotates each week through six computer facilities. Past experience reveals that the time T_i required to complete the task at the ith facility is a random variable having the p.d.f. $f_i(t)$, $i = 1, 2, 3, 4, 5$, and 6. The mean time spent at each facility may be found from the entries in Table 6.1. One may note that the expected time at the facilities ranges from 1.5 to 10 hr, though there are considerable variance differences; namely, though the expected time spent by the maintenance team is the same (10 hr) at installations 3 and 6, there is considerably more variability from week to week in the duration of the work requirement at installation 3.

Of special interest is the actual workweek of the crew $H = \sum_{i=1}^{6} T_i$. One may note that the random variable H has mean equal to the sum of the expected task durations; that is, $E(H) = 38.5$ hr. Consequently, the team might reasonably be expected to complete their rotation through the six facilities within a standard 40-hr workweek. If we presume, however, that there is no correlation between the successive task requirements at the facilities, then the variance of the team's workweek becomes:

$$\mathrm{Var}(H) = \sum_{i=1}^{6} \mathrm{Var}(T_i) = 37.25.$$

In other words, the standard deviation of the work requirement each week is $(37.5)^{1/2} \cong 6.1$ hr, so that a number of weeks may require overtime work and the necessary premium rates in wages.

The team's workweek was simulated for 2 yr (actually, 100 independent weeks) by performing 100 Monte Carlo determinations of $H = \sum_{i=1}^{6} T_i$. The resulting 100 H-values were placed in a histogram, as depicted in Fig. 6.1. The arithmetic mean of the 100 H-values was 39.0 hr, reasonably close to the theoretical expectation of 38.5 hr. In fact, the histogram of Fig. 6.1 can be seen to be a reasonable approximation to a Gaussian distribution. Since the random variable H is the direct sum of six independent random variables T_i, $i = 1, 2, 3, 4, 5$, and 6, the approach to normality for the distribution of H might well have been anticipated, especially upon reflection of the preceding discussion of the Central Limit Theorem (cf. Section 2.12).

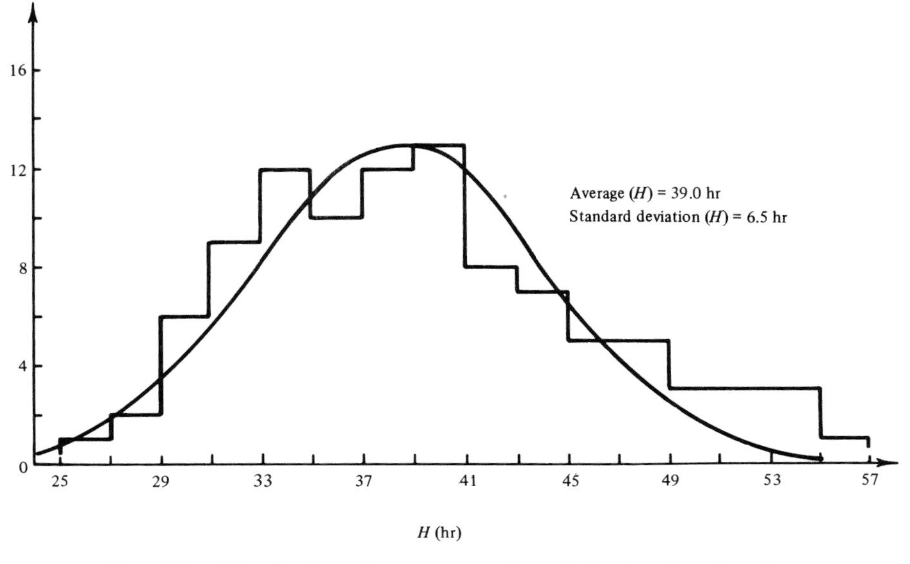

FIG. 6.1. Distribution of customer engineers' simulated workweek.

6.3.4. The Simulation of the Computer Facility Interservice Time

A computer system comprises 10 units, five of which are tape units of the same manufacturing series, each tape unit having a random operating time T between successive maintenance requirements (failures included). The distribution of each of these five random variables is presumed to be $f_1(t)$, as given in Table 6.1. The remaining five units in the computer system have random interrepair times T_i, with p.d.f. $f_i(t)$, $i = 2, 3, 4, 5$, and 6, respectively.

Whenever any one of the 10 units fails, repair service is presumed to begin immediately and to require exactly 1 hr. The system's user requires that at least two of the five tape units, together with all of the five non-tape units, be operational in order that his *system* be operational; in any other event, the system is considered "down" (or "to have failed") until the necessary repairs are completed. Therefore, once the system has, by definition, failed, each operating unit in the system is left idle until the system itself is returned to the operational state, at which time each such unit continues to operate for the time remaining in its previously determined interrepair time (T_i).

The simulation of this computer system was conducted over a 2-month period (actually, 60 consecutive 24-hr days). For simplicity, no queues were permitted to form for units which might require repair simultaneously; alternatively, an adequate number of service personnel was deemed present at all times to handle even the most unlikely crises. At the service completion time for any failed unit, a new interrepair time was generated for the unit from the appropriate p.d.f., $f_i(t)$.

If, as a result of a particular repair completion, the *system* itself became again operational, then all operational units were examined for their residual interrepair times and the *smallest* of these then used to schedule the next unit failure. At that simular time, it is determined whether the *system* itself has failed, and, if so, the time duration (since the most recent time at which the system became operational) is computed as a system interservice time (S).

Typical results of the simulation are depicted by the histogram of S-values, Fig. 6.2. One may note immediately the asymmetry of the sampling distribution of the computer system's interservice times (S). Indeed, the shape of the histogram is remarkably similar to that of a Weibull distribution possessing a shape parameter $b < 1$. When one recalls that each interservice time is a random variable that is essentially the minimum value among a set of random variables, the observed histogram appears in

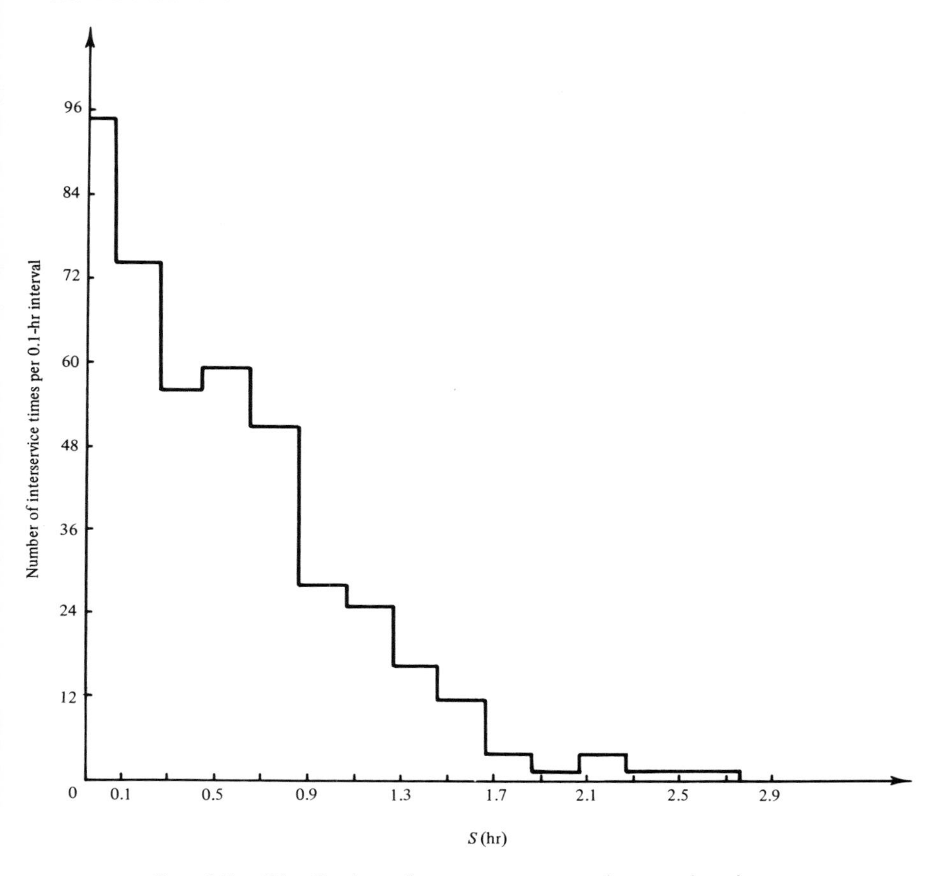

FIG. 6.2. Distribution of computer system interservice times.

consonance with the Fisher–Tippett limiting result for minima, as discussed earlier. Of note in passing is the arithmetic mean of the 736 system interservice times, $\bar{S} \cong 0.80$ hr.

6.3.5. SUMMARY

In this section two exemplary simulation models have been presented, whose responses, or outputs, are random variables. Of primary importance to the user of such models is knowledge of the distributional properties of the random variables so arising.

The Central Limit Theorem was reviewed, and its applicability to the prediction of the distribution of simular responses was noted whenever the simulation produced a response that was essentially the result of a *summation* of contributory effects. This result was contrasted with the limiting distributional form predicted by Fisher and Tippett (1928); namely, that a simular response, which is essentially the *minimum* of a sufficiently large number of positive-valued random variables, has the asymmetrical Weibull distribution.

Examples of simulation models that would lead to each of the limiting distributional forms have been presented. The first, a Monte Carlo analysis of a computer repair team's workweek, produces a normally distributed simular response, in accordance with the Central Limit Theorem, since the simular response is the direct sum of six independent random variables. The second, a stochastic simulation designed to determine a particular computer system's interservice time-distribution, produces a simular response with density function in accord with the Fisher–Tippett theorem, since the system's interservice time is essentially the minimum of its components' interrepair times.

In conclusion, one should note that model complexity is not sufficient reason to anticipate that a simular response be normally distributed; for, of the two exemplary simulations, the second is of far greater complexity. (For example, the first simulation requires only an implicit clockworks, the second demands an explicit timing mechanism for its successful simulation. In addition, entities and attributes need to be defined for the second simulation, but are not strictly necessary in the first instance.)

Hence, a more profound analysis of the intrinsic nature of a stochastic simulation model must be undertaken before predictions of the distributional properties of its simular response can be formulated. The author would generally speculate that simulation models of productivity systems, in which goods are produced over successive time intervals and are accumulated through simular time, shall produce responses which are normally distributed random variables [see, e.g., Mihram (1970)]. On the other hand, stochastic PERT models, which determine the critical path and the minimum project completion time, might be found to generate simular responses whose probability density functions are in consonance with the Fisher–Tippett limiting forms. Other distribution forms may well arise from considerations of other limiting laws of probability theory.

EXERCISES

1. Generate 100 independent, random samples of size 10 from the exponential distribution of mean 5. For each sample, determine the minimum value among the ten variates. Form a histogram of the 100 minima and compare this empirical density with the theoretical Weibull distribution of parameters $a = 0.5$, $b = 1.0$, via the Kolmogorov–Smirnov statistic.

2. Generate 100 independent random samples of size 10 from the normal distribution of mean 5 and variance 25. For each sample, determine the maximum value among the 10 normal variates. Compare the histogram of the 100 maxima with a normal distribution having the same mean and variance as the empirical histogram. Should the two necessarily be compatible?

3. Generate 100 independent and random samples of size 10 from the Weibull density

$$g_W(\omega; a, b) = b\omega^{b-1}e^{-(\omega/a)^b}/a^b \qquad \text{for} \quad \omega > 0,$$

and for parameters $a = 3.54$, $b = 0.81$. Compute $\tilde{\mu}$ and S^2 from each sample as estimates of the true mean ($\mu \cong 4$) and variance ($\sigma^2 \cong 25$), respectively. Using the T-statistic of Section 6.2.4 (as though the data were a random sample from a normal distribution), compute for each sample the 90% confidence interval for μ, as described in that section. What proportion of these 100 confidence intervals contain the actual mean ($\mu \cong 4$)? Is this proportion unusually high or low, based on the anticipated results of 100 Bernoulli trials with $p = 0.10$?

4. For each of the samples of Exercise 3, compute the 90% confidence interval for the variance σ^2 as if the data had arisen from a normal distribution of mean 4, and variance 25. What proportion contain σ^2? (See Section 6.2.5.)

6.4. Estimation of Parameters in Simular Response Distributions

The output, or simular response $Y(T)$ of a stochastic simulation model is a random variable whose probability density function can often be assigned to a familiar class of distributional forms, the class being determined by studying the intrinsic nature of the simulation model and its anticipated effects on the simular response. If this class of distributions should be the family of normal distributions, then the particular member of this family which is applicable to the simular response may be specified, provided that the two distributional properties, the mean μ and the variance σ^2, are known.

If, however, another family of probability densities, such as the Weibull family, could be anticipated as appropriate to the description of the distribution of simular responses, these distributional properties may not be directly expressible in terms of the parameters that index the family. For example, the mean of the Weibull distribution is given by

$$\mu = a\Gamma(1 + b^{-1}),$$

and the variance is

$$\sigma^2 = a^2\{\Gamma(1 + 2b^{-1}) - \Gamma^2(1 + b^{-1})\},$$

where a and b are the scale and shape parameters, respectively, for the family. Consequently, any attempt to determine estimates of these parameters from the estimates $\tilde{\mu}$ and S^2 of the mean and variance will require the solution of a pair of transcendental equations.

This state of the parametric situation exists with nearly every family of probability density functions. Not always are the moments and distributional properties of the family expressible directly in terms of the parameters that appear in the functional form for the density. Therefore, of prime importance are simple and efficient methods for determining accurate and efficient estimates of the parameters of the distribution at hand.

As was discussed in Section 6.2.1, two desirable properties of parameter estimates are their unbiasedness and their consistency. In the remainder of this section, three general methods for estimating the parameters of probability density functions will be described: moments, least squares, and maximum likelihood. The statistical properties of the resultant estimates will also be briefly discussed.

6.4.1. The Method of Moments

A first approach to the estimation of the p (≥ 1) parameters $\vec{\theta} = (\theta_1, \theta_2, \ldots, \theta_p)$ indexing a probability density function $f_Y(y; \vec{\theta})$, which describes the distribution of simular responses $Y(T)$, is to estimate unbiasedly the first p moments of the distribution. Since the kth moment,

$$\mu_k' \equiv \int_{-\infty}^{\infty} y^k f_Y(y; \vec{\theta})\, dy = \mu_k'(\theta_1, \theta_2, \ldots, \theta_p),$$

for each $k = 1, 2, \ldots, p$, is a function of the p parameters, and since an unbiased estimate of μ_k' is always available as

$$m_k' = N^{-1} \sum_{j=1}^{N} y_j^k, \qquad k = 1, 2, \ldots, p.$$

where y_j is the simular response which arises from the jth independently seeded encounter with the model at the same environmental conditions, then the simultaneous solution of the p equations

$$\begin{aligned} m_1' &= \mu_1'(\theta_1, \theta_2, \ldots, \theta_p), \\ m_2' &= \mu_2'(\theta_1, \theta_2, \ldots, \theta_p), \\ &\vdots \\ m_p' &= \mu_p'(\theta_1, \theta_2, \ldots, \theta_p), \end{aligned}$$

produces the estimates

$$\begin{aligned} \tilde{\theta}_1 &= \phi_1(m_1', m_2', \ldots, m_p'), \\ \tilde{\theta}_2 &= \phi_2(m_1', m_2', \ldots, m_p'), \\ &\vdots \\ \tilde{\theta}_p &= \phi_p(m_1', m_2', \ldots, m_p'). \end{aligned}$$

In general, there are no significant difficulties in solving the p equations for $\theta_1, \theta_2, \ldots, \theta_p$. The solutions are denoted by $\tilde{\theta}_1, \tilde{\theta}_2, \ldots, \tilde{\theta}_p$, since these quantities are clearly transformations of the first p sample moments, and are therefore random variables themselves, their values likely being different for successive sets of N simular responses.

In the event that the distribution of simular responses is presumably normal and therefore dependent upon two parameters ($\theta_1 = \mu, \theta_2 = \sigma^2$), the method of moments leads to the equations

$$m_1' = \bar{y} = \theta_1 \qquad \text{and} \qquad m_2' = \theta_2 + \theta_1^2,$$

solution for which are

$$\tilde{\theta}_1 = \bar{y} = N^{-1}\Big[\sum_{j=1}^{N} Y_j\Big] = \tilde{\mu},$$

and

$$\tilde{\theta}_2 = m_2' - (\bar{y})^2 = N^{-1}\Big[\sum_{j=1}^{N} y_j^2 - N(\bar{y})^2\Big],$$

or

$$\tilde{\theta}_2 = (N-1)S^2/N,$$

where $\tilde{\mu}$ and S^2 are the unbiased estimates of the mean (which, in this case, is μ) and the variance (which here is σ^2) as had been indicated in Section 6.2.

One may note immediately that the method of moments may not always provide unbiased estimates of the parameters, for

$$E(\tilde{\theta}_2) = [(N-1)/N]E(S^2) = (N-1)\sigma^2/N.$$

However, one may also note that the *bias*, defined generally by

$$\beta(\tilde{\theta}) = E(\tilde{\theta}) - \theta,$$

is, for the estimator $\tilde{\theta}_2$ of σ^2,

$$\beta(\tilde{\theta}_2) = (1 - N^{-1})\sigma^2 - \sigma^2 = -\sigma^2/N,$$

so that, as the number of simular responses increases, the bias diminishes in inverse proportion. In addition, the variance of $\tilde{\theta}_2$ can be shown to be (Gaussian case)

$$\mathrm{Var}(\tilde{\theta}_2) = 2\sigma^4(N-1)/N^2,$$

so that asymptotically $\tilde{\theta}_2$ approaches σ^2 with increasing certainty. Thus, $\tilde{\theta}_2$ is a consistent, though biased, estimate of $\theta_2(=\sigma^2)$.

The estimator $\tilde{\theta}_1$ of θ_1 $(=\mu)$ is the arithmetic mean of the N simular responses, and is therefore an unbiased and consistent estimate for μ, as had been shown in Section 6.2.1. It is more generally true that estimators arising from the method of moments may or may not be unbiased, but, if biased, they are usually consistent. [The reader is referred to Cramér (1946) for further details.]

In the case of the Weibull distribution, the probability density function for which is

$$g_W(w; a, b) = bw^{b-1}[\exp -(w/a)^b]/a^b, \qquad \text{for} \quad w > 0,$$

where the scale $(\theta_1 = a)$ and shape $(\theta_2 = b)$ parameters are positive, the method of moments leads to the equations

$$m_1' = a\Gamma(1 + b^{-1}), \qquad \text{and} \qquad m_2' = a^2\Gamma(1 + 2b^{-1}),$$

where $\Gamma(v) = \int_0^\infty x^{v-1}e^{-x}\,dx$ is the Gamma function of Euler and where $m_k' = N^{-1}\sum_{j=1}^N y_j^k$ for $k = 1$ and $k = 2$ are computed from the N simular responses $(y_1, y_2, \ldots, y_N)$. Noting that these equations can be solved with the aid of the function

$$R(b^{-1}) = \Gamma^2(1 + b^{-1})/\Gamma(1 + 2b^{-1}) = b^{-1}B(b^{-1}, b^{-1})/2$$

where

$$B(u, v) = \int_0^1 x^{u-1}(1 - x)^{v-1}\,dx = \Gamma(u) \cdot \Gamma(v)/\Gamma(u + v)$$

is the Euler Beta function, the present author [see Mihram (1965)] recommended the shape parameter estimator $\tilde{b}$ which arises as the solution for b of the equation

$$R(b^{-1}) = (m_1')^2/(m_2').$$

For this purpose, the Table 6.2 of the function $R(d)$ is provided for $d = b^{-1} = 0.02(.02)2.50(.06)4.00$.

The remaining Weibull parameter $\theta_1 = a$ can then be estimated by

$$\tilde{a} = m_1'/\Gamma(1 + \tilde{b}^{-1}) = \bar{y}/\Gamma(1 + \tilde{b}^{-1}).$$

Presumably this statistic would be found to be biased, although it should probably be consistent for the estimation of $\theta_1 = a$.

EXERCISES

1. Suppose that a simular response $Y(T)$ can be presumed to have the exponential distribution,

$$g_Y(y; \theta) = \theta^{-1}e^{-y/\theta}, \qquad \text{for} \quad y > 0,$$

and for the single positive parameter θ (which is equal to the mean simular response). Show then that the method of moments produces the estimator $\tilde{\theta} = \bar{y}$ which is unbiased and consistent.

2. Generate 8000 standardized normal random variables, from which select:

(*a*) 100 mutually exclusive samples of size 10 each,
(*b*) 100 mutually exclusive samples of size 20 each,
(*c*) 100 mutually exclusive samples of size 40 each,
(*d*) 100 mutually exclusive samples of size 80 each.

For each sample, compute the method-of-moments estimator for the variance ($\theta_2 = \sigma^2 = 1$) as

$$\tilde{\theta}_2 = N^{-1}\left[\sum_{j=1}^{N} y_j - N(\bar{y})^2\right],$$

TABLE 6.2

$$R(d) = \{E(X)\}^2/E\{X^2\} = \Gamma^2(1 + d)/\Gamma(1 + 2d)$$

d	$R(d)$	d	$R(d)$	d	$R(d)$	d	$R(d)$	d	$R(d)$	d	$R(d)$
0.02	0.999	0.52	0.773	1.02	0.490	1.52	0.288	2.02	0.163	2.56	0.086
0.04	0.998	0.54	0.761	1.04	0.480	1.54	0.282	2.04	0.159	2.62	0.080
0.06	0.995	0.56	0.749	1.06	0.471	1.56	0.276	2.06	0.155	2.68	0.074
0.08	0.991	0.58	0.737	1.08	0.461	1.58	0.269	2.08	0.152	2.74	0.069
0.10	0.986	0.60	0.725	1.10	0.452	1.60	0.264	2.10	0.148	2.80	0.064
0.12	0.980	0.62	0.713	1.12	0.443	1.62	0.258	2.12	0.145	2.86	0.059
0.14	0.974	0.64	0.700	1.14	0.434	1.64	0.252	2.14	0.141	2.92	0.055
0.16	0.966	0.66	0.688	1.16	0.425	1.66	0.246	2.16	0.138	2.98	0.051
0.18	0.959	0.68	0.677	1.18	0.416	1.68	0.241	2.18	0.135	3.04	0.048
0.20	0.950	0.70	0.665	1.20	0.407	1.70	0.235	2.20	0.132	3.10	0.044
0.22	0.941	0.72	0.653	1.22	0.399	1.72	0.230	2.22	0.129	3.16	0.041
0.24	0.932	0.74	0.641	1.24	0.390	1.74	0.225	2.24	0.126	3.22	0.038
0.26	0.922	0.76	0.630	1.26	0.382	1.76	0.220	2.26	0.123	3.28	0.035
0.28	0.912	0.78	0.618	1.28	0.374	1.78	0.215	2.28	0.120	3.34	0.033
0.30	0.901	0.80	0.607	1.30	0.366	1.80	0.210	2.30	0.117	3.40	0.030
0.32	0.891	0.82	0.596	1.32	0.358	1.82	0.205	2.32	0.114	3.46	0.028
0.34	0.880	0.84	0.584	1.34	0.351	1.84	0.200	2.34	0.111	3.52	0.026
0.36	0.868	0.86	0.573	1.36	0.343	1.86	0.196	2.36	0.109	3.58	0.024
0.38	0.857	0.88	0.563	1.38	0.336	1.88	0.192	2.38	0.106	3.64	0.023
0.40	0.845	0.90	0.552	1.40	0.329	1.90	0.187	2.40	0.104	3.70	0.021
0.42	0.833	0.92	0.541	1.42	0.321	1.92	0.183	2.42	0.101	3.76	0.019
0.44	0.822	0.94	0.530	1.44	0.315	1.94	0.179	2.44	0.099	3.82	0.018
0.46	0.810	0.96	0.520	1.46	0.308	1.96	0.175	2.46	0.097	3.88	0.016
0.48	0.798	0.98	0.510	1.48	0.301	1.98	0.171	2.48	0.094	3.94	0.015
0.50	0.785	1.00	0.500	1.50	0.295	2.00	0.167	2.50	0.092	4.00	0.014

where N is the applicable sample size. In each of Cases (*a*)–(*d*), build a frequency histogram for the 100 estimates and compute the sample mean, standard deviation, and the variance appropriate to each histogram. Comment on the apparent nature of $\tilde{\theta}_2$ as concerns unbiasedness and consistency with the increasing sample sizes.

3. Generate 200 random samples of size 10 from the Weibull distribution $g_W(w; 1, 1)$. Using the procedure suggested in the text [linearly interpolating in Table 6.2 for $R(d)$, as necessary], compute for each sample the method-of-moments estimate for $\theta_1 = a\ (=1)$. Construct a frequency histogram for the 200-scale parameter estimates and compute their mean, variance, and standard deviation.

6.4.2. The Method of Least Squares

A second procedure for obtaining estimates of parameters θ which index the probability density function $f_Y(y; \theta)$ of simular responses $Y(T)$ depends upon the analyst's willingness to describe the simular response at c environmental conditions denoted by $(x_1, x_2, \ldots, x_c)$ as

$$Y(T) = f(x_1, x_2, \ldots, x_c;\ \vec{\theta}) + \varepsilon(s),$$

where f is some known functional form of the c environmental conditions and of p parameters $(\theta_1, \theta_2, \ldots, \theta_p)$, and $\varepsilon(s)$ is an unknown error term dependent upon the seed s for the encounter. For example, frequently one is willing to assume that the function f is a linear combination of the $c\ (=p)$ environmental conditions:

$$f(x_1, x_2, \ldots, x_p;\ \vec{\theta}) = \theta_1 \cdot x_1 + \cdots + \theta_p \cdot x_p;$$

in this case, the assumption of the linear dependence of the response on the p environmental conditions is generally an approximation to the unknown functional form

$$Y(T) = Y(x_1, x_2, \ldots, x_p;\ s),$$

which describes the simular response.

Therefore, the error term $\varepsilon(s)$ consists of contributions from two sources:

(*a*) the inadequacy of the (linear) approximation for the actual response function, and

(*b*) the inherent variability existing in the simular response from a stochastic simulation model.

It is usual, but not necessary, to assume that the error term is a random variable with mean zero, variance equal to an unknown constant σ_ε^2, and Gaussian probability density function. This implies then that the simular response is a normally distributed random variable of mean

$$E[Y(T)] = f(x_1, x_2, \ldots, x_p; \vec{\theta})$$

and variance

$$\text{Var}[Y(T)] = \sigma_\varepsilon^2,$$

so that the Gaussian probability density function for $Y(T)$ is indexed by the parameters $\vec{\theta}$ and by the error variance σ_ε^2.

Regardless of the distributional assumptions made regarding $\varepsilon(s)$ and, therefore concerning $Y(T)$, the method of least squares attempts to utilize N independent simular responses $(y_1, y_2, \ldots, y_N)$ so as to minimize the sum of the squared errors:

$$Q = \sum_{j=1}^{N} \varepsilon^2(s_j) = \sum_{j=1}^{N} [y_j - f(x_1, \ldots, x_p; \vec{\theta}_p)]^2.$$

The usual procedure is to apply the differential calculus by taking the first partial derivatives of Q with respect to each of the p parameters, setting these expressions equal to zero, then by solving the resulting p equations to obtain the p estimates

$$\begin{aligned}
\theta_1^* &= t_1(y_1, \ldots, y_N; \vec{x}),\\
\theta_2^* &= t_2(y_1, \ldots, y_N; \vec{x}),\\
&\vdots\\
\theta_p^* &= t_p(y_1, \ldots, y_N; \vec{x}),
\end{aligned}$$

noticeably dependent upon the N simular responses and upon the environmental conditions (input specifications) which produced them.

For example, if one presumes that N simular responses are recorded at the same environmental conditions (say, the standard operating conditions $\vec{x}_0$), then the expected simular response would be

$$E[Y(T)] = f(x_{1,0}, x_{2,0}, \ldots, x_{p,0}; \vec{\theta}).$$

If, in particular

$$f(x_1, x_2, \ldots, x_p) = \theta_1 \cdot x_1 + \theta_2 \cdot x_2 + \cdots + \theta_p \cdot x_p,$$

then

$$E[Y(T)] = \theta_1 \cdot x_{1,0} + \theta_2 \cdot x_{2,0} + \cdots + \theta_p \cdot x_{p,0} \equiv \theta,$$

and the least-squares expression becomes

$$\partial Q/\partial\theta = -2 \sum_{j=1}^{N} (y_j - \theta),$$

which, when set equal to zero, yields the solution:

$$\theta^* = N^{-1}\left(\sum_{j=1}^{N} y_j\right),$$

equal to the arithmetic mean of the simular responses.

For the general case (i.e., for general functions f), the properties of the least-squares estimators are not known, even when the error term $\varepsilon(s)$ is assumed to be normally distributed. A class of exceptions to this general rule may be noted, however: Whenever f is a linear function of $\theta_1, \theta_2, \ldots, \theta_p$, it is possible to construct unbiased estimates for the parameters, the estimates having the least possible variance for unbiased estimates which are themselves linear combinations of the N simular responses. This result will be developed in a subsequent section, but for the present, the reader should be aware of the method of least squares as a fundamental technique for estimating parameters in the probability density function for the simular response. This technique will be the basis for many of the statistical techniques to be discussed later.

Of note with respect to the method of least squares is the estimate of σ_ε^2, which is always of the form

$$\sigma_\varepsilon^{2*} = N^{-1} \sum_{j=1}^{N} (y_j - f_j^*)^2,$$

where

$$f_j^* = f(x_{1,j}, x_{2,j}, \ldots, x_{p,j};\ \theta_1^*, \ldots, \theta_p^*), \qquad j = 1, 2, \ldots, N.$$

Thus, the method of least squares estimates the error variance by the mean squared deviation of responses y_j from the estimated response function f—the latter being evaluated in each case at the environmental conditions which produced y_j and at the parametric estimates θ_i^*, $i = 1, 2, \ldots, p$.

6.4.3. Maximum Likelihood Estimation

An alternative procedure for determining estimates of the p parameters $\vec{\theta} = (\theta_1, \theta_2, \ldots, \theta_p)$ of the probability density function $f_Y(y; \vec{\theta})$ applicable to the description of the distribution of simular responses, is the method of maximum likelihood. For a given random sample $(y_1, y_2, \ldots, y_N)$ of N simular responses (each arising from an independently seeded encounter with the model at the same environmental conditions), the method seeks to determine those values of $\theta_1, \theta_2, \ldots, \theta_p$ which will maximize

$$L(y_1, y_2, \ldots, y_N;\ \theta_1, \theta_2, \ldots, \theta_p) = \prod_{j=1}^{N} f_Y(y_j; \vec{\theta}).$$

The resulting values, labeled $\hat{\theta}_1, \hat{\theta}_2, \ldots, \hat{\theta}_p$ are referred to as the *maximum likelihood estimators*.

The motivation for this procedure can be found in most elementary textbooks on mathematical statistics, including the more advanced book of Cramér (1946). The usual approach to the maximization of the likelihood function L is the partial differentiation of this function (or, equivalently, of its natural logarithm $\lambda = \ln L$) with respect to θ_i, $i = 1, 2, \ldots, p$; the resulting p equations are then solved simultaneously to obtain the estimates $\hat{\theta}_1, \hat{\theta}_2, \ldots, \hat{\theta}_p$.

For example, if the simular response $Y(T)$ is a normally distributed random variable of unknown mean $(\theta_1 = \mu)$ and unknown variance $(\theta_2 = \sigma^2)$, then the likelihood function for N simular responses $(y_1, y_2, \ldots, y_N)$ becomes

$$L = (2\pi\theta_2)^{-N/2} \exp - \left\{ \sum_{j=1}^{N} (y_j - \theta_1)^2 / 2\theta_2 \right\},$$

and the log-likelihood function is

$$\lambda = -N(\ln 2\pi)/2 - \sum_{j=1}^{N} (y_j - \theta_1)^2/2\theta_2 - N(\ln \theta_2)/2.$$

Differentiating the latter with respect to θ_1 and then θ_2 yields

$$\partial\lambda/\partial\theta_1 = \sum_{j=1}^{N} (y_j - \theta_1)/\theta_2,$$

and

$$\partial\lambda/\partial\theta_2 = \sum_{j=1}^{N} (y_j - \theta_1)^2/(2\theta_2^2) - N/(2\theta_2).$$

Setting these expressions equal to zero, the simultaneous solution becomes

$$\hat{\theta}_1 = N^{-1} \sum_{j=1}^{N} y_j = \bar{y} \quad (=\tilde{\theta}_1)$$

and

$$\hat{\theta}_2 = N^{-1} \sum_{j=1}^{N} (y_j - \hat{\theta}_1)^2 \quad (=\tilde{\theta}_2),$$

which, as denoted by the parenthetical equalities, are the same estimators that arose from the method of moments.

Thus, one sees that use of the maximum likelihood method, like the method of moments, will not guarantee the unbiasedness of the resulting estimates. However, under quite general conditions (i.e., for virtually all cases of practical interest), the maximum likelihood estimate of a parameter θ_k is consistent.

Also of particular interest is the fact that the method of maximum likelihood produces the same parameter estimates as the method of least squares whenever the simular reponse is considered as

$$\begin{aligned} Y(T) &= Y(x_1, \ldots, x_p; s) \\ &= \theta_0 + \theta_1 x_1 + \cdots + \theta_p x_p + \varepsilon(s), \end{aligned}$$

where $\varepsilon(s)$ is a normally distributed random variable of mean zero, and unknown variance σ_ε^2. Even more generally, the two methods produce the same estimates, under the normality assumption for the errors $\varepsilon(s)$ *and* the assumption that the simular response is a linear combination of the *parameters*: $\theta_0, \theta_1, \ldots, \theta_p$. Since we shall be primarily concerned with essentially least-squares estimation procedures in the remainder of the book, it is of special worth to note that the assumptions [of the normality of the distribution of errors and of the linearity of the approximating response function $f(x_1, \ldots, x_p; \vec{\theta})$ in terms of the parameters $\vec{\theta}$] are sufficient to imply that any resulting least-squares estimators have the important property of consistency. In fact, one will almost always be able to ensure that these estimators be unbiased as well.

6.4.4. Summary

In this section, three procedures have been outlined for estimating the parameters required to describe the probability density functions appropriate to a particular simular response: (*a*) the method of moments, (*b*) the method of least squares, and (*c*) the method of maximum likelihood. The estimates will be required in order to select among the family of densities appropriate to the simular response, whether that family be the normal (Gaussian) densities, the Weibull densities, the exponential densities, or otherwise.

The desirability of producing unbiased and consistent estimates was noted and it was stated that none of the three methods can, in general, guarantee unbiased estimates, although the methods of maximum likelihood and moments almost always produce consistent estimators. The method of least squares, applicable whenever one feels that he has *a priori* knowledge of the functional form assumed by the simular response, can, under reasonably general conditions, be made to produce unbiased estimates identical to those of the method of maximum likelihood whenever appropriate conditions of normality and linearity are assumed. Therefore, under these conditions, least-squares estimators are consistent as well, and, as will be shown, possess the minimum variance among an important class of estimators (namely, those which are unbiased linear combinations of the simular responses $y_1, y_2, \ldots, y_N$).

The techniques displayed in this section will be encountered again during the remainder of the book. It should be noted nonetheless, that these techniques should be especially useful in the validation stage of a simulation model's development, especially so when the model represents a large-scale operating system whose actual response has been observed for some time. Even if this modeled system's response is a random variable X at its standard operating conditions, its frequent observation in the past should have provided some feeling for the nature of the distribution of responses (e.g., normal, Gamma, Weibull); hence, via a few iterations of the simulation model at the corresponding simular environmental conditions, parameter estimates can be obtained and compared with the known results for the simulated system. Criteria for assessing these comparisons will be an application of the subject matter of the next section.

6.5. Statistical Comparisons

6.5.1. Introduction

At several stages in the development of a computerized simulation model, occasions will arise that call for comparisons of the responses emanating from the model. For example, the essence of the verification and validation stages is comparison: In the first, the simular responses are contrasted with responses which would be anticipated were the model (or some module thereof) generating its intended responses; in the second, the simular responses are contrasted with known observations of the behavior of the simulated system itself. As will be shown in this and the next four chapters, the analysis stage of model development is a comparative undertaking for the purpose of detecting significant differences among simular responses arising from sets of encounters defined at diverse environmental specifications.

The very fact that a stochastic simulation model yields as its response a random variable $Y(T)$ at the end of T simular time units implies that response differences will probably exist, regardless of the stage at which comparisons are made. The critical feature of such differences, then, is their relative degree of significance. However, whether or not an absolute difference in two numerical quantities is significant is not readily apparent. For example, if 1000 tosses of a gold coin had resulted in the proportion 0.25 of heads, whereas a similar number of tosses of a silver coin had resulted in 70% heads, the difference $| 0.70 - 0.25 | = 0.45$ would be deemed by most persons quite significant. The same numerical difference in one's weight reduction after six months of strenuous daily exercise would, however, be regarded by most adults as insignificant, even if the unit of measure were the kilogram.

A measure of the intrinsic variability of any random variable is its variance σ^2 (or its standard deviation σ). Because of this important property of the variance, many of the tests for statistical significance of observed differences rely on comparisons of the magnitude of these differences with estimates of either the variance or its positive square root, the standard deviation. In this manner, recorded differences in random variables, such as simular responses, can be compared with a measure of the inherent differences which could be anticipated in any event.

In this section will be presented a representative sampling of fundamental tests for the significance of differences observed in sets of responses from a stochastic simulation model. The tests are applicable to comparative analyses of both model versus modeled responses (validation) and model versus model responses (analysis), although, for the most part, the discussion itself will surround comparisons of the latter type.

6.5.2. A Pair of Simular Responses

The most elementary comparison likely to arise in the analysis of responses, $Y(T)$, from a stochastic simulation model is the contrasting of a single pair of such responses. For the purposes of discussion, one may presume that each of the two simular responses has arisen from the specification of differing environmental conditions as well as the assignment of a different and independently selected random number seed for each model encounter; i.e.,

$$Y_1 = Y(x_{1,1}, x_{2,1}, \ldots, x_{p,1}; s_1)$$

and

$$Y_2 = Y(x_{1,2}, x_{2,2}, \ldots, x_{p,2}; s_2),$$

with $x_{i,1} \neq x_{i,2}$ for at least one index i and with $s_1 \neq s_2$ (cf. Section 5.6.2).

Quite naturally, the following question should arise. Is the observed difference $D^* = Y_1 - Y_2$, or more usually, is the observed absolute difference $D = | Y_1 - Y_2 |$, significantly different from zero? A positive reply would imply that the change of environmental conditions, from those used in the first encounter to yield Y_1 to those assigned in the second to yield Y_2, would indeed provoke a significant change in the simular response. Assuming that the model is an adequate representation of the modeled system, one might then, on the basis of such a finding, proceed to implement a corresponding change in the operating conditions or policies of the actual system.

Intuitively, one should feel that the absolute magnitude of the observed difference should play a major role in assessing its significance. A smaller difference would not tend to promote a declaration of its significance so much as would a comparatively larger difference. Thus, the decision rule for declaring the significance or nonsignificance of the responses' difference will require the specification of two regions of the real numbers:

(*a*) an *acceptance region* $\mathscr{A}$, probably a set of real numbers of the form, $\{x \mid 0 \leq x \leq c$, for c some critical constant yet to be specified$\}$, and

(*b*) *a rejection region* $\mathscr{R}$, the complement of the acceptance region; viz., $\{x \mid x > c\}$, also termed the *critical region.*

The terms "acceptance" and "rejection" refer to the disposition of some proffered, or null, hypothesis as a result of the observations recorded. In the present context, the appropriate null hypothesis is that the expected difference Δ given by

$$\Delta = E(Y_1 - Y_2) = E[Y(\vec{x}_1)] - E[Y(\vec{x}_2)]$$

is zero, where $Y(\vec{x}_i)$ denotes the simular response (random variable) arising from model encounters specified by the environmental condition $\vec{x}_i = (x_{1,i}, x_{2,i}, \ldots, x_{p,i})$, $i = 1$ and 2. That is, one tenders the hypothesis, H_0: $\Delta = 0$, intending to accept or reject it depending upon whether or not the observed absolute difference falls in $\mathscr{A}$ or in $\mathscr{R} = \bar{\mathscr{A}}$.

To specify the region $\mathscr{A}$ (and therefore $\mathscr{R}$ as well), one must determine a critical constant c, beyond the value of which an observed absolute difference will lead to the rejection of the null hypothesis (or, alternatively, to the acceptance of the existence of a significant difference in the simular responses). Furthermore, since the observed absolute difference D is a transformation of the (independently distributed) random variables Y_1 and Y_2, it is itself also a random variable; the selection of the value of c should then be based upon the unlikelihood of observing a value of D so large or larger.

The declaration of the improbability of any single sample from a probability density function rests, however, upon two factors:

(*a*) the functional form, or shape, of the random variable's probability density function, and
(*b*) a somewhat subjective assignment of a probability size α below which one presumes that any event which would occur so relatively infrequently could be categorized as "improbable" or "unlikely."

The standard practice in scientific methodology of this century has been to locate the critical constant c in correspondence with a size $\alpha = 0.05$ or $\alpha = 0.01$. Thus, if an experimental result has occurred, under the presumption that a particular null hypothesis H_0 was indeed true, with a probability as low as 0.05 then the result is termed *significant*; if the result, under the same hypothetical condition, occurs with a probability of 0.01 or less, the event is termed *highly significant.*

The selection of these particular probability levels, either of which is referred to as the *size* of a test for an hypothesis, has apparently been somewhat an historical accident, though the essential motivation should be clear: One selects the size α of a test commensurate with one's own feeling of the improbability of an event. Thus, it is very unusual to select $\alpha = 0.50$, say, because events with a probability of occurrence of this magnitude would presumably happen with great relative frequency (e.g., the tossing of a coin). Similarly, the assignment of the size $\alpha = \frac{1}{6}$ would still appear to most persons as unusually high, as reflection on the relative frequency of the occurrence of an ace in the tossing of a fair die should reveal.

Nonetheless, *prior to the experimentation*, a (necessarily subjective) determination of the size α must be assigned, thereby affixing the critical constant c (and/or region $\mathscr{R}$) *a priori*. The subsequent performance of the experiment then produces the test statistic (D, in the present case) which either assumes a value in the critical region (whence, the null hypothesis is rejected) or takes a value in the acceptance region (in which case, one assumes that the evidence, as summarized by the test statistic, is insufficient for the purpose of rejecting the null hypothesis).

For the present situation, then, it will be necessary to determine a critical constant c beyond which an observed value of $D = | Y_1 - Y_2 | = [(Y_1 - Y_2)^2]^{1/2}$ would arise strikingly infrequently. Since one may write

$$P[D > c] = P[| Y_1 - Y_2 | > c] = P\{[(Y_1 - Y_2)^2]^{1/2} > c\} \leq \alpha$$

equivalently as

$$P[(Y_1 - Y_2)^2 > c^2 \equiv C] \leq \alpha,$$

the determination of c may be transferred to the determination of C, the point at which the cumulative distribution function for the random variable $\tau = D^2 = (Y_1 - Y_2)^2$ assumes the value $(1 - \alpha)$. The determination of the critical region for testing the null hypothesis, H_0: $\Delta \equiv E(D^*) = 0$, may then be transferred to the location of a specific point on the cumulative distribution function of the positive random variable (an alternative test statistic) τ as depicted in Fig. 6.3.

The particular cumulative distribution function appropriate to the determination of the critical region is that of the test statistic τ. But this statistic is a transformation of the observed independent random variables, which are in the present exemplary situation the simular responses Y_1 and Y_2. Therefore, the determination of the probability density function which is appropriate to the simular response $Y(T)$ is of the essence, as indicated in the previous section of this chapter on the limiting forms attainable for the distributions of simular responses.

Presuming, then, that the simular response $Y(T)$ can be expected to be a normally distributed random variable of the same mean μ and the same variance σ^2 at each of the two environmental specifications, then the random variable,

$$(\tau/\sigma^2) = (Y_1 - Y_2)^2/\sigma^2 = \{[(Y_1 - \mu)/\sigma] - [(Y_2 - \mu)/\sigma]\}^2$$

is the square of the difference of two standardized and independent normal random variables. Since such a difference, that of two independent,

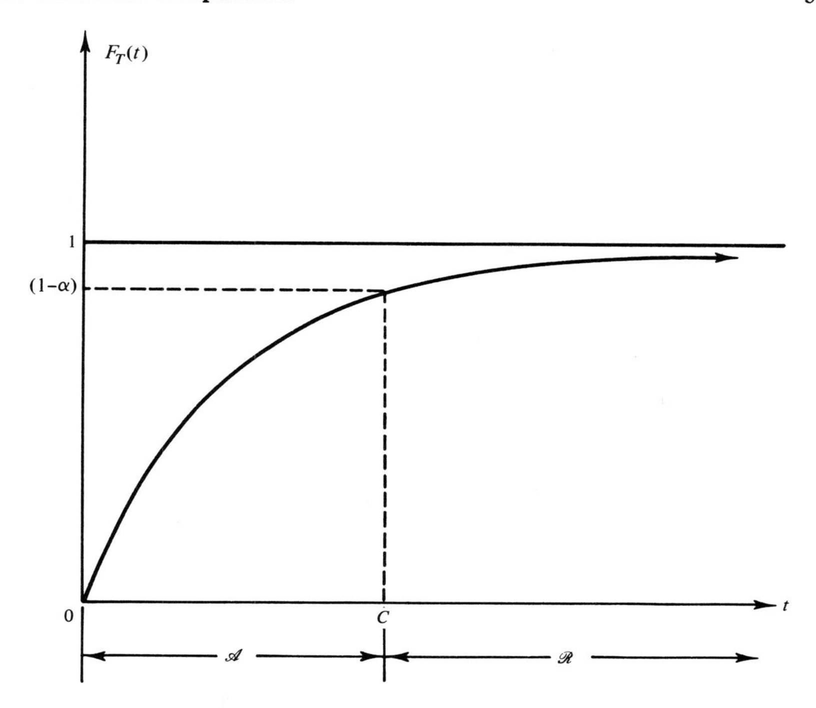

FIG. 6.3. Determination of the critical region.

normally distributed random variables, is also normally distributed with mean zero and variance two [recall that the variance of $Z_1 - Z_2$ is $\text{Var}(Z_1) + \text{Var}(Z_2) = 2$ whenever Z_1 and Z_2 are independent random variables of unit variance], then $(\tau/2\sigma^2)$, being the square of a single standardized normal variate, has the Chi-squared distribution of one degree of freedom.

The decision, whether to reject or accept the hypothesis that there is no difference between the simular responses arising at the environmental condition $\vec{x}_1$ and those arising at the environmental condition $\vec{x}_2$ is based then on the value of t at which a Chi-squared distribution of a single degree of freedom accumulates 95% (or 99%) of its distribution. From tables of this distribution, such as those of Harter (1964), one sees that the critical values for $(\tau/2\sigma^2)$ are $C = 3.84$ (for $\alpha = 0.05$), or $C = 6.63$ ($\alpha = 0.01$); that is, one would reject the null hypothesis H_0: $\varDelta = 0$, if

$$\tau = (Y_1 - Y_2)^2 > 2\sigma^2 \cdot (3.84) \qquad (\text{for} \quad \alpha = 0.05),$$

or if

$$\tau = (Y_1 - Y_2)^2 > 2\sigma^2 \cdot (6.63) \qquad (\text{for} \quad \alpha = 0.01).$$

One may note then that knowledge of the variance σ^2 of the simular responses must be available in order to perform the actual test. If this is indeed the case, the test of the hypothesis H_0: $\Delta = 0$ would proceed in the straightforward manner just indicated: i.e., by computation of $\tau = (Y_1 - Y_2)^2$ and subsequent comparison with the critical constant, say $2\sigma^2 \cdot (3.84)$. One should note that the statistic τ has expectation

$$E(\tau) = E(Y_1 - Y_2)^2 = \mathrm{Var}(Y_1 - Y_2) + \Delta^2 = 2\sigma^2 + \Delta^2$$

since $E(Y_1 - Y_2) \equiv \Delta$; thus, under the null hypothesis, when $\Delta = 0$, the expectation becomes $E(T) = 2\sigma^2$. Consequently, the ratio $(\tau/2\sigma^2)$ has expectation unity whenever the null hypothesis is true, so that it is only whenever this ratio exceeds 3.84 (or 6.63) that the hypothesis is rejected and the difference D^* is declared significant (or highly significant).

6.5.3. The Test with Unknown Variance

More usually, however, the variance σ^2 will not be known but will need to be estimated from the available data (responses). It is a general rule that such an estimate, if it is to be independent of τ, is available only when the experimentation plan involves *replication*, or repetition, of responses at the same environmental conditions. Presuming thus that m simular encounters are defined at each of the two environmental conditions $\vec{x}_1$ and $\vec{x}_2$, yet the requisite $2m$ seeds are independently selected with repetitions forbidden, then the $2m$ simular responses,

$$Y_{1,k}, \qquad k = 1, 2, \ldots, m \qquad (\text{at} \quad \vec{x}_1)$$

and

$$Y_{2,k}, \qquad k = 1, 2, \ldots, m \qquad (\text{at} \quad \vec{x}_2),$$

constitute $2m$ independent random variables.

One may take as the simular response at $\vec{x}_j$,

$$\bar{y}_j = m^{-1} \sum_{k=1}^{m} Y_{j,k}$$

and, as the statistically independent and unbiased estimate of the variance:

$$S_j^2 = (m-1)^{-1}\left[\sum_{k=1}^{m} Y_{j,k}^2 - m(\bar{y}_j)^2\right], \qquad j = 1 \text{ and } 2.$$

Since S_1^2 and S_2^2 are each statistics computed from an independent random sample, they are independent random variables; they may be pooled, or averaged, to obtain the unbiased estimate of σ^2:

$$S^2(\text{pooled}) = (S_1^2 + S_2^2)/2.$$

Furthermore, whenever the simular responses may be presumed to be normally distributed, $(m - 1)S_j^2/\sigma^2$ has the Chi-squared distribution with $(m - 1)$ degrees of freedom (cf. Section 6.2.3); thus $2(m - 1)S^2/\sigma^2$ has the Chi-squared distribution of $(2m - 2)$ degrees of freedom.

Each of the mean responses $\bar{y}_j$ has the normal distribution of mean μ and variance (σ^2/m), so that, due to their independence, the statistic

$$D^* = (\bar{y}_1 - \bar{y}_2)$$

has the normal distribution with mean 0, variance $(2\sigma^2/m)$, whenever the null hypothesis H_0: $\Delta = 0$ [i.e., $E(Y_1) = E(Y_2)$], is true. Furthermore, D^* and S^2 are independent random variables, since each is a linear combination of the mutually independent random variables $\bar{y}_1$, $\bar{y}_2$, S_1^2, and S_2^2. One sees then that

$$T = \frac{D^*/(2\sigma^2/m)^{1/2}}{\{2(m - 1)S^2/[\sigma^2(2m - 2)]\}^{1/2}} = \frac{D^*}{(2S^2/m)^{1/2}}$$

has Student's t distribution of $(2m - 2)$ degrees of freedom. Therefore, an appropriate test for the null hypothesis H_0: $\Delta = 0$ is to reject it whenever the test statistic $|T|$ exceeds t_α, that value such that the probability of a Student's t variate of $(2m - 2)$ degrees of freedom exceeds (in absolute value) t_α is α (the size of the test; conventionally, $\alpha = 0.05$ or $\alpha = 0.01$).

One should note the essential similarity of the tests proposed in the presence, then the absence, of knowledge of the inherent variance σ^2 of the simular responses. In each case, the test criterion depends upon the magnitude of the absolute difference of the simular responses relative to the intrinsic variance (or, essentially equivalently in the second instance, its estimated standard deviation). This result will recur in tests to be applied in the remainder of this book.

6.5.4. A Note Aside on the Power of a Test

In the tests so far proposed, no mention has been made of the fact that the random sample of simular responses might well lead to the computation of an insignificant test statistic T even though a nonzero difference

$\Delta = E(Y_1 - Y_2)$ indeed exists. In such a case, one would accept the null hypothesis H_0: $\Delta = 0$ even though it is false.

For any given, existent, nonzero difference Δ one may compute the probability that the null hypothesis will be rejected (e.g., $P[|T| > t_\alpha]$). As a function of Δ, the resulting curve represents the *power* of the test, and measures the ability of the test statistic to reject a false null hypothesis. An exemplary power curve for the tests of the null hypothesis H_0: $\Delta \equiv E(Y_1 - Y_2) = 0$, is given in Fig. 6.4. For the most part, the power of a test becomes important whenever there be several candidate test statistics vying for prominence as the best, or most powerful, test statistic.

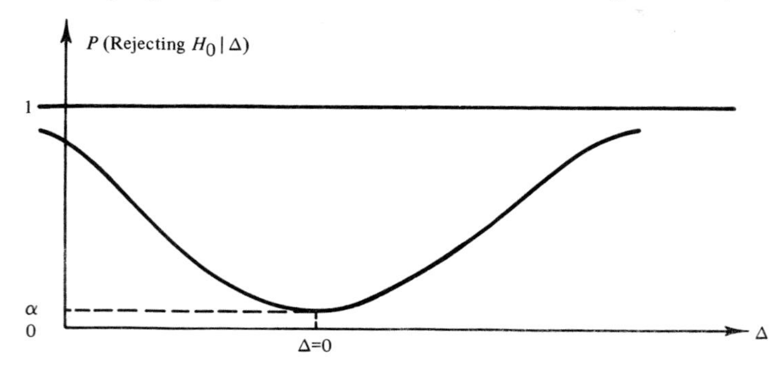

FIG. 6.4. A typical power curve.

Since we shall not be especially concerned in the present text with comparison of alternative test statistics, further discussion of power curves will be curtailed. The reader may rest assured, however, that the test statistics to be used in the remainder of the book will be of proven relative power.

EXERCISES

1. Perform a Monte Carlo analysis of the test of the hypothesis H_0: $\Delta = 0$, by first generating 100 independent pairs of normally distributed random variables of mean 4, variance 25. Computing, for each pair Y_1 and Y_2 the quantity

$$\tau/(2\sigma^2) = (Y_1 - Y_2)^2/(50)$$

form the sample cumulative distribution function and compare it with tabulations of the cumulative distribution function for the Chi-squared distribution of one degree of freedom (as tabulated in the Appendix).

By means of the Kolmogorov–Smirnov test statistic, what proportion of the 100 test statistics you have computed would lead to rejection of the null hypothesis at the significance level, $\alpha = 0.05$? Is this proportion acceptably different from α?

2. Repeat the Monte Carlo analysis of the preceding exercise, but generate the normally distributed random variables such that $\Delta \equiv E(Y_1 - Y_2)$ is actually 5.0; that is, generate 100 normally distributed random variables (Y_1) of mean 9 ($=4 + 5$) and variance 25, plus 100 normally distributed random variables (Y_2) of mean 4 and variance 25. Comment on the resulting proportion of the 100 test statistics $\tau/2\sigma^2 = (Y_1 - Y_2)^2/50$ that will exceed the critical constant, as defined for the preceding exercise.

3. The preceding tests of the null hypothesis have arisen from the presumption that the simular response $Y(T)$ is a normally distributed random variable. As has been indicated, other distributional forms may well arise in the simulation context. Perform then the Monte Carlo analysis of the first exercise by generating the pairs of simular responses (random variables) from the Weibull distribution of mean 4 and variance 25, corresponding to scale parameter value $a \cong 3.55$, shape parameter value $b \cong 0.80$. Compute the same test statistic $\tau/(2\sigma^2) = (Y_1 - Y_2)^2/50$ for each of the 100 pairs and prepare its sample cumulative distribution function therefrom. By means of the Kolmogorov–Smirnov test, test this sample cumulative distribution function (c.d.f.) with the one for a Chi-squared variate of 1 degree of freedom. Furthermore, one should note the proportion of test statistics that would lead to rejection of the null hypothesis H_0: $\Delta \equiv E(Y_1 - Y_2) = 0$ if the critical constant of Exercise 1 were employed. Would you suggest an alternative critical constant for $\alpha = 0.05$? for $\alpha = 0.01$?

6.5.5. Validation Tests

During the validation stage of model development, one generally tries to determine the compatibility of responses generated by the model with observations made on the modeled system. If the model is indeed an accurate replica of the modeled, then the probability density function of simular responses should be the same as that of the actual system's response, provided that the environmental specifications producing the simular responses are directly comparable to the operating conditions yielding the system's responses.

If, at some particular set of operating conditions, a random sample of N system responses is available and can be formed into a frequency histogram or a sample cumulative distribution function, then a suitable test for the validation of the model arises from the definition of N simular encounters at the corresponding simular environmental specifications, say $\vec{x}_0$, and the formation of either a frequency histogram or a sample cumulative distribution function from the resultant N simular responses: $Y_1, Y_2, \ldots, Y_N$. The two histograms (sample cumulative distribution functions) can then be compared by means of the usual Chi-squared test (two-sample Kolmogorov–Smirnov test), as discussed in Section 2.9 for testing random number generators [in Lindgren (1968)].

Frequently, however, one may not have, or may find it too expensive to obtain, sufficient information regarding the distribution of responses of the modeled system itself, yet at some particular set of operating conditions for the system, the expected response is reasonably well known. For example, the system's response at this set of operating conditions may have been recorded on m occasions and the arithmetic average and sample variance of these responses recorded. Denoting this average by $\bar{y}_1$, and the sample variance by S_1^2, then the same number (m) of simular encounters may be defined at the corresponding environmental conditions (say, $\vec{x}_0$) and the m resulting simular responses used to compute their mean ($\bar{y}_2$) and sample variance S_2^2. If both the system and the simular responses can be presumed to arise from a normal distribution of the same variance, then a test dependent upon the magnitude of the Student's t statistic

$$T = (\bar{y}_1 - \bar{y}_2)/(2 \cdot S^2/m)^{1/2}$$

is in order.

In many cases, however, the modeled system may not be readily accessible for such experimentation, as would often occur in the case of a simulation model constructed during the development of the system itself. In such cases, one may nonetheless possess accurate knowledge of the theoretical mean response M anticipated from the system once it is operable. One technique for aiding the validation of a stochastic simulation model of such a system is a test performed from an analysis of N simular encounters, each defined at the environmental conditions corresponding to the modeled system's (standard) operating conditions (at which, presumably, the theoretical mean response should be M).

Therefore, by offering the null hypothesis,

$$H_0: \quad \mu = M,$$

for the mean simular response (μ), one may proceed to test this hypothesis with the assumption of an appropriate distributional form for the simular response $Y(T)$. With the assumption that $Y(T)$ is a normally distributed random variable of mean μ and variance σ^2, the specification of N simular encounters, each at the same environmental conditions $\vec{x}_0$ (save for the independent and nonrepetitive selection of the random number seeds), will provide the random sample of simular responses: $Y_1, Y_2, \ldots, Y_N$. Under the null hypothesis (i.e., assuming that indeed $\mu = M$), the statistic

$$(\bar{Y} - M)/(\sigma^2/N)^{1/2}$$

is a random variable having the standardized normal distribution. Its square, the statistic

$$T = (\bar{Y} - M)^2/(\sigma^2/N)$$

is then a Chi-squared random variable of a single degree of freedom. Notable is the fact that T would tend to be large whenever $\bar{Y}$ deviates significantly from M, so that an appropriate critical region for the rejection of the null hypothesis H_0: $\mu = M$ is the set of real numbers

$$\mathscr{R} = \{t \mid t > c, \quad \text{where} \quad P[\chi_1^2 > c] = \alpha\},$$

with α specified *a priori* by the analyst (conventionally, $\alpha = 0.05$ or $\alpha = 0.01$), and where χ_1^2 is a Chi-squared variate of one degree of freedom.

The test may be performed equivalently by computing $(\bar{Y} - M)^2$ and comparing this quantity with $(c \cdot \sigma^2/N)$ for the c-value appropriate to the size α of the test. In either form, knowledge of the value of σ^2 is presumed.

If, as is more likely, the value of the standard deviation σ of the responses is unknown, the quantity (or, more usually, its square, the variance σ^2) may be estimated unbiasedly from the N simular responses by

$$S^2 = (N - 1)^{-1}\left[\sum_{j=1}^{N} y_j^2 - N(\bar{y})^2\right],$$

where again it may be recalled that $(N - 1)S^2/\sigma^2$ has the Chi-squared distribution with $(N - 1)$ degrees of freedom and that S^2 and $\bar{Y}$ are statistically independent whenever the Y_j are independently and normally distributed. Hence, an appropriate test of the null hypotheses H_0: $\mu = M$ may be conducted from the cumulative distribution function for the test statistic:

$$T = \frac{(\bar{Y} - M)^2/(\sigma^2/N)}{[(N-1)S^2/(N-1)\sigma^2]} = \frac{(\bar{Y} - M)^2}{(S^2/N)},$$

which has Snedecor's F-distribution of 1 and $(N-1)$ degrees of freedom whenever the null hypothesis is indeed true.

Again, large values of the test statistic could be presumed to be due to a significant deviation of the observed $\bar{Y}$ from the anticipated mean response M. Hence, the critical region is defined as the set of real numbers, $\mathscr{R} = \{t \mid t > c, \text{ where } P[F_{1,N-1} > c] = \alpha\}$, and the null hypothesis is rejected whenever the computed T exceeds c.

6.5.6. A Test for the Equality of Variance

Whenever two sets of data (i.e., two random samples) are to be compared for the equality of their means, by means of a test for a null hypothesis of the form H_0: $\Delta \equiv E(Y_1) - E(Y_2) = 0$, we have proceeded to suggest a test which has been predicated on the normality of the two random variables. More accurately, we have insisted not only on their normality but also on their homogeneity of variance; namely, $\sigma_1^2 = \sigma_2^2 = \sigma^2$. Since the test proposed for the null hypothesis regarding the equality of the means involves the computation of the unbiased estimates of these variances

$$S_j^2 = (N-1)^{-1}\left[\sum_{k=1}^{N} Y_{j,k}^2 - N(\bar{Y}_j)^2\right], \qquad j = 1 \quad \text{and} \quad 2,$$

one might well ask about the significance of their difference *before* proceeding to the test regarding the hypothetical equality of the means.

An appropriate null hypothesis for consideration then is H_0: $\sigma_1^2 = \sigma_2^2$, or equivalently H_0: $(\sigma_1^2/\sigma_2^2) = 1$. Since the random variables used to compute S_1^2 are presumed to be independent of those used in S_2^2, and since the presumption of normality of these $2N$ random variables implies that $(N-1)S_j^2/\sigma_j^2$ has the Chi-squared distribution of $(N-1)$ degrees of freedom, then the statistic

$$\frac{(N-1)S_1^2/\sigma_1^2}{(N-1)S_2^2/\sigma_2^2}$$

has Snedecor's F distribution of $(N-1)$ and $(N-1)$ degrees of freedom.

Therefore, if the null hypothesis of the equality of variances is true, then the statistic

$$T = (N - 1)S_1{}^2\sigma_2{}^2/(N - 1)S_2{}^2\sigma_1{}^2 = S_1{}^2/S_2{}^2$$

will have the Snedecor's F distribution of $(N - 1)$ and $(N - 1)$ degrees of freedom.

If either $S_1{}^2 \gg S_2{}^2$ (in which case, T will be quite large) or $S_1{}^2 \ll S_2{}^2$ (in which case T will be quite small), then the credibility (of the hypothesis that $\sigma_1{}^2 = \sigma_2{}^2$) would be in serious doubt. Therefore, in this case, one locates two critical constants c_1 and c_2 such that, for the preassigned significance level α and for the Snedecor's F statistic of $(N - 1)$ and $(N - 1)$ degrees of freedom,

$$P[c_1 < F_{N-1,N-1} < c_2] = \alpha.$$

Usually, one selects c_1 such that

$$P[F_{N-1,N-1} \leq c_1] = \alpha/2$$

and c_2 such that

$$P[F_{N-1,N-1} \geq c_2] = \alpha/2.$$

Therefore, presuming that the simular *and* simulated response $Y(T)$ at simular time T is normally distributed, one can employ N simular responses, as they arise from N simular encounters defined at the same environmental conditions $(\vec{x}_0)$, yet with differing and independently selected random number seeds, in conjunction with the same number of independent responses from the simulated system, in two ways during the validation stage of model development:

(*a*) Test for the equality of variances by comparing $S_1{}^2$ with $S_2{}^2$.
(*b*) Test for the equality of means, presuming that the ratio of $S_1{}^2$ to $S_2{}^2$ is deemed statistically insignificant, by pooling $S_1{}^2$ and $S_2{}^2$ and comparing $(\bar{Y}_1 - \bar{Y}_2)^2$ with this pooled estimate S^2 of the variance.

If both the means and the variances of the simular and simulated systems' responses are not deemed to differ significantly, one may assume that he has contributed to the completion of a major milestone in model development (the validation stage).

6.5.7. Summary

Of course, the preceding conclusion regarding the equality of means and variances can also be attained in the analysis phase of model development. Clearly, the tests for the homogeneity of variance and the equality of means can be applied to comparative studies of simular responses arising at two different environmental conditions, represented by the vectors $\vec{x}_1$ and $\vec{x}_2$. (By now, the reader should have noted the importance of the variance and/or its unbiased estimate in all of the tests which have been discussed in this chapter.)

In the subsequent chapters, techniques will be presented for the simultaneous comparison of simular responses that arise at many (more than two) environmental conditions. The tests for the significance of observed differences in the average responses at these environmental conditions will not only be predicated on the normality of the response but also be concerned with the comparison of variance estimators. Thus, a primary distribution in the remaining chapters of this book will be the family of Snedecor's F variates (cf. Section 2.15.5).

In concluding this section, two remarks are in order. First, the reader may wish to recall that a random variable T having the Student's t distribution of k degrees of freedom has the property that its square is a random variable having Snedecor's F distribution of one and k degrees of freedom. Hence, any test based on the absolute magnitude of a statistic which, under the null hypothesis, possesses the Student's t distribution, could be equally well performed by using its square and by determining the critical constant from the appropriate table of Snedecor's F statistics.

Second, it may well be that, in comparing two means $\bar{y}_1$ and $\bar{y}_2$ the first is naturally expected to be larger than the second. Such cases might well arise in simulation experiments wherein the environmental conditions producing the average simular response $\bar{y}_1$ would be expected to increase the average simular response which would arise at another environmental condition. In such a case, it shall be appropriate to concern oneself with the test statistic T rather than its absolute magnitude $|T|$; namely, with $T = (\bar{Y}_1 - \bar{Y}_2)/(2S^2/N)$. Clearly, a negative value of T, regardless of its absolute magnitude, could not lend credence to the hypothesis that the mean response μ_1 at $\vec{x}_1$ is greater than the mean response μ_2 at $\vec{x}_2$. The critical constant in this case would be that real number c such that

$$P[T_{2(m-1)} > c] = \alpha,$$

in accordance with the preselected size α of the test. Such a test is called a one-sided test and endeavors to acknowledge the anticipation that a particular one of the two means will surely be greater than the other.

EXERCISES

1. Generate 100 pairs of independent random samples, each of size 5 from the same normal distribution: of mean 4 and variance 25. For each of the 200 samples, compute its sample mean and sample variance (S^2).

(*a*) For each of the 100 pairs of sample variances, S_1^2 and S_2^2, compute the ratio $F = S_1^2/S_2^2$ and compile the sample cumulative distribution for this statistic. Compare, by means of the conventional Chi-squared test, this sample cumulative distribution function with that of the Snedecor's F variate of 4 and 4 degrees of freedom. What proportion of the 100 computed F-statistics exceeds the $\alpha = 0.05$ critical value in a test for the homogeneity of variance? Is this proportion acceptable?

(*b*) For each of the 100 pairs compute $D^* = \bar{Y}_1 - \bar{Y}_2$ and S^2(pooled) $= (S_1^2 + S_2^2)/2$, then form the statistic $T = D^*/(2S^2)^{1/2}$. Construct a frequency histogram of the 100 resulting t-values and compare with the appropriate theoretical distribution. What proportion of computed T values exceeds the $\alpha = 0.10$ critical value for testing the equality of means from which the samples arose? Is this an acceptable proportion?

2. Repeat Exercise 1, except for the generation of the variates, which now should arise from the Weibull distribution of mean 4 and variance 25 (i.e., with scale parameter $a = 3.55$, shape parameter $b = 0.80$). Discuss any observed differences in the results of the two exercises.

Chapter 7

EXPERIMENTAL DESIGNS FOR QUALITATIVE FACTORS

Every experiment is like a weapon which must be used in its particular way.

—PHILIPPUS AEROLUS PARACELSUS

7.1. Introductory Remarks

In the preceding chapter, discussion has centered about statistical techniques applicable to the analysis of responses $Y(T)$ arising from a stochastic simulation model. Methods for the point and interval estimations of the distributional properties and other parameters have been delineated, and some fundamental comparative statistical tests have been introduced.

The statistical tests have been presented as decision rules for deciding whether to accept or reject the null hypothesis about the distributional properties of simular responses. Much of the accompanying discussion has focused on the application of these statistical test procedures to the validation stage of model development, at which stage, it was shown, they are indeed applicable. Additionally, it has been indicated that these same tests are also applicable to the analysis stage of model development, wherein comparisons of simular responses that arise due to different environmental specifications (i.e., to differing input conditions) are of the essence.

However, for this latter application, the tests thus far discussed are overly restrictive, since they apply only to comparison of simular responses at *two* different environmental conditions. Since it is likely that an effective simulation model will incorporate several ($p \gg 2$) pertinent environmental conditions (i.e., input parameters) likely to contribute significantly to the simular response, the presentation of valid statistical methods for determining simultaneously their relative importance in affecting the simular response $Y(T)$ is in order.

For the present, $Y(T)$ will continue to be univariate, and will represent the condition of some pertinent simular state variable at the end of T units of simular time. Since the simulation model will be presumed stochastic (requiring the specification of a random number seed), the simular response will be deemed a random variable that is dependent upon the p pertinent environmental conditions $(x_1, x_2, \ldots, x_p)$, hereafter referred to as *factors*.

These factors can be categorized as either

(*a*) *qualitative*—discrete variables (environmental specifications) whose set of permissible values would not be normally associated with any continuum of real numbers;

(*b*) *quantitative*—continuous variables (environmental conditions) whose set of assignable values, though possibly constrained to a large, finite set of rational numbers when the model is implemented on the digital computer, is conceptually a continuum of real numbers.

As examples of qualitative factors, one can cite policy specifications, such as alternative queue disciplines in a model of a queueing network, alternative reorder policy specifications for a stochastic inventory control model, discrete environmental conditions, such as one of several exogenous event tapes (lists), or one of two, three, or four seasonal weather inputs. Other model input parameters that might be considered as discrete are those such as an input parameter whose assigned values would indicate that either the, say, Poisson or the Pascal distribution is to be applied to describe a certain stochastic aspect of the model.

Quantitative factors are exemplified by input parameters that can usually be thought of as continuous variates; for example, the simular time required to complete some activity in the model would serve as such a quantitative factor.

Another feature distinguishing between qualitative and quantitative factors is contextually related to their respective effects on the simular response at the end of T simular time units:

$$Y(T) = Y(x_1, x_2, \ldots, x_p;\ s).$$

For any fixed seed s, one would presume Y to be a continuous function of any quantitative factor, but not so for any qualitative factor. It is for this reason, that one need not include the input random number seed as an "environmental condition," for, although the seed is essentially a continuous variate, the simular response $Y(T)$ will probably not be a continuous function of it.

Presumably, the random number seed is essentially unique among quantitative input parameters in this respect. The simular response is then written as

$$Y(T) = Y(x_1, x_2, \ldots, x_p) + \varepsilon(s),$$

where $\varepsilon(s)$ is a random effect dependent upon the random number seed. The distribution of $\varepsilon(s)$ will be presumed independent of the environmental specifications $(x_1, x_2, \ldots, x_p)$ and will be assumed to have mean zero and variance σ_ε^2. Errors $\varepsilon(s_k)$ and $\varepsilon(s_m)$ will be presumed to be uncorrelated; i.e.,

$$\text{Cov}[\varepsilon(s_k), \varepsilon(s_m)] = 0 \qquad \text{for all seeds} \quad s_k \neq s_m,$$

regardless of the particular environmental specifications, provided that the Principium of Seeding is acknowledged.

In the present chapter will be presented a methodology for comparing simular responses that arise from encounters defined at environmental conditions, none of which represents the alteration of quantitative factors; i.e., the simular responses will be those from successive encounters for which only qualitative factors are altered. One may note that, as a consequence of the distributional assumptions made on the error term $\varepsilon(s)$,

$$E[Y(T)] = Y(x_1, x_2, \ldots, x_p);$$

it is the nature of this function $Y(x_1, x_2, \ldots, x_p)$, termed the *simular response function*, that the analyst should want to be able to describe. In the remainder of this, as well as the two subsequent chapters, the analytical techniques will focus on this goal as primary.

In order to accomplish this task with efficiency, a set of simulation encounters will need to be defined, presumably at as many different environmental conditions as are necessary for an adequate description of the simular response function. Since the number of such conditions will be barely finite for most complicated simulation models, an efficient procedure for the assignment of values to the environmental conditions must be undertaken. Since such a procedure constitutes an experimental plan, it is referred to as an *experimental design*.

It will be the purpose of this chapter to discuss experimental designs appropriate to simular experimentation in which the successive encounters with the model differ only in changes of qualitative factors. In Chapter 8, the definition of experimental designs in which successive encounters are specified by changes in quantitative factors will be undertaken.

In either case, the experimental design is specified with an ultimate goal in mind—the analysis of simular responses for the purpose of determining significant contributors to the simular response function. Thus, the next two chapters will be concerned with both the design and the analysis of simulation experiments. A straightforward extension of these techniques, valuable to the approximate description of the simular response function, will then be developed in Chapter 9.

7.2. The Factorial Design

A simulation experiment, then, is usually concerned with comparisons of simular responses as they arise from alternative environmental specifications for successive encounters with the model. The environmental specifications themselves may be either qualitative or quantitative factors. For the present discussion, those factors which are to be considered prime candidates as significant contributors to the simular response will be assumed qualitative. In addition, each factor will be presumed to have a finite number of possible states, or *levels*, which may be assigned to it. For example, the number of policies appropriate to the queue discipline in the simulation of a queueing network might be two: last in, first out (LIFO) or first in, first out (FIFO); or, the number of seasonal weather inputs might be four.

The more fundamental experimental designs for determining the effects of qualitative factors upon the simular response $Y(T)$ are concerned with factors having only a pair of levels, which might be appropriately labeled 0 and 1. Conventionally, the 0 level of a factor would correspond to the standard operating condition for the simulated system (or to the environmental status or policy which would likely be used in the actual system), whereas the 1 level would represent the alternative environmental specification. The binary assignments are, however, merely conventional, and may be assigned by the analyst as he so desires.

If the analyst wishes to determine the effects of the two alternative levels of a single qualitative factor, say x_1, then minimally a pair of simulation encounters must be defined, one at each of the environmental conditions:

(I) $(0, x_{2,0}, x_{3,0}, \ldots, x_{p,0})$ with seed s_1

(II) $(1, x_{2,0}, x_{3,0}, \ldots, x_{p,0})$ with seed $s_2 \neq s_1$.

The respective simular responses Y_1 and Y_2 can be then compared in quite the same manner suggested for the testing of the null hypothesis H_0: $\Delta = E(Y_1 - Y_2) = 0$ as described earlier (cf. Section 6.5.2). That is, one may test for the existence of an effect Δ in the simular response resulting, on the average, from altering only the qualitative factor x_1 between its two levels.

As the reader may recall, such a test requires explicit knowledge of the underlying variance σ^2 of the simular responses; furthermore, this variance will be presumed the same for both the distribution of Y_1 and Y_2. Under these assumptions, the test for the difference becomes a Chi-squared test, one employing the extreme (large) values in a Chi-squared distribution in order to indicate the significance in the difference in the two responses.

If, however, the variance σ^2 is not known, it may be estimated unbiasedly by replicating the experiment a total or m times; i.e., one selects $2m$ independent and nonrepetitious random number seeds, using these to specify m simulation encounters (with environmental condition I) and m encounters (with environmental condition II). The estimate of σ^2 becomes

$$S^2 = (S_{\mathrm{I}}^2 + S_{\mathrm{II}}^2)/2,$$

where

$$S_j^2 = (m-1)^{-1}\left[\sum_{k=1}^{m} y_{j,k}^2 - m(\bar{y}_j)^2\right],$$

with $y_{j,k}$ denoting the kth recorded simular response at the jth environmental specification, $k = 1, 2, \ldots, m$ and with $\bar{y}_j = m^{-1}\sum_{k=1}^{m} y_{j,k}$ denoting the mean response there, for $j = \mathrm{I}$ and II. The reader may recall that the test for the comparison of $\bar{y}_{\mathrm{I}}$ and $\bar{y}_{\mathrm{II}}$ relies upon the magnitude of a Student's t statistic of $(2m - 2)$ degrees of freedom, or, alternatively, upon its square, a Snedecor's F statistic of 1 and $(2m - 2)$ degrees of freedom.

An alternative viewpoint is to consider that these simular responses can be represented as:

$$Y_j(T) = \mu + \Delta + \varepsilon_{j,k}$$

where μ is a mean effect, where Δ is the effect of the qualitative factor (equal to zero if $j = \mathrm{I}$, and to a positive constant Δ if $j = \mathrm{II}$), and where $\varepsilon_{j,k}$ is the random error arising from the kth encounter at the jth environmental specification. Assuming that the $\varepsilon_{j,k}$ are uncorrelated random variables of mean 0 and common variance σ^2 then

$$E(Y_{\mathrm{I},k}) = \mu + 0 + 0 = \mu,$$

whereas

$$E(Y_{\mathrm{II},k}) = \mu + \Delta + 0 = \mu + \Delta,$$

yet

$$\mathrm{Var}(Y_{\mathrm{I},k}) = \mathrm{Var}(Y_{\mathrm{II},k}) = \sigma^2, \qquad k = 1, 2, \ldots, m.$$

Therefore, Δ does indeed represent the average difference to be anticipated between responses at the two levels of the qualitative factors of interest; hence, a test of the null hypothesis,

$$H_0: \Delta = 0,$$

is quite appropriate to the determination of the significance of any observed difference between $\bar{y}_{\mathrm{I}}$ and $\bar{y}_{\mathrm{II}}$.

7.3. The 2^2 Factorial Design

Most likely, a stochastic simulation model will not only be constructed for the purpose of investigating the effects of a single qualitative factor, but also in order to examine the relative effects of two or more factors. In the most fundamental case, one may presume that two qualitative factors, say x_1 and x_2, each of two levels, are each suspected of providing a significant contribution to the simular response. The simular response might then be written

$$Y(T) = \mu + \theta_1 x_1 + \theta_2 x_2 + \varepsilon,$$

where μ is a mean effect, θ_i is an effect due to qualitative factor i, and x_i is 0, if the ith qualitative factor is at its 0 level and 1, if it is at its 1 level, for $i = 1$ and 2. Here, ε is an error effect which again may be presumed to be uncorrelated among independently seeded encounters and which will be presumed to be a random variable of mean 0 and variance σ^2.

7.3.1. Analysis of the Single Replicate

Suppose then that four simular encounters are defined, one at each of the four ($= 2^2$) possible permutations of the two levels of the two qualitative factors, x_1 and x_2. Denoting by y_{qr} the recorded simular response at the qth level of factor x_1 and rth level of factor x_2, then the response becomes

$$y_{qr} = \mu + \theta_1 \cdot q + \theta_2 \cdot r + \varepsilon_{qr}.$$

For example, the simular response

$$y_{00} = \mu + \varepsilon_{00}$$

represents that arising from an encounter whose environmental conditions were specified by $(0, 0, x_{3,0}, x_{4,0}, \ldots, x_{p,0})$ plus a random number seed, say s_{00}; whereas the simular response

$$y_{01} = \mu + \theta_2 + \varepsilon_{01},$$

arising from environmental conditions $(0, 1, x_{3,0}, x_{4,0}, \ldots, x_{p,0})$ plus an independently selected and different seed, incorporates not only the mean and random effects, but also a presumed effect due to the specification of the "1" level for the second factor. Similarly,

$$y_{10} = \mu + \theta_1 + \varepsilon_{10}$$

is the sum of the three effects: mean, first factor, and randomness, yet

$$y_{11} = \mu + \theta_1 + \theta_2 + \varepsilon_{11},$$

the simular response resulting from the specification $(1, 1, x_{3,0}, x_{4,0}, \ldots, x_{p,0})$ contains effects due to both the "1" levels for the two factors. The reader should note the presumed additivity of these effects.

Therefore a single replication of the 2^2 factorial design consists of the following four $(= 2^2)$ simular environmental specifications, producing the corresponding responses as indicated in Table 7.1. The table lists

TABLE 7.1

THE 2^2 FACTORIAL DESIGN

Environmental condition	Code	Response	Expectation
$(0, 0, y_{3,0}, y_{4,0}, \ldots, y_{p,0})$	00	y_{00}	μ
$(0, 1, y_{3,0}, y_{4,0}, \ldots, y_{p,0})$	01	y_{01}	$\mu + \theta_2$
$(1, 0, y_{3,0}, y_{4,0}, \ldots, y_{p,0})$	10	y_{10}	$\mu + \theta_1$
$(1, 1, y_{3,0}, y_{4,0}, \ldots, y_{p,0})$	11	y_{11}	$\mu + \theta_1 + \theta_2$

the environmental conditions for each of the four encounters, introduces a standardized code for this specification, and denotes the resultant simular response and its expectation. The reader should note that, despite the differing representations for the means of the response, their variances are all presumed equal to σ^2, the error variance [i.e., $\text{Var}(Y_{qr}) = \text{Var}(\varepsilon_{qr})$].

One may note that the responses y_{00} and y_{10} are identically the Y_1 and Y_2 of Section 7.2, since they represent the responses from a pair of encounters, one of which is at the 0 level of a qualitative factor (x_1), the other at the 1 level of the same factor, while each is at the 0 (or standard) level of the second (third, etc.) factors. Therefore, the contrast

$$y_{10} - y_{00} = \mu + \theta_1 + \varepsilon_{10} - \mu - \varepsilon_{00} = \theta_1 + (\varepsilon_{10} - \varepsilon_{00})$$

is a random variable of mean θ_1, but variance $2\sigma^2$, since

$$\text{Var}(Y_{10} - Y_{00}) = \text{Var}(\varepsilon_{10} - \varepsilon_{00}) = \text{Var}(\varepsilon_{10}) + \text{Var}(\varepsilon_{00}) = 2\sigma^2.$$

Similarly, a second contrast can be formed from the two simular responses that arise from encounters defined at the 1 level of the second factor; namely,

$$y_{11} - y_{01} = \mu + \theta_1 + \theta_2 + \varepsilon_{11} - (\mu + \theta_2 + \varepsilon_{01}) = \theta_1 + (\varepsilon_{11} - \varepsilon_{01}),$$

a contrast also having expectation

$$E(Y_{11} - Y_{01}) = \theta_1$$

and variance

$$\text{Var}(Y_{11} - Y_{01}) = \text{Var}(\varepsilon_{11} - \varepsilon_{01}) = 2\sigma^2,$$

since the random variables ε_{qr} are presumed to be uncorrelated, of mean zero, and of the same variance σ^2 regardless of the environmental conditions at which the four simular encounters are defined. (Recall that the lack of correlation among errors is presumed to be a direct consequence of the independence of seed selections for the four encounters.)

Therefore, one has available in the responses from a 2^2 factorial design a pair of uncorrelated and unbiased estimates of θ_1, which is the effect of altering the qualitative factor x_1 from its 0 level; viz.,

$$(Y_{10} - Y_{00}) \quad \text{and} \quad (Y_{11} - Y_{01}).$$

The two contrasts are uncorrelated since

$$E[(Y_{10} - Y_{00} - \theta_1) \cdot (Y_{11} - Y_{01} - \theta_1)] = E[(\varepsilon_{10} - \varepsilon_{00}) \cdot (\varepsilon_{11} - \varepsilon_{01})]$$

is the same as

$$E(\varepsilon_{10} \cdot \varepsilon_{11} - \varepsilon_{10} \cdot \varepsilon_{01} - \varepsilon_{00} \cdot \varepsilon_{11} + \varepsilon_{00} \cdot \varepsilon_{01}) = 0,$$

because the errors are presumed to have zero means and covariances. A single unbiased estimate of the effect θ_1 is then

$$\tilde{\theta}_1 = \tfrac{1}{2}[Y_{10} - Y_{00} - Y_{01} + Y_{11}],$$

having variance

$$\text{Var}(\tilde{\theta}_1) = \tfrac{1}{4}[4\sigma^2] = \sigma^2.$$

The effect θ_2 of the second factor's being altered from its 0 level to its 1 level may similarly be estimated unbiasedly by either of the random variables,

$$(Y_{01} - Y_{00}) \qquad \text{or} \qquad (Y_{11} - Y_{10}),$$

the first being a contrast of the two responses from encounters defined at the same (0) level of the first factor (x_1), the second being the contrast of the remaining two responses [each deriving from encounters at the other (1) level of the first factor]. One may note that each contrast represents the difference between the simular responses at the 1 and 0 levels of the factor (x_2) of interest, yet at the same level of the remaining factor (x_1).

Again, the two contrasts may be readily shown to be unbiased for θ_2 and uncorrelated with each other. Indeed, they may be combined to form the estimate

$$\tilde{\theta}_2 = \tfrac{1}{2}[Y_{01} - Y_{00} - Y_{10} + Y_{11}]$$

having variance

$$\text{Var}(\tilde{\theta}_2) = \tfrac{1}{4}[4\sigma^2] = \sigma^2.$$

Of special importance is the fact that these two straightforward estimates of θ_1 and θ_2 are themselves uncorrelated, since

$$\begin{aligned} \text{Cov}(\tilde{\theta}_1, \tilde{\theta}_2) &= \tfrac{1}{4}\,\text{Cov}[\varepsilon_{10} - \varepsilon_{00} - \varepsilon_{01} + \varepsilon_{11}, \varepsilon_{01} - \varepsilon_{00} - \varepsilon_{10} + \varepsilon_{11}] \\ &= \tfrac{1}{4}[-\text{Var}(\varepsilon_{10}) + \text{Var}(\varepsilon_{00}) - \text{Var}(\varepsilon_{01}) + \text{Var}(\varepsilon_{11})] \\ &= \tfrac{1}{4}[-\sigma^2 + \sigma^2 + \sigma^2 - \sigma^2] = 0. \end{aligned}$$

Furthermore, each estimate is uncorrelated with the mean response, $\bar{y} = \frac{1}{4}(y_{00} + y_{01} + y_{10} + y_{11})$, as can be seen from the entries of Table 7.2, which presents the coefficients for the simular responses as required

TABLE 7.2

COEFFICIENTS OF RESPONSES FOR ESTIMATORS FROM THE 2^2 FACTORIAL DESIGN

Estimator	y_{00}	y_{01}	y_{10}	y_{11}	Expectation	Variance
$\tilde{\mu}$	$\frac{1}{4}$	$\frac{1}{4}$	$\frac{1}{4}$	$\frac{1}{4}$	$\mu + (\theta_1 + \theta_2)/2$	$\sigma^2/4$
$\tilde{\theta}_1$	$-\frac{1}{2}$	$-\frac{1}{2}$	$\frac{1}{2}$	$\frac{1}{2}$	θ_1	σ^2
$\tilde{\theta}_2$	$-\frac{1}{2}$	$\frac{1}{2}$	$-\frac{1}{2}$	$\frac{1}{2}$	θ_2	σ^2
$\tilde{\varepsilon}$	$-\frac{1}{2}$	$\frac{1}{2}$	$\frac{1}{2}$	$-\frac{1}{2}$	0	σ^2

in the estimates, in addition to the estimator's expectations and variances. The reader may note that the first three rows of coefficients in the table are orthogonal. In fact, a fourth row, orthogonal to each of the others, may be added as shown in the table (and denoted by $\tilde{\varepsilon}$, since it estimates zero, the mean error). Recalling that a set of k orthogonal linear combinations of k uncorrelated random variables becomes itself a set of k uncorrelated random variables, then the preceding statements regarding the null covariances of the estimates follow more immediately (cf. Section 2.15).

Until now, no assumptions have been made about the distributional properties of the random variables ε_{qr} (and hence, Y_{qr}), save the implicit assumption of the existence of moments of order one and two. The reader might note that if the additional assumption of normality for the distribution of the ε_{qr} is made, then the earlier assumption of their being uncorrelated implies their independence (and, hence, the independence of the simular responses Y_{qr} themselves). As a consequence, the estimators $\tilde{\mu}$, $\tilde{\theta}_1$, $\tilde{\theta}_2$, $\tilde{\varepsilon}$, being uncorrelated linear combinations of the simular responses, become independently and exactly normally distributed with mean and variance given by Table 7.2.

However, the approximate normality of these four estimators can be assumed, even without the presumption that the simular responses have Gaussian distribution. This approximate normality is reasonably well assured when one notes the likely applicability of the Central Limit Theorem, since each estimator is the direct sum of four uncorrelated random variables. These random variables of the summand, though not necessarily from the same distribution (those with negative coefficients may have probability densities different from those with positive coefficients whenever the coefficient is "absorbed into" the random variable before summing), do possess the same variance in each case. Consequently, although the simular responses may not be normally distributed, the estimates $\tilde{\theta}_1$ and $\tilde{\theta}_2$ of a pair of qualitative factors may tend to have a Gaussian distribution.

The estimates $\tilde{\theta}_1$ and $\tilde{\theta}_2$ can also be viewed as the *incremental responses* associated with the change of the respective factors x_1 and x_2 from their 0 to 1 levels. Given these estimates, one might well ask about their relative significance; i.e., one might seek to test null hypotheses of the form:

$$H_0\colon \theta_1 = 0 \qquad \text{or} \qquad H_0\colon \theta_2 = 0.$$

By presuming the (near) normality of the estimates $\tilde{\theta}_1$ and $\tilde{\theta}_2$, their lack of correlation with one another and with $\tilde{\varepsilon}$ implies the (approximately) stochastic independence of these three statistics, so that independent tests of these two hypotheses can be formulated. Unless otherwise stated, it will be assumed that the various estimators are normally distributed random variables.

The rationale for these tests can be developed by noting that from Table 7.2,

$$E[(\tilde{\theta}_1)^2] = \operatorname{Var}(\tilde{\theta}_1) + [E(\tilde{\theta}_1)]^2 = \sigma^2 + \theta_1^2,$$
$$E[(\tilde{\theta}_2)^2] = \operatorname{Var}(\tilde{\theta}_2) + [E(\tilde{\theta}_2)]^2 = \sigma^2 + \theta_2^2,$$

and

$$E[(\tilde{\varepsilon})^2] = \operatorname{Var}(\tilde{\varepsilon}) + [E(\tilde{\varepsilon})]^2 = \sigma^2.$$

Hence, if the null hypothesis $H_0\colon \theta_1 = 0$ is indeed true, the statistic

$$(\tilde{\theta}_1)^2 = (Y_{10} - Y_{00} - Y_{01} + Y_{11})^2/4$$

will be an unbiased estimate of σ^2, independent of another unbiased estimate given by

$$(\tilde{\varepsilon})^2 = (Y_{10} + Y_{01} - Y_{00} - Y_{11})^2/4.$$

Independence is assured, since $\tilde{\theta}_1$ and $\tilde{\varepsilon}$ are presumed to be normally and independently distributed; hence their squares will be independent random variables.

Furthermore, the statistic

$$(\tilde{\theta}_1)^2/\sigma^2 = (Y_{10} - Y_{00} - Y_{01} + Y_{11})^2/(4\sigma^2)$$

will have the Chi-squared distribution of a single degree of freedom *if* the null hypothesis that $\theta_1 = 0$ is true. Regardless of the veracity of the null hypothesis, the statistic

$$(\tilde{\varepsilon})^2/\sigma^2 = (Y_{10} + Y_{01} - Y_{00} - Y_{11})^2/(4\sigma^2)$$

will then have the Chi-squared distribution of 1 degree of freedom. The independence of these two statistics then implies that the easily computed ratio

$$T_1 \equiv (\tilde{\theta}_1)^2/(\tilde{\varepsilon})^2 = (Y_{10} - Y_{00} - Y_{01} + Y_{11})^2/(Y_{10} + Y_{01} - Y_{00} - Y_{11})^2$$

has Snedecor's F distribution of 1 and 1 degree of freedom whenever the incremental response θ_1 is zero. Therefore, since an extremely large value of this statistic not only would be unlikely, but also would presumably arise whenever the null hypothesis were invalid (i.e., whenever $\theta_1 \neq 0$), one may compare the computed ratio T_1 with the extreme cumulative points for the Snedecor's F distribution of 1 and 1 degree of freedom. Specifically, for a size $\alpha = 0.05$, one would reject the null hypothesis whenever the computed statistic T_1 exceeds 161.

A similar test for the significance of the estimate $\tilde{\theta}_2$ of the incremental response due to the alteration of the second factor from its 0 to its 1 level is based upon the readily computed statistic:

$$T_2 = (\tilde{\theta}_2)^2/(\tilde{\varepsilon})^2 = (Y_{01} - Y_{00} - Y_{10} + Y_{11})^2/(Y_{10} + Y_{01} - Y_{00} - Y_{11})^2$$

which shall also possess the Snedecor's F distribution with 1 and 1 degree of freedom whenever the null hypothesis H_0: $\theta_2 = 0$ is true. Therefore, upon observing the four simular responses Y_{00}, Y_{01}, Y_{11}, and Y_{01} the observed incremental effect of the levels of the second factor, as measured by $\tilde{\theta}_2$, can be declared highly significant whenever the size α of the test is, e.g., 0.01, and the computed T_2 exceeds 4050.

7.3.2. The Replicated 2^2 Factorial Design

One may note then that a single replication of the 2^2 factorial design leads to a test statistic (T_1 or T_2), which is the ratio of the squared estimate of the effect to an unbiased estimate of the error variance σ^2. Only whenever the squared estimate (of θ_1 or θ_2) is especially larger (161 times as great for $\alpha = 0.05$, 4050 times as great for $\alpha = 0.01$), will the incremental response (θ_1 or θ_2) be declared to exist. Equivalently, only when the absolute value of the estimate, say $|\tilde{\theta}_1|$, is especially larger than the estimated standard deviation (in this case $|\tilde{\varepsilon}|$), will the null hypothesis

H_0: $\theta_1 = 0$ be rejected; at the size $\alpha = 0.05$, one would require that

$$|\tilde{\theta}_1| > [161(\tilde{\varepsilon})^2]^{1/2} \cong 12.7 |\tilde{\varepsilon}|$$

before declaring that one of the levels of factor x_1 is a significant contributor to the simular response.

A more reassuring alternative is to obtain a more accurate estimate of the error variance, which may be achieved by replicating the 2^2 design a total of m (>1) times; i.e., at each of the four environmental conditions (coded 00, 01, 10, and 11 in Table 7.1), one defines m independently seeded simular encounters. The response of the kth encounter at coded environmental condition (qr) can be denoted by

$$Y_{qr(k)} \equiv y_{qr(k)} = \mu + \theta_1 \cdot q + \theta_2 \cdot r + \varepsilon_{qr(k)},$$

where the additive components are as defined in the preceding section, with the $\varepsilon_{qr(k)}$ again presumed to be uncorrelated random variables of mean zero and common variance σ^2. The simular response at environmental condition (qr) can then be taken to be the arithmetic average of the m simular responses obtained at this environmental specification; namely,

$$\bar{Y}_{qr} \equiv \bar{y}_{qr} = m^{-1} \sum_{k=1}^{m} y_{qr(k)}, \qquad q = 0 \text{ and } 1, \quad r = 0 \text{ and } 1.$$

Furthermore, an independent and unbiased estimate of the variance of the simular response at this environmental specification (qr) becomes

$$S_{qr}^2 = (m-1)^{-1} \left[\sum_{k=1}^{m} y_{qr(k)}^2 - m(\bar{y}_{qr})^2 \right],$$

and since the variances are presumed to be equivalent at all environmental specifications, and since each S_{qr}^2 is based on m simular responses that are independent of the other three sets of m responses, one may pool the four estimates to obtain the unbiased estimate of σ^2:

$$S^2 = (S_{00}^2 + S_{01}^2 + S_{10}^2 + S_{11}^2)/4.$$

Each of the four independent estimates S_{qr}^2 is related to a Chi-squared variate; namely, the statistic

$$(m-1)S_{qr}^2/\sigma^2 = \left[\sum_{k=1}^{m} y_{qr(k)}^2 - m(\bar{y}_{qr})^2 \right] \Big/ \sigma^2$$

has the Chi-squared distribution of $m - 1$ degrees of freedom. Owing to their independence then, the statistic $4(m - 1)S^2/\sigma^2$, being the sum of the four Chi-squared variates, also has the Chi-squared distribution, but of $4m - 4$ degrees of freedom.

Now, the unbiased estimates of θ_1 and θ_2 become, analogous to the estimates obtained from the simular responses for a single replication of a 2^2 factorial design:

$$\tilde{\theta}_1 = (\bar{Y}_{10} - \bar{Y}_{00} - \bar{Y}_{01} + \bar{Y}_{11})/2,$$

and

$$\tilde{\theta}_2 = (\bar{Y}_{01} - \bar{Y}_{00} - \bar{Y}_{10} + \bar{Y}_{11})/2,$$

each now having a variance equal to

$$\mathrm{Var}(\tilde{\theta}_1) = \mathrm{Var}(\tilde{\theta}_2) = \sigma^2/m.$$

Again, the estimates $\tilde{\theta}_1$ and $\tilde{\theta}_2$, being orthogonal linear combinations of the uncorrelated (and independent, if normality is assumed for the simular responses) average responses $\bar{y}_{qr}$, are uncorrelated (independent) with (of) one another.

If the null hypothesis, say H_0: $\theta_1 = 0$, is true, then

$$E[(\tilde{\theta}_1)^2] = \sigma^2/m,$$

so that the presumption of normality for the distribution of the simular responses implies that under the null hypothesis, $(\tilde{\theta}_1)^2/(\sigma^2/m)$ has the Chi-squared distribution of 1 degree of freedom. Since this statistic is distributed independently of S^2, the ratio

$$T_1 = \frac{(\tilde{\theta}_1)^2/(\sigma^2/m)}{(m - 1)4S^2/[(4m - 4)\sigma^2]} = \frac{(\tilde{\theta}_1)^2}{S^2/m}$$

has the Snedecor's F distribution of 1 and $(4m - 4)$ degrees of freedom whenever θ_1 is indeed zero.

Now, the statistic T_1 would tend to be larger whenever $\theta_1 \neq 0$ than when the null hypotheses was true, since its numerator $(\tilde{\theta}_1)^2$ has expectation, in general,

$$E[(\tilde{\theta}_1)^2] = \theta_1^2 + (\sigma^2/m).$$

Therefore, significantly large values of T_1 tend to be associated with the existence of a difference θ_1 between responses observed at the 1 and 0 levels of the qualitative factor (x_1). The estimated incremental response, $\tilde{\theta}_1$, is declared significant then whenever T_1 is improbably large.

The test for the significance of the computed estimate,

$$\tilde{\theta}_2 = \tfrac{1}{2}[\bar{y}_{01} - \bar{y}_{00} - \bar{y}_{10} + \bar{y}_{11}],$$

is conducted similarly by comparing the statistic

$$T_2 = (\tilde{\theta}_2)^2/(S^2/m)$$

with an appropriate extremal cumulative point in the Snedecor's F distribution of 1 and $4m - 4$ degrees of freedom.

One may observe that again the test of the null hypothesis, say, H_0: $\theta_1 = 0$, depends upon the relative sizes of $(\tilde{\theta}_1)^2$ and (S^2/m), both being unbiased estimates of the same quantity (σ^2/m) whenever the null hypothesis is indeed true. Equivalently, the test depends upon the relative sizes of $|\tilde{\theta}_1|$ and $(S^2/m)^{1/2}$, yet now the size of $|\tilde{\theta}_1|$ is declared significant $(\alpha = 0.05)$ whenever

$$|\tilde{\theta}_1| > \sqrt{7.71}\, S/\sqrt{2} \cong 1.96 \cdot S \qquad \text{and} \qquad m = 2,$$

or whenever

$$|\tilde{\theta}_1| > \sqrt{5.32}\, S/\sqrt{3} \cong 1.33 \cdot S \qquad \text{and} \qquad m = 3.$$

The gains of replication become then seemingly real, in the sense that a smaller observed ratio of $|\tilde{\theta}_1|$ to the estimated standard deviation S is required in order to reject the null hypothesis H_0: $\theta_1 = 0$.

The superficiality of this result becomes apparent, however, when one recalls that, for $m = 2$ replications, $\tilde{\theta}_i$ is a linear combination of 8 $(2^2 \times 2)$ simular responses, whereas for $m = 3$, $\tilde{\theta}_i$ is a linear combination of 12 $(2^2 \times 3)$ such responses and that S^2 also changes in the number of simular responses used in its computation as well. Of primary importance, however, is the fact that the *power* of the test improves with increased replication; i.e., if a small, but nonzero incremental response θ_i does indeed exist, the probability of its being detected increases as the number m of replications increases.

One further advantage of replication should be noted. The tests proposed in this section rely theoretically on the normality of the distribution of the simular response; however, since the test statistics T_1 and T_2 become based on squares of sums which possess more and more independent random variables (simular responses) as the number of replications is augmented, then one may anticipate that increased replication leads to a less approximate test (whenever the simular responses are not normally distributed from the start). If, on the other hand, the simular responses are Gaussian variates, then the tests, as given, are exact.

EXERCISES

1. Generate 100 independent random variables from each of four normal distributions:

Distribution	Mean	Variance
00	5	25
01	55	25
10	15	25
11	65	25

Presume then that each of the 100 quadruples is a set of four responses arising from simular encounters defined in accordance with a 2^2 factorial design, where apparently the actual $\theta_1 = 10$, the actual $\theta_2 = 50$. For each of the 100 experiments, compute the statistics T_1 and T_2 necessary for testing the null hypotheses H_0: $\theta_1 = 0$ and H_0: $\theta_2 = 0$, respectively. Determine the proportion of the 100 tests in which each, either, or both of the two null hypotheses are rejected.

2. Repeat Exercise 1 by generating 200 independent random variables from each of the four normal distributions there. Pairing the 200 normal variates of each distribution, presume that each of the 100 quadruples of pairs represents the simular responses arising in accordance with a twice-replicated 2^2 factorial design. For each set of eight (2×2^2) responses, perform the recommended tests of the hypothesis H_0: $\theta_1 = 0$ and H_0: $\theta_2 = 0$, noting the proportion of the 100 tests in which rejection of each, either, or both hypotheses occurs. Explain any differences in the directly comparable results of the first exercise.

3. Repeat Exercise 1, but generate the simular responses from Weibull distributions of the given means and variances. (The shape parameter b can be computed using the tables of the function $R(b^{-1}) = [E(Y)]^2/[E(Y^2)]$, as presented in Section 6.4; the shape parameter a becomes then $a = E(Y)/\Gamma(1 + b^{-1})$.) Compare the results with those of Exercise 1.

4. Repeat Exercise 2, generating the simular responses, however, as Weibull variates of the given mean and variance. Compare the results with those of the first three exercises.

7.3.3. Interaction

The reader may recall that in the discussion of the design and analysis of a single replication of a 2^2 factorial design, an "error estimate", the random variable

$$\tilde{\varepsilon} = (-Y_{00} + Y_{01} + Y_{10} - Y_{11})/2$$

of mean zero and variance σ^2, was defined. This estimate was then subsequently used in tests for both of the null hypotheses H_0: $\theta_i = 0$, $i = 1$ and 2. Nonetheless, in the discussion of the design and analysis of the multiply replicated 2^2 factorial design, the error variance was estimated by pooling the four estimates of σ^2 (one estimate from the responses of each of the four environmental conditions, or *design points*: 00, 01, 10, 11). Since in the second case, an obviously analogous error estimate

$$\tilde{\varepsilon} = (-\bar{Y}_{00} + \bar{Y}_{01} + \bar{Y}_{10} - \bar{Y}_{11})/2$$

certainly exists, one might ask the reason for not using *its* square to estimate unbiasedly (σ^2/m).

The answer lies in the following maxim: The error variance σ^2 is properly the variance of the simular response $Y(T)$ at any particular environmental specification $(x_1, x_2, \ldots, x_p)$, and therefore is properly estimated only by replicating the simular response at the same environmental conditions (except, of course, for independently and nonrepetitively selected random number seeds). Our discussions of the 2^2 factorial design, whether they referred to the existence of replication or not, have assumed that the error variances σ^2 are the same, regardless of the specification of the environmental conditions. That is, the variance of the simular response obtained whenever the environmental specification $(x_{1,0}, x_{2,0}, \ldots, x_{p,0})$ is used becomes the same as that for the simular response obtained whenever the environmental specification is otherwise.

Nonetheless, any estimate of error variance in the absence of replication can actually be an estimate of some additional term not included in the representation of the simular response. For example, the earlier presumption was that

$$Y_{qr} = \mu + \theta_1 \cdot q + \theta_2 \cdot r + \varepsilon_{qr},$$

indicating that any simular response is composed of the sum of a mean term (μ), a random term (ε_{qr}), and the conditional terms (θ_1 and θ_2), which are present if the qualitative variates x_1 and x_2 assume their 1 level in the environmental specification for the corresponding model encounters.

However, one might conceive of an additional effect arising whenever *both* qualitative variates are present at their 1 levels. Such an additional effect is called an *interaction*. If existent, an interaction between two qualitative factors must be included in the expression for the simular response:

$$Y_{qr} = \mu + \theta_1 q + \theta_2 r + \theta_{12} \cdot q \cdot r,$$

where the quantity θ_{12} (the interaction) is an additional effect present only when both q and r equal 1 (i.e., present only when both x_1 and x_2 are at their 1 levels). For example, although the qualitative variate (which describes, say, queue discipline in a simulation model) might prove to augment the system's productivity if it is altered from FIFO (0 level) to LIFO (1 level), and although a second qualitative factor (which describes, say, the use of exogenous weather conditions for the simulation) might also tend to augment the simular response if summer, rather than winter, conditions are employed, the combination of LIFO queue discipline and summer weather need not result in an improvement that is exactly the sum of the improvements of their individual effects. Rather, their combined presence might either enhance or retard the expected simular response; in either event, the interaction (θ_{12}) is present.

The error estimate

$$\tilde{\varepsilon} = (-Y_{00} + Y_{01} + Y_{10} - Y_{11})/2$$

becomes in the presence of interaction:

$$\tilde{\varepsilon} = [(-\mu) + (\mu + \theta_2) + (\mu + \theta_1) - (\mu + \theta_1 + \theta_2 + \theta_{12})]/2$$

or

$$\tilde{\varepsilon} = \theta_{12}/2,$$

half the interaction itself. The same result obtains whenever the 2^2 factorial design is replicated a total of m times; namely, the error estimate,

$$\tilde{\varepsilon} = [-\bar{Y}_{00} + \bar{Y}_{01} + \bar{Y}_{10} - \bar{Y}_{11}]/2 = \theta_{12}/2.$$

Therefore, if one intends to use only a single replicate of the 2^2 factorial design, he should be relatively certain of the nonexistence of interaction

(θ_{12}), since otherwise his estimated error variance will not estimate this quantity (zero) at all. Instead,

$$E[(\tilde{\varepsilon})^2] = \text{Var}(\tilde{\varepsilon}) + [E(\tilde{\varepsilon})]^2 = (\sigma^2/m) + (\theta_{12}/4),$$

so that for $m = 1$ (replication), no unbiased estimate of σ^2 will exist at all. Consequently, in the presence of interaction, no test for the significance of the estimated $\tilde{\theta}_1$ or $\tilde{\theta}_2$ exists after a single replication has been performed.

One may note, however, that after m replications of the 2^2 factorial design have been conducted, not only the tests for the existence of θ_1 and of θ_2 may be performed but also a test for the presence of interaction becomes available. To wit, the statistic

$$T_{\text{I}} = (\tilde{\varepsilon})^2/(S^2/m)$$

may be compared with tabulations of the cumulative distribution of Snedecor's F distribution of one and $(4m - 4)$ degrees of freedom; the null hypothesis, H_0: $\theta_{12} = 0$ is then rejected whenever T_{I} is significantly large (greater than, conventionally, the point at which 95% of this F distribution would have been accumulated).

In conclusion then, one should be reminded of the importance of replication, the independent repetition of a simular experimental design; not only is a more precise estimate (S^2) of the underlying error variance so obtained, but also this estimate is no longer confounded with any existent interaction θ_{12}. Furthermore, the replication allows testing of the presence of both interaction (θ_{12}) and main effects (or incremental effects θ_1 and θ_2). Whenever costs are not exorbitant, properly performed replications are in order, especially so in the earlier analyses with a simulation model (prior to the attainment of firm conclusions about the presence of interaction).

EXERCISES

1. Refer to Exercise 1 of Section 7.3.2, and repeat it, except for the fourth (11) distributional form in which you should use a mean of 45 (instead of 65). Compare the proportion of tests that reject the null hypotheses H_0: $\theta_1 = 0$ and H_0: $\theta_2 = 0$, with the proportions obtained previously; comment on their differences based on the presence of interaction $(\theta_{12} = -20)$ in the current exercise.

2. Repeat Exercises 2, 3, and 4 of Section 7.3.2, changing only the mean value of the fourth probability density functions. Compare the corresponding Monte Carlo analyses and comment on their differences. (Note that $\theta_{12} \neq 0$.)

7.4. The 2^3 Factorial Design

Whenever there are K qualitative factors, each possessing two assignable conditions (referred to as the 0 and 1 levels of the factor), one must define at least 2^K simulation encounters in order to ensure that every possible subset of factors has been selected for use at the 1 level in the absence *and* in the presence of the remaining factors' being at that level. For $K = 2$ qualitative factors, the minimal number of simular encounters for a single replication of the 2^2 design was seen to be 4. In the current section, the case for investigating $K = 3$ qualitative factors will be presented. The resulting experimentation plan will be referred to as the 2^3 experimental design.

One may suppose that an hypothetical incremental effect θ_i exists between the two levels of the ith of the three qualitative factors, so that the simular response may be written as

$$Y_{qrs} = \mu + \theta_1 \cdot q + \theta_2 \cdot r + \theta_3 \cdot s + \varepsilon_{qrs},$$

where μ is a mean effect, θ_i is an incremental effect for the ith factor, $i = 1, 2, 3$, and ε_{qrs} is an error term (a random variable), of mean zero, of variance σ^2, and uncorrelated with error terms arising at other environmental specifications. Again, the assignable values of q, r, and s are the binary digits, 0 and 1, corresponding to the levels of the first, second, and third qualitative factors, respectively; usually, the level 0 will be taken as that of the standard condition, or usual policy applicable to the simulated system, and the level 1 as its alternate.

Without loss of generality, one can assume that the three environmental conditions are the first three in the vector of environmental specifications $(x_1, x_2, \ldots, x_p)$ that are required as input data for the simulation model. The remaining environmental conditions $(x_4, x_5, \ldots, x_p)$ will be assigned fixed values, probably those corresponding to the standard operating conditions for these parameters in the simulated system: $(x_{4,0}, x_{5,0}, \ldots, x_{p,0})$. One can then tabulate and codify the eight environmental conditions of the 2^3 factorial design and delineate the response and the expectations, as in Table 7.3.

TABLE 7.3

THE 2^3 FACTORIAL DESIGN

Environmental condition	Code	Response	Expectation
$(0, 0, 0, x_{4,0}, \ldots, x_{p,0})$	000	Y_{000}	μ
$(0, 0, 1, x_{4,0}, \ldots, x_{p,0})$	001	Y_{001}	$\mu + \theta_3$
$(0, 1, 0, x_{4,0}, \ldots, x_{p,0})$	010	Y_{010}	$\mu + \theta_2$
$(1, 0, 0, x_{4,0}, \ldots, x_{p,0})$	100	Y_{100}	$\mu + \theta_1$
$(0, 1, 1, x_{4,0}, \ldots, x_{p,0})$	011	Y_{011}	$\mu + \theta_2 + \theta_3$
$(1, 0, 1, x_{4,0}, \ldots, x_{p,0})$	101	Y_{101}	$\mu + \theta_1 + \theta_3$
$(1, 1, 0, x_{4,0}, \ldots, x_{p,0})$	110	Y_{110}	$\mu + \theta_1 + \theta_2$
$(1, 1, 1, x_{4,0}, \ldots, x_{p,0})$	111	Y_{111}	$\mu + \theta_1 + \theta_2 + \theta_3$

Each of the eight encounters will require its own, independently selected, random number seed, with repetitions precluded. This will ensure the independence of the error terms, which have variance σ^2 regardless of the environmental conditions selected. Consequently, the variances of the simular responses Y_{qrs} will be equal to σ^2, and their correlations (and covariances) assumed equal to 0.

The reader may see that there exist four direct, paired contrasts for the purpose of estimating any particular incremental, or marginal, effect θ_i. For example, any one of the four contrasts:

$$(Y_{100} - Y_{000}), \qquad (Y_{101} - Y_{001}), \qquad (Y_{110} - Y_{010}), \qquad (Y_{111} - Y_{011}),$$

has mean θ_1, as subtraction of the appropriate entries of the final column of Table 7.3 reveals. Their arithmetic average can be written as

$$\tilde{\theta}_1 = (-Y_{000} - Y_{001} - Y_{010} - Y_{011} + Y_{100} + Y_{101} + Y_{110} + Y_{111})/4,$$

or

$$\tilde{\theta}_1 = [(Y_{100} + Y_{101} + Y_{110} + Y_{111})/4] - [(Y_{000} + Y_{001} + Y_{010} + Y_{011})/4]$$

In the latter form, the unbiased estimate $\tilde{\theta}_1$ of θ_1 becomes evidently the difference between the average response observed at the 1 level of the first factor and the average of the four responses recorded at the 0 level of this factor.

Proceeding analogously, one may determine unbiased estimates of θ_2 and θ_3, as presented in Table 7.4. One may note that these estimates are mutually orthogonal, linear combinations of the eight simular re-

TABLE 7.4

COEFFICIENTS OF RESPONSES FOR ESTIMATORS FROM THE 2^3 FACTORIAL DESIGN[a]

Estimator	Y_{000}	Y_{001}	Y_{010}	Y_{011}	Y_{100}	Y_{101}	Y_{110}	Y_{111}	Mean
$\tilde{\mu}$	$+\frac{1}{4}$	$+\frac{1}{4}$	$+\frac{1}{4}$	$+\frac{1}{4}$	$+\frac{1}{4}$	$+\frac{1}{4}$	$+\frac{1}{4}$	$+\frac{1}{4}$	$(2\mu + \theta_1 + \theta_2 + \theta_3)$
$\tilde{\theta}_1$	$-\frac{1}{4}$	$-\frac{1}{4}$	$-\frac{1}{4}$	$-\frac{1}{4}$	$+\frac{1}{4}$	$+\frac{1}{4}$	$+\frac{1}{4}$	$+\frac{1}{4}$	θ_1
$\tilde{\theta}_2$	$-\frac{1}{4}$	$-\frac{1}{4}$	$+\frac{1}{4}$	$+\frac{1}{4}$	$-\frac{1}{4}$	$-\frac{1}{4}$	$+\frac{1}{4}$	$+\frac{1}{4}$	θ_2
$\tilde{\theta}_3$	$-\frac{1}{4}$	$+\frac{1}{4}$	$-\frac{1}{4}$	$+\frac{1}{4}$	$-\frac{1}{4}$	$+\frac{1}{4}$	$-\frac{1}{4}$	$+\frac{1}{4}$	θ_3
$\tilde{\varepsilon}_1$	$+\frac{1}{4}$	$+\frac{1}{4}$	$-\frac{1}{4}$	$-\frac{1}{4}$	$-\frac{1}{4}$	$-\frac{1}{4}$	$+\frac{1}{4}$	$+\frac{1}{4}$	0
$\tilde{\varepsilon}_2$	$+\frac{1}{4}$	$-\frac{1}{4}$	$+\frac{1}{4}$	$-\frac{1}{4}$	$-\frac{1}{4}$	$+\frac{1}{4}$	$-\frac{1}{4}$	$+\frac{1}{4}$	0
$\tilde{\varepsilon}_3$	$+\frac{1}{4}$	$-\frac{1}{4}$	$-\frac{1}{4}$	$+\frac{1}{4}$	$+\frac{1}{4}$	$-\frac{1}{4}$	$-\frac{1}{4}$	$+\frac{1}{4}$	0
$\tilde{\varepsilon}_4$	$+\frac{1}{4}$	$-\frac{1}{4}$	$-\frac{1}{4}$	$+\frac{1}{4}$	$-\frac{1}{4}$	$+\frac{1}{4}$	$+\frac{1}{4}$	$-\frac{1}{4}$	0

[a] All variances are $\sigma^2/2$.

sponses, that have themselves been presumed to be uncorrelated random variables. Hence, the three estimates $\tilde{\theta}_1$, $\tilde{\theta}_2$, and $\tilde{\theta}_3$ of the incremental effects of the three factors, are themselves uncorrelated (and are independently and normally distributed if the additional assumption is made presuming the normality of the error random variables, ε_{qrs}).

Furthermore, the three estimates are each orthogonal to the sum of the eight responses,

$$4\tilde{\mu} = (Y_{000} + Y_{001} + Y_{010} + Y_{011} + Y_{100} + Y_{101} + Y_{110} + Y_{111}),$$

and one may derive four additional contrasts which are mutually orthogonal to one another and to the estimates $\tilde{\mu}$, $\tilde{\theta}_1$, $\tilde{\theta}_2$, and $\tilde{\theta}_3$. These four additional contrasts, selected so that their means are zero and their variances are the same $(\sigma^2/2)$, are labeled $\tilde{\varepsilon}_1$, $\tilde{\varepsilon}_2$, $\tilde{\varepsilon}_3$, and $\tilde{\varepsilon}_4$ in Table 7.4; each, because of its mean and variance, may be taken as an *error estimate*, its square having expectation equal to $\sigma^2/2$:

$$E[(\tilde{\varepsilon}_k)^2] = \text{Var}(\tilde{\varepsilon}_k) + [E(\tilde{\varepsilon}_k)]^2 = \sigma^2/2, \qquad k = 1, 2, 3, \quad \text{and} \quad 4.$$

7.4.1. ANALYSIS OF THE SINGLE REPLICATE

Since each $\tilde{\varepsilon}_k$ is a linear combination of uncorrelated simular responses, each is normally distributed whenever the simular response $Y(T)$ can be presumed to have the normal distribution. Even if another distribution is applicable to the simular responses, each error estimate is essentially the sum of eight independently distributed simular responses, so that, by use of the Central Limit Theorem, one might reasonably anticipate the approximate normality of the error estimates $\tilde{\varepsilon}_k$, $k = 1, 2, 3$, and 4.

We shall presume then the normal distribution for the $\tilde{\varepsilon}_k$, and that they are mutually independent random variables of mean zero, and variance $\sigma^2/2$. Therefore, the square of any one of the $\tilde{\varepsilon}_k$ will be related to a Chi-squared variate; namely, $2(\tilde{\varepsilon}_k)^2/\sigma^2$ has the Chi-squared distribution of 1 degree of freedom, the four variates being independently distributed $k = 1, 2, 3$, and 4. Hence, the sum

$$2[(\tilde{\varepsilon}_1)^2 + (\tilde{\varepsilon}_2)^2 + (\tilde{\varepsilon}_3)^2 + (\tilde{\varepsilon}_4)^2]/\sigma^2$$

has the Chi-squared distribution of 4 degrees of freedom.

A test for the significance of the computed estimate $\tilde{\theta}_i$ of the incremental effect of the ith factor, may be performed by noting that, under the assumptions of the normality and the independence of the simular responses, the random variable $\tilde{\theta}_i$ is normally distributed of mean $\tilde{\theta}_i$, variance $\sigma^2/2$. Therefore,

$$E[(\tilde{\theta}_i)^2] = \mathrm{Var}(\tilde{\theta}_i) + [E(\tilde{\theta}_i)]^2 = (\sigma^2/2) + \theta_i^2,$$

for each $i = 1, 2$, and 3; and, if the null hypothesis H_0: $\theta_i = 0$ is valid, $\tilde{\theta}_i$ has mean zero, variance $\sigma^2/2$, and the random variable $(\tilde{\theta}_i)^2$ has expectation $\sigma^2/2$. Furthermore, whenever the null hypothesis is certain, $2(\tilde{\theta}_i)^2/\sigma^2$ has the Chi-squared distribution of 1 degree of freedom.

The statistic

$$T_i = \frac{2(\tilde{\theta}_i)^2/\sigma^2}{2[(\tilde{\varepsilon}_1)^2 + (\tilde{\varepsilon}_2)^2 + (\tilde{\varepsilon}_3)^2 + (\tilde{\varepsilon}_4)^2]/(\sigma^2 \cdot 4)} = \frac{4(\tilde{\theta}_i)^2}{[\sum_{k=1}^{4} (\tilde{\varepsilon}_k)^2]}$$

will have the Snedecor's F distribution of 1 and 4 degrees of freedom, since it is formed as the ratio of two independent Chi-squared variates, each divided by its own number of degrees of freedom. The independence of the Chi-squared variates is assured, since the $\tilde{\theta}_i$ and $\tilde{\varepsilon}_k$ are orthogonal linear combinations of the independently and normally distributed simular responses. Consequently, one tests the null hypothesis by comparing the computed T_i with the cumulative distribution function for Snedecor's F distribution. For size $\alpha = 0.05$, the null hypothesis is rejected whenever $T_i > 7.71$; with $\alpha = 0.01$, one declares the estimate $|\tilde{\theta}_i|$ significant whenever $T_i > 21.2$, $i = 1, 2$, or 3.

7.4.2. Tests for Interactions

Again, one would correctly speculate that the $\tilde{\varepsilon}_k$, presented herein as estimates of error, are confounded with interaction effects. Indeed, by defining $\theta_{ii'}$ as the interaction between the levels of qualitative factor i and those of factor i', $i \neq i'$, the simular response at coded condition (qrs) can be written

$$Y_{qrs} = \mu + (\theta_1 \cdot q) + (\theta_2 \cdot r) + (\theta_3 \cdot s) + (\theta_{12} \cdot qr) + (\theta_{13} \cdot qs) + (\theta_{23} \cdot rs) + \varepsilon_{qrs}.$$

One may note that the interaction term θ_{23}, for example, appears only in the expression for a simular response Y_{qrs} for which both r and s are 1, regardless of the environmental specification for the first factor.

Employing the preceding representation for the simular responses arising in a single replication of the 2^3 factorial design, the reader can verify that the entries for the expectations (and variances) of $\tilde{\theta}_1$, $\tilde{\theta}_2$, $\tilde{\theta}_3$, and $\tilde{\varepsilon}_4$ remain unchanged in Table 7.4. However, the expectations of $\tilde{\varepsilon}_1$, $\tilde{\varepsilon}_2$, and $\tilde{\varepsilon}_3$ become

$$E(\tilde{\varepsilon}_1) = \theta_{12}/2, \qquad E(\tilde{\varepsilon}_2) = \theta_{13}/2, \qquad \text{and} \qquad E(\tilde{\varepsilon}_3) = \theta_{23}/2,$$

so that indeed these estimates are confounded with any extant interaction.

The estimate $\tilde{\varepsilon}_4$ remains nonetheless an appropriate error estimate (of mean 0) and, since it is orthogonal to the other seven estimates of Table 7.4, one may formulate tests of hypotheses by comparing an appropriate test statistic with the cumulative distribution function of a Snedecor's F-distribution of one and one degrees of freedom. For example, to test the null hypothesis H_0: $\theta_{13} = 0$, a statement that no interaction exists between the first and third factors, one notes that

$$E[(\tilde{\varepsilon}_2)^2] = (\sigma^2/2) + (\theta_{13}^2/4),$$

and that, whenever the null hypothesis is true, the random variable $2(\tilde{\varepsilon}_2)^2/\sigma^2$ has the Chi-squared distribution of 1 degree of freedom, and is independently distributed of $2(\tilde{\varepsilon}_4)^2/\sigma^2$, a random variable also having the same Chi-squared distribution. The statistic given by the ratio

$$T_{13} = (\tilde{\varepsilon}_2)^2/(\tilde{\varepsilon}_4)^2$$

can then be compared with the cumulative distribution function for a Snedecor's F variate of 1 and 1 degrees of freedom.

Other tests for interaction and the presence of main effects are performed analogously whenever a single replication of the 2^3 factorial design is implemented. Whenever m (greater than or equal to 2) replications are performed, the error estimate is properly computed as the average of the eight unbiased estimates of σ^2, with an S^2_{qrs} computable at each of the eight design points. Of course, at each design point, the simular response is taken to be

$$\bar{Y}_{qrs} = m^{-1} \sum_{k=1}^{m} Y_{qrs(k)},$$

where $Y_{qrs(k)}$ is the kth simular response to arise at the coded environmental condition (qrs), $k = 1, 2, \ldots, m$. Tests of hypotheses regarding the nullity of main effects or interactions become based on the comparison of appropriate test statistics with the cumulative distribution function of Snedecor's F statistic of 1 and $8 \cdot (m - 1)$ degrees of freedom. For further details, the reader is referred to Mendenhall (1968, Chapter 5), Graybill (1961, Vol. 1), or to Kempthorne (1952).

7.5. Generalized Interactions and Partial Replicates

The reader is also referred to these sources for an introduction to two additional topics associated with factorial designs: l-factor interactions (for $l > 2$) and the partially replicated factorial design. The first concept is related to the fact that a single replication of a 2^K factorial design cannot provide true error estimates unless certain interactions are known to be absent. For example, in our discussion of the 2^2 factorial design, the error estimate $\tilde{\varepsilon}$ was seen to estimate unbiasedly $\theta_{12}/2$, half the interaction between the two factors (unless, of course, this interaction does not exist, in which case $\tilde{\varepsilon}$ would represent an estimate of zero and thereby qualify as an error estimate); in the discussion of the 2^3 factorial design, the initially proposed error estimates $\tilde{\varepsilon}_1$, $\tilde{\varepsilon}_2$, and $\tilde{\varepsilon}_3$ were seen to estimate unbiasedly $\theta_{12}/2$, $\theta_{13}/2$, $\theta_{23}/2$, respectively. By introducing the three-factor interaction θ_{123}, assumed to exist whenever all three of the qualitative factors are specified at their "1" levels, one would find that $E(\tilde{\varepsilon}_4) = \theta_{123}/4$, and would therefore no longer qualify as an error estimate if the null hypothesis H_0: $\theta_{123} = 0$ were not valid.

These results generalize readily to the case of K (greater than 3) qualitative factors. Of primary import is the fact that generalized l-factor interactions may always be conceptually introduced, so that, unless one or more of these can be dismissed automatically by the simulation investigator, replication of the entire design becomes the only sure method for obtaining unbiased estimates of the error variance.

Fortunately, however, a good deal of experimental evidence and experience has been accumulated in the social, agricultural, medical, biological, and physical sciences; this experience indicates that significant interactions between three or more factors seldom exist. If this can be safely presumed for simular experimentation as well, then one might be able to use single replicates whenever $K > 3$ and still be able to obtain valid error estimates.

In fact, if the simulation analyst is willing to assume the nonexistence of all l-factor interactions for, say, $l \geq 3$, then experimentation requiring a large number K of qualitative factors to be examined may be accomplished by using only part (say $\frac{1}{2}$, $\frac{1}{4}$, or $\frac{1}{8}$) of the full replicate of the 2^K factorial design. A full discussion of the concept of *partial replication* would unnecessarily delay the presentation of other important analytical tools, such as regression analyses, response surface methodology, and the analyses of time series. The reader is again referred to the works of Graybill (1961) and especially Kempthorne (1952) for a more complete presentation of the ideas of, and relationships between *generalized l-factor interactions* and *partial replications* of 2^K factorial designs.

EXERCISES

1. Show that, for the 2^3 factorial design, the representation of the simular response [which arises from the coded environmental condition (qrs)] by

$$Y_{qrs} = \mu + (\theta_1 \cdot q) + (\theta_2 \cdot r) + (\theta_3 \cdot s) + (\theta_{12} \cdot qr) + (\theta_{13} \cdot qs) + (\theta_{23} \cdot rs) + (\theta_{123} \cdot qrs)$$

leaves no contrast, among the eight linear combinations of Table 7.4, which has mean zero.

2. Presume that four qualitative factors x_1, x_2, x_3, and x_4 each having two assignable levels, are to be examined for their significance in contributing to the simular response. By assuming that neither two-factor nor three-factor interactions exist, the simular response at the coded environmental condition $(qrst)$ may be written

$$Y_{qrst} = \mu + (\theta_1 \cdot q) + (\theta_2 \cdot r) + (\theta_3 \cdot s) + (\theta_4 \cdot t) + \varepsilon_{qrst},$$

with the ε_{qrst} presumed to be uncorrelated random variables of mean 0 and variance σ^2. Show that, by using only the half-replicate, consisting of the eight simular responses arising one each from the coded environmental specifications (0000), (0011), (0110), (1100), (1001), (0101), (1010), and (1111), the parameters (incremental effects) θ_1, θ_2, θ_3, and θ_4 can be estimated unbiasedly by

$$\begin{aligned}\tilde{\theta}_1 = {} & [(Y_{1111} + Y_{1010} + Y_{1100} + Y_{1001})/4] \\ & - [(Y_{0110} + Y_{0101} + Y_{0011} + Y_{0000})/4],\end{aligned}$$

$$\begin{aligned}\tilde{\theta}_2 = {} & [(Y_{1111} + Y_{0110} + Y_{0101} + Y_{1100})/4] \\ & - [(Y_{1010} + Y_{1001} + Y_{0011} + Y_{0000})/4],\end{aligned}$$

$$\begin{aligned}\tilde{\theta}_3 = {} & [(Y_{1111} + Y_{1010} + Y_{0110} + Y_{0011})/4] \\ & - [(Y_{1100} + Y_{1001} + Y_{0101} + Y_{0000})/4],\end{aligned}$$

and

$$\begin{aligned}\tilde{\theta}_4 = {} & [(Y_{1111} + Y_{1001} + Y_{0101} + Y_{0011})/4] \\ & - [(Y_{1100} + Y_{1010} + Y_{0110} + Y_{0000})].\end{aligned}$$

Furthermore, show that these four contrasts are mutually orthogonal and that each is orthogonal to the estimator,

$$\tilde{\mu} = [Y_{0000} + Y_{0011} + Y_{0110} + Y_{0101} + Y_{1001} + Y_{1010} + Y_{1100} + Y_{1111}]/4,$$

itself an unbiased estimator of $(2\mu + \theta_1 + \theta_2 + \theta_3 + \theta_4)$. Finally, determine whether the contrast,

$$\begin{aligned}\tilde{\varepsilon}_1 = {} & [(Y_{0000} + Y_{0011} + Y_{1111} + Y_{1100}) \\ & - (Y_{0110} + Y_{1001} + Y_{0101} + Y_{1010})]/4\end{aligned}$$

is an unbiased estimate of zero, and if so, determine its variance. Could you define three other error estimators, so that the eight estimators would be mutually orthogonal?

3. Repeat Exercise 2, using instead the other half-replicate of responses deriving from encounters defined by the coded environmental specifications: (0001), (0010), (0100), (0111), (1000), (1011), (1101), and (1110). Alter the estimates accordingly, noting, e.g., that $\tilde{\theta}_i$ is the contrast between the average of the four responses at the "1" level of factor i and that of the four responses at its "0" level, $i = 1, 2, 3$, and 4.

7.6. Factors at Three Levels

7.6.1. The Single Replicate

The 2^K factorial design provides an excellent experimental plan for determining whether a simular response $Y(T)$ after T units of simular time is affected significantly differently by either of the two levels of any one of the K qualitative factors. Frequently, however, a qualitative factor may possess three assignable states, or *levels*, which may be denoted by the ternary symbols (0, 1, and 2). For example, simulating a queueing network, an important queue discipline might be structured to give higher priority to earliest arrivals, or to the queue members of highest rank as based on some other attribute (such as age); still a third di scipline might structure priorities by allowing the most recent arrival to proceed for service unless an earlier elder is present in the queue (in which case, the earliest arrived elder has priority).

Often, an environmental condition that would normally be thought of as quantitative, except for its restricted assignment to only one of three values, might be considered qualitative. For example, simulating a specific queueing network, the only practicable alternatives for specifying the number of servers might be 1, 2, or 3. These specifications could be directly related in a one-to-one correspondence to the ternary symbols: 0, 1, and 2.

Presuming then that a single factor, say x_1, having only three assignable levels (coded 0, 1, and 2), is to be examined for the importance of its effect on the simular response $Y(T)$, one might begin by assigning all other environmental conditions $(x_2, x_3, \ldots, x_p)$ to some standard specification [say, $(x_{2,0}, x_{3,0}, \ldots, x_{p,0})$], by selecting independently three random-number seeds without repetitions, and by assigning the factor x_1 its three levels, thereby defining three successive encounters with the simulation model. The three resultant responses can then, without loss of generality, be labeled: Y_0, Y_1, and Y_2.

One elementary approach to the determination of significant differences in these three simular responses is the representation of the simular response at the qth level by

$$Y_q = \mu + \theta_1 \cdot (q - 1) + \theta_2 \cdot (q - 1)^2 + \varepsilon_q, \quad q = 0, 1, \text{ and } 2,$$

so that

μ is an average response (equal to the expected response at level 1),
θ_2 is an incremental response resulting from *any* change away from level 1,

θ_1 is an additional incremental response resulting from a change to level 2, yet a decremental response for a change to level 0, and

ε_q is a random, or error term, that will be considered as a set of uncorrelated random variables of mean zero and variance σ^2.

From this representation, an unbiased estimate of μ would be related to the simular responses arising at level 1 of the factor, an unbiased estimate of $(\mu + \theta_1 + \theta_2)$ would be provided by simular responses at level 2, and an unbiased estimate of $(\mu - \theta_1 + \theta_2)$ would be provided by means of simular responses at level 0. These results follow from noting the expectations of Y_1, Y_2, and Y_0, respectively; one can also note that $\text{Var}(Y_q) = \sigma^2$, regardless of the assigned level q.

As unbiased estimates of θ_1 and θ_2, one can then consider the respective statistics

$$\tilde{\theta}_1 = (Y_2 - Y_0)/2$$

and

$$\tilde{\theta}_2 = (Y_0 + Y_2 - 2Y_1)/2,$$

with variances

$$\text{Var}(\tilde{\theta}_1) = (\sigma^2 + \sigma^2)/4 = \sigma^2/2$$

and

$$\text{Var}(\tilde{\theta}_2) = (\sigma^2 + \sigma^2 + 4\sigma^2)/4 = 3\sigma^2/2,$$

and with covariance

$$\text{Cov}(\tilde{\theta}_1, \tilde{\theta}_2) = [(0 + \sigma^2 - 0) + (-\sigma^2 + 0 + 0)]/4 = 0.$$

Consequently, whenever the simular responses are jointly normally distributed, the estimates $\tilde{\theta}_1$ and $\tilde{\theta}_2$ have the bivariate normal distribution and are also independently distributed as well (cf. Section 2.15).

If a null hypothesis of the form H_0: $\theta_1 = 0$ is proposed, it can be tested by noting that

$$E[(\tilde{\theta}_1)^2] = \text{Var}(\tilde{\theta}_1) + [E(\tilde{\theta}_1)]^2 = (\sigma^2/2) + \theta_1^2,$$

so that exceedingly large values of the statistic $(\tilde{\theta}_1)^2$ would tend to make the null hypothesis rather untenable. Indeed, if this null hypothesis were true, $\tilde{\theta}_1$ would be a normally distributed random variable (of mean 0, variance $\sigma^2/2$), so that

$$T_1 \equiv (\tilde{\theta}_1)^2/(\sigma^2/2) = 2(\tilde{\theta}_1)^2/\sigma^2$$

would represent a Chi-squared variable of 1 degree of freedom. Critically large values of T_1 would then correspond to values in the extreme positive tail of this Chi-squared density function; i.e., for a given size α, one would declare the observed $\tilde{\theta}_1$ significantly different from zero whenever T_1 exceeded c_α, that real number such that

$$P[\chi_1^2 > c_\alpha] = \alpha.$$

Similarly, a null hypothesis of the form H_0: $\theta_2 = 0$ can be tested by noting that

$$E[(\tilde{\theta}_2)^2] = \operatorname{Var}(\tilde{\theta}_2) + [E(\tilde{\theta}_2)]^2 = (3\sigma^2/2) + \theta_2^2.$$

Excessively large values of $\tilde{\theta}_2$ would serve to reject this null hypothesis. Again, the random variable

$$(\tilde{\theta}_2)^2 = (Y_0 + Y_2 - 2Y_1)^2/4$$

is proportional to a Chi-squared variate of 1 degree of freedom; i.e., $T_2 = 2(\tilde{\theta}_2)^2/(3\sigma^2)$ is so distributed whenever the null hypothesis is true. Rejection of this null hypothesis depends then upon the magnitude of T_2 relative to c_α, as defined in the preceding paragraph.

If either (or both) of the statistics $\tilde{\theta}_1$ and $\tilde{\theta}_2$ is (are) deemed significant as a result of these tests, one would declare the existence of significant differences in the simular responses arising because of alternative specifications of the levels of the qualitative factor (x_1). Attributing this significance to a particular level of the factor can usually be accomplished by taking note of the relative magnitudes of the three responses.

Recalling the definitions of θ_2 (the incremental change for either deviation away from the "1" level for the factor) and θ_1 (an additional incremental change to differentiate between responses that arise from deviating to the "0" level as opposed to the "2" level), the usual procedure is to test first for the significance of the estimate $\tilde{\theta}_2$ by means of the Chi-squared statistic T_2. If T_2 is found significant, then the test for the significance of $\tilde{\theta}_1$ can be omitted, because the significance of T_2 implies the likelihood of true differences in the simular responses at the three levels of the qualitative factor (x_1). However, if the magnitude of T_2 is not found significant, the Chi-squared test for the magnitude of T_1 may be applied. If, as a result, this test results in the rejection of the null hypothesis, then apparently significant differences among the responses at the three levels do exist (see also Exercise 4 at the end of this section).

7.6.2. Testing Three Levels under Replication

The testing procedure just described requires explicit knowledge of the underlying error variance σ^2. Since this knowledge cannot always be presumed extant, estimates of this variance will need to be made. By replicating the experimental design exactly m times [implying the selection of $3m$ independent, nonrepetitive random number seeds and their assignment to the $3m$ simular encounters, m of which are defined at each of the environmental conditions $(q, x_{2,0}, x_{3,0}, \ldots, x_{p,0})$, for $q = 0$, 1, and 2], one can obtain the simular responses as the averages

$$\bar{Y}_q = m^{-1} \sum_{k=1}^{m} Y_{q(k)},$$

where $Y_{q(k)}$ is the kth simular response at level q, for $k = 1, 2, \ldots, m$, and $q = 0$, 1, and 2. Unbiased estimates of σ^2 become

$$S_q^2 = (m-1)^{-1}\left[\sum_{k=1}^{m} Y_{q(k)}^2 - m(\bar{Y}_q)^2\right], \qquad q = 0, 1, \quad \text{and} \quad 2,$$

so that the pooled estimate,

$$S^2 = (S_0^2 + S_1^2 + S_2^2)/3,$$

is also unbiased for σ^2. Furthermore, the random variables $\bar{Y}_0$, $\bar{Y}_1$, $\bar{Y}_2$, S_0^2, S_1^2, and S_2^2 are all statistically independent: the mean responses being normally distributed with variances (σ^2/m), the S_q^2 being related to Chi-squared variates; namely, $(m-1)S_q^2/\sigma^2$ has the Chi-squared distribution of $m - 1$ degrees of freedom.

The test for the null hypothesis H_0: $\theta_2 = 0$, rests upon computation of the statistic

$$T_2 = \frac{2(\tilde{\theta}_2)^2/(3\sigma^2/m)}{\{(3m-3)S^2/[(3m-3)\sigma^2]\}} = \frac{2(\tilde{\theta}_2)^2}{3S^2/m},$$

where the random variable

$$(\tilde{\theta}_2)^2 = (\bar{Y}_0 + \bar{Y}_2 - 2\bar{Y}_1)^2/4,$$

under the stated null hypothesis, is an unbiased estimate of $3\sigma^2/(2m)$, a Chi-squared variate of 1 degree of freedom, and independently distributed of S^2. The null hypothesis H_0: $\theta_2 = 0$ is therefore rejected whenever the test statistic T_2 exceeds k_α, that real number such that a

Snedecor's F distribution of 1 and $3m - 3$ degrees of freedom has accumulated $(1 - \alpha) \cdot 100\%$ of its density.

If this test does not prove significant, then the additional test, using the statistic

$$T_1 \equiv \frac{2(\tilde{\theta}_1)^2/(\sigma^2/m)}{\{(3m - 3)S^2/[(3m - 3)\sigma^2]\}} = \frac{2(\tilde{\theta}_1)^2}{S^2/m},$$

where

$$(\tilde{\theta}_1)^2 = (\bar{Y}_2 - \bar{Y}_0)^2/4,$$

may be invoked, again requiring a comparison with the tabulations of the extremes of the cumulative distribution function for Snedecor's F variate of 1 and $3m - 3$ degrees of freedom. Whenever neither of the test statistics T_1 and T_2 proves to be significant, one would conclude that there be insignificant evidence to warrant a statement that the (coded) levels 0, 1, and 2 of the factor x_1 produce significantly different simular responses (see also Exercise 4).

EXERCISES

1. For the simular responses Y_0, Y_1, and Y_2 arising from three independently seeded simular encounters, one at each of the three levels of a qualitative factor, show that $Y_1 - Y_0$ is uncorrelated with $Y_0 + Y_1 - 2Y_2$, and that the second contrast has three times the variance of the first whenever the responses are presumed to be independent random variables of the same variance (σ^2).

2. Show that, for three, independent simular responses Y_0, Y_1, and Y_2 (all homoscedastic), the statistic

$$\tilde{\sigma}^2 \equiv \sum_{q=0}^{2} (Y_q - \bar{Y})^2/(3 - 1),$$

where $\bar{Y} = (Y_0 + Y_1 + Y_2)/3$, is unbiased for σ^2 whenever the means $E(Y_q)$ are all equal, and can be written alternatively as

$$\tilde{\sigma}^2 \equiv [(2Y_0 - Y_1 - Y_2)^2 + (2Y_1 - Y_0 - Y_2)^2 + (2Y_2 - Y_0 - Y_1)^2]/18.$$

3. By presuming the normality of the simular responses as well, suggest a test for the equality of the means [i.e., a test for the null hypothesis H_0: $E(Y_0) = E(Y_1) = E(Y_2)$] by noting that $\tilde{\sigma}^2/\sigma^2$ has the Chi-squared distribution of 2 degrees of freedom. [*Hint*: In the second expression in Exercise 2, the three contrasts,

$$2Y_0 - Y_1 - Y_2, \qquad -Y_0 + 2Y_1 - Y_2, \qquad \text{and} \qquad -Y_0 - Y_1 + 2Y_2$$

are linearly related such that the sum of any two is the negative of the third. However, the difference of the same pair of contrasts is orthogonal to the third, so that two linearly independent contrasts exist.]

4. Referring to Exercises 2 and 3, show that

$$\tilde{\sigma}^2 \equiv [(2Y_0 - Y_1 - Y_2)^2 + (2Y_1 - Y_0 - Y_2)^2 + (2Y_2 - Y_0 - Y_1)^2]/18$$

can also be expressed as

$$\tilde{\sigma}^2 = [3(Y_0 - Y_2)^2 + (2Y_1 - Y_0 - Y_2)^2]/12$$

or

$$\tilde{\sigma}^2 = (\tilde{\theta}_1)^2 + [(\tilde{\theta}_2)^2/3],$$

so that one can devise a *single* test for the significance of observed differences in Y_0, Y_1, and Y_2—the test being based on the extremal values in the Chi-squared distribution of 2 degrees of freedom. Describe the test whenever the simular responses are taken to be $\bar{Y}_0$, $\bar{Y}_1$, and $\bar{Y}_2$, each an arithmetic average of an independent and random sample of m simular responses deriving from the same environmental specifications. (Note the generality of this result, applicable to the comparison of responses, whether the three environmental conditions are the three levels of a single qualitative factor or not.)

5. Generate 100 random triplets of independent random variables, the ordered elements of each triplet coming from the respective normal distributions

Y_0: mean 4, variance 25,
Y_1: mean 10, variance 25,
Y_2: mean 16, variance 25.

Presuming that each triplet is the recorded result of three simular encounters, each independently seeded yet arising due to the specification of different levels for a single qualitative factor, show that, in accordance with the definitions in the text, $\mu = 10$, $\theta_2 = 6$, and $\theta_1 = 0$. Perform the 100 tests of each of the hypothesis H_0: $\theta_1 = 0$ and H_0: $\theta_2 = 0$, and comment on the relative frequency with which you reject each (both) of these hypotheses. (Presume that $\sigma^2 = 25$ is known to the simulation analyst.)

7.7. The 3^2 Factorial Design

7.7.1. The Single Replicate

Whenever the analyst wants to examine the effects of two qualitative factors, each having three assignable levels, he will need to define $3 \times 3 = 9$ encounters with the simulation model, in order to ensure that each level of the first factor appears with each level of the second in exactly one encounter. Using the ternary symbols 0, 1, and 2 to denote the three levels of either factor, one can define two-place ternary codes (qr) for the environmental specifications of these nine encounters, as presented in Table 7.5.

Denoting the simular response at (qr) by

$$Y_{qr} = \mu + \theta_1 \cdot (q-1) + \theta_2 \cdot (q-1)^2 + \theta_3 \cdot (r-1) + \theta_4 \cdot (r-1)^2 + \varepsilon_{qr},$$

where μ is a mean effect present in all responses, especially representative of the response Y_{11}, θ_2 is the incremental effect arising from a shift from level 1 to *any* other level of the first factor, θ_1 is the additional incremental (decremental) effect associated with a shift from level 1 to level 2 (level 0) of the first factor, θ_4 is the incremental effect arising from a shift from level 1 to either of the other two levels of the second factor, and θ_3 is the additional incremental (decremental) effect associated with a shift from level 1 to level 2 (level 0) of the second factor, one may prepare the column of expectations (means) in Table 7.5.

TABLE 7.5

The 3^2 Factorial Design

Environmental condition	Code	Response	Expectation
$(0, 0, x_{3,0}, x_{4,0}, \ldots, x_{p,0})$	00	Y_{00}	$\mu - \theta_1 + \theta_2 - \theta_3 + \theta_4$
$(0, 1, x_{3,0}, x_{4,0}, \ldots, x_{p,0})$	01	Y_{01}	$\mu - \theta_1 + \theta_2$
$(0, 2, x_{3,0}, x_{4,0}, \ldots, x_{p,0})$	02	Y_{02}	$\mu - \theta_1 + \theta_2 + \theta_3 + \theta_4$
$(1, 0, x_{3,0}, x_{4,0}, \ldots, x_{p,0})$	10	Y_{10}	$\mu - \theta_3 + \theta_4$
$(1, 1, x_{3,0}, x_{4,0}, \ldots, x_{p,0})$	11	Y_{11}	μ
$(1, 2, x_{3,0}, x_{4,0}, \ldots, x_{p,0})$	12	Y_{12}	$\mu + \theta_3 + \theta_4$
$(2, 0, x_{3,0}, x_{4,0}, \ldots, x_{p,0})$	20	Y_{20}	$\mu + \theta_1 + \theta_2 - \theta_3 + \theta_4$
$(2, 1, x_{3,0}, x_{4,0}, \ldots, x_{p,0})$	21	Y_{21}	$\mu + \theta_1 + \theta_2$
$(2, 2, x_{3,0}, x_{4,0}, \ldots, x_{p,0})$	22	Y_{22}	$\mu + \theta_1 + \theta_2 + \theta_3 + \theta_4$

Since each of the three levels of either factor appears with any one of the three levels of the other factor, one can estimate unbiasedly the effect θ_1, say, by any of the three different contrasts: $(Y_{20} - Y_{00})/2$, $(Y_{21} - Y_{01})/2$, and $(Y_{22} - Y_{02})/2$. Therefore, a single unbiased estimate for the incremental effect θ_1 is their arithmetic mean, or

$$\tilde{\theta}_1 = \tfrac{1}{2}[(Y_{20} + Y_{21} + Y_{22})/3] - \tfrac{1}{2}[(Y_{00} + Y_{01} + Y_{02})/3],$$

which, in this form, reveals that the difference in the means of all responses at levels 2 and 0 of the first factor is important for computing the estimate. In a similar fashion, one would obtain the unbiased estimator for θ_3 as the contrast

$$\tilde{\theta}_3 = \tfrac{1}{2}[(Y_{02} + Y_{12} + Y_{22})/3] - \tfrac{1}{2}[(Y_{00} + Y_{10} + Y_{20})/3].$$

Unbiased estimates of θ_2 and θ_4 are presented in Table 7.6, which provides the proper coefficients for the simular responses, as required by the particular estimators. Using the preceding table, the reader should verify that the expectations of the contrasts are correct as given, and should verify that the four statistics, $\tilde{\varepsilon}_1$, $\tilde{\varepsilon}_2$, $\tilde{\varepsilon}_3$, and $\tilde{\varepsilon}_4$, indeed estimate unbiasedly zero with variances as shown in the table.

Tests of hypotheses regarding any one of the four parameters θ_i, $i = 1, 2, 3$, and 4 may be performed by comparing appropriately weighted squares of the estimates, $\tilde{\theta}_i$, with an appropriate linear combination of the squared error estimates, $\tilde{\varepsilon}_k$, $k = 1, 2, 3$, and 4.

For convenience, some typical hypotheses and their test statistics are tabulated in Table 7.7. These can be derived in the usual way.

For example, to derive a test for the null hypothesis, H_0: $\theta_1 = 0$, one notes that

$$E[(\tilde{\theta}_1)^2] = \text{Var}(\tilde{\theta}_1) + [E(\tilde{\theta}_1)]^2 = (\sigma^2/6) + \theta_1^2,$$

so that the statistic,

$$(\tilde{\theta}_1)^2 = (Y_{20} + Y_{21} + Y_{22} - Y_{00} - Y_{01} - Y_{02})^2/36,$$

if exceedingly large, reflects the possibility of the existence of a nonzero actual value of θ_1. Now, the linear combination of squared error estimates denoted by

$$\tilde{\sigma}^2 \equiv [4(\tilde{\varepsilon}_1)^2 + \tfrac{4}{3}(\tilde{\varepsilon}_2)^2 + \tfrac{4}{3}(\tilde{\varepsilon}_3)^2 + (\tilde{\varepsilon}_4)^2]/4,$$

is an unbiased estimate of σ^2.

TABLE 7.6

COEFFICIENTS OF RESPONSES FOR ESTIMATORS FROM THE 3^2 FACTORIAL DESIGN

Estimator	Y_{00}	Y_{01}	Y_{02}	Y_{10}	Y_{11}	Y_{12}	Y_{20}	Y_{21}	Y_{22}	Mean	Variance
$\tilde{\mu}$	$+\frac{1}{9}$	$+\frac{1}{9}$	$+\frac{1}{9}$	$+\frac{1}{9}$	$+\frac{1}{9}$	$+\frac{1}{9}$	$+\frac{1}{9}$	$+\frac{1}{9}$	$+\frac{1}{9}$	$\mu + (6\theta_2 + 6\theta_4)/9$	$\sigma^2/9$
$\tilde{\theta}_1$	$-\frac{1}{6}$	$-\frac{1}{6}$	$-\frac{1}{6}$	0	0	0	$+\frac{1}{6}$	$+\frac{1}{6}$	$+\frac{1}{6}$	θ_1	$\sigma^2/6$
$\tilde{\theta}_2$	$+\frac{1}{6}$	$+\frac{1}{6}$	$+\frac{1}{6}$	$-\frac{2}{6}$	$-\frac{2}{6}$	$-\frac{2}{6}$	$+\frac{1}{6}$	$+\frac{1}{6}$	$+\frac{1}{6}$	θ_2	$\sigma^2/2$
$\tilde{\theta}_3$	$-\frac{1}{6}$	0	$+\frac{1}{6}$	$-\frac{1}{6}$	0	$+\frac{1}{6}$	$-\frac{1}{6}$	0	$+\frac{1}{6}$	θ_3	$\sigma^2/6$
$\tilde{\theta}_4$	$+\frac{1}{6}$	$-\frac{2}{6}$	$+\frac{1}{6}$	$+\frac{1}{6}$	$-\frac{2}{6}$	$+\frac{1}{6}$	$+\frac{1}{6}$	$-\frac{2}{6}$	$+\frac{1}{6}$	θ_4	$\sigma^2/2$
$\tilde{\varepsilon}_1$	$+\frac{1}{4}$	0	$-\frac{1}{4}$	0	0	0	$-\frac{1}{4}$	0	$+\frac{1}{4}$	0	$\sigma^2/4$
$\tilde{\varepsilon}_2$	$-\frac{1}{4}$	$+\frac{2}{4}$	$-\frac{1}{4}$	0	0	0	$+\frac{1}{4}$	$-\frac{2}{4}$	$+\frac{1}{4}$	0	$3\sigma^2/4$
$\tilde{\varepsilon}_3$	$-\frac{1}{4}$	0	$+\frac{1}{4}$	$+\frac{2}{4}$	0	$-\frac{2}{4}$	$-\frac{1}{4}$	0	$+\frac{1}{4}$	0	$3\sigma^2/4$
$\tilde{\varepsilon}_4$	$+\frac{1}{6}$	$-\frac{2}{6}$	$+\frac{1}{6}$	$-\frac{2}{6}$	$+\frac{4}{6}$	$-\frac{2}{6}$	$+\frac{1}{6}$	$-\frac{2}{6}$	$+\frac{1}{6}$	0	σ^2

TABLE 7.7

TESTS OF HYPOTHESES ASSOCIATED WITH THE 3^2 FACTORIAL DESIGN

Hypothesis	Test statistics	Degrees of freedom	Critical constants	
			$\alpha = 0.05$	$\alpha = 0.01$
H_0: $\theta_1 = 0$	$T_1 \equiv 6(\tilde{\theta}_1)^2/(\tilde{\sigma}^2)$	(1, 4)	7.71	21.2
H_0: $\theta_2 = 0$	$T_2 \equiv 2(\tilde{\theta}_2)^2/(\tilde{\sigma}^2)$	(1, 4)	7.71	21.2
H_0: $\theta_3 = 0$	$T_3 \equiv 6(\tilde{\theta}_3)^2/(\tilde{\sigma}^2)$	(1, 4)	7.71	21.2
H_0: $\theta_4 = 0$	$T_4 \equiv 2(\tilde{\theta}_4)^2/(\tilde{\sigma}^2)$	(1, 4)	7.71	21.2
H_0: $\theta_1^2 + \theta_2^2 = 0$[a]	$T_5 \equiv 3S_1^2/(2\tilde{\sigma}^2)$[c]	(2, 4)	6.94	18.0
H_0: $\theta_3^2 + \theta_4^2 = 0$[b]	$T_6 \equiv 3S_2^2/(2\tilde{\sigma}^2)$[d]	(2, 4)	6.94	18.0

[a] An equivalent statement would be $|\theta_1| + |\theta_2| = 0$; in words, the hypothesis states that none of the three levels of the *first* factor is a significant contributor to the simular response.

[b] Equivalently, H_0: $|\theta_3| + |\theta_4| = 0$ (cf. footnote *a*).

[c] Here, $S_1^2 \equiv \frac{1}{2} \sum_{k=0}^{2} (\bar{Y}_{k\cdot} - \tilde{\mu})^2$, for $\bar{Y}_{k\cdot} = \sum_{l=0}^{2} Y_{kl}/3$, $k = 0, 1, 2$.

[d] Here, $S_2^2 \equiv \frac{1}{2} \sum_{l=0}^{2} (\bar{Y}_{\cdot l} - \tilde{\mu})^2$, where $\bar{Y}_{\cdot l} = \sum_{k=0}^{2} Y_{kl}/3$, $l = 0, 1, 2$.

Furthermore, under the presumption that the simular responses are independently and normally distributed random variables, each $\tilde{\varepsilon}_k$, being a linear combination of these uncorrelated responses, is also normally distributed with mean 0 and variance as given in Table 7.6. In addition, note that all nine linear combinations in Table 7.6 are mutually orthogonal, so that under the presumption of normally and independently distributed simular responses, the nine contrasts are also independently and normally distributed, with means and variances as tabulated in Table 7.6. Consequently, the statistics

$$4(\tilde{\varepsilon}_1)^2/\sigma^2, \qquad \tfrac{4}{3}(\tilde{\varepsilon}_2)^2/\sigma^2, \qquad \tfrac{4}{3}(\tilde{\varepsilon}_3)^2/\sigma^2, \qquad \text{and} \qquad (\tilde{\varepsilon}_4)^2/\sigma^2$$

are independently distributed Chi-squared variates, each of 1 degree of freedom, so that the random variable $4(\tilde{\sigma}^2)/\sigma^2$ has the Chi-squared distribution of 4 degrees of freedom.

By means of a quite similar argument, one arrives at the conclusion that if the null hypothesis H_0: $\theta_1 = 0$ is true, then the statistic $6(\tilde{\theta}_1)^2/\sigma^2$ also has the Chi-squared distribution, yet with 1 degree of freedom.

Due to the independence of the nine contrasts, the test statistic

$$T_1 = \frac{6(\tilde{\theta}_1)^2/\sigma^2}{4(\tilde{\sigma}^2)/4\sigma^2} = \frac{6(\tilde{\theta}_1)^2}{(\tilde{\sigma}^2)}$$

will have the Snedecor's F distribution of 1 and 4 degrees of freedom whenever θ_1 is truly zero. The appropriate critical constant, c_α, beyond which a computed value of T_1 will lead to rejection of the null hypothesis, is given in the first row of Table 7.7. Similar critical values are given in this table for the test statistics T_i appropriate to the test of the hypothesis H_0: $\theta_i = 0$, $i = 2$, 3, and 4.

In addition, one should note particularly the tests of the hypotheses H_0: $\theta_1^2 + \theta_2^2 = 0$ and H_0: $\theta_3^2 + \theta_4^2 = 0$. The first states that none of the three levels of the first factor exerts an influence on the simular response; the second that none of the three levels of the second factor is a contribution to the simular response.

EXERCISES

1. From each of the following normal distributions, generate independently a random sample of size 100. Presume that the 900 variates so generated are the results of 100 replications of a 3^2 design, the simular responses corresponding to the ternary (qr) notation for environmental specifications of a pair of three-level factors.

Code	Mean	Variance
00	24	25
01	20	25
02	24	25
10	8	25
11	4	25
12	8	25
20	24	25
21	20	25
22	24	25

For each set of nine, compute each test statistic, T_1, T_2, T_3, T_4, T_5, T_6, and compare with the appropriate F-value ($\alpha = 0.05$). Is the proportion (of 100), for each of the six tests that end in rejection, reasonably compatible with the observation that $\theta_1 = \theta_3 = 0$, $\theta_2 = 16$, and $\theta_4 = 4$?

2. Show that, as footnoted beneath Table 7.7, the statistic

$$S_1^2 \equiv \tfrac{1}{2}\sum_{k=0}^{2}(\bar{Y}_{k\bullet} - \tilde{\mu})^2, \qquad \text{for} \quad \bar{Y}_{k\bullet} = \sum_{l=0}^{2} Y_{kl}/3, \quad k = 0, 1, 2,$$

is an unbiased estimate of $(\sigma^2/3)$ whenever $\theta_1 = \theta_2 = 0$, and of $(\sigma^2/3) + (3\theta_2 - 9\theta_1)^2/81 + (36\theta_2{}^2/81) + (3\theta_1 + 3\theta_2)^2/81$ otherwise. Then show that $S_1{}^2$ is a linear combination of $(\tilde{\theta}_1)^2$ and $(\tilde{\theta}_2)^2$, thereby establishing the appropriateness of the suggested test (using T_5) for the hypothesis H_0: $\theta_1{}^2 + \theta_2{}^2 = 0$.

7.7.2. The Replicated 3^2 Design

The 3^2 factorial design can, of course, be replicated m times, in which case the simular responses are recorded as the averages

$$\bar{Y}_{qr} = m^{-1}\sum_{k=1}^{m} Y_{qr(k)},$$

where $Y_{qr(k)}$ is the kth simular response obtained from an independently seeded encounter with the simulation model at the coded environmental condition (qr). Error estimates are computed as

$$S_{qr}^2 = (m-1)^{-1}\left[\sum_{k=1}^{m} Y_{qr(k)}^2 - m(\bar{Y}_{qr})^2\right]$$

and are pooled to form the unbiased estimate of σ^2:

$$S^2 = \sum_{q=0}^{2}\sum_{r=0}^{2} S_{qr}^2/9.$$

This estimate of σ^2 has the usual property of an error-variance estimate obtained when an experimental design is replicated; namely, the estimate is in no way confounded with estimates of interaction effects, if there are any. Indeed, the estimates $\tilde{\varepsilon}_1$, $\tilde{\varepsilon}_2$, $\tilde{\varepsilon}_3$, and $\tilde{\varepsilon}_4$ of the preceding section are true error estimates only if one is certain that interaction between the levels of the two factors is nonexistent.

Tests of hypotheses may proceed exactly as stated in Table 7.7, except that the test statistics use S^2/m instead of $\tilde{\sigma}^2$ in the denominators; furthermore, the tests become based on cumulative distribution functions of Snedecor's F variates either of 1 and $9 \cdot (m-1)$ degrees of freedom (T_1, T_2, T_3, and T_4) or of 2 and $9 \cdot (m-1)$ degrees of freedom (T_5 and

T_6). One should note that the numerator statistics of these test variates are squares of linear combinations of the $\bar{Y}_{qr}$ instead of the Y_{qr} as was discussed in the preceding section (see Table 7.7).

7.8. Other Factorial Designs

A complete discussion of the interaction between the three levels of two factors would require a somewhat lengthy presentation. Here, it is sufficient to say that these interactive effects are straightforward generalizations of the interactions discussed in the context of the 2^K factorial design and its analysis (cf. Sections 7.4.2 and 7.5).

Another generalization is also rather obvious. One might seek to explore the significance of the effects of K (greater than 2) factors, each having exactly three levels. The design and analysis of the 3^K factorial experiment can be found in the excellent reference book by Kempthorne (1952).

Moreover, it may be natural to ask about the existence of effects due to several factors, some of which do not possess, in any natural sense, the same number of levels. For example, in the simulation of a particular inventory system, it may be that the only reasonable numbers of middlemen between producer and dealer are zero and one, whereas the choice of representative one-parameter distributions of the time delays for filling the retailer's stock orders, each distribution having a mean of 2 weeks, is meaningfully restricted only to the Pareto, Rayleigh, and exponential distributions. In such a case, one would hope to design an experiment consisting of the $2 \times 3 = 6$ possible combinations of the two levels of the first factor with each of the three levels of the second factor.

In the present case, six simular encounters can indeed be defined, their environmental conditions being representable by a coded variate (qr), where q is binary (assumes only 0 or 1) and r is ternary (assuming values 0, 1, or 2 in consonance with the specification of waiting times as Pareto, Rayleigh, or exponentially distributed random variables). (Clearly, the procedure is relatively context free, and thereby applicable to general situations in which the effects of one two-level factor and one three-level factor are to be examined simultaneously.) One can denote the simular response at the coded environmental condition (qr) as

$$Y_{qr} = \mu + \theta_1 \cdot q + \theta_2 \cdot (r - 1) + \theta_3 \cdot (r - 1)^2 + \varepsilon_{qr},$$

where μ is a mean effect (also, the mean of Y_{01}), θ_1 is the incremental effect resultant from specification of factor 1 at its 1 level rather than its zero level, θ_3 is the incremental effect resulting from *any* change of the level of the second factor from its level 1, θ_2 is the additional incremental (decremental) effect arising from the change to level 2 (level 0) from level 1 of the second factor, and, ε_{qr} is an independent error term, being a random variable having mean zero, variance σ^2 not dependent on the environmental specification (qr), and being stochastically independent of the errors arising at any one of the other five design points.

The resulting set of six simular responses, being independent random variables of common variance σ^2, have means as shown in the following table.

TABLE 7.8

THE 2 × 3 MIXED FACTORIAL DESIGN

Environmental specification	Code	Response	Expectation
$(0, 0, x_{3,0}, x_{4,0}, \ldots, x_{p,0})$	00	Y_{00}	$\mu - \theta_2 + \theta_3$
$(0, 1, x_{3,0}, x_{4,0}, \ldots, x_{p,0})$	01	Y_{01}	μ
$(0, 2, x_{3,0}, x_{4,0}, \ldots, x_{p,0})$	02	Y_{02}	$\mu + \theta_2 + \theta_3$
$(1, 0, x_{3,0}, x_{4,0}, \ldots, x_{p,0})$	10	Y_{10}	$\mu + \theta_1 - \theta_2 + \theta_3$
$(1, 1, x_{3,0}, x_{4,0}, \ldots, x_{p,0})$	11	Y_{11}	$\mu + \theta_1$
$(1, 2, x_{3,0}, x_{4,0}, \ldots, x_{p,0})$	12	Y_{12}	$\mu + \theta_1 + \theta_2 + \theta_3$

Estimates of the effects θ_1, θ_2, and θ_3 are provided by the contrasts presented in Table 7.9. Each estimate is unbiased, as can be seen from the "Expectation" column, and each is a linear combination of the six simular responses. Two error estimates $\tilde{\varepsilon}_1$ and $\tilde{\varepsilon}_2$ are also presented; the reader may wish to verify their expectations and variances, as well as to establish the mutual orthogonality of all six linear forms.

An appropriately weighted sum of squares of the error estimates, $\tilde{\varepsilon}_1$ and $\tilde{\varepsilon}_2$, will provide an unbiased estimate of σ^2; namely,

$$\tilde{\sigma}^2 = \tfrac{1}{2}[9(\tilde{\varepsilon}_1)^2 + 3(\tilde{\varepsilon}_2)^2],$$

which is related to a Chi-squared variate in that $2(\tilde{\sigma}^2)/\sigma^2$ has the Chi-squared distribution of 2 degrees of freedom whenever the simular responses are presumed to be independently and normally distributed. Furthermore, under the appropriate null hypotheses, $3(\tilde{\theta}_1)^2/(2\sigma^2)$, $4(\tilde{\theta}_2)^2/\sigma^2$, and $4(\tilde{\theta}_3)^2/(3\sigma^2)$ are independently distributed as Chi-squared variates of one degree of freedom.

TABLE 7.9

COEFFICIENTS OF RESPONSES FOR ESTIMATORS FROM THE 2 × 3 MIXED FACTORIAL EXPERIMENTAL DESIGN

Estimator	Y_{00}	Y_{01}	Y_{02}	Y_{10}	Y_{11}	Y_{12}	Expectation	Variance
$\tilde{\mu}$	$+\frac{1}{6}$	$+\frac{1}{6}$	$+\frac{1}{6}$	$+\frac{1}{6}$	$+\frac{1}{6}$	$+\frac{1}{6}$	$\mu + \frac{1}{2}\theta_1 + (3\theta_3/2)$	$\sigma^2/6$
$\tilde{\theta}_1$	$-\frac{1}{3}$	$-\frac{1}{3}$	$-\frac{1}{3}$	$+\frac{1}{3}$	$+\frac{1}{3}$	$+\frac{1}{3}$	θ_1	$2\sigma^2/3$
$\tilde{\theta}_2$	$-\frac{1}{4}$	0	$+\frac{1}{4}$	$-\frac{1}{4}$	0	$+\frac{1}{4}$	θ_2	$\sigma^2/4$
$\tilde{\theta}_3$	$+\frac{1}{4}$	$-\frac{2}{4}$	$+\frac{1}{4}$	$+\frac{1}{4}$	$-\frac{2}{4}$	$+\frac{1}{4}$	θ_3	$3\sigma^2/4$
$\tilde{\varepsilon}_1$	$+\frac{1}{6}$	0	$-\frac{1}{6}$	$-\frac{1}{6}$	0	$+\frac{1}{6}$	0	$\sigma^2/9$
$\tilde{\varepsilon}_2$	$-\frac{1}{6}$	$+\frac{2}{6}$	$-\frac{1}{6}$	$+\frac{1}{6}$	$-\frac{2}{6}$	$+\frac{1}{6}$	0	$\sigma^2/3$

Tests of the null hypotheses $\theta_1 = 0$, $\theta_2 = 0$, $\theta_3 = 0$ are conducted then by means of the statistics T_1, T_2, and T_3, respectively, via comparison of their magnitudes with the improbably large values of a Snedecor's F distribution of 1 and 2 degrees of freedom:

$$T_1 = 3(\tilde{\theta}_1)^2/(4\tilde{\sigma}^2), \qquad \text{under} \quad H_0\colon \theta_1 = 0,$$

$$T_2 = 2(\tilde{\theta}_2)^2/(\tilde{\sigma}^2), \qquad \text{under} \quad H_0\colon \theta_2 = 0,$$

and

$$T_3 = 2(\tilde{\theta}_3)^2/(3\tilde{\sigma}^2), \qquad \text{under} \quad H_0\colon \theta_3 = 0.$$

Compatibly with the remarks made regarding tests of hypotheses in 3^K factorial experiments, one should test T_3 first, in order to determine the presence of a significant effect due to the three levels of the second factor; only if this test fails to declare significance of this factor need T_2 be tested (see also Exercise 4 of Section 7.6.2).

The 2 × 3 design may also be replicated a total of m times. Doing so ensures that true (unconfounded) estimates of the error variances σ^2 may be obtained. Not doing so may imply that $\tilde{\varepsilon}_1$ and/or $\tilde{\varepsilon}_2$ are estimates of an interaction effect due to the combination of the 1 level of the first factor with the levels of the second factor; i.e., these estimates may be confounded with the estimation of an existing interaction. If these interactions may be safely ignored, then replication of the design is not essential.

More generally, one may define mixed factorial designs of K (greater than 2) factors. A 2 × 3 × 4 factorial design would consist of 24 simular encounters, corresponding to the 24 possible combinations of the 2 levels of the first, the 3 levels of the second, and the 4 levels of the third. The resulting simular responses in these generalized mixed factorials could be submitted to an *analysis of variance*, similar to the analyses discussed for the 2^K, 3^K, and 2×3 factorial designs described in this chapter. Any such analysis of variance should be undertaken with the design fully replicated at least twice, but if the experimenter is willing to assume that multifactor interactions are insignificant (relative to σ^2), then a single or even a partial replication can be done. Again, the reader is referred to Kempthorne (1952) for details of these designs and their analyses.

Chapter 8

EXPERIMENTAL DESIGNS FOR QUANTITATIVE FACTORS

Experimentation is an active science.

—CLAUDE BERNARD

8.1. Introduction

In Chapters 6 and 7 a number of the more fundamental statistical techniques applicable to experimentation with simulation models have been presented. A fundamental premise has been that the stochastic simulation model produces a simular response (output) $Y(T)$ which is an univariate random variable dependent upon p pertinent environmental (input) conditions, or factors $\vec{x} = (x_1, x_2, \ldots, x_p)$, in accordance with some unknown transformation

$$Y(T) = Y(x_1, x_2, \ldots, x_p; s).$$

The random number seed s is presumed to reflect the effect of stochasticity, via an error term $\varepsilon(s)$, and we write

$$Y(T) = Y(x_1, x_2, \ldots, x_p) + \varepsilon(s),$$

where $\varepsilon(s)$ is presumed to be a random variable of mean zero and of variance σ_ε^2, regardless of the particular environmental specification $(x_1, x_2, \ldots, x_p)$, and the unknown function $Y(x_1, x_2, \ldots, x_p)$, called the *simular response function*, represents the mean simular response at that specification.

Any element of the vector $\vec{x}$ of environmental specifications has been called a *factor*, and may be categorized as qualitative or quantitative, depending upon whether it could conceptually assume values from a continuum of real numbers. In addition, the presumption has been made that the simular response function will probably be a continuous function of any quantitative factor (see Section 7.1).

The error term in the simular responses has also been presumed to represent a set of uncorrelated random variables; i.e., $\text{Cov}[\varepsilon(s_k), \varepsilon(s_m)]=0$ whenever $s_k \neq s_m$, regardless of the environmental conditions at which the corresponding simular responses were defined. In the present chapter, the error term will be seen as incorporating a second source of variation. The assumption that the simular response may be represented as

$$Y(T) = f(x_1, x_2, \ldots, x_p; \vec{\beta}) + \varepsilon(s),$$

where $f(x_1, x_2, \ldots, x_p; \vec{\beta})$ is a known functional form indexed by some vector $\vec{\beta}$ of parameters of unknown values, implies then that the error term $\varepsilon(s)$ incorporates effects due both to the inadequacy of $f(x_1, x_2, \ldots, x_p; \vec{\beta})$ as a representation for the unknown simular response function $Y(x_1, x_2, \ldots, x_p)$ and the effects intrinsically present in a stochastic simulation model. Nonetheless, the fundamental assumptions about the nature of the random variable $\varepsilon(s)$ will not change. In fact, the assumption that the $\varepsilon(s)$ are normally distributed will be invoked when required, though in many cases this assumption will be nonetheless justified as a result of the Central Limit Theorem.

The preceding chapter dealt with simular experimentation that was designed to provide information about the relative effects of alternative levels of *qualitative* factors. The procedure there was relatively straightforward: A factorial experimental design was carried out in accordance with the number of qualitative factors to be examined, the number of levels of each factor to be contrasted, and the amount of replication (or partial replication) necessary to ensure that valid comparisons of the suspected effects were conducted. Once implemented, the experimental design results in a set of simular responses, appropriate linear contrasts of which may be used to determine which qualitative environmental conditions are significant contributors to the simular response.

This chapter will treat the determination of those *quantitative* factors that are suspected to be significant contributors to the simular response. Since these quantitative factors have been presumed to be essentially continuous variates, the comparison of the effects due to changes in the values assignable to any one of them could be an enormous undertaking.

An immediate extension of the idea developed in the preceding chapter, that the change from one level of a qualitative factor to another produces an incremental change in the observed simular responses, is the corresponding concept of a *marginal simular response*: a derivative, or rate at which the simular response function changes with respect to a quantitative factor.

The relationship between these marginal simular responses and the parameters $\vec{\beta}$ that appear in the assumed approximate form $f(x_1, x_2, \ldots, x_p; \vec{\beta})$ for the simular response function can be demonstrated by employing the Taylor series expansion for $Y(\vec{x})$ about some particular environmental condition $\vec{x}_0 = (x_{1,0}, x_{2,0}, \ldots, x_{p,0})$. The selection of $\vec{x}_0$, about which $Y(x_1, x_2, \ldots, x_p)$ is expanded is not especially material, though likely it will correspond to the standard operating conditions of the modeled system, or simuland. The Taylor series expansion becomes

$$\begin{aligned} Y(\vec{x}) = {} & Y(\vec{x}_0) + [Y_1(\vec{x}_0)\cdot(x_1 - x_{1,0}) + \cdots + Y_p(\vec{x}_0)\cdot(x_p - x_{p,0})] \\ & + \left[\frac{1}{2}\sum_{k=1}^{p}\sum_{l=1}^{p} Y_{kl}(\vec{x}_0)\cdot(x_k - x_{k,0})\cdot(x_l - x_{l,0})\right] \\ & + \left[\sum_{k=1}^{p}\sum_{l=1}^{p}\sum_{m=1}^{p} Y_{klm}(\vec{x}_0)\cdot(x_k - x_{k,0})\right. \\ & \quad \left.\cdot\,(x_l - x_{l,0})\cdot(x_m - x_{m,0})/6\right] + \cdots, \end{aligned}$$

where

$$Y_k(\vec{x}_0) = \partial Y(x_1, x_2, \ldots, x_p)/\partial x_k,$$
$$Y_{kl}(\vec{x}_0) = \partial^2 Y(x_1, x_2, \ldots, x_p)/\partial x_k\,\partial x_l,$$

and

$$Y_{klm}(\vec{x}_0) = \partial^3 Y(x_1, x_2, \ldots, x_p)/\partial x_k\,\partial x_l\,\partial x_m,$$

each partial derivative being evaluated at $\vec{x}_0$, for $k, l, m = 1, 2, \ldots, p$.

Since these expressions $Y_k(\vec{x}_0)$, $Y_{kl}(\vec{x}_0)$, and $Y_{klm}(\vec{x}_0)$, when evaluated at the particular environmental condition $\vec{x}_0$, each represent some constant, the simular response function can be represented by a function of the form

$$\begin{aligned} & \beta_0 + [\beta_1\cdot(x_1 - x_{1,0}) + \beta_2\cdot(x_2 - x_{2,0}) + \cdots + \beta_p\cdot(x_p - x_{p,0})] \\ & + \left[\sum_{k=1}^{p}\sum_{l=1}^{k} \beta_{kl}\cdot(x_k - x_{k,0})\cdot(x_l - x_{l,0})\right] \\ & + \left[\sum_{k=1}^{p}\sum_{l=1}^{k}\sum_{m=1}^{l} \beta_{klm}\cdot(x_k - x_{k,0})\cdot(x_l - x_{l,0})(x_m - x_{m,0})\right] + \cdots, \end{aligned}$$

where

$$\beta_0 = Y(\vec{x}_0)$$

$$\beta_k = \partial Y(\vec{x})/\partial x_k, \qquad \text{evaluated at} \quad \vec{x}_0,$$

$$\beta_{kl} = \begin{cases} \frac{1}{2} Y_{kk}(\vec{x}_0), & \text{if} \quad k = l, \\ Y_{kl}(\vec{x}_0), & \text{if} \quad k > l, \end{cases}$$

etc.

In this form, the parameters $\vec{\beta}$ appear directly in terms of the partial derivatives of the simular response function; i.e., they are the *marginal simular responses.*

Consequently, if experimentation can be conducted in the neighborhood of $\vec{x}_0$, then statistical estimates of the $\vec{\beta}$ will provide information about the relative rates of change of the simular response function with respect to each of the quantitative factors. The essential question remaining, then, is the method by which an experimental design can be implemented so as to provide, with efficiency and some facility, acceptable statistical estimates of $\vec{\beta}$. The experimental methodology suitable for providing some answers to this question will constitute the remainder of this chapter. Unless otherwise stated, all factors $(x_1, \ldots, x_p)$ are to be assumed quantitative.

8.2. Experimentation with a Single Quantitative Factor

If all environmental specifications save one (say, x_1) are assigned particular values (say, $x_{2,0}, x_{3,0}, \ldots, x_{p,0}$), then simular experimentation may be conducted in order to ascertain the relationship between the simular response $Y(T)$ and the quantitative factor x_1. This experimentation is performed by the selection of N representative specifications $(x_{1,1}, x_{1,2}, \ldots, x_{1,N})$ for the factor x_1 and the concomitant selection of N independent and nonrepetitive random-number seeds; these specifications and seeds are pairwise and sequentially assigned as the input data for the stochastic simulation model, thereby yielding the N simular responses $(Y_1, Y_2, \ldots, Y_N)$.

With the assumption that the simular response function can be approximated in a neighborhood of $(x_{1,0}, x_{2,0}, \ldots, x_{p,0})$ by the Taylor series expansion

$$Y(x_1, x_2, \ldots, x_p) = \beta_0 + \left[\sum_{k=1}^{p} \beta_k \cdot (x_k - x_{k,0})\right] + \left[\sum_{k=1}^{p} \sum_{l=1}^{k} \beta_{kl} \cdot (x_k - x_{k,0}) \cdot (x_l - x_{l,0})\right] + \cdots,$$

one sees that the constancy of the last $p - 1$ environmental specifications $(x_{2,0}, x_{3,0}, \ldots, x_{p,0})$ throughout the N simular encounters allows the simular response function to be written as the polynomial

$$Y(x_1, x_{2,0}, \ldots, x_{p,0}) = \beta_0 + \beta_1 \cdot (x_1 - x_{1,0}) + \beta_{11} \cdot (x_1 - x_{1,0})^2 + \cdots.$$

The reader may recall that the location of $x_{1,0}$ was somewhat arbitrary, it being suggested that this specification for the factor of interest might well correspond to the standard operating specification for the factor in the modeled system itself. However, without loss of generality, one can define each $x_{k,0} = 0$, $k = 1, 2, \ldots, p$, thereby effectively shifting the origin of the "factor space" to the location $(x_{1,0}, x_{2,0}, \ldots, x_{p,0}) = (0, 0, \ldots, 0)$.

In terms of x_1, the simular response function then becomes

$$Y(x_1, 0, 0, \ldots, 0) = \beta_0 + \beta_1 x_1 + \beta_{11} x_1^2 + \beta_{111} x_1^3 + \cdots.$$

8.2.1. The Linear Approximation to the Simular Response Function

Using only the initial two terms of this Taylor series expansion provides the linear approximation for the simular response function:

$$Y(x_1, 0, 0, \ldots, 0) \cong f_1(x_1) = \beta_0 + \beta_1 x_1;$$

i.e., the simular response can be represented as

$$Y(T) = \beta_0 + \beta_1 \cdot x_1 + \varepsilon(s),$$

where $\varepsilon(s)$ is an error term comprising:

(*a*) errors, due to the "lack of fit," which arise because a linear approximation to the simular response function is employed; and,

(*b*) errors due to the intrinsic stochasticity of the simulation model.

One would usually expect, then, that the N data points:

$$(x_{1,1}, y_1), (x_{1,2}, y_2), \ldots, (x_{1,N}, y_N),$$

where y_j is the recorded simular response corresponding to the jth specification $x_{1,j}$ for the factor x_1, are to be as nearly representative of the assumed relationship $Y(T) = \beta_0 + \beta_1 \cdot x_1$ as possible. In order to accomplish this desire, the method of least squares may be invoked for the purpose of estimating β_0 and β_1; the resulting estimates, $\tilde{\beta}_0$ and $\tilde{\beta}_1$, are selected such that the sum of squared deviations

$$Q = \sum_{j=1}^{N} (y_j - \beta_0 - \beta_1 x_{1,j})^2 = \sum_{j=1}^{N} \varepsilon^2(s_j)$$

is minimized whenever $\beta_0 = \tilde{\beta}_0$ and $\beta_1 = \tilde{\beta}_1$.

In accordance with the discussion (Section 6.4) of the method of least squares, Q is minimized whenever

$$\tilde{\beta}_0 = \bar{y} - \tilde{\beta}_1 \bar{x}_1, \tag{8.2.1:1}$$

where

$$\tilde{\beta}_1 = \frac{\sum_{j=1}^{N} (x_{1,j} y_j) - (\bar{y}) \sum_{j=1}^{N} x_{1,j}}{\sum_{j=1}^{N} (x_{1,j}^2) - N(\bar{x}_1)^2}, \tag{8.2.1:2}$$

where $\bar{x}_1 = N^{-1} \sum_{j=1}^{N} x_{1,j}$ and $\bar{y} = N^{-1} \sum_{j=1}^{N} y_j$. The quantities, $\tilde{\beta}_0$ and $\tilde{\beta}_1$ are called the *least squares estimates* of the parameters β_0 and β_1 and, since they are dependent upon the simular responses $(Y_1, Y_2, \ldots, Y_N)$, they become themselves random variables. For example, the estimate $\tilde{\beta}_1$ has expectation

$$E(\tilde{\beta}_1) = \frac{\sum_{j=1}^{N} x_{1,j}(\beta_0 + \beta_1 x_{1,j}) - (\bar{x}_1) \sum_{j=1}^{N} (\beta_0 + \beta_1 x_{1,j})}{\sum_{j=1}^{N} (x_{1,j}^2) - N(\bar{x}_1)^2} = \beta_1$$

whenever the errors $\varepsilon(s_j)$ are presumed to be random variables of mean zero; and, $\tilde{\beta}_1$ has variance

$$\begin{aligned} \operatorname{Var}(\tilde{\beta}_1) &= \frac{\operatorname{Var}[\sum_{j=1}^{N} (x_{1,j} y_j)] + \operatorname{Var}[\bar{x}_1 \sum_{j=1}^{N} y_j] - 2 \operatorname{Cov}[\sum_{j=1}^{N} (x_{1,j} y_j), \bar{x}_1 \cdot \bar{y} \cdot N]}{[\sum_{j=1}^{N} x_{1,j}^2 - N(\bar{x}_1)^2]^2} \\ &= \frac{\sigma_\varepsilon^2 [\sum_{j=1}^{N} x_{1,j}^2 + N(\bar{x}_1)^2 - 2N(\bar{x}_1)^2]}{[\sum_{j=1}^{N} x_{1,j}^2 - N(\bar{x}_1)^2]^2} = \frac{\sigma_\varepsilon^2}{[\sum_{j=1}^{N} x_{1,j}^2 - N(\bar{x}_1)^2]}, \end{aligned}$$

whenever the errors $\varepsilon(s_j)$ are presumed to be uncorrelated and of common variance σ_ε^2.

Under these same distributional properties for the error terms, the statistic $\tilde{\beta}_0$ has mean

$$E(\tilde{\beta}_0) = \beta_0 + \beta_1 \bar{x}_1 - E(\tilde{\beta}_1 \bar{x}_1) = \beta_0$$

and variance

$$\operatorname{Var}(\tilde{\beta}_0) = \operatorname{Var}(\bar{Y}) + (\bar{x}_1)^2 \operatorname{Var}(\tilde{\beta}_1) - 2\bar{x}_1 \operatorname{Cov}(\tilde{\beta}_1, \bar{Y}),$$

or

$$\operatorname{Var}(\tilde{\beta}_0) = \frac{\sigma_\varepsilon^2[\sum_{j=1}^N x_{1,j}^2]}{N[\sum_{j=1}^N x_{1,j}^2 - N(\bar{x}_1)^2]}.$$

Also of interest is the covariance between the two estimators:

$$\operatorname{Cov}(\tilde{\beta}_0, \tilde{\beta}_1) = \frac{-\sigma_\varepsilon^2(\bar{x}_1)}{\sum_{j=1}^N x_{1,j}^2 - N(\bar{x}_1)^2}.$$

One notes that the least squares line, a function of x_1, becomes itself a random variable,

$$\tilde{y} = \tilde{\beta}_0 + \tilde{\beta}_1 x_1,$$

having mean (at any location x_1),

$$E(\tilde{y}) = \beta_0 + \beta_1 \cdot x_1$$

and variance

$$\operatorname{Var}(\tilde{y}) = \operatorname{Var}(\tilde{\beta}_0) + x_1^2 \operatorname{Var}(\tilde{\beta}_1) + 2x_1 \cdot \operatorname{Cov}(\tilde{\beta}_0, \tilde{\beta}_1)$$

or

$$\operatorname{Var}(\tilde{y}) = \frac{\sigma_\varepsilon^2\{x_1^2 - 2x_1(\bar{x}_1) + N^{-1}\sum_{j=1}^N x_{1,j}^2\}}{\sum_{j=1}^N x_{1,j}^2 - N(\bar{x}_1)^2}.$$

Therefore, once the coefficients β_0 and β_1 have been estimated, a line representative of the simular response function can be plotted: $\tilde{y} = \tilde{\beta}_0 + \tilde{\beta}_1 x_1$, an approximate description of the simular response function.

The reader may note that considerable simplification may result from the intentional selection of the N environmental specifications $x_{1,1}, x_{1,2}, \ldots, x_{1,N}$ for the encounters to be used in the experimentation.

For example, by selecting these specifications such that

$$\sum_{j=1}^N x_{1,j} = N(\bar{x}_1) = 0,$$

the estimates $\tilde{\beta}_0$ and $\tilde{\beta}_1$ become uncorrelated; viz.,

$$\operatorname{Cov}(\tilde{\beta}_0, \tilde{\beta}_1) = 0 \text{ under this condition.}$$

Other expressions for the variances become considerably simplified:

$$\operatorname{Var}(\tilde{\beta}_0) = \sigma_\varepsilon^2/N,$$

$$\operatorname{Var}(\tilde{\beta}_1) = \sigma_\varepsilon^2 \Big/ \Big[\sum_{j=1}^N x_{1,j}^2\Big],$$

and

$$\text{Var}(\tilde{y}) = \text{Var}(\tilde{\beta}_0 + \tilde{\beta}_1 x_1) = \sigma_\varepsilon^2 \left[1 + \frac{N \cdot x_1^2}{\sum_{j=1}^N x_{1,j}^2}\right] \Big/ N.$$

Thus, one sees that an appropriate *experimental design*, for the purpose of estimating the coefficients, β_0 and β_1, is one that selects the environmental specifications $x_{1,j}$ so that their sum is null; this is most readily assured by a symmetric pairwise selection of these conditions, each pair of conditions being the negative of one another (equivalently, each element of each pair of design points being equidistant from $x_{1,0}$, the zero locus). Under such an experimental design, one may note that the estimate $\tilde{\beta}_0$ improves (i.e., has decreasing variance) with increasing N; similarly for $\tilde{\beta}_1$, provided that the additional responses are recorded away from $x_1 = 0$, so that the denominator, $\sum_{j=1}^N x_{1,j}^2$, in the expression for the variance of $\tilde{\beta}_1$, increases. One might then also note that an additional improvement in the experimental design avails itself whenever the analyst selects the design points $x_{1,j}$ as far from the zero locus as possible. This fact is compatible with the intuitive notion that a straight line will be more firmly fixed if information regarding its position comes from remotely spaced extremes. However, since the simular response function is not likely to be well represented by a linear approximation over such a vast region, one must be somewhat constrained in the selection of design points far removed from the zero locus. One may find that the use of quadratic or other polynomial representations would be more advantageous to the description of the simular response function over such extended regions.

Before proceeding to the discussion of the use of quadratic, cubic, and other polynomial representations of the simular response function, however, one additional note is in order regarding the distributional properties of the estimates $\tilde{\beta}_0$ and $\tilde{\beta}_1$. Each estimate is a linear combination of the simular responses $Y_1, Y_2, \ldots, Y_N$ as is readily noted in the expressions, applicable whenever $\sum_{j=1}^N x_{1,j} = 0$:

$$\tilde{\beta}_0 = \bar{Y} = N^{-1} \sum_{j=1}^N Y_j$$

and

$$\tilde{\beta}_1 = \sum_{j=1}^N Y_j \cdot [x_{1,j}/S^2(\vec{x}_1)],$$

where

$$S^2(\vec{x}_1) = \left[\sum_{j=1}^N x_{1,j}^2 - N(\bar{x}_1)^2\right] = \sum_{j=1}^N x_{1,j}^2.$$

The orthogonality of these linear combinations becomes readily apparent in these expressions for $\tilde{\beta}_0$ and $\tilde{\beta}_1$, since $\sum_{j=1}^{N} x_{1,j} = 0$ is a design constraint. Thus, since the simular responses Y_j are presumed to be uncorrelated, so would be any pair of orthogonal linear combinations of them. Indeed, if the additional assumption is made that the errors $\varepsilon(s_j)$, $j = 1, 2, \ldots, N$, are independently and normally distributed, then so will be the simular responses Y_j, $j = 1, 2, \ldots, N$; and, consequently $\tilde{\beta}_0$ and $\tilde{\beta}_1$ would have the bivariate normal distribution (and would be independently and normally distributed if the design constraint, $\sum_{j=1}^{N} x_{1,j} = 0$, were imposed). Therefore, by appropriately designing the simular experiment, one can compute estimators for the linear approximation to the simular response function (in terms of any particular quantitative factor) and can subsequently construct confidence intervals for, and test hypotheses associated with these coefficients. The reader is referred to Graybill (1961) and Mendenhall (1968) for details of these procedures.

In this regard, the reader may recall (Section 6.4.4) that the least-squares estimators $\tilde{\beta}_0$ and $\tilde{\beta}_1$ are the same as the maximum likelihood estimators of these coefficients whenever the error terms are assumed to be independent random variables from a normal distribution of mean zero and variance σ_ε^2. The maximum likelihood estimator for this error variance is also available as

$$\tilde{\sigma}_\varepsilon^2 = N^{-1} \sum_{j=1}^{N} [y_j - \tilde{\beta}_0 - \tilde{\beta}_1 \cdot x_{1,j}]^2,$$

the average of the squared deviations of the observed responses (y_j) from the estimated responses ($\tilde{y}_j = \tilde{\beta}_0 + \tilde{\beta}_1 \cdot x_{1,j}$). In fact, one may show that the assumptions of normality and independence of the error terms imply that the statistic, $N\tilde{\sigma}_\varepsilon^2/\sigma_\varepsilon^2$, has the Chi-squared distribution of $N - 2$ degrees of freedom, and is independently distributed of both $\tilde{\beta}_0$ and $\tilde{\beta}_1$. Chapter 9 of Lindgren (1968) provides a more complete description of these developments. The reader should note nonetheless that the preceding statements imply that the statistics

$$T_0 \equiv \frac{(\tilde{\beta}_0 - \beta_0)^2/(\sigma_\varepsilon^2/N)}{(N\tilde{\sigma}_\varepsilon^2)/[\sigma_\varepsilon^2 \cdot (N-2)]} = \frac{(\tilde{\beta}_0 - \beta_0)^2}{\tilde{\sigma}_\varepsilon^2/(N-2)}$$

and similarly,

$$T_1 \equiv \frac{(\tilde{\beta}_1 - \beta_1)^2/[\sigma_\varepsilon^2/(\sum_{j=1}^{N} x_{1,j}^2)]}{(N\tilde{\sigma}_\varepsilon^2)/[\sigma_\varepsilon^2 \cdot (N-2)]} = \frac{(\tilde{\beta}_1 - \beta_1)^2 \cdot (\sum_{j=1}^{N} x_{1,j}^2)}{\tilde{\sigma}_\varepsilon^2/(N-2)}$$

each have, whenever the design constraint $\sum_{j=1}^{N} x_{1,j} = 0$ is applicable, the Snedecor's F distribution of 1 and $N - 2$ degrees of freedom. Alternatively, these statistics are squares of Student's t variates of $N - 2$ degrees of freedom, so that tests of null hypotheses of the form, say, H_0: $\beta_1 = 0$, may be conducted by comparing the computed statistics with the cumulative distribution function of the appropriate Snedecor's F or Student's t distribution.

The null hypothesis corresponds to a speculation that the simular response function has no slope in the x_1-direction at the zero locus $x_{1,0}$ since

$$df_1(x_1)/dx_1 = d(\beta_0 + \beta_1 x_1)/dx_1 = \beta_1.$$

Similarly, the hypothesis H_0: $\beta_0 = 0$ also has an important geometric connotation, because, if indeed $\beta_0 = 0$, the simular response must pass through the origin; i.e., since $f_1(0) = \beta_0$, this null hypothesis conjectures that $f_1(0) = 0$. If indeed the simular response function should satisfy a constraint such as this (in order to be compatible, say, with the modeled system itself), then the test of this null hypothesis by examining the significance of the estimate $\tilde{\beta}_0$ is quite in order, and thus serves as an additional test applicable in the validation stage of model development.

EXERCISES

1. Design a simulation experiment that will require only two simular responses, Y_1 and Y_2, yet be capable of estimating the linear approximation,

$$f_1(x_1) = \beta_0 + \beta_1 x_1,$$

for the response surface. Show that (*a*) the least-square estimates of β_0 and β_1 are

$$\tilde{\beta}_0 = (x_{1,1} y_2 - x_{1,2} y_1)/(x_{1,1} - x_{1,2})$$

and

$$\tilde{\beta}_1 = (y_1 - y_2)/(x_{1,1} - x_{1,2});$$

(*b*) that

$$\tilde{y} = \tilde{\beta}_0 + \tilde{\beta}_1 \cdot x_1$$

passes exactly through not only the point $(\bar{x}_1, \bar{y})$ as usual, but also the data points $(x_{1,1}, y_1)$ and $(x_{1,2}, y_2)$ themselves, so that the estimate $\tilde{\sigma}_\varepsilon^2 = 0$; (*c*) that the selection of $x_{1,2} = -x_{1,1}$ so that $\sum_{j=1}^{2} x_{1,j} = 0$, provides the estimates

$$\tilde{\beta}_0 = (y_1 + y_2)/2$$

and

$$\tilde{\beta}_1 = (y_1 - y_2)/(2x_{1,1}),$$

which have moments

$$E(\tilde{\beta}_0) = \beta_0, \qquad \mathrm{Var}(\tilde{\beta}_0) = \sigma_\varepsilon^2/2,$$

$$E(\tilde{\beta}_1) = \beta_1, \qquad \mathrm{Var}(\tilde{\beta}_1) = \sigma_\varepsilon^2/(2x_{1,1}^2) = \sigma_\varepsilon^2/(x_{1,1}^2 + x_{1,2}^2)$$

and $\mathrm{Cov}(\tilde{\beta}_0, \tilde{\beta}_1) = 0$; but, (*d*) that, unless $x_{1,1} = -x_{1,2}$, $\mathrm{Cov}(\tilde{\beta}_0, \tilde{\beta}_1) \neq 0$.

2. The linear approximation,

$$f_1(x_1) = \beta_0 + \beta_1 x_1,$$

is to be fitted to the data arising from three simular encounters: $(x_{1,1}, y_1)$ $(x_{1,2}, y_2)$, and $(x_{1,3}, y_3)$. Show that, for $x_{1,1} = -x_{1,2}$ and $x_{1,3} = 0$,

(*a*) the design constraint, $\sum_{j=1}^{3} x_{1,j} = 0$, is satisfied;
(*b*) the coefficients are estimated by

$$\tilde{\beta}_0 = (y_1 + y_2 + y_3)/3,$$

and

$$\tilde{\beta}_1 = (x_{1,1}y_1 + x_{1,2}y_2)/(x_{1,1}^2 + x_{1,2}^2) = (y_1 - y_2)/(2x_{1,1});$$

(*c*) the error variance becomes estimable by

$$\tilde{\sigma}_\varepsilon^2 = \sum_{j=1}^{3} (y_j - \tilde{\beta}_0 - \tilde{\beta}_1 x_{1,j})^2/3 = (y_1 + y_2 - 2y_3)^2/18;$$

(*d*) the contrast $(y_1 + y_2 - 2y_3)$ is an unbiased estimate of zero and is indeed orthogonal to both $\tilde{\beta}_0$ and $\tilde{\beta}_1$, so that if the simular responses are independently and normally distributed, then the following statistics are independent: $\tilde{\beta}_0$, $\tilde{\beta}_1$, and σ_ε^2;
(*e*) the statistic $(3\tilde{\sigma}_\varepsilon^2)/[\sigma_\varepsilon^2(3 - 2)]$ has the Chi-squared distribution of 1 degree of freedom.

3. Generate 100 independent random variables from each of the three following normal distribution:

$$\text{I:} \qquad \mu = 1, \quad \sigma^2 = 25$$

$$\text{II:} \qquad \mu = 7, \quad \sigma^2 = 25$$

$$\text{III:} \qquad \mu = 4, \quad \sigma^2 = 25.$$

Presume that each of the 100 triads (Y_1, Y_2, Y_3) constitutes the recorded responses from a simular experiment in which the first encounter had an environmental specification of $x_{1,1} = -1$, the second had $x_{1,2} = +1$, and the third used $x_{1,3} = 0$. For each triad, estimate $\tilde{\beta}_0, \tilde{\beta}_1$ and $\tilde{\sigma}_\varepsilon^2$ as described in the preceding exercise. Form a frequency histogram for each, as well as a frequency histogram for the 100 values of $\tilde{y} = \tilde{\beta}_0 + \frac{1}{2}\tilde{\beta}_1$, the estimate of the response surface at $x_1 = \frac{1}{2}$. Are any of these histograms incompatible with the theoretical normal distributions which should be obtained in each case?

4. In Exercise 3, test the null hypotheses H_0: $\beta_0 = 4$, and H_0: $\beta_1 = 3$ for each of the 100 simular experiments. In what proportion of instances was each hypothesis rejected? Is this proportion reasonable in each case?

5. Referring to Exercise 3, repeat the 100 simular experiments and answer the questions associated with both Exercise 3 and Exercise 4, excepting now you are to have generated the second random variable of each triad from a normal distribution of mean 9 (instead of 7). Compare your results with those of Exercises 3 and 4.

6. Verify Eqs. (8.2.1:1) and (8.2.1:2) by the method of least squares, as given in Section 6.4.

8.2.2. The Quadratic Approximation to the Simular Response Function

Thus far in this section, we have presumed that experimentation with a stochastic simulation model is to be conducted to determine the behavior of the simular response $Y(T)$ at the end of T simular time units as a function of a single quantitative factor or environmental condition x_1. Consequently, the experimental designs, as recommended, have presumed the experimentation to be conducted in the neighborhood of some particular environmental specification $\vec{x}_0 = (x_{1,0}, x_{2,0}, \ldots, x_{p,0})$, which, in many simular experimental contexts, would likely correspond to the standard operating conditions for the simulated system. The factors $x_2, x_3, \ldots, x_p$ during the present experimentation have been held constant at the respective values $x_{2,0}, x_{3,0}, \ldots, x_{p,0}$, whereas the factor of interest x_1 has assumed N different values $x_{1,1}, x_{1,2}, \ldots, x_{1,N}$, one value being used as input specification for each independently seeded encounter with the simulation model.

As a function of x_1, the simular response function was presumed to be expressible as a Taylor series:

$$Y(x_1, x_{2,0}, \ldots, x_{p,0}) = \beta_0 + \beta_1 x_1 + \beta_{11} x_1^2 + \beta_{111} x_1^3 + \cdots.$$

By truncating this expression at the linear term, we have presumed that any simular response $Y(T)$ can be approximately expressed as a linear function of x_1:

$$Y(T) = \beta_0 + \beta_1 x_1 + \varepsilon(s),$$

where $\varepsilon(s)$ is a random variable composed of variation due both to the effect of the truncating the Taylor series at the linear term and to the inherent stochastic variation in the simular responses. The coefficients β_0 and β_1 were then seen to be estimable, via an application of the method of least squares employing the pairs of input–output data: $(x_{1,1}, y_1)$, $(x_{1,2}, y_2), \ldots, (x_{1,N}, y_N)$.

The resulting estimates were seen to be unbiased for β_0 and β_1, provided that $E(\varepsilon(s)) = 0$; moreover, with the additional assumptions of homogeneity of σ_ε^2 at varying x_1-locations [i.e., $\text{Var}(Y(T)) = \sigma_\varepsilon^2$, regardless of the input conditions x_1] and of the independence and normality of the distributions of the $\varepsilon(s_j)$, $j = 1, 2, \ldots, N$, these least-squares estimators, it may be recalled, are the equivalent of the maximum likelihood estimators. Under these assumptions, then, the statistics $\tilde{\beta}_0$ and $\tilde{\beta}_1$ were unbiased and consistent estimators of β_0 and β_1, respectively.

An immediate generalization of the experimental design under discussion is the quadratic approximation for the simular response function:

$$Y(x_1, x_{2,0}, \ldots, x_{p,0}) \cong \beta_0 + \beta_1 x_1 + \beta_{11} x_1^2 \equiv f_2(x_1),$$

so that any simular response is now presumed to be related to the environmental condition x_1 by

$$Y(T) = \beta_0 + \beta_1 x_1 + \beta_{11} x_1^2 + \varepsilon(s),$$

where again $\varepsilon(s)$ is a random variable comprising the two types of variation: intrinsic and "lack-of-fit." After the selection of N environmental conditions $(x_{1,1}, x_{1,2}, \ldots, x_{1,N})$ as input specifications for N independently seeded encounters with the model, the resulting set of input–output data pairs,

$$(x_{1,1}, y_1),\ (x_{1,2}, y_2), \ldots, (x_{1,N}, y_N),$$

can be used to determine the least-squares estimates of β_0, β_1, and β_{11}.

Procedurally, the estimates are obtained as the quantities $\tilde{\beta}_0$, $\tilde{\beta}_1$, and $\tilde{\beta}_{11}$, which will minimize:

$$Q = \sum_{j=1}^{N} (y_j - \beta_0 - \beta_1 x_{1,j} - \beta_{11} x_{1,j}^2)^2 = \sum_{j=1}^{N} \varepsilon^2(s_j).$$

Partially differentiating Q with respect to β_0, β_1, and β_{11} and equating to zero yields the three equations

$$\sum_{j=1}^{N} y_j = N\beta_0 + \beta_1 \cdot \sum_{j=1}^{N} x_{1,j} + \beta_{11} \cdot \sum_{j=1}^{N} x_{1,j}^2,$$

$$\sum_{j=1}^{N} \{y_j \cdot x_{1,j}\} = \beta_0 \cdot \sum_{j=1}^{N} x_{1,j} + \beta_1 \cdot \sum_{j=1}^{N} x_{1,j}^2 + \beta_{11} \cdot \sum_{j=1}^{N} x_{1,j}^3, \quad (8.2.2:1)$$

and

$$\sum_{j=1}^{N} \{y_j \cdot x_{1,j}^2\} = \beta_0 \cdot \sum_{j=1}^{N} x_{1,j}^2 + \beta_1 \cdot \sum_{j=1}^{N} x_{1,j}^3 + \beta_{11} \cdot \sum_{j=1}^{N} x_{1,j}^4,$$

having solutions which one may denote $\tilde{\beta}_0$, $\tilde{\beta}_1$, and $\tilde{\beta}_{11}$.

The solutions may be more neatly displayed in matrix notation, appropriate to which are the vectors and matrices:

$\vec{Y}^{\mathrm{T}} = [Y_1, Y_2, \ldots, Y_N]$, the $(1 \times N)$ vector of simular responses,

$\vec{\beta}^{\mathrm{T}} = [\beta_0, \beta_1, \beta_{11}]$, the (1×3) vector of coefficients,

$\vec{\varepsilon}^{\mathrm{T}} = [\varepsilon(s_1), \varepsilon(s_2), \ldots, \varepsilon(s_N)]$, the $(1 \times N)$ vector of errors,

and

$$\mathbf{X} = \begin{bmatrix} 1 & x_{1,1} & x_{1,1}^2 \\ 1 & x_{1,2} & x_{1,2}^2 \\ \vdots & \vdots & \vdots \\ 1 & x_{1,N} & x_{1,N}^2 \end{bmatrix},$$

the $(N \times 3)$ *experimental design matrix*.

Of note is the relationship between the N simular responses and these vectors and matrices; viz.,

$$\vec{Y} = (\mathbf{X} \cdot \vec{\beta}) + \vec{\varepsilon},$$

as the reader may verify by examining the jth element of each vector on the right and the left side of the vector equation. [*Note*: $(\mathbf{X} \cdot \vec{\beta})$ is also an $(N \times 1)$ vector.] The sum of squared errors becomes representable then as

$$Q = \sum_{j=1}^{N} \varepsilon^2(s_j) = \vec{\varepsilon}^{\mathrm{T}}\vec{\varepsilon} = (\vec{Y} - \mathbf{X} \cdot \vec{\beta})^{\mathrm{T}} \cdot (\vec{Y} - \mathbf{X} \cdot \vec{\beta}),$$

a scalar quantity that is a quadratic expression in terms of β_0, β_1, and β_{11}. Specifically,

$$Q = (\vec{Y}^{\mathrm{T}}\vec{Y}) - (\vec{Y}^{\mathrm{T}}\mathbf{X}\vec{\beta}) - (\vec{\beta}^{\mathrm{T}}\mathbf{X}^{\mathrm{T}}\vec{Y}) + \vec{\beta}^{\mathrm{T}}(\mathbf{X}^{\mathrm{T}}\mathbf{X})\vec{\beta}$$

or

$$Q = (\vec{Y}^{\mathrm{T}}\vec{Y}) - 2(\vec{Y}^{\mathrm{T}}\mathbf{X}\vec{\beta}) + \vec{\beta}^{\mathrm{T}} \cdot (\mathbf{X}^{\mathrm{T}}\mathbf{X}) \cdot \vec{\beta}.$$

The matrix $(\mathbf{X}^{\mathrm{T}}\mathbf{X})$ is the (3×3) symmetric matrix:

$$(\mathbf{X}^{\mathrm{T}}\mathbf{X}) = \begin{bmatrix} N & \sum_{j=1}^{N} x_{1,j} & \sum_{j=1}^{N} x_{1,j}^2 \\ \sum_{j=1}^{N} x_{1,j} & \sum_{j=1}^{N} x_{1,j}^2 & \sum_{j=1}^{N} x_{1,j}^3 \\ \sum_{j=1}^{N} x_{1,j}^2 & \sum_{j=1}^{N} x_{1,j}^3 & \sum_{j=1}^{N} x_{1,j}^4 \end{bmatrix},$$

so that differentiation of Q with respect to each of the elements of $\vec{\beta}$ and the subsequent setting of these three equal to zero leads to the matrix equations (called the *normal equations*):

$$\mathbf{X}^{\mathrm{T}} \cdot \vec{Y} = (\mathbf{X}^{\mathrm{T}} \cdot \mathbf{X}) \cdot \tilde{\vec{\beta}}$$

identically those presented in summation notation above (8.2.2:1).

Any solution of these three simultaneous equations may be called a set of least-squares estimators for $\vec{\beta}^{\mathrm{T}} = (\beta_0, \beta_1, \beta_{11})$. If the matrix, $\mathbf{X}^{\mathrm{T}} \cdot \mathbf{X}$, is nonsingular, however, the solution is unique, being

$$\tilde{\vec{\beta}} = (\mathbf{X}^{\mathrm{T}} \cdot \mathbf{X})^{-1} \cdot \mathbf{X}^{\mathrm{T}} \cdot \vec{Y}.$$

One of the criteria for the experimental design matrix $\mathbf{X}$ (whose second column, the reader should note, comprises elements whose assignment is at the discretion of the analyst) is that the resulting matrix $(\mathbf{X}^{\mathrm{T}}\mathbf{X})$ have an inverse. With a minimum of caution, this result will obtain; a sufficient condition for this purpose is that the rank of the matrix $\mathbf{X}$ be 3, for the proof of which the reader is referred to Graybill (1961).

Therefore, in order to obtain unique estimates of the coefficients β_0, β_1, and β_{11} for the quadratic approximation to the simular response function, the analyst need only select N (≥ 3) values of the environmental condition x_1 (each to be used as the input specification for an independently seeded encounter with the simulation model), in such a manner that the matrix $\mathbf{X}$ has rank 3.

The resulting estimates are unbiased, since

$$E(\vec{\tilde{\beta}}) = E[(\mathbf{X}^{\mathrm{T}}\mathbf{X})^{-1} \cdot (\mathbf{X}^{\mathrm{T}}\vec{Y})] = E[(\mathbf{X}^{\mathrm{T}}\mathbf{X})^{-1}(\mathbf{X}^{\mathrm{T}}\mathbf{X})\vec{\beta}\,] = \vec{\beta};$$

i.e., since the jth element of $\vec{\tilde{\beta}}$ has expectation β_j, $j = 0$, 1, and 11. Furthermore, the variance–covariance matrix for the random variables, $\vec{\tilde{\beta}}$, becomes

$$\begin{aligned}\mathbf{\Sigma}_{\tilde{\beta}} &= E\{[(\mathbf{X}^{\mathrm{T}}\mathbf{X})^{-1} \cdot \mathbf{X}^{\mathrm{T}} \cdot \vec{Y} - \vec{\beta}\,] \cdot [(\mathbf{X}^{\mathrm{T}}\mathbf{X})^{-1} \cdot \mathbf{X}^{\mathrm{T}} \cdot \vec{Y} - \vec{\beta}\,]^{\mathrm{T}}\} \\ &= E[(\mathbf{X}^{\mathrm{T}}\mathbf{X})^{-1} \cdot \mathbf{X}^{\mathrm{T}} \cdot \vec{\varepsilon} \cdot \vec{\varepsilon}^{\,\mathrm{T}} \cdot \mathbf{X} \cdot (\mathbf{X}^{\mathrm{T}}\mathbf{X})^{-1}],\end{aligned}$$

since $\vec{Y} = \mathbf{X}\vec{\beta} + \vec{\varepsilon}$ and since $(\mathbf{X}^{\mathrm{T}}\mathbf{X})$ is a symmetric matrix which, if nonsingular, possesses a symmetric inverse $(\mathbf{X}^{\mathrm{T}}\mathbf{X})^{-1}$; thus

$$\mathbf{\Sigma}_{\tilde{\beta}} = (\mathbf{X}^{\mathrm{T}}\mathbf{X})^{-1}\mathbf{X}^{\mathrm{T}}(\sigma_\varepsilon^2\mathbf{I})\mathbf{X}(\mathbf{X}^{\mathrm{T}}\mathbf{X})^{-1},$$

since the errors are presumed to be uncorrelated and of common variance σ_ε^2.

The variance–covariance matrix for the statistics (estimators) $\tilde{\beta}_0$, $\tilde{\beta}_1$, and $\tilde{\beta}_{11}$ then becomes

$$\mathbf{\Sigma}_{\tilde{\beta}} = \sigma_\varepsilon^2 \cdot (\mathbf{X}^{\mathrm{T}}\mathbf{X})^{-1}.$$

There are two important points to be made with regard to this (3×3) variance–covariance matrix. First, the estimates $\tilde{\beta}_0$, $\tilde{\beta}_1$, and $\tilde{\beta}_{11}$ are uncorrelated only if $(\mathbf{X}^{\mathrm{T}}\mathbf{X})^{-1}$ is a diagonal matrix, or equivalently, only whenever $(\mathbf{X}^{\mathrm{T}}\mathbf{X})$ is diagonalized. Second, a judicious choice of the elements of $\mathbf{X}$ should lead to a matrix $(\mathbf{X}^{\mathrm{T}}\mathbf{X})^{-1}$ which is not only diagonal, but also one that has diagonal elements made as small as or as equal as practicable.

The matrix $(\mathbf{X}^{\mathrm{T}}\mathbf{X})^{-1}$ will assuredly be diagonalized if the three columns of the design matrix $\mathbf{X}$ are defined orthogonally, for the element e_{ij} of the matrix $(\mathbf{X}^{\mathrm{T}}\mathbf{X})$ is the scalar product of the ith row of $\mathbf{X}^{\mathrm{T}}$ with the jth column of $\mathbf{X}$; i.e., e_{ij} is the scalar product of the ith and jth columns of $\mathbf{X}$. The orthogonality of these columns would then ensure that $(\mathbf{X}^{\mathrm{T}}\mathbf{X})$, and hence $(\mathbf{X}^{\mathrm{T}}\mathbf{X})^{-1}$, would be a diagonal matrix. Presuming that this could be achieved, and, if, the assumptions of the independence and the normality of the errors $\varepsilon(s_j)$ are made, then, the estimates $\tilde{\beta}_0$, $\tilde{\beta}_1$, and $\tilde{\beta}_{11}$ would also be independently and normally distributed random variables.

However, as the reader may note, the first and third columns of the design matrix, $\mathbf{X}$, cannot possibly be orthogonal unless $x_{1,j}^2 = 0$, $j = 1, 2, \ldots, N$. But, even in this degenerate case the second and third columns of $\mathbf{X}$ would be identical and the rank of $\mathbf{X}$ could not equal 3, as required. Therefore, the present situation does not admit a design matrix that will lead to mutually independent estimates. The reader should note, however, the conditions under which the estimate $\tilde{\beta}_1$ will be uncorrelated with $\tilde{\beta}_0$ and $\tilde{\beta}_{11}$:

$$\sum_{j=1}^{N} x_{1,j} = 0 \qquad \text{and} \qquad \sum_{j=1}^{N} x_{1,j}^3 = 0,$$

respectively. Thus, a design criterion which would produce the null covariances,

$$\operatorname{Cov}(\tilde{\beta}_0, \tilde{\beta}_1) = 0 = \operatorname{Cov}(\tilde{\beta}_1, \tilde{\beta}_{11}),$$

is a choice of *pairs* of design points, $x_{1,j}$ that are symmetrically placed about $x_1 = 0$. In this case the least-squares unbiased estimates become [see Eqs. (8.2.2:1)]

$$\tilde{\beta}_0 = \bar{y} - \tilde{\beta}_{11}\left(\sum_{j=1}^{N} x_{1,j}^2/N\right),$$

$$\tilde{\beta}_1 = \left(\sum_{j=1}^{N} y_j \cdot x_{1,j}\right) \Big/ \left(\sum_{j=1}^{N} x_{1,j}^2\right),$$

and

$$\tilde{\beta}_{11} = \frac{\left(\sum_{j=1}^{N} y_j x_{1,j}^2\right) - \bar{y} \cdot \left(\sum_{j=1}^{N} x_{1,j}^2\right)}{\sum_{j=1}^{N} x_{1,j}^4 - N^{-1}\left(\sum_{j=1}^{N} x_{1,j}^2\right)^2}.$$

One may note, however, that, owing to their correlation, $\tilde{\beta}_0$ and $\tilde{\beta}_{11}$ cannot be independent random variables. Therefore, tests of hypotheses about either should take into account the values of both statistics; i.e., the conditional distribution of $\tilde{\beta}_0$ given $\tilde{\beta}_{11}$ should be employed either in defining confidence intervals for, or in testing hypotheses concerning β_0 and vice-versa. Tests of hypotheses concerning β_1 may of course be performed under the constrained design; the test statistic may again be shown to be related to a Snedecor's F variate of 1 and $N - 3$ degrees of freedom. For details, the reader is again referred to the text of Graybill (1961).

EXERCISE

1. Show that, for the linear representation for the response surface

$$Y(T) = \beta_0 + \beta_1 x_1 + \varepsilon(s),$$

as discussed in Section 8.2.1, the N simular responses $Y_1, Y_2, \ldots, Y_N$ may be written in vector–matrix notation as $\vec{Y} = (\mathbf{X}\vec{\beta}) + \vec{\varepsilon}$, where

$$\vec{Y}^{\mathrm{T}} = (Y_1, Y_2, \ldots, Y_N),$$
$$\vec{\beta}^{\mathrm{T}} = (\beta_0, \beta_1),$$
$$\vec{\varepsilon}^{\mathrm{T}} = [\varepsilon(s_1), \ldots, \varepsilon(s_N)],$$

and

$$\mathbf{X} = \begin{bmatrix} 1 & x_{1,1} \\ 1 & x_{1,2} \\ \vdots & \vdots \\ 1 & x_{1,N} \end{bmatrix}.$$

Show also that the estimates $\tilde{\beta}_0$ and $\tilde{\beta}_1$ are given as elements of

$$\tilde{\vec{\beta}} = (\mathbf{X}^{\mathrm{T}} \cdot \mathbf{X})^{-1} \cdot \mathbf{X}^{\mathrm{T}} \cdot \vec{Y},$$

and that these estimates are equivalent to those derived previously in summation notation. Note that $(\mathbf{X}^{\mathrm{T}}\mathbf{X})$ will be a (2×2) nonsingular matrix provided that there exist at least two distinct design points $x_{1,j}$. Describe a simple condition on the $x_{1,j}$ such that $\mathbf{X}^{\mathrm{T}}\mathbf{X}$, and hence $(\mathbf{X}^{\mathrm{T}}\mathbf{X})^{-1}$, will be diagonalized, thereby assuring that $\tilde{\beta}_0$ and $\tilde{\beta}_1$ will be uncorrelated. Show also that $\mathrm{Var}(\tilde{\beta}_0) = \mathrm{Var}(\tilde{\beta}_1)$ whenever the design points are selected in a manner such that their average square is unity, i.e., whenever $\sum_{j=1}^{N} x_{1,j}^2 = N$. Presuming that the error terms $\varepsilon(s_j)$ are independently and normally distributed, describe conditions that one can impose (and meet) on the second column of the design matrix $\mathbf{X}$ such that $\tilde{\beta}_0$ and $\tilde{\beta}_1$ are independently and identically normally distributed random variables.

8.2.3. The Polynomial Approximation to the Simular Response Function

Simular experimentation, conducted with respect to a single quantitative factor, can also be designed to allow polynomial approximations of practically any degree to be estimated. In the case of the kth-degree polynomial approximation, the procedure

$$Y(x_1, x_{2,0}, \ldots, x_{p,0}) \cong f_k(x_1) = \beta_0 + \beta_1 x_1 + \beta_{11} x_1^2 + \cdots + \beta_{11\ldots1} x_1^k,$$

is a straightforward generalization of the technique described in the preceding section. One defines the design matrix $\mathbf{X}$ to consist of $k+1$ columns, the ith of which has jth element $x_{1,j}^{i-1}$, for $i = 1, 2, \ldots, k+1$, and $j = 1, 2, \ldots, N$:

$$\mathbf{X} = \begin{bmatrix} 1 & x_{1,1} & x_{1,1}^2 & \cdots & x_{1,1}^k \\ 1 & x_{1,2} & x_{1,2}^2 & \cdots & x_{1,2}^k \\ \vdots & \vdots & \vdots & & \vdots \\ 1 & x_{1,N} & x_{1,N}^2 & \cdots & x_{1,N}^k \end{bmatrix}.$$

With $\vec{Y}$ and $\vec{\varepsilon}$ denoting $(N \times 1)$ vectors of simular responses and error variates, the jth of which is associated with a simular encounter specified by assigning the environmental condition, $x_1 = x_{1,j}$, and by independently selecting a random number seed s_j, one defines the $[1 \times (k+1)]$ vector of parameters,

$$\vec{\beta}^{\mathrm{T}} = [\beta_0, \beta_1, \beta_{11}, \beta_{111}, \ldots, \beta_{11\ldots1}],$$

the last subscript containing k ones. The vector–matrix equation for the responses becomes

$$\vec{Y} = \mathbf{X} \cdot \vec{\beta} + \vec{\varepsilon},$$

so that the least-squares estimates arise as elements of that vector $\vec{\tilde{\beta}}$ which minimizes

$$\vec{\varepsilon}^{\mathrm{T}} \vec{\varepsilon} = (\vec{Y} - \mathbf{X} \cdot \vec{\beta})^{\mathrm{T}} (\vec{Y} - \mathbf{X} \cdot \vec{\beta}).$$

Completely analogously to the procedure suggested in the preceding section, least-squares estimators are elements of the vector $\vec{\tilde{\beta}}$ which is a solution for $\vec{\beta}$ of

$$(\mathbf{X}^{\mathrm{T}} \mathbf{X}) \cdot \vec{\beta} = \mathbf{X}^{\mathrm{T}} \vec{Y}.$$

A sufficient condition for the existence of the inverse $(\mathbf{X}^{\mathrm{T}}\mathbf{X})^{-1}$ is that the design matrix $\mathbf{X}$ be of rank $(k+1)$, i.e., the number N of model encounters must equal or exceed the number $(k+1)$ of parameters, and the environmental specifications $x_{1,1}, x_{1,2}, \ldots, x_{1,N}$ for the factor x_1 must be made in a manner such that the rank of the design matrix $\mathbf{X}$ not be less than $(k+1)$. It might well be noted that the avoidance of duplication among the specifications $x_{j,1}$ will usually suffice for the purpose of achieving the full rank for the design matrix, though in any case this should be checked. If full rank is not attained, then the estimates $\tilde{\vec{\beta}}$ may still be derived from the solution of the normal equations

$$(\mathbf{X}^{\mathrm{T}}\mathbf{X})\vec{\beta} = \mathbf{X}^{\mathrm{T}}\vec{Y},$$

though the resulting estimates will be linearly dependent and thus not uniquely specified.

Presuming that $(\mathbf{X}^{\mathrm{T}}\mathbf{X})^{-1}$ exists, the estimates $\tilde{\vec{\beta}}$ are unbiased for $\vec{\beta}$ and the variance–covariance matrix is given by

$$\mathbf{\Sigma}_{\tilde{\beta}} = \sigma_{\varepsilon}^{2}(\mathbf{X}^{\mathrm{T}}\mathbf{X})^{-1}.$$

Therefore, in addition to the desirability of $\mathbf{X}$'s having rank $(k + 1)$, one would prefer that $(\mathbf{X}^{\mathrm{T}}\mathbf{X})^{-1}$ be a diagonal matrix. With this additional feature, the vector $\tilde{\vec{\beta}}$ of estimates contains $(k + 1)$ independently and normally distributed random variables whenever the error terms $\vec{\varepsilon}$ have been presumed to be distributed in a like manner.

8.3. Multiple Regression: The General Linear Model

8.3.1. Multiple Regression

These same results also obtain in still a more general context. Usually, one will not want to pursue extensively the polynomial representation of the simular response in terms of a single quantitative factor, x_1, but will likely seek to experiment with a number of such factors simultaneously. The reader may recall that the simular response function is representable as the Taylor series expansion (in p quantitative variables):

$$\begin{aligned} Y(x_1, \ldots, x_p) = {} & Y(x_{1,0}, x_{2,0}, \ldots, x_{p,0}) + \sum_{k=1}^{p} \beta_k \cdot (x_k - x_{k,0}) \\ & + \sum_{k=1}^{p} \sum_{l=k}^{p} \beta_{kl} \cdot (x_k - x_{k,0}) \cdot (x_l - x_{l,0}) + \cdots, \end{aligned}$$

where $\vec{x}_0 = (x_{1,0}, x_{2,0}, \ldots, x_{p,0})$ is some particular environmental specification about which experimentation is to be conducted and which could, without loss of generality, be considered as an origin $(0, 0, \ldots, 0)$, so that the simular response function becomes

$$Y(x_1, x_2, \ldots, x_p) = \beta_0 + \sum_{k=1}^{p} \beta_k x_k + \sum_{k=1}^{p} \sum_{l=k}^{p} \beta_{kl} x_k x_l + \cdots.$$

In this latter form, one may note a similarity of the simular response function here and that defined when only a single factor x_1 was deemed of importance; in each case, the simular response function is represented as a *linear combination of the parameters*

$$\vec{\beta}^{\mathrm{T}} = (\beta_0, \beta_1, \ldots, \beta_p, \beta_{11}, \ldots, \beta_{pp}, \ldots).$$

For example, if one were to truncate the most recent expression for the simular response function at its linear terms (in the quantitative factors x_k), simular responses would be represented as

$$Y(T) = \beta_0 + \beta_1 x_1 + \beta_2 x_2 + \cdots + \beta_p x_p + \varepsilon(s),$$

where again $\varepsilon(s)$ is presumed to be a random variable of mean 0 and variance σ_ε^2. If N environmental specifications are now delineated, say,

$$\begin{aligned} (x_{1,1}, x_{2,1}, \ldots, x_{p,1}) &\equiv \vec{x}_1, \\ (x_{1,2}, x_{2,2}, \ldots, x_{p,2}) &\equiv \vec{x}_2, \\ &\vdots \\ (x_{1,N}, x_{2,N}, \ldots, x_{p,N}) &\equiv \vec{x}_N, \end{aligned}$$

then the corresponding simular responses for the N independently seeded encounters can be written as

$$\vec{Y} \equiv \mathbf{X} \cdot \vec{\beta} + \vec{\varepsilon},$$

where $\vec{\varepsilon}^{\mathrm{T}} = [\varepsilon(s_1), \ldots, \varepsilon(s_N)]$ is a $(1 \times N)$ vector of uncorrelated errors, $\vec{\beta}^{\mathrm{T}} = (\beta_0, \beta_1, \ldots, \beta_p)$ is a $[1 \times (p + 1)]$ vector of the parameters (coefficients, $\vec{Y} = (Y_1, \ldots, Y_N)$ is the $(1 \times N)$ vector of simular responses, and

$$\mathbf{X} = \begin{bmatrix} 1 & x_{1,1} & x_{2,1} & \cdots & x_{p,1} \\ 1 & x_{1,2} & x_{2,2} & \cdots & x_{p,2} \\ \vdots & \vdots & \vdots & & \vdots \\ 1 & x_{1,N} & x_{2,N} & \cdots & x_{p,N} \end{bmatrix}$$

is the $[N \times (p + 1)]$ experimental-design matrix whose last p columns are left to the specification of the analyst. Indeed, row by row, $\mathbf{X}$ represents the environmental specifications for the N independently seeded simular encounters.

The design matrix should be structured so that, first, it is of rank $p + 1$, presumably less than N; second, if the resulting unbiased estimates

$$\tilde{\vec{\beta}} = (\mathbf{X}^T\mathbf{X})^{-1} \cdot \mathbf{X}^T\vec{Y}$$

are to be uncorrelated, then $\mathbf{X}$ should be selected such that the variance–covariance matrix,

$$\mathbf{\Sigma}_{\tilde{\beta}} = \sigma_\varepsilon^2(\mathbf{X}^T\mathbf{X})^{-1},$$

is a diagonal matrix; i.e., $(\mathbf{X}^T\mathbf{X})$ should be a diagonalized matrix.

If the errors $\varepsilon(s_j)$ are presumed to be independently and normally distributed, and the two conditions are imposed on the design matrix, then the $p + 1$ estimators $\tilde{\beta}_0, \tilde{\beta}_1, \ldots, \tilde{\beta}_p$ are also independently and normally distributed; indeed, in this case, $\tilde{\beta}_i$ is a normally distributed random variable of mean β_i and of variance

$$\mathrm{Var}(\tilde{\beta}_i) \equiv \sigma_i^2 = \begin{cases} \sigma_\varepsilon^2/N, & i = 0, \\ \sigma_\varepsilon^2 \Big/ \left(\sum_{j=1}^{N} x_{i,j}^2\right), & i = 1, 2, \ldots, N. \end{cases}$$

Hence, $(\tilde{\beta}_i - \beta_i)^2/\sigma_i^2$ will have the Chi-squared distribution of 1 degree of freedom under these conditions, for each $i = 1, 2, \ldots, p$. Moreover, the maximum likelihood estimator of σ_ε^2 becomes

$$\tilde{\sigma}_\varepsilon^2 = N^{-1} \sum_{j=1}^{N} (y_j - \tilde{\beta}_0 - \tilde{\beta}_1 x_{1,j} - \cdots - \tilde{\beta}_p x_{p,j})^2,$$

which may be written equivalently as

$$\tilde{\sigma}_\varepsilon^2 = N^{-1} \cdot (\vec{Y} - \mathbf{X}\tilde{\vec{\beta}})^T \cdot (\vec{Y} - \mathbf{X}\tilde{\vec{\beta}}).$$

The reader is referred to Graybill (1961) for the proof of the results that, under the assumed conditions for the error terms and for the design matrix $\mathbf{X}$, the random variables $\tilde{\beta}_0, \tilde{\beta}_1, \ldots, \tilde{\beta}_p$, *and* $\tilde{\sigma}_\varepsilon^2$ are mutually independent, and that the statistic $N\tilde{\sigma}_\varepsilon^2/\sigma_\varepsilon^2$ has the Chi-squared distribution of $N - p - 1$ degrees of freedom.

Consequently, tests of hypotheses of the form,

$$H_0\colon \beta_i = 0,$$

can be readily constructed via comparisons of the respective statistics

$$T_i \equiv \frac{(\tilde{\beta}_i)^2/\sigma_i^2}{N\tilde{\sigma}_\varepsilon^2/[\sigma_\varepsilon^2 \cdot (N - p - 1)]};$$

or

$$T_i = \begin{cases} (\tilde{\beta}_0)^2/[\tilde{\sigma}_\varepsilon^2/(N-p-1)], & i=0, \\ (\tilde{\beta}_i)^2 \cdot \left(\sum_{j=1}^{N} x_{i,j}^2\right) \Big/ [N\tilde{\sigma}_\varepsilon^2/(N-p-1)], & i=1,2,\ldots,p, \end{cases}$$

with the cumulative distribution function for Snedecor's F-distribution of 1 and $N-p-1$ degrees of freedom. Again, unusually large values of these test statistics are considered as evidence for the rejection of the corresponding null hypothesis, because the numerators have expectations

$$E[(\tilde{\beta}_i)^2] = (\sigma_i^2 + \beta_i^2),$$

tending to be larger whenever $\beta_i \neq 0$, $i = 0, 1, 2, \ldots, p$.

The use of the relationship,

$$Y(T) = \beta_0 + \beta_1 x_1 + \beta_2 x_2 + \cdots + \beta_p x_p + \varepsilon(s),$$

is a fundamental example of *multiple regression*, in which the simular response function is deemed to be expressible as a function of several quantitative environmental conditions. By now, the reader should have noted that fundamentally this relationship is the same functional format as the representation,

$$Y(T) = \beta_0 + \beta_1 x_1 + \beta_{11} x_1^2 + \cdots + \beta_{11\ldots1} x_1^p + \varepsilon(s),$$

useful whenever the simular experimentation is concerned with only a single factor, x_1. Indeed, in each case, the N simular responses arising in the experiment are linear functions of the unknown parameters (coefficients), $\beta_0, \beta_1, \ldots, \beta_p$; only the elements of the design matrix $\mathbf{X}$ change between the two situations, yet in either case these are represented by the vector–matrix equation $\vec{Y} = \mathbf{X} \cdot \vec{\beta} + \vec{\varepsilon}$.

Under the same assumptions of normality, independence, and homoscedasticity for the variance of the errors $\varepsilon(s_j)$, tests for the null hypotheses, H_0: $\beta_i = 0$, use statistics of the same form. In the case of multiple regression, the tests indicate the significance of the estimated rates $\tilde{\beta}_i$ by which the simular response surface varies with the ith factor; in the second case, the significance of $\tilde{\beta}_i$ implies that the term of degree i should be included. (Hence, in this case, one usually tests for the significance of $\tilde{\beta}_p$ first; if found insignificant, then one tests $\tilde{\beta}_{p-1}$ for significance, and so forth.)

One may even consider a more general representation that would mix multiple and polynomial regressions; e.g., as a function of two quantitative factors x_1 and x_2, one might propose that an appropriate representation for the simular response function be

$$Y(x_1, x_2) = \beta_0 + \beta_1 x_1 + \beta_2 x_2 + \beta_{11} x_1^2 + \beta_{12} x_1 x_2 + \beta_{22} x_2^2.$$

Here one can define the parameter vector

$$\vec{\beta}^{\mathrm{T}} = (\beta_0, \beta_1, \beta_2, \beta_{11}, \beta_{12}, \beta_{22})$$

of six elements. An appropriate experimental design, as specified by the $(N \times 6)$ design matrix,

$$\mathbf{X} = \begin{bmatrix} 1 & x_{1,1} & x_{2,1} & x_{1,1}^2 & (x_{1,1} \cdot x_{2,1}) & x_{2,1}^2 \\ 1 & x_{1,2} & x_{2,2} & x_{1,2}^2 & (x_{1,2} \cdot x_{2,2}) & x_{2,2}^2 \\ \vdots & \vdots & \vdots & \vdots & \vdots & \vdots \\ 1 & x_{1,N} & x_{2,N} & x_{1,N}^2 & (x_{1,N} \cdot x_{2,N}) & x_{2,N}^2 \end{bmatrix},$$

leads to normal equations of the form

$$(\mathbf{X}^{\mathrm{T}}\mathbf{X}) \cdot \tilde{\vec{\beta}} = \mathbf{X}^{\mathrm{T}} \vec{Y},$$

and to estimates $\tilde{\vec{\beta}}$ having the statistical properties already described in general. One may also note in this present context that the simular response function can, itself, be estimated by the quadratic surface,

$$\tilde{y}(x_1, x_2) = \tilde{\beta}_0 + \tilde{\beta}_1 x_1 + \tilde{\beta}_2 x_2 + \tilde{\beta}_{11} x_1^2 + \tilde{\beta}_{12} x_1 x_2 + \tilde{\beta}_{22} x_2^2.$$

Indeed, this interpretation will be of prime importance in the next chapter, where the simular response function will be viewed as a response surface, the locus of whose maximal (or optimal) response will be sought in an endeavor to determine the environmental conditions under which the simular (and, presumably, the simulated) system will respond optimally.

8.3.2. The General Linear Model

For the present, however, the current section can be concluded by observing that a *general linear model* for the simular response can be written as

$$Y(T) = \beta_0 + \beta_1 z_1 + \beta_2 z_2 + \cdots + \beta_k z_k + \varepsilon(s).$$

Again, $\varepsilon(s)$ represents an error term which comprises both the variations inherent in the model's structure and that due to the inadequacy of the representation; the errors arising from independently seeded simular encounters are, as usual, presumed to be uncorrelated, of mean 0, and of common variance σ_ε^2. The variables $z_1, z_2, \ldots, z_N$ no longer need be quantitative environmental conditions, but can represent squares or cubes of such factors; in fact, they may be assigned numerical values corresponding to levels of *qualitative* factors as well. For example, one may presume that an appropriate representation for the simular response is

$$Y(T) = \mu + \gamma_1 x_1 + \gamma_2 x_2 + \gamma_{22} x_2^2 + \theta_1 + \varepsilon(s),$$

where μ is a mean effect (β_0); θ_1 is an effect present whenever a particular qualitative factor is specified at its 1 level for an encounter (say, $\beta_4 z_4$, where $z_4 = i - 1$, if the qualitative factor is set at level i, $i = 1$ and 2); γ_i is the rate at which the simular response function varies with respect to the ith of 2 quantitative factors $(\beta_i z_i$, $i = 1$ and 2); and γ_{22} is a coefficient for a quadratic term in this second factor: $(\beta_3 z_3)$.

The general linear model for the simular response in this instance becomes then

$$Y(T) = \beta_0 + \beta_1 z_1 + \beta_2 z_2 + \beta_3 z_3 + \beta_4 z_4 + \varepsilon(s),$$

which is linear in the β_i and therefore amenable to the analysis suggested in this section.

Of special note, in the general linear representation, is the Gauss–Markov property which exists for the resulting estimators $\tilde{\beta}_i$, $i = 1, 2, \ldots, k$; namely, of all the unbiased estimates β_i^* of β_i that are linear combinations of the N simular responses $(Y_1, Y_2, \ldots, Y_N)$, the least-squares estimator β_i, which arises from the normal equations

$$(\mathbf{Z}^{\mathrm{T}}\mathbf{Z}) \cdot \vec{\tilde{\beta}} = \mathbf{Z}^{\mathrm{T}}\vec{Y},$$

has smallest variance. The estimators $\vec{\tilde{\beta}}$ are thus referred to as the *best linear unbiased estimates* (*BLUE*) of $\vec{\beta}$. This important property does not require conditions about the distribution of the error terms; only the condition of their mutual lack of correlation, their nullity of mean, and homoscedasticity are assumed in order to derive the BLUE property. For the proof, the reader is referred to the compendium by Kendall and Stuart (1963).

Therefore, estimators for the coefficients $\vec{\beta}$ in the general linear representation for the responses arising from N independently seeded simular encounters $\vec{Y} = \mathbf{Z} \cdot \vec{\beta} + \vec{\varepsilon}$, where $\mathbf{Z}$ is the experimental-design matrix whose rows correspond to the designated environmental conditions, whether qualitative or quantitative, are in at least one sense optimal estimators.

8.3.3. Selection of the Design Points

The reader may recall the two primary purposes for undertaking a simulation experiment:

(*a*) to determine the behavior of the simular response function; and
(*b*) to determine the relative importance of p pertinent environmental conditions (factors).

Since the general linear representation $\vec{Y} = \mathbf{Z} \cdot \vec{\beta} + \vec{\varepsilon}$, permits one to incorporate simultaneously experimentation with both qualitative and quantitative factors, a question may naturally arise as to the choice of levels and values to be used in this experimentation for either of these two purposes. The reader may recall that the estimate $\tilde{\theta}$ of the incremental effect of altering a two-level qualitative factor from its 0 to its 1 level is not interpreted as a rate of change for that factor; however, the estimate $\tilde{\beta}$ of the coefficient associated with a quantitative factor is indeed a rate estimator. Therefore, since null hypotheses of the forms, H_0: $\theta = 0$ or H_0: $\beta = 0$, are tested by essentially the same procedure (viz., a comparison of $(\tilde{\theta})^2$ and $(\tilde{\beta})^2$ with $\tilde{\sigma}_\varepsilon^2$), the rejection of either hypothesis must be somehow reckoned in the light of knowledge regarding the two factors' effects on the simulated system itself.

For example, the change from one queueing discipline to a second discipline (two-level qualitative factor) may produce an estimate $\tilde{\theta}$, which, relative to $\tilde{\sigma}_\varepsilon^2$, is declared significantly nonzero; yet, the estimate $\tilde{\beta}$ of the coefficient β for a pumping-rate (quantitative) factor may prove relatively insignificant *over the range of pumping rates* used in the simulation experiment. If one were to presume that the simular response *is* a linear function of this pumping rate, however, clearly there would exist rates beyond which the effects on the simular response would exceed those effects due to the change in queue disciplines.

Thus, the comparison of alternative factors should be undertaken on some externally equivalent basis. For this purpose, the author would recommend that design points be selected for simular experimentation such that:

(*a*) the design matrix **Z** be of full rank (i.e., the number N of encounters must exceed the number of parameters in $\vec{\beta}$, yet the N environmental conditions (design points) must be selected so that the design matrix **Z** has rank equal to this number of parameters);

(*b*) the design matrix **Z** be selected so that $(\mathbf{Z}^T\mathbf{Z})$ have as many zero off-diagonal elements as possible (i.e., as many of the columns of **Z** as possible should be defined orthogonally);

(*c*) the N environmental conditions (design points) should cover an equicost range for each of the factors studied in the experiment (i.e., the range of the changes in the factors should represent changes which, if imposed on the *simulated* system, would involve approximately the same costs for implementation, per factor).

It may not always be possible to accomplish design objective (*c*). As a goal, however, it is indeed recommended, because if attained, one knows that the relative significance of the estimated coefficients is indeed relatable on a common (equicost) basis.

In the next chapter, techniques will be presented for accomplishing the remaining experimental objective—that of determining the behavior of the simular response function, primarily so that one can ascertain the environmental specification(s) under which the simular response is optimal. Since the techniques there will use the methodology suggested in the present discussion on experimental designs, the reader is urged to work through as many of the following exercises as possible before proceeding to Chapter 9. This chapter will end with a discussion of analytical techniques, applicable whenever the simular response is itself of a multivariate nature.

EXERCISES

1. A simular response is presumed to be primarily dependent upon a pair of quantitative environmental conditions x_1 and x_2. It has been assumed that an adequate representation for the simular response function is the planar approximation

$$Y(T) = \beta_0 + \beta_1 x_1 + \beta_2 x_2 + \varepsilon(s).$$

(*a*) Show that the experimental design using four simular responses (Y_1, Y_2, Y_3, and Y_4), one arising from each of the four environmental conditions $(-5, +4)$, $(+5, +4)$, $(-5, -4)$, and $(+5, -4)$, respectively, leads to the vector–matrix equations $\vec{Y} = \mathbf{X} \cdot \vec{\beta} + \vec{\varepsilon}$, where $\vec{\varepsilon}$ are the independently and normally distributed random errors of mean zero and common variance σ_ε^2, and where the design matrix becomes

$$\mathbf{X} = \begin{bmatrix} 1 & -5 & +4 \\ 1 & +5 & +4 \\ 1 & -5 & -4 \\ 1 & +5 & -4 \end{bmatrix}.$$

(*b*) Show then that the columns of the design matrix are mutually orthogonal, so that $(\mathbf{X}^\mathrm{T}\mathbf{X})$ becomes the 3×3 diagonal matrix

$$(\mathbf{X}^\mathrm{T}\mathbf{X}) = \begin{bmatrix} +4 & 0 & 0 \\ 0 & +100 & 0 \\ 0 & 0 & +64 \end{bmatrix}.$$

(*c*) Show that the best linear unbiased estimates of β_0, β_1, and β_2 are, from $\tilde{\vec{\beta}} = (\mathbf{X}^\mathrm{T}\mathbf{X})^{-1}\mathbf{X}^\mathrm{T}\vec{Y}$:

$$\tilde{\beta}_0 = (+y_1 + y_2 + y_3 + y_4)/4,$$
$$\tilde{\beta}_1 = 5(-y_1 + y_2 - y_3 + y_4)/100,$$

and

$$\tilde{\beta}_2 = 4(+y_1 + y_2 - y_3 - y_4)/64,$$

noting that the three estimators are orthogonal linear combinations of the recorded simular responses: y_1, y_2, y_3, and y_4.

(*d*) Show then that the variance–covariance matrix for the three estimators is given by

$$\mathbf{\Sigma}_{\tilde{\beta}} = \sigma_\varepsilon^2(\mathbf{X}^\mathrm{T}\mathbf{X})^{-1} = \sigma_\varepsilon^2 \begin{bmatrix} +\frac{1}{4} & 0 & 0 \\ 0 & +\frac{1}{100} & 0 \\ 0 & 0 & +\frac{1}{64} \end{bmatrix}.$$

(*e*) Show that the maximum likelihood estimator for σ_ε^2 is:

$$\tilde{\sigma}_\varepsilon^2 = (\vec{Y} - \mathbf{X} \cdot \tilde{\vec{\beta}})^\mathrm{T} \cdot (\vec{Y} - \mathbf{X} \cdot \tilde{\vec{\beta}})/4 = c(-y_1 + y_2 + y_3 - y_4)^2;$$

i.e., $\tilde{\sigma}_\varepsilon^2$ is proportional to the square of a linear combination of the simular responses, the contrast being orthogonal to each of the independent estimates $\tilde{\beta}_0$, $\tilde{\beta}_1$, and $\tilde{\beta}_2$, so that $\tilde{\sigma}_\varepsilon^2$ is distributed independently of these estimates. Conclude then that $\tilde{\sigma}_\varepsilon^2$ is proportional to a Chi-squared variate of 1 degree of freedom; viz., $4\tilde{\sigma}_\varepsilon^2/[\sigma_\varepsilon^2(4 - 2 - 1)]$ has the Chi-squared distribution of $(N - p - 1) = 1$ degree of freedom.

(f) Show that the contrast,

$$\tilde{\varepsilon} = (-y_1 + y_2 + y_3 - y_4)/4$$

has mean 0 and variance $\sigma_\varepsilon^2/4$, and that $\tilde{\sigma}_\varepsilon^2 = (\tilde{\varepsilon})^2$, so that the constant $c = \frac{1}{16}$ for part (e) of this exercise.

(g) Show that a proper test of the hypothesis H_0: $\beta_1 = 0$ is based on the comparison of the statistic,

$$T_1 = \frac{(\tilde{\beta}_1)^2/(\sigma_\varepsilon^2/100)}{(4\tilde{\sigma}_\varepsilon^2)/(\sigma_\varepsilon^2 \cdot 1)} = \frac{25(\tilde{\beta}_1)^2}{(\tilde{\varepsilon})^2},$$

with the cumulative distribution function for Snedecor's F variate of 1 and 1 degrees of freedom.

(h) Show that a test of the hypothesis H_0: $\beta_2 = 0$ may depend on the test statistic,

$$T_2 = \frac{(\tilde{\beta}_2)^2/(\sigma_\varepsilon^2/64)}{(4\tilde{\sigma}_\varepsilon^2)/(\sigma_\varepsilon^2 \cdot 1)} = \frac{16(\tilde{\beta}_2)^2}{(\tilde{\varepsilon}^2)}.$$

(i) Comment on the relative magnitudes of $\tilde{\beta}_1$ and $\tilde{\beta}_2$ in order to declare each statistically different from zero.

(j) Approximating the simular response function at the point (x_1, x_2) by $\tilde{y} = \tilde{\beta}_0 + \tilde{\beta}_1 x_1 + \tilde{\beta}_2 x_2$, show that

$$E(\tilde{y}) = \beta_0 + \beta_1 x_1 + \beta_2 x_2 = Y(x_1, x_2),$$

and

$$\text{Var}(\tilde{y}) = \text{Var}(\tilde{\beta}_0) + x_1^2 \,\text{Var}(\tilde{\beta}_1) + x_2^2 \,\text{Var}(\tilde{\beta}_2) + (2) \cdot (0)$$

or

$$\text{Var}(\tilde{y}) = \sigma_\varepsilon^2(1600 + 64x_1^2 + 100x_2^2)/6400.$$

Note that the *contours* of equal variance for the estimated planar approximation are ellipses.

2. (a) Show that the three parameters, β_0, β_1, and β_2 may be estimated unbiasedly from only three simular responses Y_1, Y_2, and Y_3 which arise from independently seeded encounters specified by the environmental conditions $(x_1, x_2) = (0, 1)$, $(-\sqrt{3}/2, -\frac{1}{2})$, and $(+\sqrt{3}/2, -\frac{1}{2})$, respectively, assuming that the simular responses may be represented by $Y(T) = \beta_0 + \beta_1 x_1 + \beta_2 x_2 + \varepsilon(s)$. Show that the design matrix is

$$\mathbf{X} = \begin{bmatrix} 1 & 0 & 1 \\ 1 & -\sqrt{3}/2 & -\frac{1}{2} \\ 1 & +\sqrt{3}/2 & -\frac{1}{2} \end{bmatrix}$$

leading to the matrix

$$(\mathbf{X}^{\mathrm{T}}\mathbf{X}) = \begin{bmatrix} 3 & 0 & 0 \\ 0 & \frac{3}{2} & 0 \\ 0 & 0 & \frac{3}{2} \end{bmatrix},$$

and to the estimates as the orthogonal contrasts

$$\hat{\beta}_0 = (y_1 + y_2 + y_3)/3,$$

$$\hat{\beta}_1 = (-y_2 + y_3)/\sqrt{3},$$

and

$$\hat{\beta}_2 = (2y_1 - y_2 - y_3)/3.$$

(*b*) Show that $\mathrm{Var}(\hat{\beta}_0) = \sigma_\varepsilon^2/3$, $\mathrm{Var}(\hat{\beta}_1) = 2\sigma_\varepsilon^2/3$, and $\mathrm{Var}(\hat{\beta}_2) = 2\sigma_\varepsilon^2/3$ and that the three estimators are uncorrelated.

(*c*) Show that the random variable

$$\hat{y} = \hat{\beta}_0 + \hat{\beta}_1 x_1 + \hat{\beta}_2 x_2,$$

the estimate of the response function at the environmental condition (x_1, x_2), has mean

$$E(\hat{y}) = \beta_0 + \beta_1 x_1 + \beta_2 x_2,$$

and variance

$$\begin{aligned} \mathrm{Var}(\hat{y}) &= \mathrm{Var}(\hat{\beta}_0) + x_1^2 \,\mathrm{Var}(\hat{\beta}_1) + x_2^2 \,\mathrm{Var}(\hat{\beta}_2) + (2) \cdot (0) \\ &= \sigma_\varepsilon^2(1 + 2x_1^2 + 2x_2^2)/3. \end{aligned}$$

Note that the *contours* of $\mathrm{Var}(\hat{y})$ are circles in the Euclidean plane of (x_1, x_2) specifications.

(*d*) Show that $\hat{y}$ passes exactly through $(0, 1, y_1)$.

3. Presuming that the simular response is representable as

$$Y(T) = \beta_0 + \beta_1 x_1 + \beta_2 x_2 + \varepsilon(s),$$

where x_1 and x_2 are quantitative variates, the following experimental design (matrix) was defined in order to estimate $\vec{\beta}^{\mathrm{T}} = (\beta_0, \beta_1, \beta_2)$:

$$\mathbf{X} = \begin{bmatrix} 1 & 0 & 1 \\ 1 & -\sqrt{3}/2 & -\frac{1}{2} \\ 1 & +\sqrt{3}/2 & -\frac{1}{2} \\ 1 & 0 & 0 \end{bmatrix}.$$

(*a*) Locate by points, in the x_1, x_2-plane, the loci at which the four simular responses are to be recorded; note their positions at the vertices and center of an equilateral triangle.

(*b*) Show that

$$(\mathbf{X}^{\mathrm{T}}\mathbf{X}) = \begin{bmatrix} 4 & 0 & 0 \\ 0 & \frac{3}{2} & 0 \\ 0 & 0 & \frac{3}{2} \end{bmatrix},$$

so that the unbiassed estimates of the elements of $\vec{\beta}$ become

$$\tilde{\beta}_0 = (y_1 + y_2 + y_3 + y_4)/4,$$

$$\tilde{\beta}_1 = (0 - y_2 + y_3 + 0)/\sqrt{3},$$

and

$$\tilde{\beta}_2 = (2y_1 - y_2 - y_3 + 0)/3,$$

having variances as in the preceding exercise, save for

$$\mathrm{Var}(\tilde{\beta}_0) = \sigma_\varepsilon^2/4.$$

(*c*) Show that, unlike the preceding exercise, an estimate of σ_ε^2 now becomes

$$\tilde{\sigma}_\varepsilon^2 = (\vec{Y} - \mathbf{X}\tilde{\vec{\beta}})^{\mathrm{T}}(\vec{Y} - \mathbf{X}\tilde{\vec{\beta}})/4 = C \cdot (y_1 + y_2 + y_3 - 3y_4)^2,$$

for constant C ($= \frac{1}{48}$). Note that $\tilde{\sigma}_\varepsilon^2$ is proportional to the square of a linear contrast that is mutually orthogonal to $\tilde{\beta}_0$, $\tilde{\beta}_1$, and $\tilde{\beta}_2$.

(*d*) Denoting the contrast by

$$\tilde{\varepsilon} = (y_1 + y_2 + y_3 - 3y_4),$$

show that $\tilde{\beta}_0$, $\tilde{\beta}_1$, $\tilde{\beta}_2$, and $\tilde{\varepsilon}$ are independent and normally distributed random variables whenever the $\varepsilon(s_j)$ are likewise distributed.

(*e*) Show that $E(\tilde{\varepsilon}) = 0$ and that $\mathrm{Var}(\tilde{\varepsilon}) = 12\sigma_\varepsilon^2$.

(*f*) Conclude that $(\tilde{\varepsilon})^2/(12\sigma_\varepsilon^2)$ is a Chi-squared variate of 1 degree of freedom, under the assumptions of normality and independence for the error variates.

(*g*) Show that the appropriate statistic for testing the null hypothesis, H_0: $\beta_1 = 0$, is

$$T_1 = \frac{(\tilde{\beta}_1)^2/(2\sigma_\varepsilon^2/3)}{4\tilde{\sigma}^2/(\sigma_\varepsilon^2)} = \frac{3(\tilde{\beta}_1)^2}{8\tilde{\sigma}^2},$$

and that, for testing the null hypothesis H_0: $\beta_2 = 0$, one uses

$$T_2 \equiv \frac{(\tilde{\beta}_2)^2/(2\sigma_\varepsilon^2/3)}{(4\tilde{\sigma}_\varepsilon^2)/(\sigma_\varepsilon^2)} = \frac{3(\tilde{\beta}_2)^2}{8(\tilde{\sigma}_\varepsilon^2)},$$

in each case the cumulative distribution function for Snedecor's F distribution of 1 and 1 degrees of freedom being appropriate for the comparison.

(*h*) Denoting by $\tilde{y} = \tilde{\beta}_0 + \tilde{\beta}_1 x_1 + \tilde{\beta}_2 x_2$ the estimated simular response for the environmental conditions (x_1, x_2), show that

$$E(\tilde{y}) = \beta_0 + \beta_1 x_1 + \beta_2 x_2,$$

and that

$$\text{Var}(\tilde{y}) = \sigma_\varepsilon^2(3 + 32x_1^2 + 32x_2^2)/48,$$

contours of the latter of which are circles in the x_1, x_2-plane.

4. Repeat Exercise 3, substituting the design matrix,

$$\mathbf{Z} = \begin{bmatrix} 1 & 0 & -1 \\ 1 & -\sqrt{3}/2 & +\frac{1}{2} \\ 1 & +\sqrt{3}/2 & +\frac{1}{2} \\ 1 & 0 & 0 \end{bmatrix},$$

for $\mathbf{X}$. Determine the estimates $\tilde{\beta}_0, \tilde{\beta}_1, \tilde{\beta}_2, \tilde{\varepsilon}$, and $\tilde{y}(x_1, x_2)$; comment on any differences that arise in the properties (expectations, variances) as compared with the results of Exercise 3.

5. Consider the simular response function,

$$Y(T) = \beta_0 + \beta_1 z_1 + \beta_2 z_2 + \varepsilon(s),$$

where z_1 and z_2 are both quantitative factors that serve as environmental conditions for a stochastic simulation model.

(*a*) Show that the design matrix, for the experiment that will yield simular responses Y_1, Y_2, Y_3, and Y_4 at $(-1, +1)$, $(+1, +1)$, $(-1, -1)$, and $(+1, -1)$ in z_1, z_2-coordinates, respectively, is

$$\mathbf{Z} = \begin{bmatrix} 1 & -1 & +1 \\ 1 & +1 & +1 \\ 1 & -1 & -1 \\ 1 & +1 & -1 \end{bmatrix},$$

a matrix with mutually orthogonal columns.

(*b*) Show that the matrix $(\mathbf{Z}^T\mathbf{Z})$ is diagonal:

$$(\mathbf{Z}^T\mathbf{Z}) = \begin{bmatrix} 4 & 0 & 0 \\ 0 & 4 & 0 \\ 0 & 0 & 4 \end{bmatrix}.$$

(*c*) Show that the least-squares estimates of β_0, β_1, and β_2 are

$$\tilde{\beta}_0 = (+y_1 + y_2 + y_3 + y_4)/4;$$
$$\tilde{\beta}_1 = (-y_1 + y_2 - y_3 + y_4)/4 = \tfrac{1}{4}[(y_2 + y_4) - (y_1 + y_3)];$$

and

$$\tilde{\beta}_2 = (+y_1 + y_2 - y_3 - y_4)/4 = \tfrac{1}{4}[(y_1 + y_2) - (y_3 + y_4)].$$

(*d*) Show that an unbiased estimate of zero is given by

$$\tilde{\varepsilon} = (-y_1 + y_2 + y_3 - y_4)/4,$$

and that $\tilde{\beta}_0$, $\tilde{\beta}_1$, $\tilde{\beta}_2$, and $\tilde{\varepsilon}$ are mutually orthogonal, each with variance $(\sigma_\varepsilon^2/4)$.

(*e*) Show that the maximum-likelihood estimator for σ_ε^2, derivable once the error terms have been presumed normally and independently distributed as

$$\tilde{\sigma}_\varepsilon^2 = \sum_{j=1}^{4} (y_j - \tilde{\beta}_0 - \tilde{\beta}_1 z_{1,j} - \tilde{\beta}_2 z_{2,j})^2/4,$$

may be written

$$\tilde{\sigma}_\varepsilon^2 = (\tilde{\varepsilon})^2.$$

(*f*) Show that $4(\tilde{\sigma}_\varepsilon^2)/(\sigma_\varepsilon^2)$ has the Chi-squared distribution of 1 $(=4 - 2 - 1)$ degree of freedom.

(*g*) Show that an appropriate test of the hypothesis H_0: $\beta_i = 0$ is the comparison of the statistic,

$$T_i = \frac{(\tilde{\beta}_i)^2/(\sigma_\varepsilon^2/4)}{(4\tilde{\sigma}_\varepsilon^2/\sigma_\varepsilon^2)} = \frac{(\tilde{\beta}_i)^2}{(\tilde{\varepsilon})^2},$$

with the cumulative distribution function for Snedecor's F variate of 1 and 1 degrees of freedom, for $i = 0$, 1, or 2.

(*h*) The resulting estimate of the approximate simular response function becomes, at any point (z_1, z_2):

$$\tilde{y}(z_1, z_2) = \tilde{\beta}_0 + \tilde{\beta}_1 z_1 + \tilde{\beta}_2 z_2.$$

Show that the random variable $\tilde{y}$ has moments:

$$E(\tilde{y}) = \beta_0 + \beta_1 z_1 + \beta_2 z_2$$

and

$$\text{Var}(\tilde{y}) = \sigma_\varepsilon^2(1 + z_1^2 + z_2^2)/4.$$

(*i*) Compare $\text{Var}(\tilde{y})$ with the result obtained in Exercise 3, presuming that $z_1 = x_1$ and $z_2 = x_2$. Show that the augmented triangular design (Exercise 3) estimates the simular response function with lower variance than does the square design of this exercise if and only if the simular response function is estimated within the circle:

$$(x_1^2 + x_2^2 = 0.45).$$

(*j*) Compare the results of this exercise with those of Exercise 1, noting that the designs are the same if

$$z_1 = x_1/5 \qquad \text{and} \qquad z_2 = x_2/4.$$

Review again your comments on part (*i*) of the first exercise.

6. The simular response function for a stochastic simulation model is presumed to be dependent upon the levels of a pair of two-level qualitative factors x_1 and x_2. Designating these levels by 0 and 1 for each factor, an experiment was conducted by defining four independently seeded encounters at the points (0, 0), (0, 1), (1, 0), and (1, 1). The simular response being denoted by

$$Y(T) = \beta_0 + \beta_1 x_1 + \beta_2 x_2 + \varepsilon(s),$$

the four responses Y_1, Y_2, Y_3, and Y_4 can be represented in vector–matrix notation by $\vec{Y} = \mathbf{X}\cdot\beta + \vec{\varepsilon}$, where the $\vec{\varepsilon}$ are normally and independently distributed random variables, each of mean 0 and variance σ_ε^2.

(*a*) Show that the design matrix can be represented by

$$\mathbf{X} = \begin{bmatrix} 1 & 0 & 0 \\ 1 & 0 & 1 \\ 1 & 1 & 0 \\ 1 & 1 & 1 \end{bmatrix},$$

and that the matrix $(\mathbf{X}^T\mathbf{X})$ becomes

$$(\mathbf{X}^T\mathbf{X}) = \begin{bmatrix} 4 & 2 & 2 \\ 2 & 2 & 1 \\ 2 & 1 & 2 \end{bmatrix}.$$

(*b*) Show that the estimates of the *incremental effects*, β_1 and β_2, become, from the solution of $(\mathbf{X}^T\mathbf{X})\vec{\beta} = \mathbf{X}^T\vec{Y}$:

$$\tilde{\beta}_1 = (-y_1 - y_2 + y_3 + y_4)/2,$$

and

$$\tilde{\beta}_2 = (-y_1 + y_2 - y_3 + y_4)/2.$$

Note: The matrix $(\mathbf{X}^T\mathbf{X})$ has inverse

$$(\mathbf{X}^T\mathbf{X})^{-1} = \tfrac{1}{4}\begin{bmatrix} 3 & -2 & -2 \\ -2 & 4 & 0 \\ -2 & 0 & 4 \end{bmatrix}.$$

(*c*) Note that $\tilde{\beta}_1$ and $\tilde{\beta}_2$ are orthogonal and hence independent, but that $\tilde{\beta}_0 = (3y_1 + y_2 + y_3 - y_4)/4$ is not orthogonal to either of the incremental effect estimators.

(*d*) Show that, if the zero levels of the qualitative factors were denoted by (-1), then the estimates for the incremental effects would be the "same" contrasts as those given in Exercise 5.

(*e*) Derive the normal equations, and solutions thereof, whenever *one* of the two factors of this exercise is quantitative and is to be experimented with at values ± 1. (Assume that the two-level qualitative factor has levels denoted by ± 1.)

7. A simular experiment is being conducted to represent the simular response as functionally dependent upon a quadratic formulation of two quantitative variates:

$$Y(T) = \beta_0 + \beta_1 x_1 + \beta_2 x_2 + \beta_{11} x_1^2 + \beta_{22} x_2^2 + \beta_{12} x_1 x_2.$$

Seven simular encounters are to be defined at the loci:

$$\underset{Y_1}{(0, 1)}, \quad \underset{Y_2}{(-\sqrt{3}/2, -\tfrac{1}{2})}, \quad \underset{Y_3}{(+\sqrt{3}/2, -\tfrac{1}{2})}, \quad \underset{Y_4}{(0, -1)},$$

$$\underset{Y_5}{(-\sqrt{3}/2, +\tfrac{1}{2})}, \quad \underset{Y_6}{(+\sqrt{3}/2, +\tfrac{1}{2})}, \quad \underset{Y_7}{(0, 0)}$$

(*a*) Show that, in the Euclidean plane of coordinates (x_1, x_2), these loci correspond to the center and vertices of a regular hexagon inscribed inside the unit circle.

(*b*) Show that, for $\beta^{T} = (\beta_0, \beta_1, \beta_2, \beta_{11}, \beta_{22}, \text{and } \beta_{12})$, the design matrix becomes

$$\mathbf{X} = \begin{bmatrix} 1 & 0 & 1 & 0 & 1 & 0 \\ 1 & -\sqrt{3}/2 & -\frac{1}{2} & \frac{3}{4} & \frac{1}{4} & +\sqrt{3}/4 \\ 1 & +\sqrt{3}/2 & -\frac{1}{2} & \frac{3}{4} & \frac{1}{4} & -\sqrt{3}/4 \\ 1 & 0 & -1 & 0 & 1 & 0 \\ 1 & -\sqrt{3}/2 & +\frac{1}{2} & \frac{3}{4} & \frac{1}{4} & -\sqrt{3}/4 \\ 1 & +\sqrt{3}/2 & +\frac{1}{2} & \frac{3}{4} & \frac{1}{4} & +\sqrt{3}/4 \\ 1 & 0 & 0 & 0 & 0 & 0 \end{bmatrix},$$

and the (6×6) matrix $(\mathbf{X}^{T}\mathbf{X})$ is given by

$$(\mathbf{X}^{T}\mathbf{X}) = \begin{bmatrix} 7 & 0 & 0 & 3 & 3 & 0 \\ 0 & 3 & 0 & 0 & 0 & 0 \\ 0 & 0 & 3 & 0 & 0 & 0 \\ 3 & 0 & 0 & \frac{9}{4} & \frac{3}{4} & 0 \\ 3 & 0 & 0 & \frac{3}{4} & \frac{9}{4} & 0 \\ 0 & 0 & 0 & 0 & 0 & \frac{3}{4} \end{bmatrix},$$

having inverse

$$(\mathbf{X}^{T}\mathbf{X})^{-1} = \begin{bmatrix} 1 & 0 & 0 & -1 & -1 & 0 \\ 0 & \frac{1}{3} & 0 & 0 & 0 & 0 \\ 0 & 0 & \frac{1}{3} & 0 & 0 & 0 \\ -1 & 0 & 0 & \frac{3}{2} & \frac{5}{6} & 0 \\ -1 & 0 & 0 & \frac{5}{6} & \frac{3}{2} & 0 \\ 0 & 0 & 0 & 0 & 0 & \frac{4}{3} \end{bmatrix}.$$

(*c*) Show then that the unbiased estimators of the elements of β are obtained as $\mathbf{A} \cdot \vec{Y}$, where $\mathbf{A}$ is the (6×7) matrix

$$\mathbf{A} = \begin{bmatrix} 0 & 0 & 0 & 0 & 0 & 0 & 1 \\ 0 & -\sqrt{3}/6 & +\sqrt{3}/6 & 0 & -\sqrt{3}/6 & +\sqrt{3}/6 & 0 \\ +\frac{1}{3} & -\frac{1}{6} & -\frac{1}{6} & -\frac{1}{3} & +\frac{1}{6} & +\frac{1}{6} & 0 \\ -\frac{1}{6} & +\frac{1}{3} & +\frac{1}{3} & -\frac{1}{6} & +\frac{1}{3} & +\frac{1}{3} & -1 \\ +\frac{1}{2} & 0 & 0 & +\frac{1}{2} & 0 & 0 & -1 \\ 0 & +\sqrt{3}/3 & -\sqrt{3}/3 & 0 & -\sqrt{3}/3 & +\sqrt{3}/3 & 0 \end{bmatrix},$$

i.e.,

$$\begin{aligned}
\tilde{\beta}_0 &= y_7, \\
\tilde{\beta}_1 &= -\sqrt{3}\,(-y_2 + y_3 - y_5 + y_6)/6, \\
\tilde{\beta}_2 &= (2y_1 - y_2 - y_3 - 2y_4 + y_5 + y_6)/6, \\
\tilde{\beta}_{11} &= (-y_1 + 2y_2 + 2y_3 - y_4 + 2y_5 + 2y_6 - 6y_7)/6, \\
\tilde{\beta}_{22} &= (+y_1 + y_4 - 2y_7)/2,
\end{aligned}$$

and

$$\tilde{\beta}_{12} = \sqrt{3}\,(y_2 - y_3 - y_5 + y_6)/3.$$

(*d*) Show that

$$\text{Cov}(\tilde{\beta}_0, \tilde{\beta}_{11}) = -\sigma_\varepsilon^2 = \text{Cov}(\tilde{\beta}_0, \tilde{\beta}_{22}),$$

and

$$\text{Cov}(\tilde{\beta}_{11}, \tilde{\beta}_{22}) = 5\sigma_\varepsilon^2/6,$$

yet all other pairs of estimates are uncorrelated.

(*e*) Defining, at any point (x_1, x_2) the random variable

$$\tilde{y}(x_1, x_2) \equiv \tilde{\beta}_0 + \tilde{\beta}_1 x_1 + \tilde{\beta}_2 x_2 + \tilde{\beta}_{11} x_1^2 + \tilde{\beta}_{22} x_2^2 + \tilde{\beta}_{12} x_1 x_2,$$

show that

$$E(\tilde{y}) = \beta_0 + \beta_1 x_1 + \beta_2 x_2 + \beta_{11} x_1^2 + \beta_{22} x_2^2 + \beta_{12} x_1 x_2,$$

and

$$\begin{aligned}
\text{Var}(\tilde{y}) = {} & \text{Var}(\tilde{\beta}_0) + x_1^2 \text{Var}(\tilde{\beta}_1) + x_2^2 \text{Var}(\tilde{\beta}_2) + x_1^4 \text{Var}(\tilde{\beta}_{11}) \\
& + x_2^4 \text{Var}(\tilde{\beta}_{22}) + x_1^2 x_2^2 \text{Var}(\tilde{\beta}_{12}) + 2x_1^2 \text{Cov}(\tilde{\beta}_0, \tilde{\beta}_{11}) \\
& + 2x_2^2 \text{Cov}(\tilde{\beta}_0, \tilde{\beta}_{22}) + 2x_1^2 x_2^2 \text{Cov}(\tilde{\beta}_{11}, \tilde{\beta}_{22}) + 2\cdot(0),
\end{aligned}$$

or

$$\text{Var}(\tilde{y}) = \sigma_\varepsilon^2[6 - 10(x_1^2 + x_2^2) + 9(x_1^2 + x_2^2)^2]/6.$$

8.4. The Multivariate Simular Response

Throughout this chapter, discussion has centered about techniques for analyzing a univariate simular response. Most sophisticated stochastic simulation models produce a variety of state variables, or system attributes, that would serve quite well to describe the state of the simular system at the end of T simular time units. Thus, the simular response might more properly be described as a vector $V_1, V_2, \ldots, V_k$ of random variables having some joint k-dimensional probability density function $f_V(v_1, v_2, \ldots, v_k)$. Each element of the vector would correspond to an univariate simular response, as has been discussed at some length; i.e., any V_i would represent, at the end of T simular time units, a random variable dependent upon the environmental conditions specified for the simular encounter:

$$V_i(T) = V_i(x_1, x_2, \ldots, x_p; s),$$

where $x_1, x_2, \ldots, x_p$ represent the p environmental conditions save the random number seed s. Consequently, for any specified environmental condition $\vec{x}$, stipulation of a single seed leads to a vector $\vec{V} = (V_1, V_2, \ldots, V_p)$ of responses for the corresponding encounter; repetition of the model at the same environmental condition, yet over all possible seeds, would lead to a barely finite collection of multivariate simular responses which, when placed in a multidimensional frequency histogram, could be considered as a random vector.

Though a considerable statistical literature currently exists for the analysis of multiple experimental responses, this section will present only one of the most basic statistical methods: *correlation analysis*. The correlation between a pair of random variables (or, between the elements of a bivariate simular response) U and V is given, as in Section 2.11.4, by

$$\varrho_{U,V} \equiv \mathrm{Cov}(U, V)/[\mathrm{Var}(U)\,\mathrm{Var}(V)]^{1/2},$$

where $\mathrm{Cov}(U, V)$ is the covariance between the simular responses U and V.

Hence, in general, knowledge of the correlation between two random variables requires knowledge of their joint probability density function, $f_{U,V}(u, v)$, since

$$\mathrm{Cov}(U, V) = \int_{-\infty}^{\infty}\int_{-\infty}^{\infty} (u - \mu_U)\cdot(v - \mu_V) f_{U,V}(u, v)\, du\, dv.$$

However, one may recall that some correlation values are associated with situations that are essentially distribution free. For example, if the random variables are independently distributed, then their covariance, and hence their correlation, is zero, although the converse is not necessarily true. Also, the correlation coefficient $\varrho_{U,V} = \pm 1$ if and only if the random variables are strictly linearly related; i.e., if and only if $U = a + bV$

for some constants a and b ($\neq 0$). This last condition obtains whenever the joint probability density function degenerates and exists only along the line $u - bv = a$.

Therefore, if a pair of simular response variables (U, V) arises from a simulation encounter, computation of an estimate of the correlation coefficient may tend to reveal the existence of a near-linear relationship between the two variates. Though such a finding need not imply causation between the two variables, one would often seek to find possible "cause-and-effect" relationships whenever an estimated correlation coefficient is deemed significantly nonzero.

A second reason for the use of correlation analysis is the possibility of reducing the number of response variates for subsequent analysis. Indeed, if an estimated correlation coefficient for a pair of response variates revealed a significantly high degree of linearity between the two variates, then further simular experimentation might well be concerned with only one of the two variates.

The report of the author's experience with such an eventuality appears in *Transportation Science* (1970). A large-scale stochastic simulation model of a transportation–queueing network had been designed primarily to ascertain the number of tons of goods that the system could deliver to the sink nodes during, say, a 30-day period. It became readily apparent, however, that in addition to this measure of system productivity, other responsible authorities would ask about the average number of hours which the fleet of vehicles (cargo aircraft) experienced daily in productive movement. The two variates, U (aircraft utilization rate) and V (aircraft productivity), were observed after each of a number of independently seeded simulation encounters, each defined by its own environmental specifications. The high estimated correlation between the resulting data pairs indicated indeed that the original univariate measure (system productivity) was apparently quite adequate for describing the simular response.

The estimated correlation coefficient $r_{U,V}$ between two random variables U and V is given by

$$r_{U,V} \equiv \frac{\sum_{j=1}^{N} (u_j - \bar{u}) \cdot (v_j - \bar{v})}{[\sum_{j=1}^{N}(u_j - \bar{u})^2 \cdot \sum_{j=1}^{N}(v_j - \bar{v})^2]^{1/2}},$$

where u_j, v_j is the recorded simular response vector for the jth of N independently seeded simular responses. In general, one might propose that the actual correlation coefficient should vary with the environmental specifications, but if one is willing to assume "homoscedasticity" of covariance as well as of the variances, then the computation of $r_{U,V}$ with response pairs from differing environmental specifications can be undertaken.

The estimated correlation coefficient $r_{U,V}$ is clearly a random variable itself, yet not much is known regarding its exact probability density function, even for some of the more elementary bivariate distributions $f_{U,V}(u, v)$. Fisher (1921) showed that, if $f_{U,V}(u, v)$ is the bivariate normal distribution, then the random variable,

$$Z = \tfrac{1}{2} \ln[(1 + r)/(1 - r)],$$

where $r = r_{U,V}$, is approximately normally distributed with mean

$$E(Z) = \tfrac{1}{2} \ln[(1 + \varrho)/(1 - \varrho)],$$

where $\varrho = \varrho_{U,V}$, and with variance

$$\mathrm{Var}(Z) = 1/(N - 3),$$

for N sufficiently large. Therefore, under these conditions, tests of hypotheses of the form, H_0: $\varrho_{U,V} = \varrho_0$, can be conducted by comparisons of the computed Z-statistic with the extremes of the appropriate cumulative normal distribution.

If the existence of a significant correlation between two simular response variates is apparent, then the reader might well ask if it is permissible to plot the N simular response vectors (u_j, v_j) and then draw a (or, fit the) least-squares line among the N data points. One may indeed plot the N points (called a *scattergram*) and may indeed use the regression methodology (described earlier) in order to pass the least-squares line $u = \tilde{\beta}_0 + \tilde{\beta}_1 v$ through the N recorded simular response pairs. However, strictly speaking, many of the properties of the estimated coefficients $\tilde{\beta}_0$ and $\tilde{\beta}_1$ no longer remain those exposed earlier. The reader may recall that only the dependent variable [$Y(T)$, the simular response] was considered to be a random variable, whereas the independent variable (x_1, an environmental condition, or quantitative factor) was presumed to be fixed in each data pair $(x_{1,j}, y_j)$. The complications that arise because both elements of the data points in a scattergram are random variables requires far too lengthy a discussion to be entertained seriously here; the reader is referred to the second volume of Kendall and Stuart (1963) for a quite complete discussion of the matter.

For the N-variate simular response ($N > 2$), the developing techniques of multivariate statistical analyses will prove appropriate. The reader is referred to the work of Anderson (1958) as well as to the more recent text of Morrison (1967) for a discussion of these procedures.

EXERCISE (*A Test for Independence in Random Number Chains*)

Denote by $W_1, W_2, \ldots, W_{501}$ a chain of 501 successive random numbers produced by the power residue method. The 500 successive pairs (W_1, W_2), (W_2, W_3), $(W_3, W_4), \ldots$, (W_{500}, W_{501}) of random variates would presumably constitute pairs of independent and uniformly distributed random variables, each variate being between 0 and 1. The log-and-trig method for the production of pairs of independently and normally distributed random variates would produce the 500 data pairs, (U_i, V_i):

$$\begin{aligned} U_i &= [\ln(1 - W_i)^{-2}]^{1/2} \cdot [\cos(2\pi W_{i+1})], \\ V_i &= [\ln(1 - W_i)^{-2}]^{1/2} \cdot [\sin(2\pi W_{i+1})], \end{aligned} \qquad i = 1, 2, \ldots, 500.$$

For 500 such variates, compute the estimated correlation coefficient, $r = r_{U,V}$, as described in the preceding section; then compute Fisher's Z-statistic,

$$Z = \tfrac{1}{2}\{\ln[(1 + r)/(1 - r)]\},$$

and compare with the normal distribution of mean 0 and variance 1/497. Is the computed Z-statistic too great (in its absolute value) to warrant a statement that the U_i, V_i are uncorrelated normal variates?

Repeat the test for uniform variates generated from an additive congruential technique; from a mixed congruential technique; from published tables of random digits.

8.5. A Scholium on Simular Experimentation

The three preceding chapters have dealt with techniques for the analysis of computerized simulation experiments. It has been pointed out that many of these techniques are especially applicable to three of the stages in the development of a simulation model: verification, validation, and analysis. Whenever simular responses can be generated under environmental conditions whose probabilistic effects on the response can be predicted beforehand, many of the techniques of Chapter 6 may be indeed appropriate to the verification stage of the model's development. [See Mihram (1972b).]

Once the analyst is reasonably well convinced that the model is producing the kinds of responses its intended programmed structure should provide, other techniques of that chapter, as well as those of Chapters 7 and 8, will probably prove applicable to the comparison of simular responses with those of the simulated system itself. In this validation stage, as many comparisons as possible (of environmental conditions and their resulting simular responses with the simulated system's operating conditions and its corresponding observed responses, or yields) should be made; in many cases, this is readily accomplished by considering the simular responses and the actual system's responses as random variables resulting from the assignment of one of the two levels of a "factor" (real versus simular responses). If the effect of these "levels" is significant, then the model remains somewhat invalidated. Thus, simple factorial designs can prove applicable in the validation stage as well.

As soon as the analyst has verified and validated his model, he will undertake the analyses for which the model was constructed initially. Two categories, or analytical frameworks, have been suggested as the primary motivations for building a simulation model; hence, they become the two primary analytical goals, although it will not be necessary that both goals are relevant to a particular simulation model. To reiterate, these are

(*a*) the determination of the relative significance of likely contributors to the simular response, whether these suspected contributors be of a qualitative or a quantitative nature; and,

(*b*) the determination of the environmental specifications for which the simular response attains its optimum.

Chapters 6–8 have been concerned strictly with the first category of analyses. Factorial designs have been stressed because these provide an excellent, rapid, and systematic procedure for isolating both quantitative and qualitative factors which are significant contributors to the simular response function. Moreover, the use of regression analyses itself suggests the likely nature of the simular response function, since a planar or quadratic estimate of this function is usually available at the conclusion of the experimentation. As a matter of fact, in the case of quantitative variates, some of the estimated coefficients, $\tilde{\beta}_i$, are estimates of the derivative of the simular response function with respect to a particular quantitative environmental condition (x_i). Therefore, one may hope by the use of factorial designs and regression analyses to eliminate a large number of suspected contributors (factors) before undertaking search procedures designed to locate the optimal simular response. (These search procedures will be delineated in the next chapter.)

The usually large number of suspected contributors to the simular response function makes imperative the need to reduce the dimensionality of the factor space. If only five simular environmental conditions are considered prime suspects as being significant contributors to the simular response, then, even if each factor is explored at only two values (or levels), a complete replicate of the possible combinations will require $2^5 = 32$ encounters with the simulation model. The use of a one-half or one-quarter (i.e., partial) replicate would assist greatly in reducing the required number of simulation encounters, but one undertakes such simular experimentation with the knowledge that any significant interactive effects on the part of the factors will be undetectable; in fact, they will contribute directly to the error estimate, possibly making more difficult the isolation of significant effects of the factors themselves.

In order to reduce this problem of dimensionality, one might find feasible the lumping together of the total number p of environmental conditions into subsets of factors. Usually, any such partition should consist of subsets containing factors which:

(*a*) if altered in the same direction, would each be expected to provide similar effects as concerns the direction taken by the simular response; and,

(*b*) factors in any subset should have relatively equivalent marginal costs for unit alterations in their specifications (in the simulated system, or simuland).

In addition, each subset should, if possible, require a collective marginal cost for its unit alteration that equals the cumulative marginal cost for unit alterations in all the factors of any other subset.

Thus, each initial subset of factors may be presumed to be a two-level qualitative factor (though unit increases in the various factors constituting a given subset may vary widely in their numerical specification for the model's input data). Presuming that one has as few as three initial subsets of suspected significant factors, a 2^3 factorial experiment may be designed and analyzed. If one of these subsets does not prove to be significant, its factors may be "set aside" and the factors from the significant subsets may be assigned new groupings, or subsets, in accordance with the criteria delineated above (insofar as possible). A new 2^K factorial design can be defined, and the subsequent analysis used to eliminate other subsets of factors. The procedure may be applied until the individual, significant contributors may be singled out, at which time more intensive analyses (e.g., regression) may be conducted with respect to their apparent effects on the simular response function.

Such a heuristic procedure is not without its latent difficulties, since some of the factors within a particular subset may have a significant interactive effect that would tend to conceal the true significance of other factors in the subset. The attempt to reduce the dimensionality of the simular experimentation shall often require, however, that initial analyses proceed in a manner similar to that suggested above. The procedure has been applied with considerable success by the author (see Mihram, 1970a), though at this writing sufficient experimentation has not yet been undertaken to reveal the presence of negative interactive effects which might have camouflaged significant factors.

The preceding three chapters have indicated some of the primary statistical techniques that are directly applicable to simular experimentation. This experimentation, when conducted during the analysis stage of model development, generally presumes a "stable" simulation model; i.e., the simulation model has been verified and validated and it is assumed that the model shall not be itself altered during the experimentation. Of course, as in any scientific inquiry, when a model is found to be somehow inadequate, every effort should be made to improve it. Indeed, one would be remiss in his duty if he failed to make alterations to provide a better mimic for the simulated system. However, if significant alterations are to be implemented during the simular experimentation, then a return to the verification and/or validation stages will usually be necessary before recommencing the analysis stage.

The simular experimentation techniques suggested in this chapter are quite analogous to those of the standard statistical methodology: The simular system (simulation model) corresponds to the *experimental unit*, the specified environmental conditions correspond to the *treatments* (which are *randomized* by the concomitant assignment of independently selected and nonrepetitive random number seeds for the encounters in the *experimental design*), the simular responses correspond to the usual *yields*, and the *experimental error* is, as usual, represented by the failure of two simular responses to be the same, even though the environmental specifications for the two successive (though independently seeded) encounters remain identical.[†]

[†] See footnote, page 258.

In this regard, one may define *blocks* of treatments by proper manipulation of the elements of a model's juxtaposed seed (see page 235). One selects a particular subset (say, $s_1, s_2, \ldots, s_k$) of k from among the model's required K seeds; the initial values of the elements of this subset are selected randomly from their admissible sets of values, yet are repeatedly employed by each of the b treatment specifications which constitute a block. Additional blocks can then be defined by randomly selecting additional sets of k seeds, each set then repeated in the b encounters constituting a block. Of course, in accordance with the Principium of Seeding, the remaining $K - k$ seeds are selected randomly and nonrepetitively for every encounter in the experimental design. One must note, however, that the resulting *block effects* become random quantities and therefore require analysis in terms of *variance components* (see footnote, page 258). In addition, one might note that the repetitive use of a particular, random, exogenous event chain [such as a recorded sequence of actual, yet random, customer arrival instants, as suggested by Conway (1963)] can also be construed as a specification of a random block of treatments in simular experimentation.

Determining the significance of factors (treatments) contributing to a simular response is usually accomplished by means of the analysis of variance (or regression analysis). This methodology reveals whether there are significant differences among the treatment means, but does not indicate by how much they differ. Of course, one can inspect the estimated treatment means for their comparative magnitudes; more properly, one should employ in such instances *multiple comparisons*, or *multiple ranking techniques*, for the application of which the reader is referred to Kendall and Stuart (1966).

Also omitted from the statistical tools have been discussions of, notably, sequential experiments and analysis [cf. Davies (1954)] and of multivariate techniques [cf. Anderson (1958)]. The discussions of other topics, such as partial replication of factorial experiments and the power of the suggested statistical tests of proffered hypotheses, have been necessarily brief. The reader is again referred to the cited references, however, for further details on the applications of these techniques.

Chapter 9

THE SEARCH FOR OPTIMAL SYSTEM CONDITIONS

Get thee up into the high mountains.

ISAIAH 40: 9

9.1. Introduction

The simulation model is usually constructed in order to determine either:

(*a*) the relative importance of alternative policies, of different environmental conditions, or of differing parametric specifications as they affect the simular response at some point T in simular time; or,

(*b*) that set (or combination) of policies, environmental conditions, and parametric specifications which will provide, in some sense, the optimal simular response at time T.

In the three preceding chapters, the statistical procedures available for, and pertinent to, the first of these aims have been investigated. The analysis of variance is particularly applicable whenever comparisons of the simular responses $Y(T)$ (as they arise under a set of changes in qualitative input parameters, such as alterations of policies governing the simular system), are to be made; it has also been seen to be appropriate to comparisons of the simular responses which derive from alternative specifications of input parameters, the values of which more appropriately arise from a continuum. In the latter case, exemplified by the dispensing rate of a simulated fuel pump, the analysis of variance might be more usually referred to as regression analysis.

Nonetheless, in either event, the aim of the analysis is the determination of the relative contribution of the alternative input parameters, whether these parameters define governing policies or other environmental conditions for the simular system. These analyses of variance, and regression analyses, are conducted from relatively straightforward computations involving the simular responses $Y_j(T)$, $j = 1, 2, \ldots, N$, even though these responses are random variables, and therefore possess intrinsic stochasticity. The reader may recall that analyses are based on comparisons, of the variance contributions arising from the particular effects of interest, with the underlying, inherent variance in the simular response.

In a similar fashion, the attempt to analyze simular responses, in order to ascertain the optimal operating conditions for the simular (and simulated) system, must be conducted in the presence of the intrinsic variability of the simular response $Y(T)$. In the absence of this stochasticity, optimum-seeking methods and strategies, such as those discussed by Wilde (1964), can be used in the search for the optimal response from a deterministic simulation model. However, this chapter is concerned with techniques appropriate to the location of the optimal operating, or environmental, conditions even in the presence of randomness, such as exists in a stochastic simulation model.

9.2. The Response Surface

A simulation model is usually not constructed to evaluate the effect of just one, but rather of several, environmental conditions upon the simular response. This section will discuss an orderly procedure for isolating and determining the location of the optimal simular response whenever a *pair* of environmental conditions is presumed pertinent.

Let x_1 and x_2 represent two continuously variable (alternatively, quantitative) environmental conditions (or input parameters) associated with a stochastic simulation model. The response, or output, at the end of a simular time period of T units is

$$Y(T) = Y(x_1, x_2; s),$$

a function of the inputs x_1, x_2, and the seed s. For any fixed location (x_1, x_2) in the Euclidean representation of the environmental conditions, selection of all permissible values of s leads to a collection of simular responses $Y(T)$ which will therefore be distributed according to some probability density function $f_Y(y)$; that is, $Y(T)$ in the usual manner may be deemed to be a random variable having its mean and variance each dependent, of course, on the selection of (x_1, x_2):

$$E[Y(T)] = \mu(x_1, x_2) = \int_{-\infty}^{\infty} y f_Y(y)\, dy$$

and

$$\mathrm{Var}[Y(T)] = \sigma^2(x_1, x_2) = \int_{-\infty}^{\infty} (y - \mu)^2 f_Y(y)\, dy.$$

The expected simular response $\mu(x_1, x_2)$ is then a function of the two simular environmental conditions. Presuming that this function is continuous, its graph may be conceivably sketched and referred to as the simular response surface, or simply the *response surface*. A set of exemplary response surfaces is depicted in Fig. 9.1.

Of course, the varieties of such response surfaces are infinite. Those shown are representative of the types of response surfaces of considerable interest to the analyst who searches for the maximal (i.e., optimal) simular response. An alternative presentation of the simular response surface is the two-dimensional plot of its *isohypses*, or lines of contour, each depicting the locations (x_1, x_2) at which the response surface assumes a constant height.

Figure 9.2 displays contours for the four types of response surfaces shown in Fig. 9.1:

(a) a unique maximum,
(b) a ridge of maxima,
(c) a rising ridge,
(d) maximin.

The reader may note that, in cases (a) and (b), the maximal simular response exists, although in (b) there exists a number of environmental conditions [that is, input data pairs (x_1, x_2)] at which the maximal response is attained. The loci of these maxima are referred to as a *stationary ridge.*

A second type of ridge is depicted in part (c). Here, no search will locate the maximal simular response, yet if the nature of the rising ridge could be determined, environmental conditions could be quickly altered to improve the simular response.

In part (d), a maximin exists: That is, if one were to conduct a search in the x_2-direction along a line running vertically through the contour labels (1 and 3 of the figure), he would declare that a maximal response

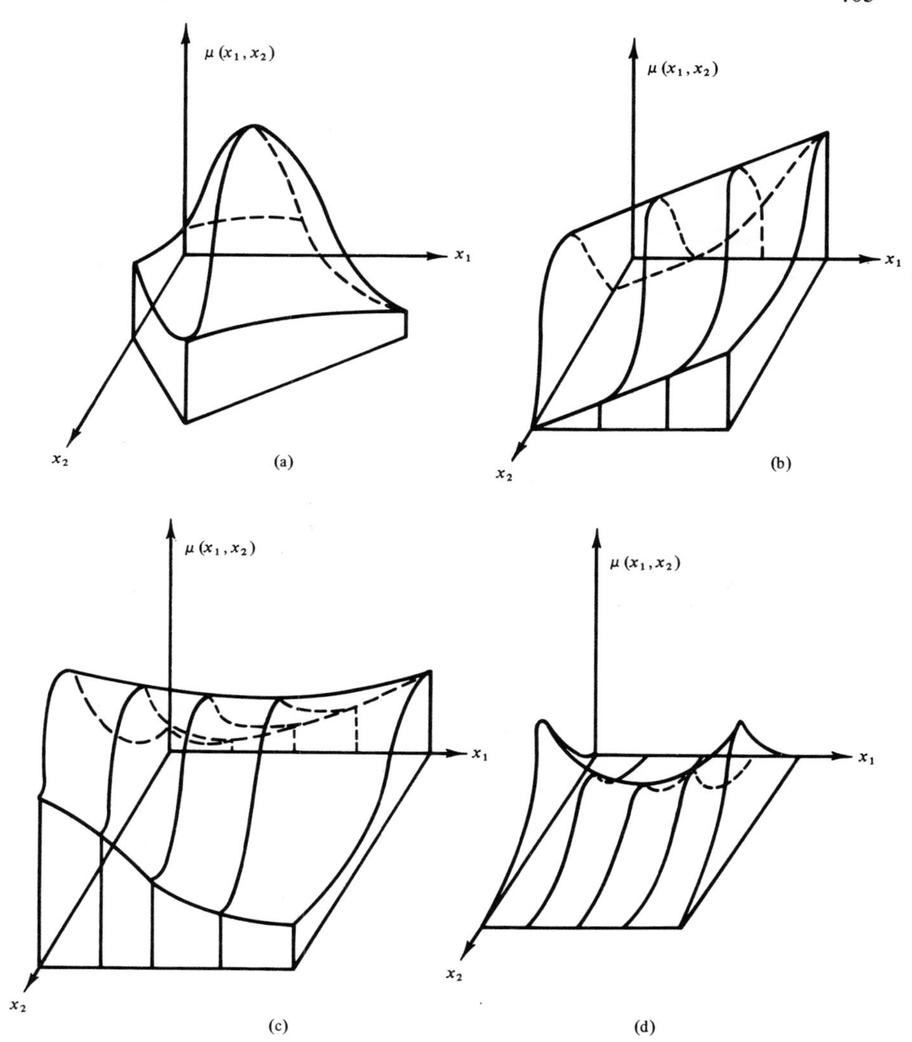

FIG. 9.1. Exemplary simular response surfaces.

somewhat less than 4 would exist. If a similar search were conducted in the x_2-direction, but a few x_1-units to the right of left, an inevitable conclusion that the maximal response be, say, 4, 6, or 8 (depending upon the x_1-situs of the search) would be derived. Of course, were a search to be conducted along some fixed x_2-value, one would likely discover an apparent minimal (but no maximal) simular response. Hence, the term "maximin" is applied, since the "saddle point" constitutes either

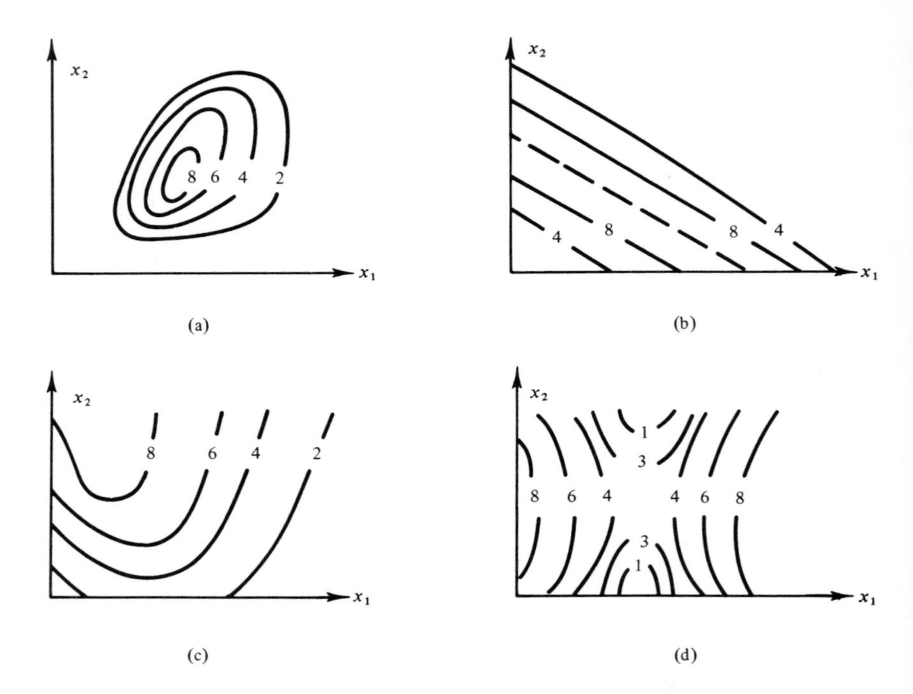

FIG. 9.2. Isohypses of typical simular response surfaces.

a maximum or a minimum, depending upon the direction along the surface from which it is approached.

In general, then, the response surface is seen to be a set of response curves; i.e., for any fixed value of the second environmental condition (say, x_2^*), the loci of the intersection of the plane $x_2 = x_2^*$ and the response surface $\mu(x_1, x_2)$ corresponds to a *response curve* $\mu(x_1)$. With this in mind, the preceding paragraph serves as a cautionary note to the reader: The attempt to locate the maximal simular response by varying one environmental condition at a time (holding the second condition constant) implies that the search for the optimum is conducted only along one of the response curves of the response surface. This procedure should be avoided in order to ensure coverage of the greatest area of the response surface with the least encounters of the simulation model as necessary.

Furthermore, another typical difficulty that can arise in this regard is a search conducted over a surface containing a rising ridge; altering only one variable at a time may lead to a false conclusion that a maximum exists.

However, one may presume that the response surface $\mu(x_1, x_2)$, which represents the mean simular response at the end of T simular time units whenever the environmental conditions (i.e., input data) were specified by (x_1, x_2), can be expressed as a Taylor series about some pertinent environmental condition, which one may denote by $(x_{1,0}, x_{2,0})$ and which one may label the *initial search center*. Hence,

$$\begin{aligned}\mu(x_1, x_2) = \mu(x_{1,0}, x_{2,0}) &+ \mu_1(x_{1,0}, x_{2,0})(x_1 - x_{1,0}) \\ &+ \mu_2(x_{1,0}, x_{2,0})(x_2 - x_{2,0}) + \mu_{11}(x_{1,0}, x_{2,0})(x_1 - x_{1,0})^2/2 \\ &+ \mu_{22}(x_{1,0}, x_{2,0})(x_2 - x_{2,0})^2/2 \\ &+ \mu_{12}(x_{1,0}, x_{2,0})(x_1 - x_{1,0})(x_2 - x_{2,0}) + \cdots,\end{aligned}$$

where μ_{ab} denotes partial differentiation of $\mu(x_1, x_2)$ with respect to arguments x_a and x_b, for a and b equal to 1 or 2.

If one assumes that an appropriate scale S_i exists for the ith experimental condition, then a pair of dimensionless variates,

$$z_1 = (x_1 - x_{1,0})/S_1 \qquad \text{and} \qquad z_2 = (x_2 - x_{2,0})/S_2,$$

can be defined so that the Taylor series representation for the simular response surface may be written:

$$\mu(z_1, z_2) = \beta_0 + \beta_1 z_1 + \beta_2 z_2 + \beta_{11} z_1^2 + \beta_{12} z_1 \cdot z_2 + \beta_{22} z_2^2 + \cdots,$$

where

$$\begin{aligned}\beta_0 &= \mu(x_{1,0}, x_{2,0}) \\ \beta_i &= S_i \mu_i(x_{1,0}, x_{2,0}), \qquad i = 1, 2, \\ \beta_{ii} &= 2S_i^2 \mu_{ii}(x_{1,0}, x_{2,0}), \qquad i = 1, 2,\end{aligned}$$

and

$$\beta_{12} = S_1 S_2 \mu_{12}(x_{1,0}, x_{2,0}), \quad \text{etc.}$$

The choice of the scales S_1 and S_2 is not obvious; a skillful selection of these units of measure may imply a considerable reduction in the effort required in the subsequent search procedure. Such a selection implies some *a priori* knowledge of the behavior of the response surface, however; presumably only a minimal amount is known, else the search for the optimum response would be somewhat a game on the part of the analyst.

Nonetheless, if such *a priori* information is available, the analyst should indeed use it to his benefit, just as the choice of the initial search center

$(x_{1,0}, x_{2,0})$ might be based on prior knowledge (or an educated guess) about the approximate location of the optimal simular response. In any case, the search technique (to be discussed presently) requires that the analyst use his judgment as to the relative scales of measure S_1 and S_2, bearing in mind that the search technique will require successive movements, or changes, in environmental conditions to be essentially at right angles to estimated contour lines.

9.3. The Search for the Optimal Response

The search procedure for locating the optimal response along a simular response surface is conducted in two phases:

(*a*) an *isolation phase* designed to situate the optimum within a quite restricted range of environmental conditions;

(*b*) an *intensive search*, or local exploration, phase designed to fix precisely the location of the optimal simular response.

The isolation phase of the search is begun by presuming that a linear approximation to the response curve will be adequate to determine the slope in a neighborhood of a search center; therefore, the signs of the regression estimator of the slopes of this approximating plane are suggested as the mechanism for indicating the direction in which to conduct the subsequent search. After sufficient iterations of this procedure, when the optimal simular response could be deemed isolated within a smaller region of environmental conditions, the local exploration phase is then conducted by employing a polynomial of degree two or more as an approximation to the response surface in the isolated range. A sufficient number of encounters with the model are then defined, for values of the search variates (i.e., environmental conditions), so as to permit an accurate estimate of the response surface in the isolated range. The maximal (optimal) value on this estimated response surface is then assumed to be (or to be near) the optimal simular response.

For the purpose of this discussion, one can presume that the response surface is continuous and may take as a working assumption that the surface has a single unique maximum over the region R of environmental conditions (x_1, x_2) of interest. Cases in which ridges or a maximin exist are not exceedingly more difficult to treat, and the reader is referred to the excellent presentations of Davies (1954), Box and Hunter (1958), and Box (1954) for a more complete description than appears here.

9.3.1. The Initial Search

Let us begin with the presumption that the response surface $E[Y(T)]$ may be expressed as the expected simular response at the end of T simular time units and that the surface is a function of two search variates z_1 and z_2:

$$E[Y(T)] = \mu(z_1, z_2).$$

The search variates represent scaled, dimensionless, environmental conditions (x_1, x_2), in such a manner that the latter may be unraveled from the former by means of the relationships:

$$x_i = x_{i,0} + S_i \cdot z_i, \qquad i = 1 \quad \text{and} \quad 2,$$

where the S_i are positive scaling quantities to be discussed momentarily and where $(x_{1,0}, x_{2,0})$ is the *initial search center.*

The initial search center should be chosen either near the anticipated location of the optimal response or quite near the usual environmental conditions of the simulated system itself. In a neighborhood of this initial search center, corresponding to the location (0, 0) for the search variates (z_1, z_2), the response surface may be approximately expressed as the plane:

$$\mu(z_1, z_2) = \beta_0 + \beta_1 z_1 + \beta_2 z_2,$$

so that the observation of simular encounters defined by environmental conditions which are situated near the initial search center (and by an independently selected seed for each encounter, of course) should provide responses lying on or near such a plane.

The reader may note that this approximating plane corresponds to the (linear) truncation of the Taylor series expansion for the response surface, as used in Section 9.2. The plane is described completely with a specification of the three coefficients, or parameters, β_0, β_1, and β_2, and one should note the physical interpretation applicable to each of these parameters:

$$\beta_0 = \mu(x_{1,0}, x_{2,0}),$$

$$\beta_i = S_i \mu_i(x_{1,0}, x_{2,0}), \qquad i = 1 \quad \text{and} \quad 2.$$

That is, β_0 is the expected (or average) simular response at the initial search center, whereas β_1 and β_2 represent the slopes of the proximate plane at this search center, in the direction of the z_1 and z_2 axes, respectively.

Consequently, statistical estimates of β_1 and β_2 yield information concerning both the "tilt" of the approximating plane and, hopefully, the direction in which the response surface is rising most rapidly from this search center. Since three parameters are required to describe the approximating plane, at least three simular responses are required. These responses may arise from simulation encounters defined by three environmental conditions $(x_1, x_2)_j$, $j = 1, 2$, and 3, provided these three points [or, equivalently, that the three corresponding points in terms of the search variates: $(z_1, z_2)_j$, $j = 1, 2$, and 3] are not collinear. (The reader should note that, in general, the collinearity of these points would not uniquely specify an estimated plane. Indeed, Hunter (1959) suggests that the vertices of an equilateral triangle, each vertex being a unit distance from the search origin (0, 0), be employed as the loci of three search points at which the three simular encounters are defined.

This procedure will provide the required estimates β_0, β_1, and β_2; in fact, the resulting estimate of the approximating plane

$$y = \beta_0 + \beta_1 z_1 + \beta_2 z_2$$

will pass directly through the three points:

$$[(z_1, z_2)_1, y_1], \quad [(z_1, z_2)_2, y_2], \qquad \text{and} \qquad [(z_1, z_2)_3, y_3],$$

where y_j is the simular response arising from the encounter defined by the *search vector* or *search pair* $(z_1, z_2)_j$ and by an independently selected random number seed s_j, $j = 1, 2$, and 3.

In order to provide a measure of the closeness of the estimated plane to the simular responses, however, it is recommended that at least four simular encounters be defined in the immediate neighborhood of the search center. The resulting data, consisting of triads of the form $(z_1, z_2)_j$, y_j for the jth such encounter, can then be used in the usual regression computations. The selection of the search vectors for this purpose is usually left to the analyst who should, however, avoid the collinearity of any three (or more) of these locations (search vectors).

The estimates for β_0, β_1, and β_2, under these essentially unconstrained selections of the simular environmental conditions, are straightforward computations of the well-known multiple-regression techniques, for which the reader is referred either to Hunter (1959) or to the lucid text by Graybill (1961). One category of these techniques, however, is especially noteworthy.

The reader may recall from the preceding chapter the 2^2 factorial design in which two levels (values) of each of two factors (variables) appear in all four (2^2) possible permutations; i.e., in the present context, four simulation encounters are defined by four specifications of environmental conditions that correspond to the following four points in terms of the search vector:

$$(z_1, z_2)_1 = (-1, -1); \qquad (z_1, z_2)_2 = (-1, +1);$$
$$(z_1, z_2)_3 = (+1, -1); \qquad (z_1, z_2)_4 = (+1, +1).$$

Thus, the location of the environmental conditions for the encounters correspond to the vertices of a rectangle with center at $(x_{1,0}, x_{2,0})$ and with dimensions $2S_1$ and $2S_2$.

With y_j denoting the simular response arising from assignment of the input conditions $(z_1, z_2)_j$ and, of course, an independently selected random number seed, the regression estimates become:

$$\hat{\beta}_0 = (y_1 + y_2 + y_3 + y_4)/4;$$
$$\hat{\beta}_1 = (-y_1 - y_2 + y_3 + y_4)/4;$$

and

$$\hat{\beta}_2 = (-y_1 + y_2 - y_3 + y_4)/4.$$

The computational simplicity of these estimators derives from the symmetry of the experimental design about the search center [equal to (0, 0), in terms of the search vector]. One may note the alternative forms for the estimates of the coefficients β_1 and β_2:

$$\hat{\beta}_1 = \{[(y_3 + y_4)/2] - [(y_1 + y_2)/2]\}/2$$

and

$$\hat{\beta}_2 = \{[(y_4 + y_2)/2] - [(y_3 + y_1)/2]\}/2,$$

in which the estimators are half the difference between the means computed at each of the levels of the corresponding search variate.

The properties of these estimators can be exhibited if one presumes that the responses themselves may be represented as

$$Y(T) = \beta_0 + \beta_1 z_1 + \beta_2 z_2 + \varepsilon$$

where ε is a random variable usually taken to be normally distributed of mean 0 and variance σ_ε^2 and with succeeding ε's presumed to be independent. The random variable ε is presumed to represent a combination of errors arising from

(*a*) the inadequacy of the planar approximation for the response surface; and
(*b*) the intrinsic variability in the simular response itself.

Therefore, the estimators have the properties:

(*a*) $E(\hat{\beta}_i) = \beta_i$, $i = 0, 1$, and 2;
(*b*) $\text{Var}(\hat{\beta}_i) = \sigma_\varepsilon^2/4$, $i = 0, 1$, and 2;
(*c*) $\text{Cov}(\hat{\beta}_i, \hat{\beta}_j) = 0$, $i \neq j = 0, 1$, and 2;
(*d*) $\hat{\beta}_i$ are normal and independent random variables, as were developed in the discussion of multiple regression techniques (cf. Section 8.3).

The estimate of the simular response in a square neighborhood of the search center $(-1 < z_1 < 1, \ -1 < z_2 < 1)$ becomes then

$$\hat{y} = \hat{\beta}_0 + \hat{\beta}_1 z_1 + \hat{\beta}_2 z_2,$$

a planar surface whose contours can be drawn as straight lines in the space of (z_1, z_2) variates; i.e., a given contour line is the loci of all points (z_1, z_2) such that

$$\hat{\beta}_0 + \hat{\beta}_1 z_1 + \hat{\beta}_2 z_2 = K \quad \text{(const)},$$

as depicted in Fig. 9.3. The reader should note that, of all the directions one can move away from the search center (0, 0), movement perpendicular to these parallel contours provides the greatest promise for improvement in the simular response. Since the slopes of the estimated response surface $\hat{y}$ are given by $\hat{\beta}_1$ in the z_1-direction and by $\hat{\beta}_2$ in the z_2-direction, one sees that marginal increases in the simular responses, arising due to unit increase in the variates z_1 and z_2, are in the ratio $\hat{\beta}_1 : \hat{\beta}_2$. Alternatively, movement in the z_1- and z_2-directions in these proportions will provide the greatest marginal increase in $\hat{y}$.

Consequently, it is recommended that a new search center be located at the search variate coordinates

$$z_1 = k\hat{\beta}_1, \qquad z_2 = k\hat{\beta}_2,$$

for some constant k at the discretion of the simulation analyst. Another 2^2 factorial experiment can be conducted about this new search center and the four resulting simular responses used to estimate the response surface as a plane in a neighborhood of this search center. The new

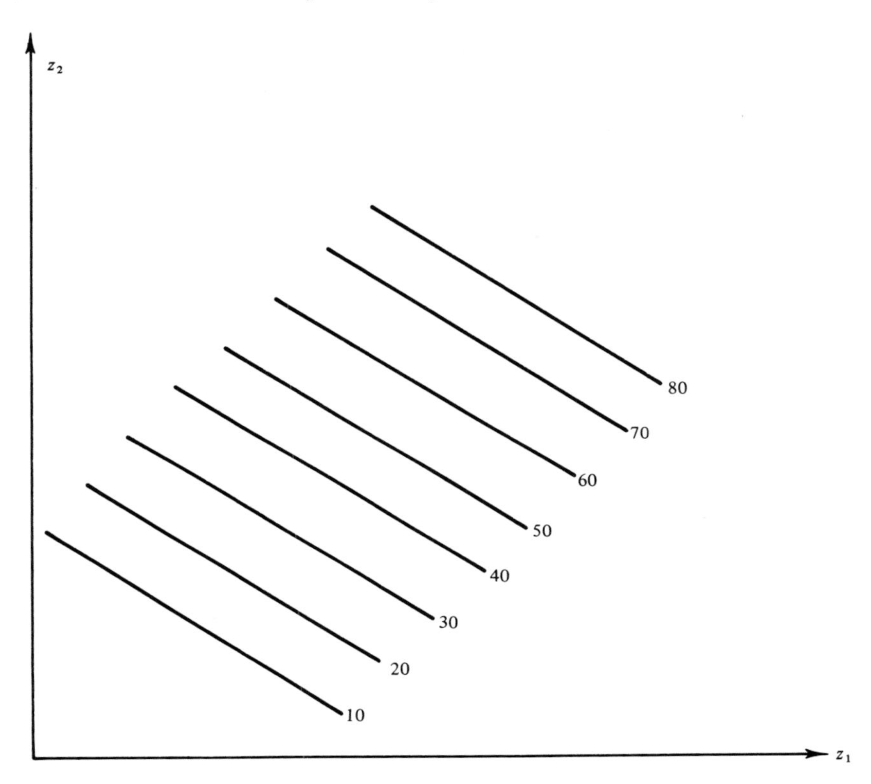

FIG. 9.3. Contours of the estimated planar responses.

estimates $\hat{\beta}_1$ and $\hat{\beta}_2$ of the slopes of the planar estimate can in turn be used to direct proportional changes in the variates z_1 and z_2 to locate a third search center.

It is anticipated that such a search will eventually lead to estimates, $\hat{\beta}_1$ and $\hat{\beta}_2$, which are not significantly different from zero, in which case the planar estimate $\hat{y} = \hat{\beta}_0 + \hat{\beta}_1 z_1 + \hat{\beta}_2 z_2$, which is valid in the square neighborhood $-1 < z_1 < +1$, $-1 < z_2 < +1$, would be relatively constant; i.e., $y \cong \hat{\beta}_0$. Since this condition would likely obtain for a search center located at, or near a stationary point of the simular response surface, one would usually presume that the first phase (isolation) of the search process has been completed.

In order to determine whether the estimated coefficients, $\hat{\beta}_1$ and $\hat{\beta}_2$, are significantly different from zero, one must recall that the simular responses $Y(T)$ are random variables with variances presumed to be given by σ_ε^2, equal to the variance of the errors as described previously.

At any given search center, the four simular responses (y_1, y_2, y_3, y_4) have arisen from the 2^2 factorial experimental design, so that under the usual assumption of the stochastic independence of the errors in these responses, the three estimators $\hat{\beta}_0$, $\hat{\beta}_1$, and $\hat{\beta}_2$ may be seen to be uncorrelated random variables, since they are orthogonal linear transformations of the uncorrelated random variables (Y_1, Y_2, Y_3, and Y_4):

$$\hat{\beta}_0 = (Y_1 + Y_2 + Y_3 + Y_4)/4,$$

$$\hat{\beta}_1 = (-Y_1 - Y_2 + Y_3 + Y_4)/4,$$

and

$$\hat{\beta}_2 = (-Y_1 + Y_2 - Y_3 + Y_4)/4.$$

A fourth linear combination of the responses which is orthogonal to each of these three is

$$\hat{\varepsilon} = (Y_1 - Y_2 - Y_3 + Y_4)/4,$$

which is written as an estimate of ε, the response error, since $E(\hat{\varepsilon}) = 0$ and $\mathrm{Var}(\hat{\varepsilon}) = \sigma_\varepsilon^2/4$ whenever the simular response is given by

$$Y(T) = \beta_0 + \beta_1 z_1 + \beta_2 z_2 + \varepsilon,$$

as just discussed. Therefore, since

$$\mathrm{Var}(\hat{\varepsilon}) = E(\hat{\varepsilon}^2) - [E(\hat{\varepsilon})]^2 = E(\hat{\varepsilon}^2),$$

one sees that an unbiased estimate of σ_ε^2 is provided by

$$\hat{\sigma}_\varepsilon^2 = 4(\hat{\varepsilon})^2 = (Y_1 - Y_2 - Y_3 + Y_4)^2/4.$$

Analogous to the method used in the preceding chapter, one may show that

$$E[4(\hat{\beta}_0)^2] = E[(Y_1 + Y_2 + Y_3 + Y_4)^2/4] = \sigma_\varepsilon^2 + 4\beta_0^2,$$

$$E[4(\hat{\beta}_1)^2] = E[(-Y_1 - Y_2 + Y_3 + Y_4)^2/4] = \sigma_\varepsilon^2 + 4\beta_1^2,$$

and

$$E[4(\hat{\beta}_2)^2] = E[(-Y_1 + Y_2 - Y_3 + Y_4)^2/4] = \sigma_\varepsilon^2 + 4\beta_2^2,$$

so that the tests for the significance of the estimated slopes $\hat{\beta}_1$ and $\hat{\beta}_2$ are the straightforward F-tests based on Snedecor's F-distribution with 1 and 1 degrees of freedom. The statistics computed for this purpose are

$$F_1^* = (\hat{\beta}_1)^2/(\hat{\varepsilon})^2$$

used to test for the significance of $\hat{\beta}_1$, and

$$F_2^* = (\hat{\beta}_2)^2/(\hat{\varepsilon})^2,$$

used to test for the significance of $\hat{\beta}_2$. The significance of either is declared whenever the corresponding F^* is unusually large, as reflected by tabulations of Snedecor's F distribution of 1 and 1 degrees of freedom. (The accepted procedure is to declare the slope estimate *significant* whenever the corresponding $F^* > 161$, or *highly significant* whenever F^* exceeds 4050; often it is accepted in response surface methodology to declare $\hat{\beta}_i$ significantly nonzero whenever F_i^* exceeds 39.9, but the reader should note that according to this criterion, he will declare significance in about 10% of those cases in which the true slope is zero.)

Actually, the simulation analyst is more interested in determining whether the *maximal* slope at the search center is nonzero. Since this maximal slope (gradient) is estimated by $[(\hat{\beta}_1)^2 + (\hat{\beta}_2)^2]^{1/2}$, an appropriate test of the hypothesis,

$$H_0\colon \beta_1^2 + \beta_2^2 = 0,$$

can be conducted by means of the single test statistic

$$F^{**} = [(\hat{\beta}_1)^2 + (\hat{\beta}_2)^2]/2(\hat{\varepsilon})^2.$$

Since $\hat{\beta}_1$ and $\hat{\beta}_2$ are independently and normally distributed random variables whenever the simular responses are likewise distributed, and since these slope estimators have mean 0 and variance $\sigma_\varepsilon^2/4$ whenever H_0 is true, then F^{**} will have the Snedecor's F distribution of 2 and 1 degrees of freedom. Comparison of F^{**} with the improbably large values of this Snedecor's F distribution provides then the test of the hypothesis that the surface's gradient be zero.

The reader should be aware that these proposed F-tests for the significance of $\hat{\beta}_1$ and $\hat{\beta}_2$ are more properly tests for the adequacy of a planar representation for the response surface in a neighborhood of the search center. The suggested estimate $\hat{\sigma}_\varepsilon^2$ measures the average squared deviation of the four simular responses from the fitted response plane $\hat{y} = \hat{\beta}_0 + \hat{\beta}_1 z_1 + \hat{\beta}_2 z_2$, as may be seen by substitution for $\hat{\beta}_i$ and for $(z_1, z_2)_j$ in the left-hand side of the expression:

$$\sum_{j=1}^{4} [y_j - \hat{y}(z_1, z_2)_j]^2 = 4\hat{\varepsilon}^2 = \hat{\sigma}_\varepsilon^2,$$

where $\hat{y}(z_1, z_2)_j$ is the value of $\hat{y}$ at $(z_1, z_2)_j$. In order to obtain a valid estimate of σ_ε^2, then, one should properly replicate the experimental design.

This replication would imply at least a doubling of the number of simular responses around each search center, since the replication means that at least two responses be generated at each assigned search vector value [i.e., at $(-1, -1)$, $(-1, +1)$, $(+1, -1)$, and $(+1, +1)$]. Each replicated simulation would be defined then by the same environmental conditions, save for the assignment of independently selected and different random number seeds. The subsequent analysis of the simular responses would proceed as just described, except that the arithmetic mean of the simular responses arising from encounters, which are defined at the same environmental conditions, would be employed as the simular response at that point; in addition, the previously computed $\hat{\sigma}_\varepsilon^2$ would be more properly denominated the "lack-of-fit" mean square, so that the properly denoted estimate of error would become

$$S_\varepsilon^2 = \sum_{j=1}^{4} (S_j^2),$$

where

$$S_j^2 = \sum_{k=1}^{K} (y_{j,k} - \bar{y}_j)^2/(K - 1),$$

with $y_{j,k}$ denoting the kth of the K simular responses at $(z_1, z_2)_j$, and

$$\bar{y}_j = K^{-1} \sum_{k=1}^{K} y_{j,k},$$

for each $j = 1, 2, 3$, and 4.

Since the user of response surface methodology usually wants to define as few simulation encounters as absolutely necessary, one would usually select $K = 2$; i.e., one would "duplicate" the experimental design, so as to obtain the eight simular responses: $y_{j,k}$, for $k = 1$ and 2 at each of the four search vectors: $(z_1, z_2)_j$, $j = 1, 2, 3$, and 4. In this case, one should note the somewhat simplified computation that can be used to obtain the estimate of σ_ε^2, the error variance:

$$S_\varepsilon^2 = 8^{-1} \sum_{j=1}^{4} (y_{j,1} - y_{j,2})^2,$$

which may be associated with a Chi-squared variate of 4 degrees of freedom whenever the usual assumptions concerning the normality of the error terms ε are invoked. Tests for the significance of $\hat{\beta}_1$ and $\hat{\beta}_2$ can then proceed from computations of

$$F_i = 32(\hat{\beta}_i)^2 \Big/ \left[\sum_{j=1}^{4} (y_{j,1} - y_{j,2})^2 \right],$$

where $\hat{\beta}_i$ is the estimate of the slope in the z_i-direction, as computed by using the means $\bar{y}_j = (y_{j,1} + y_{j,2})/2$, as the responses. The actual tests are then performed via Snedecor's F distribution of 1 and 4 degrees of freedom.

Similarly, a test of the hypothesis H_0: $\beta_1^2 + \beta_2^2 = 0$, may be performed by comparing the statistic,

$$F = 32[(\hat{\beta}_1)^2 + (\hat{\beta}_2)^2] \Big/ \left[\sum_{j=1}^{4} (y_{j,1} - y_{j,2})^2 \right],$$

with the improbably large values of Snedecor's F distribution of 2 and 4 degrees of freedom.

9.3.2. A Note on the Selection of the Scale Factors

Whenever one of the computed $\hat{\beta}_i$ proves not to be statistically significant, one should not conclude immediately that further movement in the corresponding z_i-direction would be fruitless; rather, the nonsignificance may well reflect the relative sizes of the scale factors S_i that relate to the actual simular environmental conditions, but that were used so that a *search area* could be conveniently defined as a square, of sides 2 units each, situated about the search center.

Hence, it is recommended, especially in the initial searches, that a nonsignificant slope estimate $\hat{\beta}_i$ be verified by significantly augmenting the corresponding scale factor S_i *after* relocating the (next) search center. If, after conducting the search there (i.e., after estimating the planar approximation to the simular response surface there), the newly computed $\hat{\beta}_i$ remains nonsignificant, one may presume that indeed little is to be gained by further variation of the search center in the corresponding z_i-direction, at least not until a relatively large movement in the other direction may have accumulated through the subsequent relocations to new search centers.

Another method for determining the scale factors is noteworthy. Presuming that the initial search center $(x_{1,0}, x_{2,0})$, in terms of the simular environmental conditions, is a locus that would correspond to standard operating conditions for the actual system being simulated, the analyst may well be aware of the relative costs associated with unit changes in the environmental conditions x_1 and x_2. For example, if x_1 represents the flow rate of a dispensing device, x_2 the speed of a manufacturing tool, then the *marginal costs* C_1 and C_2, for devices operating at a rate $(x_1 + 1)$ and for tools that perform at a speed $(x_2 + 1)$, respectively, may be known. In such event, the scale factors may well be chosen such that

$$S_1/S_2 = C_2/C_1 = K;$$

i.e., the scales may be selected in the proportion,

$$S_1 = KS_2.$$

The magnitude of the larger of the two (i.e., the larger of S_1 and S_2) is left to the discretion of the analyst, though a reasonable constraint to apply to this assignment might be related to the implicit unit cost associated with a change of that size for that variate in the simulated system.

If the approach based on relative costs can be implemented, then it probably will not be necessary to alter a scale factor S_i at the time at which the analyst finds the nonsignificant estimate $\hat{\beta}_i$. The two search variates z_1 and z_2 have become equicost variables; i.e., a unit change in either reflects the same marginal cost increase as a unit change in the other, so that if an estimated slope of the simular system's response plane is not statistically significant, then presumably a more cost-effective response could be pursued by proceeding to search only in the significant direction. Of course, if considerable movement away from the original search center occurs, the scale factors may need to be redefined in terms of the marginal costs at the new search center.

Nonetheless, the isolation phase of the search effort proceeds from one search center to another in the z_1- and z_2-directions—the movement proportional to the estimated slopes $\hat{\beta}_1$ and $\hat{\beta}_2$—until such time as the estimated gradient is no longer statistically significant. At that time, the intensified search may commence in accordance with one of the procedures to be described presently.

9.3.3. The Equilateral Triangular Design

Before proceeding, the reader should be advised that experimental designs other than the 2^2 factorial (with or without replication) can be used in the initial search phase. Indeed, the use of an equilateral triangular design is often recommended since, even if twice replicated, one would require only six simulation encounters each time the planar approximation to the response surface is estimated. This economy of effort is probably quite justified, and the only reason for not introducing the triangular design here first has been the comparative ease of presentation of the data analyses arising from the 2^2 factorial design.

In fact, if economy of effort is of the essence, one might consider a single replication of the equilateral triangular design augmented by two simular encounters defined at the search center. Such an experimental design requires only five simulation encounters for each estimated planar approximation to the response surface, yet provides 1 degree of freedom for lack of fit and another for an error variance estimate.

Because the equilateral triangular design's facile employment in the subsequent intensive phase of the search for the optimal simular response, the estimators arising from the three simular responses,

$$y_1 = \mu(0, 1) + \varepsilon_1,$$
$$y_2 = \mu(-\sqrt{3}/2, -\tfrac{1}{2}) + \varepsilon_2,$$

and

$$y_3 = \mu(3^{1/2}/2, -\tfrac{1}{2}) + \varepsilon_3,$$

are presented anew:

$$\hat{\beta}_0 = 3^{-1}(y_1 + y_2 + y_3),$$
$$\hat{\beta}_1 = -(y_2/\sqrt{3}) + (y_3/\sqrt{3}),$$
$$\hat{\beta}_2 = (2y_1/3) - (y_2/3) - (y_3/3)$$

(cf. Exercise 2, Section 8.3.3).

One may note that each estimate is unbiased for the corresponding coefficient, that the estimates are orthogonal linear combinations of the responses and are therefore uncorrelated if the responses are, although $\mathrm{Var}(\hat{\beta}_0) = \sigma_\varepsilon^2/3$, whereas $\mathrm{Var}(\hat{\beta}_1) = 2\sigma_\varepsilon^2/3 = \mathrm{Var}(\hat{\beta}_2)$.

The estimated planar approximation to the response surface, as it is derived from the equilateral triangular design, becomes

$$\hat{y} = \hat{\beta}_0 + \hat{\beta}_1 z_1 + \hat{\beta}_2 z_2,$$

deemed valid within or near the equilateral triangle. Indeed, the reader may wish to verify that the estimated plane $\hat{y}$ passes exactly through the three simular responses by showing that $\hat{y} = y_1$ when $z_1 = 0$, $z_2 = 1$, that $\hat{y} = y_2$ when $z_1 = -\sqrt{3}/2$, $z_2 = -\tfrac{1}{2}$, and that $\hat{y} = y_3$ when $z_1 = +\sqrt{3}/2$, $z_2 = -\tfrac{1}{2}$. Therefore, no lack of fit exists, nor is there any measure of the intrinsic variation by which one can test whether the estimated slopes $\hat{\beta}_1$ and $\hat{\beta}_2$ differ significantly from zero (unless σ_ε^2 is known).

For the estimated planar approximation $\hat{y}$ of the simular response surface, the contours can be plotted as illustrated in Fig. 9.3. Again, since the estimated surface $\hat{y}$ is a plane, these contours will appear as parallel straight lines when plotted in terms of those coordinates (z_1, z_2) such that

$$\hat{\beta}_0 + \hat{\beta}_1 z_1 + \hat{\beta}_2 z_2 = c \text{ (const).}$$

Thus, the apparently optimal direction in which to move so as to locate a new search center will be in the direction perpendicular to these contour lines. Movement in this direction is the equivalent of movement in the z_1 and z_2 directions in proportion to $\hat{\beta}_1$ and $\hat{\beta}_2$, respectively, so that the next search center may be located at search vector $(k \cdot \hat{\beta}_1, k \cdot \hat{\beta}_2)$ for some constant k at the discretion of the analyst. It would be suggested, however, that k be chosen so that the new search center be located not too far outside the present search area, as defined by the sides of the equilateral triangle; such a recommendation may be more apparent whenever one considers that the current estimates of the simular response surface are based on a planar approximation only.

About the newly located search center, which may be relabeled (0, 0), a new equilateral triangular design can be defined and a triplet (y_1, y_2, y_3) of simular responses recorded. The estimated slopes $\hat{\beta}_1$ and $\hat{\beta}_2$ may again be employed to move the search to another search center. Such a procedure may continue until such time as the estimated slopes are no longer significantly different from zero.

Determination of the significance of the magnitudes of $\hat{\beta}_1$ and $\hat{\beta}_2$ must depend upon knowledge of the intrinsic variability in the simular responses, usually measured by the error variance σ_ε^2. There are several methods by which the error variance may be estimated, three of which shall be discussed briefly here.

The first approach would be the addition of a fourth design point at the search center (0, 0). Denoting the simular response arising there by y_4, one may solve the associated normal equations to obtain the parameter estimates:

$$\hat{\beta}_0 = (y_1 + y_2 + y_3 + y_4)/4,$$

$$\hat{\beta}_1 = -(y_2/\sqrt{3}) + (y_3/\sqrt{3}),$$

and

$$\hat{\beta}_2 = (2y_1/3) - (y_2/3) - (y_3/3),$$

which remain as unbiased, orthogonal (uncorrelated) estimates of the coefficients for the planar approximation to the simular response surface. An estimate of error (i.e., an unbiased estimate of zero) is given by

$$\hat{\varepsilon} = (2\sqrt{3})^{-1}(y_1 + y_2 + y_3 - 3y_4),$$

a linear combination of the four simular responses, which is orthogonal to each of the three estimates $\hat{\beta}_0$, $\hat{\beta}_1$, and $\hat{\beta}_2$. Furthermore, $\mathrm{Var}(\hat{\varepsilon}) = \sigma_\varepsilon^2 = 3[\mathrm{Var}(\hat{\beta}_1)]/2 = 3[\mathrm{Var}(\hat{\beta}_2)]/2$; under the assumptions of normality and independence for the response Y_1, Y_2, Y_3, and Y_4, the random variables (statistics) $\hat{\beta}_0$, $\hat{\beta}_1$, $\hat{\beta}_2$, and $\hat{\varepsilon}$ are normally and independently distributed with

$$E[(\hat{\beta}_1)^2] = (2\sigma_\varepsilon^2/3) + \beta_1^2,$$

$$E[(\hat{\beta}_2)^2] = (2\sigma_\varepsilon^2/3) + \beta_2^2,$$

and

$$E(\hat{\varepsilon}^2) = \sigma_\varepsilon^2.$$

Hence, in order to test an hypothesis of the form H_0: $\beta_i = 0$, one may compute the statistic

$$F_i^* = \{\tfrac{3}{2}\} \cdot \{(\hat{\beta}_i)^2/(\hat{\varepsilon})^2\}, \qquad i = 1 \quad \text{and} \quad 2,$$

rejecting the hypothesis that the corresponding slope is zero whenever F_i^* is too large, as determined from an examination of the tabulations of the Snedecor F-distribution of 1 and 1 degrees of freedom (cf. Exercise 3, Section 8.3.3).

The estimate $\hat{\varepsilon}$ of error in this case, however, is an estimate of the combined variability arising from the intrinsic variation of the simular responses and that arising from the failure of the plane to pass exactly through the four simular responses. Actually, one can show that $(\hat{\varepsilon})^2 = \sum_{j=1}^{4} (\hat{y}_j - y_j)^2$, the sum of the squared deviations of the responses y_j from the respective points $\hat{y}_j$ on the estimated planar surface (e.g., $\hat{y}_4$ is the height of the estimated plane $\hat{y} = \hat{\beta}_0 + \hat{\beta}_1 z_1 + \hat{\beta}_2 z_2$, at (0, 0); viz., $\hat{y}_4 = \hat{\beta}_0$). One would then conclude that $(\hat{\varepsilon})^2$ represents a measure of the lack of fit of the planar surface to the four responses and is therefore not a true measure of the intrinsic variability in the simular responses.

Nonetheless, its use in the F-test, in order to determine the significance of the magnitudes of $\hat{\beta}_1$ and $\hat{\beta}_2$, is widespread. The reader may wish to refer to the analogous situation, which arises in using the 2^2 factorial design for the purpose of estimating the planar approximation, as in Section 9.3.1.

A second approach to the determination of an estimate of the intrinsic variability is the use of replication at the search center. Minimally, this implies the definition of two simulation iterations with the same environmental conditions (those input conditions corresponding to $z_1 = 0$, $z_2 = 0$), yet with independently selected and different random number seeds.

Denoting these two simular responses by y_4 and y_5, one will find that the appropriate parameter estimates become

$$\hat{\beta}_0 = (y_1 + y_2 + y_3 + y_4 + y_5)/5,$$

$$\hat{\beta}_1 = -(y_2/\sqrt{3}) + (y_3/\sqrt{3}),$$

and

$$\hat{\beta}_2 = (2y_1 - y_2 - y_3)/3 \qquad \text{(cf. Section 8.3).}$$

Two additional orthogonal linear combinations of the simular responses are given by

$$\hat{\varepsilon}_1 = (2y_1 + 2y_2 + 2y_3 - 3y_4 - 3y_5)/(3\sqrt{5}),$$

and

$$\hat{\varepsilon}_2 = (y_4 - y_5)/\sqrt{3},$$

each representing an unbiased estimate of zero and each having variance $2\sigma_\varepsilon^2/3$. The first of these estimates $\hat{\varepsilon}_1$ still provides a measure of the lack of fit in the sense that

$$(\hat{\varepsilon}_1)^2 = 20 \sum_{j=0}^{3} (y_j - \hat{y}_j)^2/21,$$

where $y_0 \equiv (y_4 + y_5)/2$, $\hat{y}_0 \equiv \hat{\beta}_0$; whereas the second is a true measure of the inherent variability in the simular responses. Therefore, the tests for the significance of the $\hat{\beta}_i$ would proceed by using the statistics $F_i^* = (\hat{\beta}_i)^2/(\hat{\varepsilon}_2)^2$ and by comparing the result with tabulations of the F distribution of 1 and 1 degrees of freedom, $i = 1$ and 2.

Analogously, additional replication at the search center would provide an appropriate estimate of σ_ε^2, so that the significance of the magnitudes of the estimated slopes, $\hat{\beta}_1$ and $\hat{\beta}_2$, could be tested before moving to a new search center. In these cases, the estimates of β_1 and β_2 remain the same, but the estimate of σ_ε^2 becomes the usual mean square estimate:

$$\hat{\sigma}_\varepsilon^2 = (r - 1)^{-1} \sum_{k=1}^{r} (\eta_k - \bar{\eta})^2,$$

where η_k is the kth of r simular responses observed at the search center, and $\bar{\eta} = \sum_{k=1}^{r} \eta_k/r$. In this case, however, $\hat{\sigma}_\varepsilon^2$ has $r - 1$ degrees of freedom so that the F-statistics

$$F_i^* = 3(\hat{\beta}_i)^2/(2\hat{\sigma}_\varepsilon^2), \qquad i = 1 \quad \text{and} \quad 2,$$

must be compared with the tabulations of Snedecor's F distribution of 1 and $r - 1$ degrees of freedom.

Usually, the search for the optimal simular response is conducted with a minimal amount of effort (with as few simulation encounters as necessary to locate the optimal response). Consequently, one usually uses the equilateral triangular design and, if estimates of the variability in simular responses are required (as they should be, in order for a valid determination of a stopping procedure for the isolation phase of the search to be available), one may augment the design with one (or, preferably, two) design points at the current search center.

A third approach to the estimation of the variation inherent in the simular responses is to replicate the equilateral triangular design itself. Minimally, this requires the definition of six simular encounters, two at each of the three design points corresponding to the vertices of the equilateral triangle. In this case, one defines three pairs of simulation encounters, each encounter of a pair requiring the same environmental conditions as input, save for the assignment of independently selected and different random number seeds.

At each design point, one may record the mean simular response there as

$$\bar{y}_j = \sum_{k=1}^{r} y_{j,k}/r,$$

where $y_{j,k}$ is the kth of r responses at the point. Additionally, an unbiased estimate of the intrinsic variation of the simular responses at this design point becomes

$$S_j^2 = (r - 1)^{-1} \sum_{k=1}^{r} (y_{j,k} - \bar{y}_j)^2.$$

In the case for $r = 2$, often more frequently employed, this error variance estimate becomes

$$S_j^2 = (y_{j,1} - y_{j,2})^2/2.$$

One should note that these estimates need not be construed as lack of fit for the resulting estimated planar approximation to the simular response surface, but rather as proper estimates of σ_ε^2. The three estimates may themselves be averaged to provide an unbiased estimate of σ_ε^2, one having $3(r-1)$ degrees of freedom:

$$S_\varepsilon^2 = \sum_{j=1}^{3} S_j^2/3.$$

The estimates of the coefficients for the planar approximation become the orthogonal contrasts:

$$\hat{\beta}_0 = (\bar{y}_1 + \bar{y}_2 + \bar{y}_3)/3,$$

$$\hat{\beta}_1 = (-\bar{y}_2 + \bar{y}_3)/\sqrt{3},$$

and

$$\hat{\beta}_2 = (2\bar{y}_1 - \bar{y}_2 - \bar{y}_3)/3,$$

notably the same forms as arise in estimating these coefficients for the triangular design whenever there is no replication (except that $\bar{y}_j$ is used instead of y_j, $j = 1$, 2, and 3). Hence, one may note that the resulting planar estimate of the simular response,

$$\hat{y} = \hat{\beta}_0 + \hat{\beta}_1 z_1 + \hat{\beta}_2 z_2,$$

passes exactly through the locations of the mean responses: $\bar{y}_1$, $\bar{y}_2$, and $\bar{y}_3$.

If the replication of the entire equilateral triangular design is performed by the analyst, he should note that the test for the significance of the estimated slopes $\hat{\beta}_1$ and $\hat{\beta}_2$ rests on comparisons of the statistics

$$F_i^* = 3r(\hat{\beta}_i)^2/(2S_\varepsilon^2), \qquad i = 1 \quad \text{and} \quad 2,$$

with tabulations of Snedecor's F-statistic with 1 and $3(r-1)$ degrees of freedom.

In addition, one should note that the use of replication (whether at the search centers or at the vertices or the successive triangles used in the search) provides independent estimates of σ_ε^2 at the successive search centers. One may indeed pool (i.e., take the arithmetic mean of) these estimates as they accumulate during the isolation phase of the search. Of course, the number of degrees of freedom accumulates as well, so that the statistics F_i^* become Snedecor F-variables with an accumulating number of "denominator" degrees of freedom.

9.3.4. A Comparison of the Factorial and Triangular Designs

In summary, one sees that the equilateral triangular design is adequate to estimate the planar approximation for the simular response surface. By augmenting this basic simplex design with a simular response at the triangle's center, an estimate of the lack of fit of the estimated plane to the four responses becomes available. Since the basic 2^2 (square, simplex) design of Section 9.3.1 also requires exactly four model encounters and also provides a degree of freedom for the measurement of the lack of fit, one might understandably ask about possible comparisons between the 2^2 and the (augmented) equilateral triangular design.

Notable in this regard is the fact that for each of the two designs, the estimate $\hat{\beta}_0$ is of the same form; namely, it is the arithmetic mean of the four simular responses. However, the remaining coefficients $\hat{\beta}_1$ and $\hat{\beta}_2$ of the planar approximation are not so directly comparable in the two cases. One might expect then that each design estimates the planar approximation to the response surface with a differing accuracy, as measured by $\text{Var}(\hat{y})$. [These two variances are of course dependent upon the location (z_1, z_2) at which the response is estimated.]

For the equilateral triangular design with augmentation at its design (the search) center,

$$\begin{aligned}
\text{Var}(\hat{y}) &= \text{Var}(\hat{\beta}_0 + \hat{\beta}_1 z_1 + \hat{\beta}_2 z_2) \\
&= \text{Var}(\hat{\beta}_0) + z_1^2 \text{Var}(\hat{\beta}_1) + z_2^2 \text{Var}(\hat{\beta}_2) \\
&\quad + 2z_1 \text{Cov}(\hat{\beta}_0, \hat{\beta}_1) + 2z_2 \text{Cov}(\hat{\beta}_0, \hat{\beta}_2) + 2z_1 z_2 \text{Cov}(\hat{\beta}_1, \hat{\beta}_2) \\
&= \sigma_\varepsilon^2(\tfrac{1}{4} + \tfrac{2}{3}z_1^2 + \tfrac{2}{3}z_2^2 + 0 + 0 + 0),
\end{aligned}$$

since the estimates $\hat{\beta}_i$ are uncorrelated, $i = 0$, 1, and 2; i.e.,

$$\text{Var}(\hat{y}) = \sigma_\varepsilon^2(3 + 8z_1^2 + 8z_2^2)/12.$$

For the estimate $\hat{y}$ arising from the 2^2 factorial design, one finds that

$$\text{Var}(\hat{y}) = \sigma_\varepsilon^2(1 + z_1^2 + z_2^2)/4,$$

or

$$\text{Var}(\hat{y}) = \sigma_\varepsilon^2(3 + 3z_1^2 + 3z_2^2)/12.$$

Hence, at the design center, the two variances are identical. But at search vectors (z_1, z_2) away from the search center, the augmented equilateral triangular design appears to produce a less accurate estimate of the planar approximation to the simular response surface.

Nonetheless, the equilateral triangular design has one compelling advantage: Once the isolation phase of the search for the optimal simular response has been completed, the most recent triangular design is conveniently incorporated within a simplex design ideally suited to undertaking the second phase of the search: the local exploration for the optimum.

EXERCISES

1. For the equilateral triangular design, show that the design matrix $\mathbf{Z}$ is given by

$$\mathbf{Z} = \begin{bmatrix} 1 & 0 & 1 \\ 1 & -\sqrt{3}/2 & -\frac{1}{2} \\ 1 & +\sqrt{3}/2 & -\frac{1}{2} \end{bmatrix}$$

and

$$(\mathbf{Z}^{\mathrm{T}}\mathbf{Z})^{-1} = \begin{bmatrix} \frac{1}{3} & 0 & 0 \\ 0 & \frac{2}{3} & 0 \\ 0 & 0 & \frac{2}{3} \end{bmatrix}$$

so that the unbiased estimates of β_0, β_1, and β_2 are given by $\hat{\vec{\beta}} = (\mathbf{Z}^{\mathrm{T}}\mathbf{Z})^{-1}\mathbf{Z}^{\mathrm{T}}\vec{Y}$, for $\vec{Y}^{\mathrm{T}}$ the vector of simular responses (y_1, y_2, y_3), thus showing that the estimates are those discussed in Section 9.3.3.

2. For the equilateral triangular design, augmented by a design point located at the search center (0, 0) (the center of the triangle as well), show that the design matrix becomes

$$\mathbf{Z} = \begin{bmatrix} 1 & 0 & 1 \\ 1 & -\sqrt{3}/2 & -\frac{1}{2} \\ 1 & +\sqrt{3}/2 & -\frac{1}{2} \\ 1 & 0 & 0 \end{bmatrix}$$

and that

$$(\mathbf{Z}^{\mathrm{T}}\mathbf{Z})^{-1} = \begin{bmatrix} \frac{1}{4} & 0 & 0 \\ 0 & \frac{2}{3} & 0 \\ 0 & 0 & \frac{2}{3} \end{bmatrix},$$

leading to the estimates $\hat{\vec{\beta}} = (\mathbf{Z}^{\mathrm{T}}\mathbf{Z})^{-1}\mathbf{Z}^{\mathrm{T}}\vec{Y}$, where $\vec{Y}^{\mathrm{T}} = (y_1, y_2, y_3, y_4)$ is the vector of the four simular responses, as discussed in the preceding text. Ensure that each $\hat{\beta}_i$ is that of the text, $i = 0$, 1, and 2.

3. For the equilateral triangular design, augmented by a pair of simular responses, Y_4 and Y_5, at the search center (0, 0), show that the design matrix is

$$\mathbf{Z} = \begin{bmatrix} 1 & 0 & 1 \\ 1 & -\sqrt{3}/2 & -\frac{1}{2} \\ 1 & +\sqrt{3}/2 & -\frac{1}{2} \\ 1 & 0 & 0 \\ 1 & 0 & 0 \end{bmatrix}$$

and that

$$\mathbf{Z}^{\mathrm{T}}\mathbf{Z} = \begin{bmatrix} 5 & 0 & 0 \\ 0 & \frac{3}{2} & 0 \\ 0 & 0 & \frac{3}{2} \end{bmatrix};$$

hence, that

$$(\mathbf{Z}^{\mathrm{T}}\mathbf{Z})^{-1}\mathbf{Z}^{\mathrm{T}} = \begin{bmatrix} \frac{1}{5} & \frac{1}{5} & \frac{1}{5} & \frac{1}{5} & \frac{1}{5} \\ 0 & -\sqrt{3}/3 & +\sqrt{3}/3 & 0 & 0 \\ \frac{2}{3} & -\frac{1}{3} & -\frac{1}{3} & 0 & 0 \end{bmatrix}$$

thus providing the estimates $\hat{\beta}_0$, $\hat{\beta}_1$, and $\hat{\beta}_2$ as given in the text. Show also that, if ε_1 is to be a linear combination $[a, b, c, d, d] \cdot \vec{Y}$, which is orthogonal to the three rows of $(\mathbf{Z}^{\mathrm{T}}\mathbf{Z})^{-1}\mathbf{Z}^{\mathrm{T}}$, then $a = b = c$ and $d = -3a/2$. Show then that the choice of $a = 2/(3\sqrt{5})$ leads to the estimator, $\hat{\varepsilon}_1$, of zero as given in the text.

9.4. The Intensive Search

9.4.1. The Quadratic Simular Response Surface

Section 9.3 indicated several techniques for estimating the simular response surface by a plane in some neighborhood of one of a succession of search centers, each search center being located by moving along the path of estimated steepest ascent away from the previous search center. The movement from one search center to another ceases whenever the estimated gradient of the approximating planar surface is no longer significantly different from zero (i.e., whenever weighted sum of the squares of the estimated slopes is no longer significantly greater than appropriate error variances, as estimated from the simular responses).

Whenever this condition obtains, one presumes that the most recent search center occupies a position at or near a stationary point on the simular response surface, hopefully the location of the optimal simular response. Of course, the condition of a relatively flat planar approximation to a response surface could also be attained along a ridge (stationary or rising) or at a maximin (or saddle point).

Therefore, the local exploration, or intensive search, within the isolated search area must be undertaken in a manner also conducive to the determination of the nature of the response surface in this area. Since our initial search, or isolation phase, is conducted via the use of first-order (planar) approximations to the simular response surface, an appropriate improvement in the approximation of the response surface is the consideration of a second-order (quadratic) approximation to the surface.

This approximation to the simular response surface will be deemed valid in a neighborhood surrounding the most recent search center, which one may, without loss of generality, relabel either $(x_{1,0}, x_{2,0})$, in terms of the simular environmental conditions, or (0, 0), in terms of the search variates:

$$z_i = (x_i - x_{i,0})/S_i,$$

where S_i is the scale selected for the ith environmental condition, $i = 1$ and 2. One should note that in the neighborhood of this search center, several simular responses have already been observed, their number being dependent upon the experimental design and the amount of replication used in the most recent estimation.

The quadratic approximation to the simular response surface can be derived by using those terms of the Taylor series expansion for the surface which are of at most second degree in the search variates z_i:

$$E[Y(T)] \equiv \mu(z_1, z_2) = \beta_0 + \beta_1 z_1 + \beta_2 z_2 + \beta_{11} z_1^2 + \beta_{22} z_2^2 + \beta_{12} z_1 z_2,$$

for loci (z_1, z_2) such that, say $|z_i| \leq 1$, $i = 1$ and 2. Therefore, if it be assumed that the simular response in this neighborhood can be represented by

$$Y(T) = \beta_0 + \beta_1 z_1 + \beta_2 z_2 + \beta_{11} z_1^2 + \beta_{22} z_2^2 + \beta_{12} z_1 z_2 + \varepsilon, \quad (9.4.1{:}1)$$

where ε is a random error term with the usual assumptions of normality, nullity of mean, homoscedasticity of variance, and independence, then one may attempt to estimate the coefficients in the quadratic form from the available simular responses. That these estimates are not often suitable for their intended purposes may be shown by considering the orthogonal estimates arising from a single replication of the 2^2 factorial design:

$$\hat{\beta}_0 = (y_1 + y_2 + y_3 + y_4)/4,$$
$$\hat{\beta}_1 = (-y_1 - y_2 + y_3 + y_4)/4,$$
$$\hat{\beta}_2 = (-y_1 + y_2 - y_3 + y_4)/4,$$

and

$$\hat{\varepsilon} = (y_1 - y_2 - y_3 + y_4)/4.$$

(See Section 9.3.1.)

If indeed these simular responses, Y_1, Y_2, Y_3, and Y_4, are random variables of the structure of Eq. (9.4.1:1), the expectations of the statistics become:

$$E(\hat{\beta}_0) = \beta_0 + \beta_{11} + \beta_{22},$$
$$E(\hat{\beta}_1) = \beta_1,$$
$$E(\hat{\beta}_2) = \beta_2,$$

and

$$E(\hat{\varepsilon}) = \beta_{12},$$

as may be verified by substitution of the appropriate values for z_1 and z_2 corresponding to the Y_j as they appear in each estimate. Consequently, unbiased estimates of the coefficients of the linear terms in the quadratic expression for the simular response surface are available, and are the same linear combinations of the responses used to estimate the slopes on the planar (first-order) approximation to the surface. One notes, however, that the error estimate $\hat{\varepsilon}$ no longer estimates zero, but rather the coefficient β_{12} of the cross-product (or interaction) term in the quadratic expression for the simular response surface. Furthermore, the previously unbiased estimate of β_0 is seen to confound in this context the estimation of β_0, β_{22}, and β_{11}.

9.4.2. The 3^2 Factorial Design

Of course, the estimation of the six coefficients ($\beta_0, \beta_1, \beta_2, \beta_{12}, \beta_{22}$, and β_{11}) by means of four simular responses could hardly be expected to prove entirely successful. A minimum of six simular responses, each located above a different locus in the search area, will be required to estimate the quadratic surface. Consequently, a symmetrical hexagonal design or a symmetrical pentagonal design (plus a design point located at its search center) would be in order.

If, in the isolation phase of the search, the 2^2 factorial design had been used, then it becomes difficult to define an augmented design in which all four of these recorded responses may be incorporated, using a total of *exactly* six design points symmetrically. However, one may proceed to complete a 3^2 design by defining five additional simular encounters: one at the search center (0, 0), the remaining four on the search axes, at (0, 1), (1, 0), (0, −1), and (−1, 0), as depicted in Fig. 9.4, wherein the original design loci are denoted by noughts (○), the additional five loci by crosses (×).

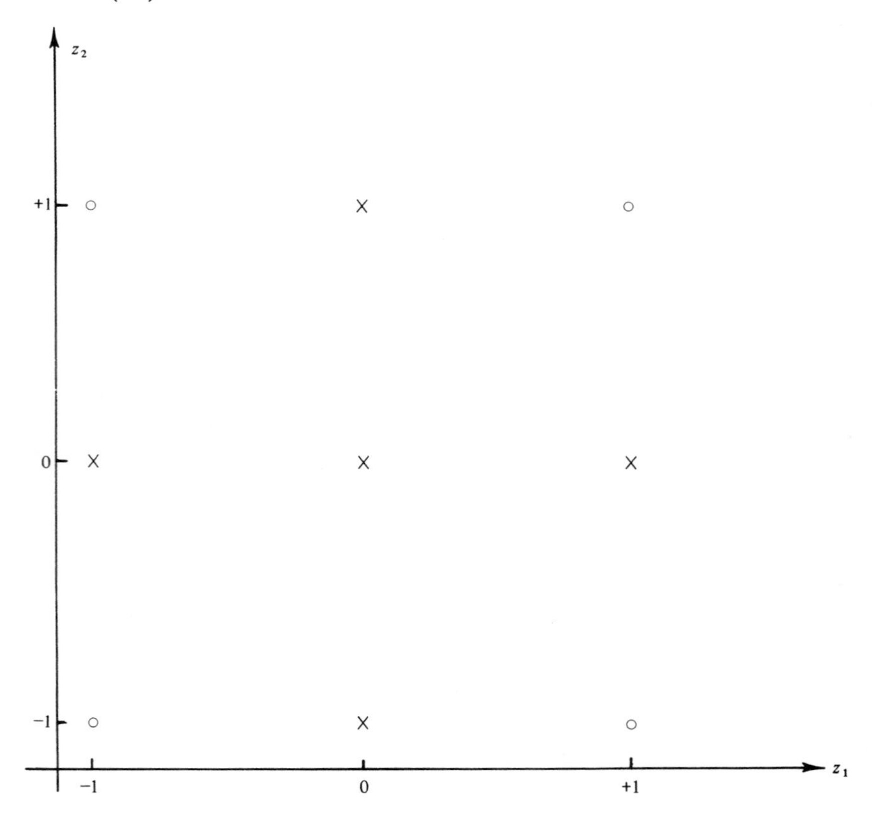

Fig. 9.4. The 3^2 factorial, with embedded 2^2 factorial, design.

The resulting set of nine simular responses, their locations in terms of the search variates, and their dependence on the coefficients in the quadratic approximation to the response surface, as given by Eq. (9.4.1:1), are listed in Table 9.1.

TABLE 9.1

LOCI AND EXPECTATIONS OF SIMULAR RESPONSES, 3^2 FACTORIAL DESIGN

Response	Locus	Expectation					
		β_0	β_1	β_2	β_{11}	β_{22}	β_{12}
y_1	$(-1, -1)$	+1	−1	−1	+1	+1	+1
y_2	$(-1, +1)$	+1	−1	+1	+1	+1	−1
y_3	$(+1, -1)$	+1	+1	−1	+1	+1	−1
y_4	$(+1, +1)$	+1	+1	+1	+1	+1	+1
y_5	$(0, 0)$	+1	0	0	0	0	0
y_6	$(0, +1)$	+1	0	+1	0	+1	0
y_7	$(+1, 0)$	+1	+1	0	+1	0	0
y_8	$(0, -1)$	+1	0	−1	0	+1	0
y_9	$(-1, 0)$	+1	−1	0	+1	0	0

From the nine simular responses, as tabulated, one can determine the following unbiased estimates of the quadratic surface's coefficients by noting that the matrix tabulated under the heading "Expectation" is the design matrix, **Z**, itself; hence,

$$\hat{\vec{\beta}} = (\mathbf{Z}^T\mathbf{Z})^{-1} \cdot \mathbf{Z}^T\vec{Y},$$

or

$$\begin{aligned}
\hat{\beta}_0 &= (-y_1 - y_2 - y_3 - y_4 + 5y_5 + 2y_6 + 2y_7 + 2y_8 + 2y_9)/9,\\
\hat{\beta}_1 &= (-y_1 - y_2 + y_3 + y_4 + y_7 - y_9)/6,\\
\hat{\beta}_2 &= (-y_1 + y_2 - y_3 + y_4 + y_6 - y_8)/6,\\
\hat{\beta}_{11} &= (y_1 + y_2 + y_3 + y_4 - 2y_5 - 2y_6 + y_7 - 2y_8 + y_9)/6,\\
\hat{\beta}_{22} &= (y_1 + y_2 + y_3 + y_4 - 2y_5 + y_6 - 2y_7 + y_8 - 2y_9)/6,
\end{aligned}$$

and

$$\hat{\beta}_{12} = (y_1 - y_2 - y_3 + y_4)/4.$$

The last five of these estimators constitute a set of orthogonal contrasts of the simular responses Y_j so that, if the Y_j are normally distributed about their respective expectations, and are independent of one another, then they will be independently and normally distributed. Each of the six estimators is unbiased for its corresponding parameter; i.e., has mean corresponding to the parameter that each estimates, yet variances:

$$\mathrm{Var}(\hat{\beta}_0) = 5\sigma_\varepsilon^2/9, \qquad \mathrm{Var}(\hat{\beta}_1) = \mathrm{Var}(\hat{\beta}_2) = \sigma_\varepsilon^2/6,$$
$$\mathrm{Var}(\hat{\beta}_{11}) = \mathrm{Var}(\hat{\beta}_{22}) = \sigma_\varepsilon^2/2, \qquad \text{and} \qquad \mathrm{Var}(\hat{\beta}_{12}) = \sigma_\varepsilon^2/4.$$

Three orthogonal and unbiased estimates of zero, themselves mutually orthogonal to the preceding six parameter estimators, are available; viz.,

$$\hat{\varepsilon}_1 = (-y_1 - y_2 + y_3 + y_4 - 2y_7 + 2y_9),$$
$$\hat{\varepsilon}_2 = (-y_1 + y_2 - y_3 + y_4 - 2y_6 + 2y_8),$$

and

$$\hat{\varepsilon}_3 = (y_1 + y_2 + y_3 + y_4 + 4y_5 - 2y_6 - 2y_7 - 2y_8 - 2y_9).$$

Though each of these error estimators has expectation 0, their variances are given by

$$\mathrm{Var}(\hat{\varepsilon}_1) = 12\sigma_\varepsilon^2,$$
$$\mathrm{Var}(\hat{\varepsilon}_2) = 12\sigma_\varepsilon^2,$$

and

$$\mathrm{Var}(\hat{\varepsilon}_3) = 36\sigma_\varepsilon^2.$$

Consequently, a single unbiased estimate of σ_ε^2 can be formed as a weighted sum of the squares of the $\hat{\varepsilon}_k$:

$$\hat{\sigma}_\varepsilon^2 = [3(\hat{\varepsilon}_1)^2 + 3(\hat{\varepsilon}_2)^2 + (\hat{\varepsilon}_3)^2]/108.$$

Under the usual normality, homocedasticity, and independence assumptions for the simular responses, each of the three terms is proportional to one-third of a Chi-squared variate with 1 degree of freedom; hence $3\hat{\sigma}_\varepsilon^2/\sigma_\varepsilon^2$ is a Chi-squared variate with 3 degrees of freedom.

Using the estimates of the six coefficients in the quadratic approximation to the response surface, one can express an estimate of the simular response surface as

$$\hat{y} = \hat{\beta}_0 + \hat{\beta}_1 z_1 + \hat{\beta}_2 z_2 + \hat{\beta}_{11} z_1^2 + \hat{\beta}_{22} z_2^2 + \hat{\beta}_{12} z_1 z_2,$$

a random variable which exists at each locus (z_1, z_2) in the search area. The random variable $\hat{y} \equiv \hat{\mu}(z_1, z_2)$ has expectation

$$E(\hat{y}) = \beta_0 + \beta_1 z_1 + \beta_2 z_2 + \beta_{11} z_1^2 + \beta_{22} z_2^2 + \beta_{12} z_1 z_2$$

and variance

$$\mathrm{Var}(\hat{y}) = (20 - 18z_1^2 - 18z_2^2 + 18z_1^4 + 18z_2^4 + 9z_1^2z_2^2) \cdot \sigma_\varepsilon^2/36.$$

One may then plot the contours of the estimated quadratic approximation to the response surface by connecting all loci (z_1, z_2) of the search variates such that $\hat{y} = c$, a constant. Varying the constant at regular intervals provides successive contours. As a practical suggestion, the initial value of c might be taken as a convenient number near $\hat{\beta}_0$, since this is the value of the estimated quadratic approximation to the simular response surface at the search center (0, 0).

Once the contours have been plotted, the nature of the simular response surface in the isolated search area can be somewhat revealed. For example, if the contours assume the form of concentric ellipses, then the optimal simular response may be presumed to lie somewhere near the center of these ellipses. A more intensive search, possibly employing a sufficient number of simular responses so as to estimate a third-order approximation to the simular response function, could then be conducted about this apparent center in order to ascertain more precisely the location of the optimal response.

However, one may find that the contours of the quadratic surface,

$$\hat{y} = \hat{\beta}_0 + \hat{\beta}_1 z_1 + \hat{\beta}_2 z_2 + \hat{\beta}_{11} z_1^2 + \hat{\beta}_{22} z_2^2 + \hat{\beta}_{12} z_1 z_2$$

are not conveniently elliptical, but rather are other conic sections such as hyperbolas, parabolas or even straight lines. These cases correspond, respectively, to situations in which the estimated quadratic approximation to the simular response surface is a maximin, a rising ridge, or a stationary ridge, as discussed previously, and as depicted in Fig. 9.1.

Hence, the analyst may take advantage of this newly found knowledge, as inferred from the contour plots. If the contours are elliptical or circular, then the location of the optimal response can be affixed by a more intensive search centered about its apparent location at the center of the ellipses. If a ridge apparently exists, then further search should be conducted in the apparent direction of the ridge to determine whether the ridge, if extant for the actual simular response surface, is of the rising or stationary category. In the remaining case, if a minimax surface apparently exists, search may need to be conducted in both directions of greatest promise away from the minimax; such a condition may imply a bimodal simular response surface, so that two stationary points will need to be isolated and their relative magnitudes subsequently compared in order to fix the location of *the* optimal simular response.

9.4.3. The Hexagonal Design

The factorial design has been shown to be especially effective in both the isolation phase and the subsequent intensified search phase of the search procedure for locating the optimal simular response. Indeed, the 2^2 factorial design, adequate for estimating the three coefficients of a planar approximation to the response surface, is appropriate to the isolation phase of the search, since it provides explicit instructions about the relocation of the search center until such time as a stationary point has been apparently isolated. At the stationary point, the 3^2 factorial design may be invoked in order to estimate the six coefficients of a quadratic (second-order) approximation to the response surface in the neighborhood of the most recent search center. Additionally, the most recently employed 2^2 factorial design is readily embedded in the 3^2 design, so that some economy of effort is extant in moving from the isolation phase to the local exploration phase.

One may recall, however, that the 3^2 design provides nine simular responses for the purpose of estimating the six coefficients of the approximating quadratic surface, thereby requiring three more simular encounters than absolutely necessary for the purpose. In this section, then, the use of the hexagonal design, for the purpose of estimating the quadratic approximation to the simular response surface, will be considered.

The hexagonal design consists of six loci, at each of which a single simular encounter is to be defined by specifying the corresponding environmental (input) conditions and by assigning an independently selected random number seed. The six loci are at the vertices of a regular hexagon which may be thought of as inscribed in the unit circle, so that the search vector (0, 0), or *search center*, is situated at the center of the hexagon.

If the isolation phase of the search procedure has used the equilateral triangular design (with or without a central design point to measure the goodness of the planar fit, or with or without replication designed to provide an estimate of the inherent variation in the simular responses), then the most recent triangular design already contains three of the design points for the hexagonal design. All that remains to complete the hexagonal design is to superimpose a second equilateral triangular design above, but rotated 60° from the first, as depicted in Fig. 9.5.

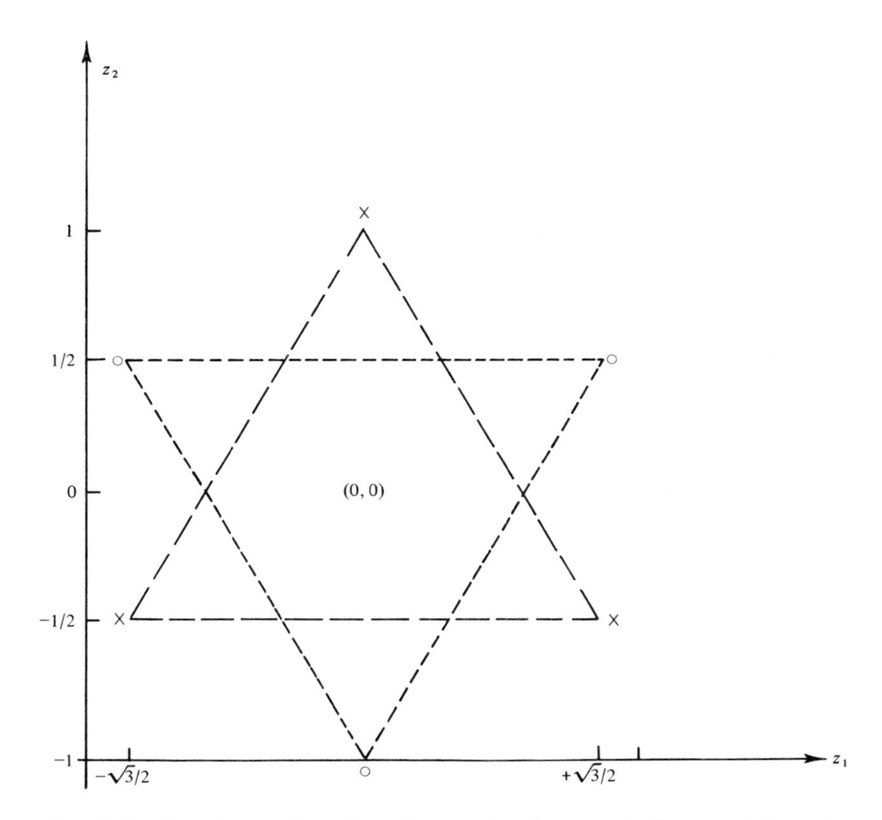

FIG. 9.5. Superimposed equilateral triangular designs (the hexagonal design).

Denoting by y_j the simular response at the jth vertex of the hexagon, one may recall the quadratic approximation for the simular response,

$$Y(T) = \beta_0 + \beta_1 z_1 + \beta_2 z_2 + \beta_{11} z_1^2 + \beta_{22} z_2^2 + \beta_{12} z_1 z_2 + \varepsilon,$$

for (z_1, z_2) in, say, the unit circle (or unit square) surrounding the search center, and for ε a random error component comprising, as usual, errors arising due both to the inadequacy of the quadratic representation for the response surface, and to the intrinsic variability in the simular responses. For each of the six simular responses of this hexagonal design, Table 9.2 provides the locus and the expectation.

One may note that the first six rows of the array beneath the heading "Expectation" in the table constitute the design matrix $\mathbf{Z}_1$ so that the vector β of coefficients has unbiased estimates that would be given by

TABLE 9.2

LOCI AND EXPECTATIONS OF SIMULAR RESPONSES, HEXAGONAL DESIGN

Response	Locus	Expectation					
		β_0	β_1	β_2	β_{11}	β_{22}	β_{12}
Y_1	(0, 1)	1	0	1	0	1	0
Y_2	$(-\sqrt{3}/2, -\frac{1}{2})$	1	$-\sqrt{3}/2$	$-\frac{1}{2}$	$\frac{3}{4}$	$\frac{1}{4}$	$+\sqrt{3}/4$
Y_3	$(+\sqrt{3}/2, -\frac{1}{2})$	1	$+\sqrt{3}/2$	$-\frac{1}{2}$	$\frac{3}{4}$	$\frac{1}{4}$	$-\sqrt{3}/4$
Y_4	(0, −1)	1	0	−1	0	1	0
Y_5	$(-\sqrt{3}/2, +\frac{1}{2})$	1	$-\sqrt{3}/2$	$\frac{1}{2}$	$\frac{3}{4}$	$\frac{1}{4}$	$-\sqrt{3}/4$
Y_6	$(+\sqrt{3}/2, +\frac{1}{2})$	1	$+\sqrt{3}/2$	$\frac{1}{2}$	$\frac{3}{4}$	$\frac{1}{4}$	$+\sqrt{3}/4$
Y_7	(0, 0)	1	0	0	0	0	0

the elements of the vector

$$\beta = (\mathbf{Z}_1{}^T\mathbf{Z}_1)^{-1}\mathbf{Z}_1{}^T\vec{Y},$$

where $\vec{Y}^T = (y_1, y_2, y_3, y_4, y_5, y_6)$, the vector of simular responses. However, one may verify that

$$\mathbf{Z}_1{}^T\mathbf{Z}_1 = \begin{bmatrix} 6 & 0 & 0 & 3 & 3 & 0 \\ 0 & 3 & 0 & 0 & 0 & 0 \\ 0 & 0 & 3 & 0 & 0 & 0 \\ 3 & 0 & 0 & \frac{9}{4} & \frac{3}{4} & 0 \\ 3 & 0 & 0 & \frac{3}{4} & \frac{9}{4} & 0 \\ 0 & 0 & 0 & 0 & 0 & \frac{3}{4} \end{bmatrix},$$

from which one may observe that the first row is equal to the sum of the fourth and fifth rows, thereby implying that $(\mathbf{Z}_1{}^T \cdot \mathbf{Z}_1)$ has no inverse. In turn, this would imply that any unbiased estimates of β_0, β_{11}, and β_{22} which were linear combinations of the six simular responses would be linearly related in the sense that $\hat{\beta}_0$ would equal some linear combination of $\hat{\beta}_{11}$ and $\hat{\beta}_{22}$.

Hence, though the hexagonal design is symmetrical, it is not really adequate for estimating all six coefficients of the quadratic approximation to the simular response surface. This inadequacy can be easily overcome by the addition of a design point at the search center (0, 0) and by the recording of the subsequent response Y_7 from a simulation encounter

TABLE 9.3

VARIANCES OF ESTIMATORS OF QUADRATIC APPROXIMATION TO SIMULAR RESPONSE SURFACE, AT SELECTED LOCI

(z_1, z_2)	3^2 Factorial	Hexagon and center
(0, 0)	$20\sigma_\varepsilon^2/36$	σ_ε^2
(0, ± 1)	$20\sigma_\varepsilon^2/36$	$30\sigma_\varepsilon^2/36$
(± 1, 0)	$20\sigma_\varepsilon^2/36$	$30\sigma_\varepsilon^2/36$
(± 1, ± 1)	$29\sigma_\varepsilon^2/36$	$132\sigma_\varepsilon^2/36$
($\pm\sqrt{3}/2$, $\pm\frac{1}{2}$)	$251\sigma_\varepsilon^2/(16 \cdot 36)$	$480\sigma_\varepsilon^2/(16 \cdot 36)$

defined there. The last row of Table 9.2 provides the locus and expectation of this response.

Denoting by $\mathbf{Z}$ the matrix (of seven rows and six columns) immediately beneath the tabular heading "Expectation," one may compute the matrix:

$$\mathbf{Z}^T\mathbf{Z} = \begin{bmatrix} 7 & 0 & 0 & 3 & 3 & 0 \\ 0 & 3 & 0 & 0 & 0 & 0 \\ 0 & 0 & 3 & 0 & 0 & 0 \\ 3 & 0 & 0 & \frac{9}{4} & \frac{3}{4} & 0 \\ 3 & 0 & 0 & \frac{3}{4} & \frac{9}{4} & 0 \\ 0 & 0 & 0 & 0 & 0 & \frac{3}{4} \end{bmatrix},$$

having inverse

$$(\mathbf{Z}^T\mathbf{Z})^{-1} = \begin{bmatrix} 1 & 0 & 0 & -1 & -1 & 0 \\ 0 & \frac{1}{3} & 0 & 0 & 0 & 0 \\ 0 & 0 & \frac{1}{3} & 0 & 0 & 0 \\ -1 & 0 & 0 & \frac{3}{2} & \frac{5}{6} & 0 \\ -1 & 0 & 0 & \frac{5}{6} & \frac{3}{2} & 0 \\ 0 & 0 & 0 & 0 & 0 & \frac{4}{3} \end{bmatrix},$$

so that

$$(\mathbf{Z}^T\mathbf{Z})^{-1}\mathbf{Z}^T$$

$$= \begin{bmatrix} 0 & 0 & 0 & 0 & 0 & 0 & 1 \\ 0 & -1/2\sqrt{3} & +1/2\sqrt{3} & 0 & -1/2\sqrt{3} & +1/2\sqrt{3} & 0 \\ +\frac{1}{3} & -\frac{1}{6} & -\frac{1}{6} & -\frac{1}{3} & +\frac{1}{6} & +\frac{1}{6} & 0 \\ -\frac{1}{6} & +\frac{1}{3} & +\frac{1}{3} & -\frac{1}{6} & +\frac{1}{3} & +\frac{1}{3} & -1 \\ +\frac{1}{2} & 0 & 0 & +\frac{1}{2} & 0 & 0 & -1 \\ 0 & +1/\sqrt{3} & -1/\sqrt{3} & 0 & -1/\sqrt{3} & +1/\sqrt{3} & 0 \end{bmatrix},$$

the estimates of the parameters becoming then $\hat{\vec{\beta}} = (\mathbf{Z}^T\mathbf{Z})^{-1} \cdot \mathbf{Z}^T\vec{Y}$, or

$$\begin{aligned}
\hat{\beta}_0 &= y_7, \\
\hat{\beta}_1 &= (-y_2 + y_3 - y_5 + y_6)/(2\sqrt{3}), \\
\hat{\beta}_2 &= (2y_1 - y_2 - y_3 - 2y_4 + y_5 + y_6)/6, \\
\hat{\beta}_{11} &= (-y_1 + 2y_2 + 2y_3 - y_4 + 2y_5 + 2y_6 - 6y_7)/6, \\
\hat{\beta}_{22} &= (y_1 + y_4 - 2y_7)/2,
\end{aligned}$$

and

$$\hat{\beta}_{12} = (y_2 - y_3 - y_5 + y_6)/\sqrt{3}.$$

With these estimates from the seven simular responses, the simular response function's quadratic approximation is estimated by

$$\hat{y} = \hat{\beta}_0 + \hat{\beta}_1 z_1 + \hat{\beta}_2 z_2 + \hat{\beta}_{11} z_1^2 + \hat{\beta}_{22} z_2^2 + \hat{\beta}_{12} z_1 z_2,$$

for which the contours may be plotted in, say, the unit circle or the unit square (in terms of the search variates z_1 and z_2). Thus, the apparent nature of the simular response surface may be ascertained and further exploration conducted, if necessary, to affix more precisely the environmental conditions giving rise to the optimal simular response.

9.4.4. Comparison of Quadratic Surface Estimates

The reader may note that the intensive search is conducted identically to the search when the quadratic approximation had been estimated by nine simular responses arising from the use of a 3^2 factorial design; i.e., advantage is taken of apparent ridges or saddle points, or an intensive search is conducted about the center of the contour ellipses. Quite naturally, the question of preferability of the hexagonal and factorial designs for this purpose should then arise.

The 3^2 factorial design, as may be recalled, uses nine simular responses, arising from environmental conditions that are arrayed in a pattern of three rows and three columns, in order to estimate the six coefficients of the quadratic approximation to the simular response surface. These six estimates are given by linear combinations of the nine responses, the weights having been explicitly given in Section 9.4.2. Each estimate is unbiased for the corresponding coefficient β_{ij} and is orthogonal to (i.e., uncorrelated with) each of the remaining estimates, so that the variance of the estimate $\hat{y}$ of the quadratic approximation to the simular response surface at (z_1, z_2) was readily computed for the 3^2 design as

$$\text{Var}(\hat{y}) = (20 - 18z_1^2 - 18z_2^2 + 18z_1^4 + 18z_2^4 + 9z_1^2z_2^2)\sigma_\varepsilon^2/36.$$

On the other hand, the regular hexagonal design was seen to be inadequate without the augmentation of the design by inclusion of a simular response arising from environmental conditions corresponding to the search center. Even so, only seven simular responses are required to provide unbiased estimates of the six coefficients in the quadratic approximation. Again, these estimates are relatively straightforward linear combinations of the seven simular responses, the weights having been discussed in Section 9.4.3. However, in this instance, the six estimates are not all pairwise orthogonal, so that some of the estimates are intercorrelated:

$$\text{Cov}(\hat{\beta}_0, \hat{\beta}_{11}) = -\text{Var}(Y_7) = -\sigma_\varepsilon^2,$$

$$\text{Cov}(\hat{\beta}_0, \hat{\beta}_{22}) = -\text{Var}(Y_7) = -\sigma_\varepsilon^2,$$

and

$$\text{Cov}(\hat{\beta}_{11}, \hat{\beta}_{22}) = +5\sigma_\varepsilon^2/6.$$

Consequently, the variance of the resulting estimate $\hat{y}$ of the quadratic approximation to the simular response surface at any point (z_1, z_2) becomes, for the augmented hexagonal design:

$$\text{Var}(\hat{y}) = \sigma_\varepsilon^2[6 - 10(z_1^2 + z_2^2) + 9(z_1^2 + z_2^2)^2]/6.$$

A comparison of this variance with that just presented for the variance of the estimate $\hat{y}$ as associated with the 3^2 factorial design, would reveal a preference for the latter estimate. Table 9.3 is provided in order to facilitate these comparisons.

Though the apparent preferability is for a response surface estimate derived from the nine simular responses of the 3^2 factorial design, the analyst may want to weigh the relative cost of using the augmented hexagonal design, especially if the simulation model itself requires extensive computer time in order to perform a single encounter.

Of course, the selection of the experimental design for the local exploration phase of the search for the optimal simular response may be somewhat dependent on the design selected in the isolation phase. For example, if an equilateral triangular design, augmented by a fourth response at the search center (rather than a 2^2 factorial design), had been used in the isolation phase, then only three additional simular responses would be required in order to implement the analysis by means of the augmented hexagonal design (as opposed to the five responses required in order to complete the 3^2 factorial design from the original 2^2 design). In other words, the *marginal* costs for the alternative implementations of the intensive search phase are roughly in a ratio of 3 : 5. (*Note*: This ratio would be 4 : 5 if an equilateral triangular design, without augmentation, were employed in the last search of the isolation phase.)

These marginal cost ratios do not presume, however, that an equal amount of effort (i.e., an equal cost) has been experienced in the isolation phase, and they should not be taken as statements which would tend to imply that a subsequently equal degree of effort will be required in the final stages of the local exploration phase (such as searching along an apparent ridge). This discussion hopefully raises an interesting decision–theoretic problem, but it will not be the present purpose to propose a strategy for its solution. The search for the optimal strategy will probably depend on the nature of the underlying (and unknown) simular response function; one might then speculate that computer simulation itself may provide some important insights into such strategies.

9.5. Summary: A Guide to Response Surface Methodology

This chapter has presented and developed methods for providing an exploration of the response surface associated with the simular response $Y(T)$ at the end of T simular time units. The techniques are appropriate whenever the response can be seen as a function of two quantitative and continuous environmental conditions x_1 and x_2. This exploration was discussed in the context of an attempt to isolate and fix the location of the optimal simular response, although the experimental designs used might well be used merely to determine the approximate nature of the simular responses in a neighborhood of some specific search center. (This center might, for example, correspond to the standard operating conditions for the system being simulated.)

For example, in the preceding chapters, the application of the analysis of variance, regression techniques, and experimental designs for the purpose of determining which of several environmental (input) conditions are significant contributors to the simular response, $Y(T)$, were discussed. It should now be evident that appropriate to several of these designs is their use to estimate the response surface in a neighborhood of the locale in which the design was implemented; e.g., in addition to performing a 2^2

or a 3^2 factorial design in order to obtain a measure of the relative significance of each environmental conditions, one may estimate the response surface itself, irrespective of the significance or nonsignificance of the two factors, by plotting the contours of

$$\hat{y} = \hat{\beta}_0 + \hat{\beta}_1 z_1 + \hat{\beta}_2 z_2 + \hat{\beta}_{11} z_1^2 + \hat{\beta}_{22} z_2^2 + \hat{\beta}_{12} z_1 z_2.$$

Thus, the response surface methodology becomes a candidate for use in model verification whenever the simular response surface might be anticipated under some restricted set of environmental conditions.

One might also speculate that since experimental designs exist for the purpose of comparing three or more factors (environmental conditions), there exist response surface methodologies for the purpose of finding the direction of the "steepest ascent" whenever so many environmental conditions are deemed important to the search. Indeed, such search techniques are available, but their exposition here would carry us unnecessarily far afield. The reader is referred to the excellent paper by Box (1954), an earlier paper by Box (1952), and the subsequent papers by Box and Hunter (1957) and by Draper (1962). Chapter 11 of Davies (1954) carries a lucid account. These multifactor analyses, it may be noted, are relatively straightforward extensions of the procedures of this chapter.

In addition to the consideration of multifactor response surface methodology, an additional generalization is possible. One may recall that the transition from the isolation phase to the local exploration phase was accompanied by the shift from a linear to a quadratic approximation for the simular response surface. The order of the polynomial approximation need not be constrained at two, but may be enlarged to provide cubic or quartic or even quintic approximations to the simular response surface. The reader would correctly anticipate that the experimental designs discussed in this chapter would not prove adequate for the task of estimating the additional polynomial coefficients. The interested reader is referred to Draper (1962), as well as papers by Gardiner *et al.* (1959) and DeBaun (1959).

For further references on the exploration of response surfaces, the reader is referred to the important bibliography of Hill and Hunter (1966), as well as to the more recent paper on the application of experimental designs to simulation experiments by Hunter and Naylor (1970) and Mihram (1970b).

Before concluding the present chapter, a note of caution is in order.

Though much of this chapter has been devoted to a discussion of procedures for locating the optimal simular response, this search, in the presence of random variation, may be somewhat akin to the proverbial quest for the philosopher's stone—elusive. As presented, the search procedures, have been directed toward locating the optimal *expected* simular response, as opposed to the actual optimum. One should note then that, even if the locus of the optimal expected simular response can be determined, there is a great probability that larger simular responses could be recorded not only at this locus but also at nearby loci as well. A measure or estimate of the intrinsic variation, such as σ_ε^2 or σ_ε or their estimates, is therefore of prime importance in any decision to undertake further exploration of the simular response function.

Chapter 10

THE SIMULATION AS A STOCHASTIC PROCESS

But times do change and move continually.

—EDMUND SPENCER

10.1. Introduction

The simulation model is a dynamic representation of an actual system, and therefore must keep track of the behavior of the system in time. The system's behavior can itself be represented by a *state vector* $\vec{Y}(t)$, each element of the vector providing information on some aspect of the system's status at any point t in simular time. In many instances, one will be interested in a particular element of the state vector, the particular element of interest being denoted then by $Y(t)$ and referred to as the *simular response* (or, more succinctly, as the *response*) at simular time t. Cases in which two or more elements of the state vector are required to describe the system's performance may be treated separately. Thus for a given encounter with a simulation model (of duration T time units), the response $Y(t)$ may be measured and recorded for $t = 1, 2, \ldots, T$, thereby producing the record of finite length:

$$[Y(1), Y(2), \ldots, Y(T)].$$

Since each encounter requires its own random number seed, a barely finite collection of such records of length T is possible. The collection of possible responses $Y(t)$ at simular time t constitutes therefore a random variable $Y(t)$ having, for each $t = 1, 2, \ldots, T$, its own probability distribution function.

Conceivably, any encounter could be allowed to run indefinitely, so that its record may be deemed of infinite length; such an infinite-length record is called a *realization*:

$$[Y(1), Y(2), \ldots, Y(T), Y(T+1), \ldots].$$

Again, since each encounter can be initiated with its own random number seed, there exists a nearly infinite collection, or *ensemble*, of these realizations. At each point t in simular time, then, the simular response is a random variable $Y(T)$ described by its own distribution function and its corresponding moments. The infinite collection $\{Y(t), t = 1, 2, \ldots\}$ of these random variables constitutes then a *stochastic process*.

In general, the notation $Y(t)$ will be reserved for instances in which reference is being made to the simular response as the random variable arising at simular time t. The notation $y(t)$ will be used to refer to an actual response at simular time t in a particular encounter. The notation $Y(t)$, $t = 1, 2, \ldots$ or, more simply $[Y(t)]$ will refer to a realization, a record of infinite length. The use of $[y(t), t = 1, 2, \ldots, T]$ or, more usually $[y(t)]$ will imply discussion of a particular record of finite length, as would be recorded for a given encounter observed over a simular time period of length T. Furthermore, the notation $\{Y(t)\}$ will refer to the stochastic process, or *time series*, defined either by the entire ensemble of random variables $Y(t)$, $t = 1, 2, \ldots$, or by the complete collection of realizations $[Y(t)]$.

Thus, one may view the response of a stochastic simulation: (a) statically; i.e., in terms of the random variable $Y(t)$ at a particular point, t (usually $t = T$) in simular time; or, (b) dynamically; i.e., in terms of the time series or stochastic process $\{Y(t)\}$. Analyses of simular responses are therefore categorized accordingly, so that one may endeavor to determine either the static effects or the dynamic effects by using statistical techniques applicable, respectively, to random samples (of a random variable) or sample records (of a time series). The statistical techniques applicable to a random sample of $Y(T)$ are discussed in Chapters 6–9. In this chapter, an examination will be made of the statistical techniques applicable whenever one wishes to analyze the dynamic effects of a simular system.

There are generally two occasions on which one seeks to determine the simulation dynamics. In the first instance (model verification), one seeks to determine whether the simular system (or, for that matter, a subsystem or subroutine) is responding in accordance with its anticipated behavior. Often, as was noted earlier, the rectitude of such simular behavior may be reasonably well determined by observing the agreement of an encounter's time series with that expected from an encounter for a given set of input conditions. In other events, a more sophisticated statistical analysis will be in order. [See Mihram (1972b).]

In the second instance (the analysis of sample records arising from encounters with the completed and validated model), statistical techniques appropriate to the determination of the effects that given input conditions produce in the dynamic simular response $[Y(t)]$, are essential. In either instance, then, the statistical techniques appropriate to time-series analysis may be invoked to study the dynamic behavior of the simular response.

In this chapter some of the more fundamental procedures are presented for the analysis of the time series arising in simulation encounters. The discussion will generally be restricted to the case in which a univariate response $y(t)$ is recorded at each point t in simular time, though in a later section cases will be discussed in which two or more state variables are recorded in simular time.

10.2. Stationarity and Ergodicity

It has been shown that theoretical developments pertinent to stochastic processes are often easier under assumptions of stationarity. Of course, stationarity is not necessary to a rigorous development of the theory surrounding a stochastic process, since the Markov processes are quite amenable to study, though they are not necessarily stationary. For example, the Poisson process $\{N(t)\}$ is an ensemble of random variables $N(t)$ such that $N(t)$ has the Poisson distribution with mean λt. For $t \neq s$, then, $N(t)$ and $N(s)$ possess different distributions. However, one may recall that the incremental Poisson process $\{N(t + \Delta t) - N(t)\}$ is stationary for any fixed Δt; i.e., the random variables $N(t + \Delta t) - N(t)$ and $N(s + \Delta t) - N(s)$ possess the same probability density function for all s and t.

Many stochastic processes that are not themselves stationary may be converted to stationary processes through relatively simple artifices such as incrementation. The Wiener–Lévy process $\{W(t)\}$, an ensemble of normally distributed random variables $W(t)$ with $E(W(t)) = 0$ and $\text{Var}(W(t)) = \sigma^2 \cdot |t|$, is not stationary since the normal distribution applicable to $W(t)$ has variance increasing linearly with time t. Again, the device of incrementation may be invoked, and the resulting incremental Wiener–Lévy process $\{W(t + \Delta t) - W(t)\}$ becomes an ensemble of normally distributed random variables with

$$E(W(t + \Delta t) - W(t)) = 0 \quad \text{and} \quad \mathrm{Var}(W(t + \Delta t) - W(t)) = \sigma^2(\Delta t).$$

Hence, since a normal distribution is completely specified by its mean value and its variance, the distributions of the random variables

$$W(t + \Delta t) - W(t) \quad \text{and} \quad W(s + \Delta t) - W(s)$$

are indeed identical.

We note that, for the Poisson process $\{N(t)\}$, both the mean-value function and the variance function increase with time; i.e.,

$$E(N(t)) = \mathrm{Var}(N(t)) = \lambda t;$$

incrementation of the process leads to a process that has mean value function and variance function each equal to $\lambda(\Delta t)$. Thus, incrementation of the Poisson process removes the linear trend present in the mean of the Poisson process; fortunately it does so in a manner that produces a common probability density function for the random variables of the incremental process.

In the case of the Wiener–Lévy process $\{W(t)\}$, used to describe the distance that a particle has moved after encountering a large number of molecular collisions in time t, no trend exists in the mean-value function itself. However, the ever-increasing variance of the process can be controlled (and made constant) by the selection of a time increment Δt and by using instead the incremental Wiener–Lévy process $\{W(t + \Delta t) - W(t)\}$.

Thus, by studying the progress that a stochastic process has made in the most recent (Δt) units of time, as opposed to the progress it has made since time 0, one may often induce stationarity as a property of the process under study. Of course, to this point, remarks have been limited to stochastic processes whose incremental processes possess the property of first-order stationarity. In the case of the Wiener–Lévy process, incrementation produces strict stationarity; i.e., the joint probability density function for the n random variables:

$$W(t_1 + \Delta t) - W(t_1),\ W(t_2 + \Delta t) - W(t_2), \ldots,\ W(t_n + \Delta t) - W(t_n),$$

is a multivariate normal distribution whose variance–covariance matrix remains invariant under a shift of the n time indices by any amount $h > 0$.

In a simulation, then, it may well be that the response $Y(t)$ need not be a random variable from an ensemble $\{Y(t)\}$ that is a stationary stochastic process. For example, in simulating a queueing situation, an appropriate measure of effectiveness is the queue length $Q(t)$ at simular time t. For many queueing situations, especially those in which the arrival rate exceeds the effective service rate, one can expect a trend to exist in $Q(t)$. By studying instead the change $Q(t + \Delta t) - Q(t)$ in the length of the queue over a fixed interval of simular time (Δt), one can attempt to remove the trend effect and thus hope to produce stationarity.

Another example of a simular response that will probably have a time trend is the total production $Y(t)$ accumulating from a simulated system organized to develop raw materials into finished products. In many such simulation models, one would anticipate that a linear time trend exists in any realization $[y(t), t = 1, 2, \ldots]$ from the simulation model. Indeed, most managerial philosophy seems centered about such a linearity principle.

A more generalized concept of productivity arises in delivery systems, such as transportation networks designed to deliver tonnages of goods to a sink (consumer, retailer, or wholesaler) from a source (depot, manufacturing plant) or from a set of sources. In simulating such systems, we are frequently concerned with measuring the total tonnage delivered to the sink (or sinks) by simular time t. Describing this tonnage as the *productivity* of the transportation system, one might again anticipate a linearly increasing simular response with time; i.e., in the ensemble of responses $\{Y(t)\}$, we have $E\{Y(t)\} = kt$.

One should note that the incremental process $\{Y(t + \Delta t) - Y(t)\}$ associated with the simular response $Y(t)$ is a difference, and, hence, a linear combination, of random variates from the stochastic process $\{Y(t)\}$. A more general method for the removal of polynomial trends, to be discussed in the following section, is also structured by using special linear combinations (called moving averages) of the elements of a sample record $[y(t), t = 1, 2, \ldots, T]$.

For example, the mean-value function of the stochastic process is, in general, $E(Y(t)) = m(t)$, some function of t. Presuming that such trends have been removed in such a way as to leave the resulting ensemble (which will be denoted by $\{Y^*(t)\}$) stationary, then the statistical properties of the trend-free, stationary time series will not depend upon simular time; e.g., for the random variable $Y^*(t)$ one would have $E(Y^*(t)) = m$ and $\mathrm{Var}(Y^*(t)) = \sigma^2$, for m and σ^2 constant.

However, this need not imply that any particular realization $[Y^*(t)]$ will have average properties which are those of the ensemble $\{Y^*(t)\}$, as may be seen by considering the elementary ensemble of three time functions

$$Y(t) = \begin{cases} +1, & \text{if the toss of a die is 1 or 2,} \\ 0, & \text{if the toss of a die is 3 or 4,} \\ -1, & \text{if the toss of a die is 5 or 6.} \end{cases}$$

In this example, the random number seed is determined by the toss of a die and the resulting realization is $+1$, 0, or -1 in accordance with the result on its uppermost face. At any time point t the random variable $Y(t)$ will be observed to assume one of these three values, depending upon the result of the tossing of the die originally. Thus, the p.d.f. for $Y(t)$ is

$$g_{Y(t)}(u) = \tfrac{1}{3}, \qquad u = -1, 0, 1,$$

and the stochastic process $\{Y(t)\}$ is thus stationary of order 1. Furthermore, $E(Y(t)) = 0$ and $\text{Var}(Y(t)) = \frac{2}{3}$, for all t. Nonetheless, if one is observing the process, already in progress after noting the result on the die, he will record a constant sequence (either $+1$, 0, or -1). Clearly, although the process itself has stationary properties, sample moments from any realization will not necessarily be the equivalent of the corresponding ensemble moments. One should note that, if the realization $y(t) = +1$ is under observation, then for a record of any length T,

$$\bar{y} = T^{-1} \sum_{t=1}^{T} y(t) = 1,$$

and

$$S_y^2 = (T-1)^{-1} \sum_{t=1}^{T} [y(t) - \bar{y}]^2 = 0,$$

both result at variance with the ensemble's mean-value and variance functions (0 and $\frac{2}{3}$, respectively).

Thus, in concluding this section, a remark on *ergodicity* is in order. One may recall that the ensemble $\{Y(t)\}$ of random variables (simular responses) might be equally well considered as a collection of realizations, $[Y(t), t = 1, 2, \ldots]$, each then conceptually a function of time, and each resulting from the specification of a particular random number seed. Quite naturally, one may ask about the adequacy of using a single simular record $[y(t), t = 1, 2, \ldots, T]$ of finite length in order to estimate properties of the entire ensemble.

The preceding example reveals that one should take some care in making inferences concerning ensemble properties from the elements in a sample record arising from a single simulation encounter. More generally, even though a simular response has been treated to remove trends and instill stationarity, one must also be reasonably sure of the presence of ergodicity, so that the sample moments computed from a particular realization can be used as estimates of the corresponding ensemble moments.

Several authors (cf. Fishman and Kiviat, 1967) note that one can obtain essentially independent observations from a single realization of a simular time series, provided that the successive observations are recorded from the *steady-state* segment of the stochastic process (where the probability laws have presumably become independent of the model's starting conditions) and provided that these recordings are themselves sufficiently separated (T time units) in simular time in order to avoid, for all practical purposes, their autocorrelation. Since their recommended procedures for determining both the duration of the *transient stage* (at the end of which the collection of the successive observations could begin) and the time duration between successive, essentially uncorrelated, observations thereafter are somewhat heuristic, the present author has not recommended this approach for obtaining multiple simular responses corresponding to a given environmental specification. Rather, the emphasis has been on recording the responses of independently seeded encounters, each defined at the same environmental conditions, since

(*a*) Independent, and therefore definitely uncorrelated, responses may be obtained.

(*b*) Many models are constructed solely for the purpose of examining their transient stage.

(*c*) Many simulation models have extremely short transient stages, especially if their initialization conditions are properly randomized at simular time 0.

(*d*) The cost of computation has become significantly reduced so that even moderately long transient periods may be simulated prior to recording the simular response $Y(T)$.

(*e*) One need not be concerned about the possibility that the realizations from a simular time series may not always be ergodic whenever independently seeded encounters are employed.

Furthermore, Gafarian and Ancker (1966) recently indicated that for a large class of weakly stationary and ergodic processes, estimates will be usually more precise if computed from independently seeded encounters, rather than from nonoverlapping portions of a single encounter.

The following exercise serves as an additional reminder of the importance of ergodicity. For further discussion of ergodicity, the reader is referred to Feller (1950, Vol. I, p. 395), though many covariance stationary processes will be ergodic (see Parzen, 1962, p. 74).

EXERCISE

Consider a simular response $Y(t)$ of the form

$$Y(t) = \cos(2\pi t + \Phi),$$

where Φ is a phase angle randomly selected from a uniform distribution over the interval $(0, \pi)$. Note that, once Φ has been selected, the realization $[Y(t),\ t = 1, 2, \ldots]$ will be a series of measures taken from a cosine wave of unit period. Show that the random variable $Y(t)$ has probability density function

$$g_{Y(t)}(u) = \begin{cases} 1/[\pi(1 - u^2)^{1/2}], & u \in (-1, 1), \\ 0, & \text{other } u, \end{cases}$$

which is independent of t, thereby showing that the stochastic process $\{Y(t)\}$ is stationary of order 1. Show that $E(Y(t)) = 0$, for all t. Show also that the stochastic process is strictly stationary. In addition, conclude that the computation, from a simular record $[y(t),\ t = 1, 2, \ldots, T]$, of

$$\bar{y} = T^{-1} \sum_{t=1}^{T} y(t) = \cos \varphi$$

reveals that the stochastic process $\{Y(t)\}$ is not ergodic.

10.3. Removal of Trends

The effects contributing to a time series $\{Y(t)\}$ are usually categorized as (a) a trend; (b) one or more oscillatory effects about the trend; and (c) a random component. In a subsequent section, methods will be discussed that are appropriate to the isolation of oscillatory effects. For the present, the isolation and removal of trends will be the topic of interest.

A trend may be defined loosely as a slow, long-term movement in a time series. It might indeed, for a given sample record $[y(t),\ t = 1, 2, \ldots, T]$, constitute a portion of an oscillatory effect of extremely long period (relative to T, the length of the sample record), so that, unless a record of sufficient length is examined, certain low-frequency oscillatory effects may escape notice, and be assumed to represent trend. The importance of this fact should not be overlooked, though in a practical sense there will always exist the possibility of undetected low-frequency oscillatory components.

One of the primary desiderata for the removal of trend effects is the wish to operate on records from a stationary time series. Stationarity is of prime importance in the analysis of time series, especially stationarity in the wide sense, since the autocorrelation function in this case is dependent upon the time lag between, but not the absolute location of, the times for which the correlation between two of the random variables $Y(t)$ and $Y(s)$ of the ensemble is computed. Denoting the *autocovariance function* by

$$\gamma_{YY}(t, s) = E((X(t) - \mu)(X(s) - \mu)),$$

where $\mu = E(X(t))$ for all t, then stationarity in the wide sense implies that, for all t and s,

$$\gamma_{YY}(t, s) = \gamma_{YY}(t - s).$$

Thus, $\gamma_{YY}(t, t) = \mathrm{Var}(Y(t)) = \gamma_{YY}(0)$, for any t, so that the *autocorrelation function* becomes the ratio

$$\varrho_{YY}(\tau) = \gamma_{YY}(\tau)/\gamma_{YY}(0),$$

for lag $\tau = (t - s)$.

The fact that the autocovariance and autocorrelation functions are both dependent upon a single variable is of immense assistance to facilitating analysis of a time series. Indeed, as we have seen, the behavior of the autocorrelation function for wide-sense stationary time series can be used, to a limited extent, to determine the underlying mechanism that may be generating the time series.

In addition, a wide-sense stationary time series has a spectral distribution function which is the Fourier transform of the series' autocorrelation function. This means, then, that a time series that is stationary in the wide sense can be decomposed into its frequency components, thus revealing the oscillatory effects present in the series. Indeed, we can use a sample record from such a series to estimate the relative strength or power of these frequency components, as will be seen in a subsequent section.

If a stochastic process is Gaussian (i.e., if all its joint probability distribution functions are multivariate normal distributions), then the process is completely described by its mean-value function and the autocovariance function. Therefore, if a Gaussian process is wide-sense stationary, the variance–covariance matrix of its n-dimensional joint probability density functions is completely specified and the densities are stationary; the wide-sense stationary Gaussian process is thus strictly stationary as well.

The time series that arise from simulation encounters may, as has been seen, frequently be Gaussian, owing to the Central Limit Theorem. Consequently, if one may assume that the responses from such simulation models are wide-sense stationary time series, then stationarity in its strict sense is also present. In this event, multidimensional probability density functions can be studied by shifting the time origin, without fear of introducing complications into the analysis.

One would hope, then, that any effects of trend might be statistically isolated and subsequently removed from the time series, leaving a stochastic process composed only of oscillatory and random effects and, therefore, a random process that is essentially stationary.

10.3.1. Removal of a Linear Trend

Let us presume that a time series, in addition to being composed of oscillatory and random elements, has a component which increases (or decreases) linearly with time:

$$Y(t) = a + bt + P(t) + \varepsilon(t),$$

where $P(t)$ is the periodic or oscillatory component and where $\varepsilon(t)$ is the random component.

For example, the productivity of a simulated delivery system might be expected to increase in such a manner with time, yet with both systematic and unsystematic fluctuations about this trend. Clearly, the time series arising from an encounter with such a simulation model will not be stationary, because the mean-value function, and hence the univariate probability density functions, will vary with time.

If we consider any time lag of fixed duration (Δt), then the incremental process $Y(t + \Delta t) - Y(t)$ assumes the form

$$Y(t + \Delta t) - Y(t) = b(\Delta t) + [P(t + \Delta t) - P(t)] + [\varepsilon(t + \Delta t) - \varepsilon(t)].$$

Since the lag (Δt) is fixed, the effect of the linear trend, $a + bt$, in the $\{Y(t)\}$ process has been replaced by the constant $b(\Delta t)$ in the incremental process. Thus, the removal of a linear trend is readily accomplished; in the event a sample record $[y(t), t = 1, 2, \ldots, T]$ is available from a simulation encounter, the effect of a linear trend may be removed by selecting a time increment Δt and proceeding to the computation of the corresponding sample record from the time series $\{Y^*(t)\}$:

$$y^*(t) = y(t + \Delta t) - y(t), \qquad t = 1, 2, \ldots, (T - \Delta t).$$

In the example cited, that of a simulation of a productivity system, this simple device for removal of trend is indeed quite natural, since the incremental process represents the productivity normalized over a fixed period of time (Δt). Equivalently, the incremental productivity is essentially a productivity *rate*, indeed constant with respect to the presence of a linear trend.

10.3.2. Removal of a Quadratic Trend

If we assume that the time series arising from encounters with a simulation model is of the form:

$$Y(t) = a + bt + ct^2 + P(t) + \varepsilon(t),$$

where again $P(t)$ and $\varepsilon(t)$ represent systematic and unsystematic fluctuations, respectively, then the quadratic trend component may be removed by selecting a time increment Δt and by considering the stochastic process $\{Y(t + 2(\Delta t)) - 2Y(t + \Delta t) + Y(t)\}$, which will be denoted $\{Y^{**}(t)\}$. One sees immediately that for constant k,

$$\begin{aligned} Y^{**}(t) = k &+ \{P[t + 2(\Delta t)] - 2P(t + \Delta t) + P(t)\} \\ &+ \{\varepsilon[t + 2(\Delta t)] - 2\varepsilon(t + \Delta t) + \varepsilon(t)\}, \end{aligned}$$

so that only the periodic and random effects remain in the time series $\{Y^{**}(t)\}$.

An alternative approach to the definition of the process $\{Y^{**}(t)\}$ is via the finite-difference operator Δ, which operates on a stochastic process $\{Y(t)\}$ to produce the stochastic process $\{Y(t + \Delta t) - Y(t)\}$; i.e.,

$$\Delta Y(t) \equiv Y(t + \Delta t) - Y(t) \equiv Y^*(t),$$

for some fixed Δt. Thus, a repeated application of the finite-difference operator yields

$$\begin{aligned} \Delta Y^*(t) &= \{Y[t + 2(\Delta t)] - Y(t + \Delta t)\} - [Y(t + \Delta t) - Y(t)] \\ &\equiv Y^{**}(t); \end{aligned}$$

i.e.,

$$\Delta[\Delta Y(t)] \equiv \Delta^2 Y(t) \equiv Y^{**}(t).$$

We see then that the process $\{\Delta^2 Y(t)\}$ will not contain any quadratic or linear effect present in the series $\{Y(t)\}$. A relatively simple linear combination of the terms of the time series $\{Y(t)\}$ will produce a stochastic process $\{Y^{**}(t)\}$ independent of such effects.

Moreover, for a given realization $[y(t), t = 1, 2, \ldots]$, one may proceed to the removal of quadratic and linear trends by considering the *moving average*:

$$y^{**}(t) = y(t+2) - 2y(t+1) + y(t), \qquad t = 1, 2, \ldots$$

with (Δt) taken for convenience to be unity. Therefore, if one anticipates linear and/or quadratic components in the record $[y(t), t = 1, 2, \ldots, T]$ arising from an encounter with a simulation model, these trend effects may be reduced by considering instead the record $[y^{**}(t), t = 1, 2, \ldots, (T-2)]$. Further statistical analyses can then proceed using the resulting record of moving averages as a trend-free and (hopefully) stationary time series.

10.3.3. Removal of More General Trends

If a stochastic process $\{Y(t)\}$ comprises a general trend component, $T(t)$, in addition to a periodic component $P(t)$ and a random component $\varepsilon(t)$, then one can often assume that a polynomial of degree p will suffice to describe the trend effect; i.e.,

$$T(t) \cong a_0 + a_1 t + \cdots + a_p t^p.$$

For a given simular record $[y(t), t = 1, 2, \ldots, T]$, one may endeavor to eliminate the polynomial trend by considering a contiguous set of an odd number $(2m + 1)$ of terms from the record $[y(t)]$. Without loss of generality, one may relabel these $2m + 1$ data elements:

$$v_{-m}, v_{-m+1}, \ldots, v_0, \ldots, v_{m-1}, v_m,$$

so that the central term of the set will possess the null subscript. A polynomial of degree p $(<2m + 1)$ is then fitted to these data by the method of least squares; i.e., one solves simultaneously the set of $(p + 1)$ equations

$$0 = (\partial/\partial a_j) \sum_{t=-m}^{m} (v_t - a_0 - a_1 t - \cdots - a_p t^p)^2, \qquad j = 0, 1, 2, \ldots, p.$$

The jth of these equations will assume the form

$$\sum_{t=-m}^{m} v_t t^j = a_0\left(\sum_{t=-m}^{m} t^j\right) + a_1\left(\sum_{t=-m}^{m} t^{j+1}\right) + \cdots + a_p\left(\sum_{t=-m}^{m} t^{p+j}\right),$$

$$j = 0, 1, \ldots, p,$$

so that the solution for the a_j's will be of the form

$$a_j^* = b_{0,j} + b_{1,j}v_{-m} + b_{2,j}v_{-m+1} + \cdots + b_{2m+1,j}v_m,$$

a linear combination of the data points $v_{-m}, \ldots, v_0, \ldots, v_m$, for each $j = 0, 1, 2, \ldots, p$.

Therefore, we may arrive at a polynomial expression for the data points:

$$v_t^* = a_0^* + a_1^*t + \cdots + a_p^*t^p,$$

approximately valid for t near zero. In fact, $v_0^* = a_0^*$, and one may show that $b_{0,0} = 0$, so that

$$v_0^* = a_0^* = b_{1,0}v_{-m} + b_{2,0}v_{-m+1} + \cdots + b_{2m1,0}v_m.$$

Assuming then that the fitted polynomial v_t^* provides an approximate expression for v_t at $t = 0$, one may use v_0^* instead of v_0. One proceeds to obtain the next estimated trend value by considering the data points

$$v_{-m+1}, v_{-m+2}, \ldots, v_0, v_1, \ldots, v_m, v_{m+1};$$

i.e., one drops the term v_{-m} and includes v_{m+1}, so that the central term is now v_1. By redefining the subscripts, one may assign the subscript 0 to the term denoted now by v_1 and then proceed to estimate the trend value at this location. Continuing in this fashion along the simular record, one obtains a moving record of estimated trend values; and, since, each estimated trend value is a linear combination of the observed simular responses up to m time units on either side of its location, the procedure is referred to as a *moving average*.

The coefficients, $b_{0,0}, b_{1,0}, \ldots, b_{2m+1,0}$, which are necessary to the determination of a_0^* (and, equivalently, of the estimated trend value v_0^*) are functions of p and m only and do not depend, for example, on the recorded values v_t. In addition, the coefficients required for fitting a polynomial of odd degree $2r + 1$ also serve as the coefficients for fitting the polynomial of even degree $2r$, so that the coefficients can be conveniently tabulated. Exemplary values are provided in Table 10.1.

TABLE 10.1

COEFFICIENTS FOR MOVING AVERAGES, AS REQUIRED IN FITTING A POLYNOMIAL OF DEGREE $2r$ OR $2r+1$ TO $2m+1$ SUCCESSIVE TIME-SERIES VALUES

$r = 1$ (*Quadratic or cubic polynomials*)

$2m+1$	Coefficients
5	$\frac{-3}{35}, \frac{12}{35}, \frac{17}{35}, \frac{12}{35}, \frac{-3}{35}$
7	$\frac{-2}{21}, \frac{3}{21}, \frac{6}{21}, \frac{7}{21}, \frac{6}{21}, \frac{3}{21}, \frac{-2}{21}$
9	$\frac{-21}{231}, \frac{14}{231}, \frac{39}{231}, \frac{54}{231}, \frac{59}{231}, \frac{54}{231}, \frac{39}{231}, \frac{14}{231}, \frac{-21}{231}$

$r = 2$ (*Quartic or quintic polynomials*)

$2m+1$	Coefficients
7	$\frac{5}{231}, \frac{-30}{231}, \frac{75}{231}, \frac{131}{231}, \frac{75}{231}, \frac{-30}{231}, \frac{5}{231}$
9	$\frac{15}{429}, \frac{-55}{429}, \frac{30}{429}, \frac{135}{429}, \frac{179}{429}, \frac{135}{429}, \frac{30}{429}, \frac{-55}{429}, \frac{15}{429}$
11	$\frac{18}{429}, \frac{-45}{429}, \frac{-10}{429}, \frac{60}{429}, \frac{120}{429}, \frac{143}{429}, \frac{120}{429}, \frac{60}{429}, \frac{-10}{429}, \frac{-45}{429}, \frac{18}{429}$

For example, in fitting a cubic trend to the five data values, v_{-2}, v_{-1}, v_0, v_1, and v_2, one obtains:

$$v_0^* = \frac{-3}{35} v_{-2} + \frac{12}{35} v_{-1} + \frac{17}{35} v_0 + \frac{12}{35} v_1 + \frac{-3}{35} v_2.$$

Similarly, one obtains the next fitted value, v_1^*, as

$$v_1^* = \frac{-3}{35} v_{-1} + \frac{12}{35} v_0 + \frac{17}{35} v_1 + \frac{12}{35} v_2 + \frac{-3}{35} v_3,$$

and proceeds to compute the fitted series v_j^*, $j = 0, 1, 2, \ldots$ as a moving average, each average employing the coefficients appropriate to the degree of the polynomial and the number of successive terms over which the fit is made. For more complete tables of the requisite coefficients, the reader is referred to the third volume of the compendium by Kendall and Stuart (1966).

Therefore, one may replace a simular record $[y(t),\ t = 1, 2, \ldots, T]$ by a trend-fitted record of moving averages $[v_t^*,\ t = (m+1), (m+2), \ldots, (T-m)]$; one reduces of course the length of the resulting record by $2m$ terms.

If a sample record is indeed exactly a polynomial of degree p, then the terms $y(t)$ will be identically equal to v_t^*, $t = (m+1), (m+2), \ldots, (T-m)$; therefore, the "trend-free" series $y(t) - v_t^*$, will be null for every $t = (m+1), (m+2), \ldots, (T-m)$. It is for this reason that one proceeds to use the record $[y(t) - v_t^*,\ t = (m+1), (m+2), \ldots, (T-m)]$ as a trend-free series to which techniques appropriate to the analysis of stationary time series may be applied.

EXERCISE

Consider the time series generated exactly by the cubic polynomial: $y(t) = (t-2)^3$, $t = 1, 2, \ldots, T$. Show that, using the moving average implied by the coefficients in the first row of Table 10.1, the fitted terms v_t^*, $t = 3, 4, \ldots, (T-2)$, will be exactly equal to the corresponding $y(t)$ values, so that the resulting trend-free record $y(t) - v_t^*$, $t = 3, 4, \ldots, (T-2)$, will not contain cubic, quadratic, or linear trend effects.

10.3.4. Effects of Trend Elimination

A word of warning is in order. For a time series $\{Y(t)\}$ of the form

$$Y(t) = T(t) + P(t) + \varepsilon(t),$$

where $T(t)$ is a trend effect, $P(t)$ a periodic effect, and $\varepsilon(t)$ a random effect, elimination of trend effects by the use of linear operations on a sample record $[y(t)]$ has implicit effects upon the components $P(t)$ and $\varepsilon(t)$ as well. If we presume that a polynomial of degree p, sufficient to eliminate the trend effect adequately, is to be employed, then the resulting record $y^*(t) = y(t) - v_t^*$ is essentially a transformation of the record $[y(t)]$; i.e.,

$$[y^*(t)] = \mathrm{Tr}[y(t)].$$

Thus, considering the application of such a transformation to each record in the complete collection, one may assume that the trend-free stochastic process is the sum

$$\{Y^*(t)\} = \{P(t) - \mathrm{Tr}[P(t)]\} + \{\varepsilon(t) - \mathrm{Tr}[\varepsilon(t)]\}.$$

Of primary concern, then, is the effect that trend removal has upon the systematically and unsystematically fluctuating components $P(t)$ and $\varepsilon(t)$.

The reader is referred to the third volume of Kendall and Stuart's statistical compendium for the justification of the following points regarding trend removal:

(*a*) The suggested methods of trend removal will probably remove periodic components of low frequency as well.

(*b*) The suggested methods of trend removal will not alter the mean value of the random component, provided that the mean of the random process $\{\varepsilon(t)\}$ was zero initially.

(*c*) The suggested methods of trend removal shall, however, introduce certain periodicities into the residual component $\{\varepsilon(t) - \mathrm{Tr}[\varepsilon(t)]\}$.

Result (*c*) is known as the Slutzky–Yule effect, whereby the removal of trend in a purely random series of uncorrelated observations from the same probability density function leaves a series with periodicities not present in the original. From this fundamental case, one infers that some caution is in order whenever trend removal is undertaken, especially whenever a harmonic or spectral analysis is to be applied to the "trend-free" series.

As for result (*a*), the fact that trend removal can eliminate low-frequency periodic effects is not too surprising in view of the earlier comments regarding the fact that many trend effects are actually realizations of cyclic effects of long duration, whose period is of sufficient length to avoid detection in a record of available length.

In summary, one notes that the effects of polynomial trends in time series may be eliminated, or at least somewhat removed, by specific moving averages defined along the time series. As was noted, finite-differencing techniques are applicable to the elimination of linear and quadratic effects. Indeed, reiterated finite-differences are applicable to the removal of higher-order polynomial effects, just as are the specific moving average techniques that have been recommended.

In the event that reiterated finite-differencing is applied to a time series, relatively simple expressions for the variance of the resulting time series is available; namely, for the time series $\{\Delta^r \varepsilon(t)\}$,

$$\mathrm{Var}(\Delta^r \varepsilon(t)) = \binom{2r}{r} \mathrm{Var}(\varepsilon(t)),$$

whenever $\{\varepsilon(t)\}$ is the purely random series. This result is the basis of a method developed by Tintner (1940) for determination of the appropriate degree of polynomial for trend removal: The estimated variances, divided by $\binom{2r}{r}$, of the rth successively differenced series may be compared until such time as reasonable compatibility of two successive calculations develops. The appropriate degree of polynomial adequate to describe the trend is then taken to be the current value of either r or $r - 1$.

The difficulties intrinsic to the use of these trend-removal techniques should not be an excuse for failing to make use of them. The procedures that will be discussed in the remaining sections of this chapter will presume the existence of stationarity (wide-sense), a condition that clearly does not apply to series containing trend effects.

10.4. The Correlogram

10.4.1. Relevance of the Correlogram

Presuming then that the time series $\{Y(t)\}$ arising from encounters with a simulation model is stationary of first order, we recall that each of the random variables $Y(t)$ has the same probability density function $F_Y(y)$. Therefore the mean μ of each of the random variables is the same, and the variance of every random variable in the ensemble $\{Y(t)\}$ is the same constant, σ^2.

In most cases, one probably does not know distributional properties such as μ and σ^2, so that he will seek to estimate these with a reasonable degree of precision by using the data from a simular record of length T: $[y(t), t = 1, 2, \ldots, T]$. Of course, one may have been required to apply a trend-removal technique, as described in the preceding section, so that the data at hand may not be the actual record resulting from the simulation itself. Nonetheless, it will be presumed that the record of data at hand is of length T, that it is trend-free, and, for the moment, that it arises from a stochastic process which is stationary of order 1. In addition, it will be assumed that moments from a simular record will, in the limit, equal corresponding moments for the stochastic process.

Under these conditions, the simular record $[y(t), t = 1, 2, \ldots, T]$ can be used to estimate the stationary mean μ by computing

$$\hat{\mu} = \bar{y} = T^{-1} \sum_{t=1}^{T} y(t).$$

One notes that $E(\hat{\mu}) = T^{-1} \sum_{t=1}^{T} E(Y(t)) = T^{-1} \cdot T \cdot \mu = \mu$, so that the estimator $\hat{\mu}$ is unbiased for μ. The variance of this estimator may be derived as

$$\mathrm{Var}(\hat{\mu}) = T^{-2} \sum_{t=1}^{T} \mathrm{Var}(Y(t)) + T^{-2} \sum_{\substack{t=1,s=1 \\ t \neq s}}^{T} \sum^{T} \mathrm{Cov}[Y(t), Y(s)].$$

Thus, knowledge of the precision of the estimate $\hat{\mu}$ requires information concerning the autocovariance function

$$R(s, t) = \mathrm{Cov}[Y(s), Y(t)].$$

In general, the probability density function for the statistic $\hat{\mu}$ will remain concealed; however, in the event that the time series $\{Y(t)\}$ is a normal process, so that any collection of n random variables $Y(t_1)$, $Y(t_2), \ldots, Y(t_n)$ has the multivariate normal distribution, the statistic $\hat{\mu}$ (being a linear combination of such random variables) has the univariate normal distribution specified by the mean and variance delineated above. One should note in this event that the expression for the variance of $\hat{\mu}$ is a linear combination of the elements of the variance–covariance matrix specific to the joint normal distribution for $Y(1), Y(2), \ldots, Y(T)$. In those cases where it is not likely that the trend-free simular response constitutes a normal process, the distribution of $\hat{\mu}$ cannot be stated so precisely; even so, the Central Limit Theorem may be somewhat applicable and the normal approximation for the distribution of $\hat{\mu}$ may prove satisfactory.

In order to estimate the variance σ^2 of the time series, one may compute from the record $[y(t), t = 1, 2, \ldots, T]$ the quantity

$$\tilde{\sigma}^2 = T^{-1} \sum_{t=1}^{T} [y(t) - \mu]^2$$

whenever the mean μ of the stochastic process $\{Y(t)\}$ may be presumed known. The expected value of this variance estimator is

$$E(\tilde{\sigma}^2) = T^{-1} \sum_{t=1}^{T} E[Y(t) - \mu]^2 = T^{-1} T \sigma^2 = \sigma^2,$$

so that the estimator is indeed unbiased. The variance of the statistic $\tilde{\sigma}^2$ generally depends upon the fourth moment of the random variable $Y(t)$ and will not be discussed further.

More likely to arise is the situation in which μ is not known, but must be itself estimated by $\hat{\mu} = \bar{y}$. The estimator for the variance σ^2 then becomes

$$\hat{\sigma}^2 = (T - 1)^{-1} \sum_{t=1}^{T} [y(t) - \bar{y}]^2,$$

where $(T - 1)^{-1}$ is used instead of T^{-1}, analogous to the procedure necessary to obtain an unbiased estimate of σ^2 whenever the random variables $Y(t)$ may be presumed to be uncorrelated. The distributional properties of $\hat{\sigma}^2$ become, however, much more complicated than those of $\tilde{\sigma}^2$. However, presuming that the sample record $[y(t)]$ is of sufficient length and that the autocovariance function for the process $\{Y(t)\}$ is sufficiently well behaved for one to assert that $\hat{\mu} = \bar{y}$ is quite near μ, then one could probably feel confident that $\hat{\sigma}^2$ is approximately unbiased for σ^2.

One sees then that the behavior of the autocovariance function is of prime importance in the estimation of statistical properties of a time series stationary of order 1. Indeed, as was noted in Chapter 3, the autocovariance function may serve in a limited capacity to characterize the nature of a time series (e.g., autoregressive, moving average), and thus may assist the analyst of a simular output in describing the apparent behavior of the dynamic model of a system.

It will be presumed that the time series $\{Y(t)\}$ deriving from a stochastic simulation model is stationary in the wide sense, so that the autocovariance function

$$R(s, t) = \text{Cov}[Y(s), Y(t)] = R(s - t)$$

is a function of the time difference, or lag, $s - t$ only. Since knowledge of the behavior of this function is of assistance in describing the behavior of a time series, as well as the precision of estimates of properties of the process, a technique for the estimation of $R(s - t)$ is of the essence.

10.4.2. Estimation of the Autocovariance Function

If the mean μ of the stationary time series $\{Y(t)\}$ is again presumed known, an estimate of the autocovariance function of lag 1 is computed from the simular record $[y(t),\ t = 1, 2, \ldots, T]$ as

$$\tilde{R}(1) = (T - 1)^{-1} \sum_{t=1}^{T-1} \{[y(t) - \mu] \cdot [y(t + 1) - \mu]\};$$

i.e., one averages the successive products of pairs of factors of the form

$[y(t) - \mu]$ and $[y(t+1) - \mu]$. One should note in passing that

$$\tilde{R}(-1) = (T-1)^{-1} \sum_{t=2}^{T} \{[y(t) - \mu] \cdot [y(t-1) - \mu]\}$$

yields the same result as $\tilde{R}(1)$. The mean value of the estimate $\tilde{R}(-1)$ or equivalently $\tilde{R}(1)$ is

$$\begin{aligned} E(\tilde{R}(1)) &= (T-1)^{-1} \sum_{t=2}^{T-1} E([Y(t) - \mu] \cdot [Y(t+1) - \mu]) \\ &= (T-1)^{-1} \cdot (T-1)R(1) = R(1), \end{aligned}$$

so that $\tilde{R}(1) = \tilde{R}(-1)$ is an unbiased estimate of the autocovariance function of lag 1.

In this context, one notes that the estimate $\tilde{\sigma}^2$ of the variance of the process $\{Y(t)\}$ is an unbiased estimate of the autocovariance function of lag 0:

$$\tilde{\sigma}^2 = \tilde{R}(0) = T^{-1} \sum_{t=1}^{T} \{[y(t) - \mu] \cdot [y(t) - \mu]\}.$$

More generally, an unbiased estimate of the autocovariance function of lag k is computed from the sample record $[y(t)]$ as

$$\tilde{R}(k) = (T-k)^{-1} \sum_{t=1}^{T-k} \{[y(t) - \mu] \cdot [y(t+k) - \mu]\},$$

again an average of lagged products, each successive product being the pair of factors $[y(t) - \mu]$ and $[y(t+k) - \mu]$ for $t = 1, 2, \ldots, (T-k)$. Precisely equivalent to this estimate is the estimator

$$\tilde{R}(-k) = (T-k)^{-1} \sum_{t=k+1}^{T} \{[y(t) - \mu] \cdot [y(t-k) - \mu]\},$$

having expectation

$$E(\tilde{R}(-k)) = E(\tilde{R}(k)) = (T-k)^{-1} \cdot (T-k)R(k) = R(k).$$

Therefore, from a sample record of length T one can compute unbiased estimates of $\sigma^2 = R(0)$ and the autocovariances $R(k)$ of lags $k = 1, 2, \ldots, (T-1)$. One notes of course that, the greater is the lag k, then the smaller is the number of lagged products being averaged to obtain the estimate $\tilde{R}(k)$; indeed, $\tilde{R}(T-1)$ will consist of the single product $[y(1) - \mu] \cdot [y(T) - \mu]$. The precision of the estimates of the autocovariance function at the longer lags is then often lacking, relative to that of the estimates for shorter lags. In fact, one frequently computes only the initial m autocovariance estimates $\tilde{R}(1), \ldots, \tilde{R}(m)$, for $m \ll T$. Since one generally presumes that the theoretical autocovariances diminish significantly in absolute value for longer lags, little is usually lost by truncating the estimation procedure at some lag m.

More usually, the mean-value of the simular response shall not be known, however, and shall need be estimated from the trend-free simular record $[y(t), t = 1, 2, \ldots, T]$. The estimates of the autocovariance function are then altered to become

$$\hat{R}(k) = (T - k)^{-1} \sum_{t=1}^{T-k} \{[y(t) - \bar{y}] \cdot [y(t + k) - \bar{y}]\},$$

for any lag $k = 0, 1, 2, \ldots, (T - 1)$. The distributional properties of the $\hat{R}(k)$ remain concealed for most time series $\{Y(t)\}$, yet for $T - k$ sufficiently large and for a well-behaved autocovariance function $R(t)$ one may expect that $E(\hat{R}(k)) \cong E[\tilde{R}(k)] = R(k)$.

The variances of the estimates $\hat{R}(k)$ are known approximately to be

$$\mathrm{Var}[\hat{R}(k)] \cong (T - k)^{-1} \sum_{r=-\infty}^{\infty} [R^2(r) + R(r + k) \cdot R(r - k)].$$

More generally, since $\hat{R}(k)$ and $\hat{R}(l)$ constitute a bivariate random vector, an approximate expression for their covariance may be obtained as

$$\mathrm{Cov}[\hat{R}(k), \hat{R}(l)] \cong T^{-1} \sum_{r=-\infty}^{\infty} [R(r) \cdot R(r + l - k) + R(r + l) \cdot R(r - k)].$$

Further details regarding these approximations may be found in the source book by Jenkins and Watts (1968, Chapter 6).

The autocovariance function $R(k)$ for a stationary time series $\{Y(t)\}$ has units of measure in terms of the square of the units of measure of the random variables $Y(t)$. Consequently, if one seeks to compare autocovariance functions of two stochastic processes, each defined on a different scale of measurement, an inherent difficulty exists. This difficulty may be somewhat overcome by employing instead the dimensionless *autocorrelation function* of lag k:

$$\begin{aligned} \varrho(k) &= R(k)/R(0) \\ &= \mathrm{Cov}[Y(t), Y(t + k)]/\mathrm{Var}[Y(t)], \end{aligned}$$

the correlation coefficient between random variables $Y(t)$ and $Y(t+k)$ separated by a time lag k.

Statistical estimates of the autocorrelation function are readily suggested by

$$\tilde{\varrho}(k) \equiv \tilde{R}(k)/\tilde{R}(0),$$

whenever the mean μ of the time series may be presumed known, or by

$$\hat{\varrho}(k) \equiv \hat{R}(k)/\hat{R}(0), \qquad \text{otherwise.}$$

The plot of such estimates, as a function of k, is called a *correlogram*. However, in neither case are the statistical properties of these estimates known generally. For example, their being biased estimates of $\varrho(k)$ would likely be a safe presumption in most applications.

In the analysis of the encounters arising from a simulation model, it is infrequent that one will need to compare simular responses of differing units of measure. For example, the productivity of a simulated delivery system would likely be expressed in the same units in the responses of each of two encounters with the model, even though those encounters be defined by different environmental (i.e., input) conditions. Therefore, the use of estimates of the autocovariance function, as opposed to the correlogram, will usually suffice in most applications arising from simulation experiments.

10.4.3. Effects of Trend Elimination

We have noted the utility of the estimated autocovariances: (a) to provide estimates of $R(k)$, so that the precision of estimators such as $\hat{\mu}$ may be somewhat assigned; and, (b) to estimate visually, via a plot of the correlogram, the nature of the behavior of the actual covariance (or autocorrelation) function of the time series $\{Y(t)\}$. Recalling that the sample record $[y(t), t = 1, 2, \ldots, T]$ may not be that actually obtained from the simular response, but rather may be the simular record treated for the removal of trends, denoted more appropriately by $[y^*(t), t = 1, 2, \ldots, T]$, one might well ask of the effect that the trend elimination techniques may have produced in the resulting autocovariance estimators.

It will be presumed initially that a simular record of length T, $[y(t), t = 1, 2, \ldots, T]$, is a finite portion of a realization from a time series $\{Y(t)\}$, which is stationary. However, let it be presumed that, despite the assumed stationarity, the sample record has been treated for elimination of a linear trend by use of the difference operator:

$$\Delta y(t) = y(t+1) - y(t) \equiv y^*(t), \qquad t = 1, 2, \ldots, (T-1).$$

[Here, for convenience, $(\Delta t) = 1$.] The resulting sample record $[y^*(t), t = 1, 2, \ldots, (T-1)]$, of length $(T-1)$, is then used to estimate the autocovariance function:

$$\hat{R}^*(k) = (T-1-k)^{-1} \sum_{t=1}^{T-1-k} \{[y^*(t) - \bar{y}^*] \cdot [y^*(t+k) - \bar{y}^*]\},$$

for $k = 0, 1, \ldots, (T-1)$, where

$$\bar{y}^* = (T-1)^{-1} \sum_{t=1}^{T-1} y^*(t).$$

In actuality,

$$\bar{y}^* = (T-1)^{-1} \sum_{t=1}^{T-1} [y(t+1) - y(t)] = (T-1)^{-1} \cdot [y(T) - y(0)],$$

so that, for T adequately large, $\bar{y}^*$ will be approximately zero. Moreover, substitution of $y(t+1) - y(t)$ for $y^*(t)$ in the expression for $\hat{R}^*(t)$ yields

$$\hat{R}^*(k) \cong (T-1-k)^{-1} \sum_{t=1}^{T-1-k} [y(t+1) - y(t)] \cdot [y(t+k+1) - y(t+k)],$$

and, with $\bar{y} = T^{-1} \sum_{t=1}^{T} y(t)$,

$$\begin{aligned} \hat{R}^*(k) &\cong (T-1-k)^{-1} \sum_{t=1}^{T-1-k} \{[y(t+1) - \bar{y} - (y(t) - \bar{y})] \\ &\qquad \cdot [(y(t+k+1) - \bar{y}) - (y(t+k) - \bar{y})]\} \\ &\cong \hat{R}(k) - \hat{R}(k-1) - \hat{R}(k+1) + \hat{R}(k), \end{aligned}$$

where the $\hat{R}(k)$ are the estimates of the autocovariance function $R(k)$ of the process $\{Y(t)\}$ and are those estimates that would have been obtained had the original simular record been used instead of the trend-fitted record $[y^*(t)]$. Thus, the autocovariance estimates for the incremental process $\{\Delta Y(t)\}$ become

$$\hat{R}^*(k) \cong -[\hat{R}(k-1) - 2\hat{R}(k) + \hat{R}(k+1)]$$

or

$$\hat{R}^*(k) \cong -\Delta^2 \hat{R}(k-1);$$

i.e., the estimates of the autocovariance function for the record $[y^*(t)]$ obtained from the first differences of the simular record $[y(t)]$ are essentially equal to the second differences in the autocovariance estimators that would have arisen from use of the original simular record itself.

Similarly, if the original simular record $[y(t),\ t = 1, \ldots, T]$ is differenced twice in an effort to eliminate both quadratic and linear effects, the resulting record, $[y^{**}(t),\ t = 1, 2, \ldots, (T-2)]$, where

$$y^{**}(t) = y(t+2) - 2y(t+1) + y(t), \qquad t = 1, 2, \ldots, (T-2),$$

can be used to estimate the autocovariance function; namely

$$\hat{R}^{**}(k) = (T-2-k)^{-1} \sum_{t=1}^{T-2-k} \{[y^{**}(t) - \bar{y}^{**}] \cdot [y^{**}(t+k) - \bar{y}^{**}]\},$$

where

$$\bar{y}^{**} = (T-2)^{-1} \sum_{t=1}^{T-2} y^{**}(t), \qquad k = 1, 2, \ldots, (T-2).$$

Noting that

$$\bar{y}^{**} = (T-2)^{-1}[y(0) - y(1) - y(T-1) + y(T)],$$

then, for T sufficiently large $\bar{y}^{**} \cong 0$. Proceeding as for $\hat{R}^*(k)$, one obtains the relationship

$$\hat{R}^{**}(k) \cong \hat{R}(k-2) - 4\hat{R}(k-1) + 6\hat{R}(k) - 4\hat{R}(k+1) + \hat{R}(k+2),$$

where the approximation is quite close if $\bar{y}^{**}$ is quite nearly null. One should note then that

$$\hat{R}^{**}(k) \cong \Delta^4 \hat{R}(k-2);$$

i.e., the estimates from the simular record $[Y^{**}(t)]$ itself formed by second differences of the original simular record, can be computed from the fourth differences of the estimated autocovariances that would have arisen had the original series been used directly.

Thus, by computing the estimates $\hat{R}(k)$ of the autocovariance directly from the simular record $[y(t),\ t = 1, 2, \ldots, T]$, one may readily obtain the estimates $\hat{R}^*(k)$ as the negative of the second differences of the $\hat{R}(k)$, without necessitating the computation of $[y^*(t),\ t = 1, 2, \ldots, T]$ and the calculation of the $\hat{R}^*(k)$ therefrom. Moreover, the estimates of $\hat{R}^{**}(k)$ are then available as the negative of the second differences of $\hat{R}^*(k)$, thereby being the fourth differences of the $\hat{R}(k)$:

$$\hat{R}^{**}(k) = -\Delta^2\hat{R}^*(k-1) = +\Delta^2[\Delta^2\hat{R}(k-2)] = \Delta^4\hat{R}(k-2).$$

One notes then that trend elimination, via the technique of finite differencing, results in a time series whose autocovariance function is a weighted average of the autocovariance function (shifted), as would be estimated for the original simular record. The attempt at trend elimination causes, then, a smoothing or averaging of the autocovariance function estimates.

By plotting these successively smoothed correlograms, one may be able to ascertain the mechanism generating the simular time series; for, comparisons of the correlograms with the theoretical autocovariance (or autocorrelation) functions of specific time-series models (e.g., autoregressive, moving average) may allow one to induce a "model" for the underlying behavior of the simular response.

10.5. Spectral Analysis

We see then that the correlogram, or, equivalently, the sample autocovariance function, can be employed with some degree of success in the analysis of the dynamics of a simular response. However, the categorization of a time series by means of the correlogram is somewhat perplexing in that two or more series, essentially different in that their mechanisms of generation differ, may have remarkably similar autocovariance functions. The reader may recall that the autocovariance function for the following two time series can be quite alike in appearance; namely, a damped oscillatory behavior:

$$X(t) = aX(t-1) + \varepsilon(t),$$

where $-1 < a < 0$ and $\{\varepsilon(t)\}$ is a purely random process; and

$$Z(t) = aZ(t-1) + bZ(t-2) + \varepsilon(t),$$

where $-\frac{1}{2} < a, b < \frac{1}{2}$ and $\{\varepsilon(t)\}$ is a purely random process. Therefore, when one takes into account the sampling variations arising in the correlogram, care must be exercised in forming any inference from them regarding the apparent nature of a simulation model.

Of course, if the nature of a simular response can be anticipated, in the sense that its autocovariance function has a predictable behavior, then the correlogram may prove especially useful. Especially in the verification stage of model development, then, can the correlogram be of value. Subroutines or modules of a simulation model can be tested by supplying to them input processes of known properties (usually white noise) and by analyzing the output processes, often by comparing visually the computed correlogram with the anticipated autocovariance function.

Furthermore, in some simulation models, such as those requiring the generation of correlated sequences of random variates, use of the correlogram may prove valuable for testing the random variate sequence itself.

The reader may recall that the *spectral density function* for a stochastic process $\{Y(t)\}$ is the Fourier transform of its autocorrelation function:

$$f(\omega) = (\pi)^{-1} \sum_{k=-\infty}^{\infty} \varrho(k) e^{-i\omega k},$$

or

$$f(\omega) = (\pi)^{-1} \sum_{k=-\infty}^{\infty} \varrho(k) \cos \omega k \qquad \text{for} \quad \omega \in (0, \pi),$$

since $e^{-i\omega k} = \cos \omega k - i \cdot \sin \omega k$, and since the autocorrelation function is an even function:

$$\varrho(-k) = \varrho(k) \qquad \text{for all} \quad k.$$

Therefore, the spectral density function may be written

$$f(\omega) = (\pi)^{-1} R^{-1}(0) \sum_{k=-\infty}^{\infty} R(k) \cos \omega k,$$

since $\varrho(k) = R(k)/R(0)$, leading one to define the *spectrum* $I(\omega)$ of the stochastic process as

$$I(\omega) = R(0) f(\omega) = \pi^{-1} \left\{ \sigma^2 + 2 \sum_{k=1}^{\infty} R(k) \cos \omega k \right\}, \qquad \omega \in (0, \pi).$$

Due to the one-to-one correspondence between autocovariance functions $R(k)$ and spectra $I(\omega)$ one might suspect that any attempt to analyze the behavior of a simular response via computations of a sample spectrum from the simular record $[y(t), t = 1, 2, \ldots, T]$ would encounter the same difficulty as was experienced in employing the correlogram; namely, the similarity of the behavior of spectra arising from stochastic processes of differing characters. Nonetheless, many instances will arise in the simulation context for the employment of spectral analyses. That is, in the verification stage of model development, the sample spectrum may be computed and examined for its compatibility with that theoretical spectrum anticipated in the event that the model had been indeed properly structured. Additionally, the sample spectrum may be computed as a test for the acceptability of generated chains of correlated (or, for that matter, uncorrelated) random variables; the presence of undesirable, yet hidden, periodicities in the chain of variates might thus be disclosed (see Coveyou and MacPherson, 1967).

10.5.1. Spectral Estimation

If one presumes that the trend-free simular responses, under the set of random number seeds, constitute a stationary stochastic process $\{Y(t)\}$ of known mean μ one may generate the simular record of length T from a given encounter with the model by subtracting the known μ from the trend-removed record, to produce the record which one may then denote by

$$[y(t),\ t = 1, 2, \ldots, T].$$

From this trend-free, mean-removed record, one may compute, as an estimate of the spectrum at any frequency ω in the interval $(0, \pi)$, the Schuster (1898) periodogram:

$$\tilde{I}(\omega) = (\pi T)^{-1}\left\{\left[\sum_{t=1}^{T} y(t) \cos \omega t\right]^2 + \left[\sum_{t=1}^{T} y(t) \sin \omega t\right]^2\right\}.$$

Strictly speaking, $\tilde{I}(\omega)$ should be computed for frequencies $\omega = \omega_l = 2\pi l/T$, for $l = 1, 2, \ldots, [[T/2]]$, where $[[a]]$ is the greatest integer less than or equal to a, whence each ω_l lies in the interval $(0, \pi)$.

One may note that

$$\tilde{I}(\omega) = (\pi T)^{-1}\left\{\sum_{t=1}^{T} y^2(t)[\cos^2 \omega t + \sin^2 \omega t] + 2 \sum_{t=1}^{T-1} \sum_{s=t+1}^{T} y(t)y(s)[\cos \omega s \cos \omega t + \sin \omega t \sin \omega s]\right\}$$

or, equivalently, the Schuster periodogram becomes

$$\tilde{I}(\omega) = (\pi T)^{-1}\left\{\sum_{t=1}^{T} y^2(t) + 2 \sum_{t=1}^{T-1} \sum_{u=1}^{T-t} y(t)y(t+u) \cos \omega u\right\},$$

since

$$\cos \omega u \equiv \cos \omega(s - t) = \cos \omega t \cos \omega s + \sin \omega s \sin \omega t.$$

Recalling then the autocovariance estimators (μ is presumed to be known, or equivalently, null):

$$\tilde{R}(k) = (T-k)^{-1} \sum_{t=1}^{T-k} y(t)y(t+k), \qquad \text{for} \quad k = 0, 1, \ldots, (T-1),$$

we may define alternative estimators of the form

$$\begin{aligned} r(k) &= (T-k)\tilde{R}(k)/T \\ &= T^{-1} \sum_{t=1}^{T-k} y(t)y(t+k), \end{aligned}$$

so that the periodogram becomes

$$\tilde{I}(\omega) = (\pi)^{-1}\left\{r(0) + 2 \sum_{k=1}^{T-1} r(k) \cos \omega k\right\},$$

which presents itself in this form as a truncated expression for the spectrum $I(\omega)$, with, however the $R(k)$, as required therein, here replaced by their biased estimates $r(k)$, $k = 0, 1, 2, \ldots, (T-1)$.

In general, then, one would not anticipate that the statistic $\tilde{I}(\omega)$ supply an unbiased estimate of $I(\omega)$. Indeed, since $E[\tilde{R}(k)] = R(k)$ for each $k = 0, 1, 2, \ldots, (T-1)$, we see that the estimates $r(k)$ have expectation

$$E[r(k)] = [1 - (k/T)]R(k)$$

which are each biased downward, save for $r(0)$; each, however, is asymptotically unbiased, at least so for $k \ll T$. Consequently, the expected value for the Schuster periodgram becomes

$$E[\tilde{I}(\omega)] = (\pi)^{-1}\left\{R(0) + 2 \sum_{k=1}^{T-1} [1 - (k/T)]R(k) \cos \omega k\right\}$$

which is biased for $I(\omega)$ unless

$$-(T)^{-1} \sum_{k=1}^{(T-1)} kR(k) \cos \omega k = \sum_{k=T}^{\infty} R(k) \cos \omega k.$$

One case in which this condition holds is that of the autocorrelation function $R_\varepsilon(k)$ for a purely random process $\{\varepsilon(t)\}$; for, in this instance, the random variables $\varepsilon(t)$ are uncorrelated, with

$$R_\varepsilon(k) = \begin{cases} R_\varepsilon(0) = \sigma_\varepsilon^2, & k = 0, \\ 0, & k \neq 0. \end{cases}$$

In this event, then,

$$E[\tilde{I}_\varepsilon(\omega)] = \pi^{-1}R_\varepsilon(0) = \sigma_\varepsilon^2/\pi, \qquad \text{for} \quad \omega \in (0, \pi),$$

which is identically $I_\varepsilon(\omega)$, the spectrum of a purely random process $\{\varepsilon(t)\}$ of random variables $\varepsilon(t)$ having $\mathrm{Var}\{\varepsilon(t)\} = \sigma_\varepsilon^2$. In general, however, one may not expect that the Schuster periodogram be an unbiased estimate of the spectrum $I(\omega)$.

In recent years, a number of alternative spectral estimators have been proposed. The reader is referred to the excellent review article by Jenkins (1961) for a discussion of these estimators, as well as the subsequent article by Parzen (1961) for a mathematical comparison of the alternative estimators and for a good bibliography on the topic of spectral analysis.

In essence, these alternative estimators of the spectrum take the form:

$$I_\lambda^*(\omega) = (\pi)^{-1}\left\{r(0) + \sum_{k=1}^{T-1} \lambda_k r(k) \cos \omega k\right\},$$

where the coefficients λ_k (or *lag windows*, as they are often called) are chosen to improve the statistical properties of the estimator with respect to the spectrum at hand. Particularly bothersome is the fact that the Schuster periodogram, corresponding to the spectral estimator having lag window,

$$\lambda_k = \begin{cases} 1, & k = 0, 1, 2, \ldots, (T-1), \\ 0, & k \geq T, \end{cases}$$

is not even a consistent estimate of the spectrum; i.e., as T becomes infinite (equivalently, as the simular encounter is lengthened), the variance of $\tilde{I}(\omega)$ does not tend to zero, so that increasing the record length and computing additional covariance estimates $\tilde{R}(k)$ does not imply a more precise estimate of the spectrum. With alternative specifications of the lag window λ_k, however, this malignancy may be overcome and consistent estimates defined.

10.5.2. Summary

Thus, one is able to compute estimates of the spectrum $I(\omega)$, associated with a stochastic process $\{Y(t)\}$ of simulation responses by employing a simular record $[y(t), t = 1, 2, \ldots, T]$ from an encounter with the model. One should note that the assumption has been made, however, that the mean μ of the stochastic process is known and, after subtraction, assumed to be zero. An alternative approach, and one that shall of necessity be invoked on many occasions, is the definition of a spectral estimator

$$\hat{I}(\omega) = \pi^{-1}\left\{\hat{r}(0) + 2 \sum_{k=1}^{T-1} \lambda_k \hat{r}(k) \cos \omega k\right\},$$

where the autocovariance estimators are given by

$$\hat{r}(k) = (T)^{-1} \sum_{t=1}^{T} \{[y(t) - \bar{y}] \cdot [y(t+k) - \bar{y}]\},$$

with

$$\bar{y} = T^{-1} \sum_{t=1}^{T} y(t)$$

used as an estimate of μ. Little is known of the effects upon the statistical properties of $\hat{I}(\omega)$ as a result of this most reasonable modification. Again it may be reiterated that, if the length of the simulation record is sufficiently great and if the autocovariance function of the process $\{Y(t)\}$ is adequately well behaved, little damage will probably result from the use of $\bar{y}$ (instead of μ) in the estimation of the spectrum.

Thus, the estimation of the spectrum of a stochastic process from a simular record is rather straightforward, requiring the computation of a sum of products of three factors:

(*a*) the autocovariance estimators, either $\hat{r}(k)$ or $r(k)$, depending whether or not μ is to be estimated by $\bar{y}$;
(*b*) the lag windows, or weights λ_k;
(*c*) the cosine weights, $\cos \omega k$ corresponding to the frequency ω at which the spectral ordinate is to be estimated.

One should note the earlier comment regarding the selection of the frequencies ω of the form:

$$\omega_l = 2\pi l/T, \qquad l = 1, 2, \ldots, [[T/2]].$$

To do so means that one has estimates of the spectrum at equally spaced points in the frequency interval $(0, \pi)$. Furthermore, under certain conditions, the spectral estimates $\tilde{I}(\omega_1), \tilde{I}(\omega_2), \ldots, \tilde{I}(\omega_m)$, for $m = [[T/2]]$, the greatest integer less than or equal to $T/2$, are random variables which are statistically independent.

Specifically, if the time series $\{Y(t)\}$ constitutes *normal white noise* (i.e., is an ensemble of normally distributed, independent random variables of mean zero and common variance σ^2), then the Schuster periodogram values $\tilde{I}(\omega_1)$, $\tilde{I}(\omega_2),\ldots$, $\tilde{I}(\omega_m)$ are independent, Chi-squared distributed random variables, each having two degrees of freedom (unless T is even, in which case all have two degrees of freedom, save $\tilde{I}(\pi)$ with only 1 degree of freedom). Though it will seldom be the case that the response of a simulation model consists of normal white noise, one should note that the Schuster periodogram estimate is proportional to the sum of two squares, each square being a sine- (or cosine-) weighted sum of the simular responses. Thus, one might anticipate that each resulting sum, before being squared, constitutes an approximately normally distributed random variable. If the autocovariance function $R(k)$ for the process $\{Y(t)\}$ dampens to zero sufficiently rapidly, one might well anticipate the relative independence of the estimators $\tilde{I}(\omega_1)$, $\tilde{I}(\omega_2)$, $\ldots$, $\tilde{I}(\omega_m)$, in which case the Chi-squared distributional properties of the estimates could be expected to be reasonably valid.

In conclusion, one should note that the general spectral estimators $I_\lambda^*(\omega_l)$, when computed for $\omega_l = 2\pi l/T$, $l = 1, 2,\ldots, m = [[T/2]]$, are essentially spectral averages that represent a weighted smoothing of the spectrum over a small range of frequency centered about ω_l. The lag windows λ_k, by means of which the estimators $I_\lambda^*(\omega_l)$ are distinguished, are a determining factor in the method by which the spectrum is averaged, or smoothed, over this small frequency range. A thorough discussion of this phenomenon would take us far beyond the scope of this work, but the interested reader is referred to the aforementioned papers of Jenkins (1961) and Parzen (1961), as well as to the reference work of Jenkins and Watts (1968).

10.6. Tests for Random Number Chains

The finite sequence of random (or pseudorandom) numbers to be used in a simulation encounter may be considered as a sample record $[u(s),\ s = 1, 2,\ldots, N]$ of a stochastic process $\{U(s)\}$, where the index s merely denotes the position of $u(s)$ in the sample record of length N. One usually assumes that the technique, by which these presumably uniformly distributed random numbers have been generated, produces uncorrelated sequences; i.e., the theoretical autocovariance function for the stochastic process $\{U(s)\}$ is

$$R_U(k) = \begin{cases} \mathrm{Var}(U(s)) = \frac{1}{12}, & k = 0 \\ 0, & k = 1, 2,\ldots \end{cases}$$

Of course, if an algorithmic generator is being used on a finite-state machine, the chain of pseudorandom numbers must have a period (of length, say, C), so that the autocorrelation function has the property:

$$\varrho_U(C) = 1 = \varrho_U(2C) = \varrho_U(3C) = \cdots$$

Ignoring this anomaly, however, the spectrum for the time series $\{U(s)\}$ becomes:

$$I_U(\omega) = \pi^{-1}\left\{R_U(0) + 2 \cdot \sum_{k=1}^{\infty} R_U(k) \cos \omega k\right\} = \pi^{-1}R_U(0),$$

for $0 \leq \omega \leq \pi$; i.e., the spectrum is constant for all frequencies ω in the interval $(0, \pi)$. In consonance with the decomposability of pure white light into contributions of equal magnitude from all light frequency sources, any generation technique which yields a sequence of uncorrelated random variables from the same probability density function is referred to as a *white-noise generator*.

10.6.1. Anderson's Test for White Noise

Tests for the whiteness of a random number generator then serve as tests for the randomness of the sequence $[u(s), s = 1, 2, \ldots, N]$. There are two categories of such tests: those that use a test statistic based directly on sample autocorrelation functions and those that use a test based on the sample spectrum.

In the first case, the following test is proposed. The sample autocorrelation function is computed:

$$\hat{\varrho}_U(k) = \hat{R}_U(k)/\hat{R}_U(0),$$

where

$$\hat{R}_U(k) = (N - k)^{-1} \cdot \sum_{s=1}^{N-k} \{[u(s) - \bar{u}] \cdot [u(s + k) - \bar{u}]\},$$

for $k = 0, 1, 2, \ldots, N - 1$, with

$$\bar{u} = N^{-1} \sum_{s=1}^{N} u(s).$$

[Usually, not *all* $(N - 1)$ autocorrelations are computed, but rather $m \ll (N - 1)$ of them.] By presuming that $\{U(s)\}$ is indeed a white-noise process, the autocorrelation estimates themselves are essentially uncorrelated with one another. Indeed, if $E(U(s))$ is known [presumably $E(U(s)) = \frac{1}{2}$] and need not be estimated by $\bar{u}$, then the autocorrelation estimates

$$\tilde{\varrho}_U(k) = \tilde{R}_U(k)/\tilde{R}_U(0),$$

where

$$\tilde{R}_U(k) = (N - k)^{-1} \cdot \sum_{s=1}^{N-k} \{[u(s) - \tfrac{1}{2}][u(s + k) - \tfrac{1}{2}]\},$$

for $k = 0, 1, 2, \ldots, (N - 1)$, are themselves uncorrelated. In addition, the distribution of the autocorrelation estimators [both $\hat{\varrho}_U(k)$ and $\tilde{\varrho}_U(k)$] is approximately normal for N large, the mean being 0, the variance $1/N$, as was shown by Anderson (1942).

Thus, one may compute the (say, 95%) confidence interval for the mean of random variables arising from a normal distribution of mean 0, variance $1/N$; viz.,

$$P[\tilde{\varrho} - 1.96/\sqrt{N} \leq \varrho \leq \tilde{\varrho} + 1.96/\sqrt{N}\,] \cong 0.95,$$

or

$$P[\hat{\varrho} - 1.96/\sqrt{N} \leq \varrho \leq \hat{\varrho} + 1.96/\sqrt{N}\,] \cong 0.95.$$

Since 95% of all the (m) autocorrelation confidence intervals should include zero, one may count the number of computed estimates that fail to do so. A comparison with tabulations of the binomial probability distribution (for which $p = 0.05$, $n = m$) provides an approximate measure of the likelihood of having observed so many (or more) failures. If this probability is significantly small (say, less than 0.05 or 0.01), one can fairly safely assume that the generation technique is not producing white noise (i.e., uncorrelated random variables).

10.6.2. A Test Employing the Integrated Spectrum

Another test for white noise, suggested by Jenkins and Watts (1968), employs the spectral estimators:

$$\hat{I}(\omega_l) = (\pi)^{-1}\left\{r(0) + 2 \cdot \sum_{k=1}^{N-1} r(k) \cos \omega_l k\right\},$$

for $\omega_l = 2\pi l/N$, $l = 1, 2, \ldots, m$, for $m = [[N/2]]$, the greatest integer less than or equal to $N/2$. Since the spectrum $I(\omega)$ for the white noise process is constant, its cumulative (or integrated) spectrum is given by:

$$\mathscr{I}(\omega) \equiv \int_0^\omega I(\omega)\, d\omega, \qquad \text{for} \quad \omega > 0.$$

Thus,

$$\mathscr{I}(\omega_l) = \pi^{-1} \int_0^{\omega_l} \sigma^2\, d\omega = \pi^{-1} \cdot \sigma^2 \cdot \omega_l, \qquad l = 1, 2, \ldots, m,$$

where $\sigma^2 = \text{Var}(U(s)) = R_U(0)$.

One can similarly accumulate the sample spectrum as

$$\tilde{\mathscr{I}}(\omega) \equiv \sum_{l:\omega_l \le \omega} \tilde{I}(\omega_l),$$

so that the sequences

$$\mathscr{I}(\omega_l), \qquad l = 1, 2, \ldots, m,$$

and

$$\tilde{\mathscr{I}}(\omega_l), \qquad l = 1, 2, \ldots, m,$$

can be compared. In essence, since $\mathscr{I}(\omega)$ is a linear function of ω for the white-noise process, one is testing the sequence $\mathscr{I}(\omega_l)$ for its departure from linearity.

Alternatively, one might consider the sequence $\tilde{\mathscr{I}}(\omega_l)/\tilde{R}(0)$ as though it were a cumulative distribution function; the sequence $\tilde{\mathscr{I}}(\omega_l)/\tilde{R}(0)$ then corresponds to a sample cumulative distribution function, so that departures of the sample sequence from the sequence $\mathscr{I}(\omega_l)/\tilde{R}(0)$ may be tested by means of the Kolmogorov–Smirnov test statistic, for the application of which the reader is referred to Section 2.9.4.

10.6.3. A Summary

One sees then that tests for white noise can use either the sample autocorrelation function or the sample spectrum. These tests are of prime importance in the testing of random (or pseudorandom) number chains. In addition, they are especially applicable to the determination of the adequacy of a presumed time series representation for a response.

For example, if experience with a simulation model (or with the actual system itself) had always indicated the appropriateness of the first-order autoregressive scheme, for a constant,

$$[Y(t) - \mu] = a[Y(t-1) - \mu] + \varepsilon(t),$$

where $\{\varepsilon(t)\}$ is a white-noise process (i.e., an ensemble of uncorrelated random variables of mean zero and common variance, σ_ε^2), and where μ is the mean of the stochastic process $\{Y(t)\}$. For a given simular record $[y(t),\ t = 1, 2, \ldots, T]$, the record

$$e(t) = y(t) - a \cdot y(t-1), \qquad t = 2, 3, \ldots, (T-1),$$

can then be computed and used as if it were a sample record from a realization of the $\{\varepsilon(t)\}$ process. The tests for the whiteness of the record $[e(t)]$ would then correspond to tests for the adequacy of the given autoregressive model.

Such tests could also be applied to higher-order autoregressive schemes. Of course, the parameters in these schemes must either be known or estimated in order for the test statistics to be computed. The reader is referred to Jenkins and Watts (1968, Chapter 5) for the estimates of the parameters of such linear processes.

10.7. Comparison of Time Series from Simular Encounters

In many simulation experiments, it will be of primary concern to observe the dynamic behavior of two or more simular responses as simular time evolves. For example, in the simulation of a production system, the productivity $\{Y(t)\}$ of the simular system constitutes a stochastic process, a realization of which may be observed from any encounter with the model. This productivity is governed somewhat by the accessibility of certain personnel and/or pieces of equipment, which may themselves be constrained in a random fashion either by exogeneous events (such as weather conditions or work schedules) or by other endogeneous events inherent in the model's structure (such as queues or limited resources). Thus, a second variate of importance in the simulation of a productivity system may likely be the utilization experience $\{X(t)\}$ of these personnel or resources.

Each of the variates, productivity $Y(t)$ and utilization $X(t)$, is usually expressed as a cumulative quantity, being, respectively, the total goods produced and the total hours of use up to simular time t. Therefore, neither of the stochastic processes $\{X(t)\}$ or $\{Y(t)\}$ could be expected to be stationary, since a positive trend would be anticipated in each. By the use of the finite first differences, as discussed previously, the resulting incremental processes $\{\Delta X(t)\}$ and $\{\Delta Y(t)\}$ would, in many instances, appear stationary, since these would represent, respectively, equipment utilization and system productivity *rates*.

In general, one would likely proceed to an inquiry of the nature of the relationship between the $\{X(t)\}$ and $\{Y(t)\}$ processes, or between the $\{\Delta X(t)\}$ and $\{\Delta Y(t)\}$ processes. Such an inquiry would appropriately be conducted by the definition of a simulation encounter and the subsequent collection of the sample records of responses:

$$[x(t),\ t = 1, 2, \ldots, T], \qquad \text{and} \qquad [y(t),\ t = 1, 2, \ldots, T].$$

For the remainder of this discussion, it will be presumed that each of these records is a recording of a realization from a weakly stationary stochastic process; e.g., trend effects have been removed, if necessary.

The enquiry into the relationship between the two processes will likely be directed so as to establish:

(*a*) the existence of a time lag between the effect of one series on the other;
(*b*) the degree of correlation between the two series; and,
(*c*) the possibility of a functional relationship that may adequately describe the connection between the two series.

10.7.1. Estimation of the Cross-Covariance Function

The cross-covariance functions between two time series, or stochastic processes $\{X(t)\}$ and $\{Y(t)\}$ or, equivalently, in the bivariate time series $\{X(t),\ Y(t),\ t = 1, 2, \ldots\}$, are expressed as

$$\gamma_{XY}(k) \equiv E\{[X(t) - \mu_X] \cdot [Y(t+k) - \mu_Y]\},$$

and

$$\gamma_{YX}(k) \equiv E\{[Y(t) - \mu_Y] \cdot [X(t+k) - \mu_X]\},$$

denominated the cross-covariance functions of lag k with $Y(t)$ leading $X(t)$ and then with $X(t)$ leading $Y(t)$, respectively. One may observe that $\gamma_{XY}(k)$ is the covariance between the random variables $X(t)$ and $Y(t+k)$, whereas $\gamma_{YX}(k)$ is the covariance between $Y(t)$ and $X(t+k)$. Therefore, one has the relationships

$$\gamma_{XY}(-k) = \gamma_{YX}(k), \qquad \text{and} \qquad \gamma_{YX}(-k) = \gamma_{XY}(k).$$

Presuming then that the bivariate time series $\{X(t),\ Y(t)\}$ is stationary, with its mean value vector presumed to be zero [i.e., $E(X(t)) =$

$E(Y(t)) = 0$, for all t], one can compute, as estimates of the cross-covariance functions of lag k, the quantities

$$\tilde{\gamma}_{XY}(k) = (T - k)^{-1} \sum_{t=1}^{T-k} [x(t) \cdot y(t + k)]$$

and

$$\tilde{\gamma}_{YX}(k) = (T - k)^{-1} \sum_{t=1}^{T-k} [y(t) \cdot x(t + k)],$$

each unbiassed for $\gamma_{XY}(k)$ and $\gamma_{YX}(k)$, respectively, $k = 0, 1, \ldots, (T - 1)$. One may note that for negative lags, we define

$$\tilde{\gamma}_{XY}(-k) \equiv (T - k)^{-1} \sum_{t=k+1}^{T} [x(t) \cdot y(t - k)] = \tilde{\gamma}_{YX}(k),$$

and

$$\tilde{\gamma}_{YX}(-k) \equiv (T - k)^{-1} \sum_{t=k+1}^{T} [y(t) \cdot x(t - k)] = \tilde{\gamma}_{XY}(k),$$

compatible with the relationships extant for the ensemble covariance properties of negative lag. Computationally, this implies that one need calculate only lagged products of positive lag, since those of negative lag are related in the straightforward manner just described.

For two different lags, k and l, the covariance between the estimators $\tilde{\gamma}_{XY}(k)$ and $\tilde{\gamma}_{XY}(l)$ may be shown to be approximately

$$\operatorname{Cov}[\tilde{\gamma}_{XY}(k), \tilde{\gamma}_{XY}(l)]$$
$$\cong T^{-1} \sum_{s=-\infty}^{\infty} [\gamma_{XX}(s)\gamma_{YY}(s + l - k) + \gamma_{XY}(s + l)\gamma_{YX}(s - k)],$$

where $\gamma_{XX}(k)$ and $\gamma_{YY}(k)$ are the autocovariance functions of lag k for the respective processes $\{X(t)\}$ and $\{Y(t)\}$. Therefore, the variance of the cross-covariance estimators may be approximated by

$$\operatorname{Var}(\tilde{\gamma}_{XY}(k)) \cong T^{-1} \sum_{s=-\infty}^{\infty} [\gamma_{XX}(s)\gamma_{YY}(0) + \gamma_{XY}(s + k)\gamma_{YX}(s - k)],$$

as may be inferred from the discussions surrounding these estimators in the text of Jenkins and Watts (1968).

Of course, whenever the means μ_X and μ_Y are nonzero, yet known, they can be subtracted from each observation in the corresponding simular records, so that the resulting records are those from a bivariate stochastic process having means 0. In the event that μ_X and μ_Y are unknown, and must be estimated by

$$\bar{x} = T^{-1} \sum_{t=1}^{T} x(t)$$

and

$$\bar{y} = T^{-1} \sum_{t=1}^{T} y(t),$$

respectively, then the simular records $[x(t)]$ and $[y(t)]$ may be prepared by subtracting $\bar{x}$ and $\bar{y}$ from the corresponding observations. The resulting cross-covariance estimators become

$$\hat{\gamma}_{XY}(k) = (T-k)^{-1} \sum_{k=1}^{T-k} \{[x(t) - \bar{x}] \cdot [y(t+k) - \bar{y}]\}$$

and analogously

$$\hat{\gamma}_{YX}(k) = (T-k)^{-1} \sum_{k=1}^{T-k} \{[y(t) - \bar{y}] \cdot [x(t+k) - \bar{x}]\}, \qquad k = 0, 1, 2, \ldots, (T-1);$$

these estimators can be expected to be biased for $\gamma_{XY}(k)$ and $\gamma_{YX}(k)$, although, in general, their nonasymptotic statistical properties remain concealed. In a practical sense, one is usually forced to compute the estimators $\hat{\gamma}_{XY}(k)$ and $\hat{\gamma}_{YX}(k)$, hoping that the introduction of $\bar{x}$ and $\bar{y}$ will not introduce undesirable properties for the cross-covariance estimators.

Once the estimators have been computed, they can be compared with the theoretical cross-covariance functions for specific bivariate time series. For this purpose, the dimensionless cross-correlation function

$$\varrho_{XY}(k) = \gamma_{XY}(k)/[\gamma_{XX}(0)\gamma_{YY}(0)]^{1/2}$$

may be estimated by

$$\tilde{\varrho}_{XY}(k) = \tilde{\gamma}_{XY}(k)/[\tilde{R}_X(0)\tilde{R}_Y(0)]^{1/2}, \qquad \text{for } k = 0, 1, \ldots, (T-1)$$

and subsequently compared with the cross-correlation functions of known bivariate serial relationships, such as bivariate autoregressive and bivariate moving average processes. The reader is referred to the discussion by Jenkins and Watts (1968) for a more thorough presentation of these bivariate stochastic processes.

10.7.2. Bivariate Spectra

Additionally, one may desire to compare two simular time series in the frequency domain; i.e., one may wish to examine the pair of realizations in order to determine the compatibility of their periodic components. For this purpose, cross-amplitude and phase-spectral estimators can be computed from the simular records $[x(t)]$ and $[y(t)]$ by means of the Fourier decomposition of the sample cross-covariance function. In fact, the square of the cross-amplitude spectral estimator can be computed as the product of the spectral estimators for the individual time series

$$\tilde{A}^2_{XY}(\omega) = \tilde{I}_X(\omega)\tilde{I}_Y(\omega).$$

Consequently, whenever, the stochastic processes $\{X(t)\}$ and $\{Y(t)\}$ are independent and both normal white-noise processes of mean 0 with respective variances $\sigma_X{}^2$ and $\sigma_Y{}^2$, the random variable

$$\alpha^2(\omega_l) = 4\tilde{A}^2_{XY}(\omega_l)/(\sigma_X{}^2\sigma_Y{}^2), \qquad \text{for} \quad \omega_l = 2\pi l, \;\; l = 1, 2, \ldots, [[T/2]]$$

is the product of two independent Chi-Squared random variables, so that the distribution of $\alpha^2(\omega_l)$ is a member of the family described by Malik (1968). Furthermore, the natural logarithm of $\alpha^2(\omega_l)$ has probability density function given by the convolution of densities of the form discussed by Mihram (1965).

The utility of the cross-amplitude spectral estimator becomes apparent whenever one recalls that the cross amplitude spectrum reveals whether the periodic contributors at frequency ω in the time series $\{X(t)\}$ are associated with great or small amplitudes at the same frequency in the series $\{Y(t)\}$. The reader should also note the remaining bivariate spectra: the phase spectrum, the cospectrum, and the quadrature spectrum, estimates of each of which can be used to determine relationships (in the frequency sense, or domain) between two observed simular records. In general, the statistical properties of these estimates remain concealed, although recently the use of Monte Carlo evaluations has proved exceptionally valuable to their determination. [See Granger and Hughes (1968).] It would appear that certain analyses of simular responses shall be guided by simulation itself!

10.8. Summary

The dynamic, symbolic simulation model of the stochastic variety can be seen as a generator of realizations $[Y(t), t = 1, 2, \ldots]$ from a time series, or stochastic process $\{Y(t)\}$. Consequently, the statistical tools of time-series analysis are applicable to the determination of the dynamic behavior of the simular response. These tools, particularly those of the correlogram and the estimation of the spectrum, are especially suitable for the analyses that should accompany the verification stage of model development, since it is in this phase that one wishes to ascertain whether the programmed behavior of the system provides the dynamic response that was intended. However, the employment of time series methodology need not be restricted to the verification of a model (or of its subroutines), but can also be applied to the analysis of the responses arising from the completed and validated model. These techniques are particularly useful in the analysis of data arising from simulations of input–output systems such as communication channels and networks.

In this latter regard, the use of the statistical estimates of properties of bivariate time series is of especial import, for, via the estimates of the cross amplitude, phase, quadrature, and cospectra, one may infer relationships between a pair of variates, $X(t)$ and $Y(t)$, arising from a simular encounter.

The focal point of these techniques, whether one is applying univariate or bivariate procedures, is the covariance function (auto or cross). The covariance function, or more properly its estimate, is of prime importance in evaluating the accuracy of estimates of the fundamental properties of time series, such as the estimators for the mean value and variance functions, as well as those for the autocovariance function itself.

In this chapter, the importance of stationarity has been noted and reiterated; procedures have also been outlined for the elimination of trend components that are likely to exist in many simular records. The elimination of trend is a necessary step toward the production of a simular record which is stationary, though we have not claimed that it is sufficient for the purpose. The technique of Tintner (1940) has been suggested for the determination of the degree of the polynomial trend extant in a simular record.

Nonetheless, a word of caution is deserving of repetition: that is, the treatment of time series by finite differencing techniques, as a technique for trend removal, implies the averaging of covariance estimators for the original series and, consequently, implies the weighting of (shifted) spectral estimators. Thus, the blind application of trend removal procedures is to be avoided, for their unnecessary application may lead to the subsequent analysis of smoothed, or averaged, covariance and spectral estimators whose analysis should not in the first place have been initiated. The reader is referred to Kendall and Stuart (1966, Vol. 3) for further discussion of the possible effects of trend removal; also, see Mihram (1971).

Despite these cautionary notes, the application of time series analyses to simular records is proving more and more an asset to the builder of the stochastic simulation model. Indeed, one can predict that *ad hoc* simulation languages will become increasingly oriented toward analyses of this type and will probably incorporate routines for the purpose of collecting and analyzing simular time series in accordance with the specifications of the model/language user.

Chapter 11

VISTAS IN SIMULATION

Intelligence ... is the faculty of manufacturing artificial objects, especially tools to make tools.

—HENRI BERGSON

11.1. Perspectives

Chapter 1 of this book presented a number of classification schemes for models. It was noted that the dynamic, stochastic simulation model constitutes the most general symbolic model category. In particular, it was argued that the desire to achieve extensive detail in a model usually requires the need to represent a time-dependent system not only as a dynamic, but also as a stochastic, simulation model. This Uncertainty Principle of Modeling implies then that the dynamic, stochastic simulation becomes a symbolic mimicry of a time series whose statistical properties shall not be generally known in advance.

In order that an investigator, or systems analyst, may attend to the incertitudes extant in nature and may find expression for these in a symbolic model, the lengthy Chapter 2 and the somewhat condensed Chapter 3 have included discussions of the conditions under which a host of random variables and stochastic processes may arise. Thus, concepts such as the transformation of random variables, the Central Limit Theorem, and the Poisson process may each be invoked not only to provide a framework for specifying random behavior in a stochastic simulation, but also to indicate efficient methods by which random variables may be readily generated from virtually any applicable distribution function or stochastic process.

The brief discussion of the Monte Carlo method in Chapter 4 noted the value of statistical sampling experiments in order to ascertain the distributional properties of random variables which arise from experimentation with random or pseudorandom numbers. From the subsequent chapter on modeling and systems analysis, it should become apparent that the dynamic, stochastic simulation model is a generalized Monte Carlo experiment designed to reproduce (in a statistical sense) a time series of essentially unknown (or, at best, partially known) behavior. The borderline between the more classical sampling experiments and the more recently developed simulation models becomes difficult to define precisely. Indeed, a PERT (Program Evaluation and Review Technique) model, of the network of interdependent activities that must be completed in order for the overall project of interest to be considered finished, provides an example of a model which can be manipulated either as a rather static Monte Carlo sampling experiment or as a quite dynamic stochastic simulation model. In the first instance, random activity times can be generated for every activity, then the standard techniques of the Critical Path Analysis (CPA) method invoked in order to ascertain the project completion time (cf. Lockyer, 1967); alternatively, a simular clockworks may be constructed, and the entire project may be simulated, from project initiation to completion. It is this difference in orientation, however, that usually distinguishes between the Monte Carlo sampling experiment and the dynamic, stochastic simulation model.

Since the dynamic, stochastic simulation model is so general, modeling will probably become increasingly concerned with randomness. Model responses will need to be analyzed accordingly. Indeed, as this book has indicated, statistical techniques are directly applicable to the responses arising from a stochastic simulation model: the importance of statistical comparisons, factorial experimental designs, multiple regression analyses, sequential estimation and experimentation, multivariate analyses, response surface methodology, and of time series analysis has been discussed in the preceding five chapters.

In this chapter, note will be taken of several activities currently related to simulation. Indications of the applicability of the simulation methodology to the "solution" of certain biological, environmental, and societal problems will be presented in the next section. The relationship of simulation to gaming will be explored in Section 11.3.

The dynamic, stochastic simulation model can also be viewed from another important perspective. The analog computer, it has been realized for some time, constitutes a parallel processor, in that several physical (or electrical) parameters can be traced simultaneously, with instantaneous feedback available for the purpose of influencing other parameters as time evolves. On the other hand, the electronic digital computer has appeared constrained in this regard, being essentially a sequential processor that handles one matter (i.e., variable) at a time. However, the software-defined simular clock appears to be a compromise allowing for *pseudoparallel processing* by the digital computer; although the digital computer proceeds sequentially from task to task in actual time, the clocking mechanism of the simulation model's executive routine holds simular time constant until all necessary feedback effects have been accounted for. Thus, the dynamic simulation model, when implemented on a digital computer, provides a computational capability not otherwise readily available on such machines.

The use of a simulation model as a pseudoparallel processor has repercussions in a number of intellectual endeavors. Especially pertinent are the current efforts in heuristics, self-organizing systems, and artificial intelligence. These efforts will be discussed in Section 11.4. In conclusion, some indication of the probable evolutions that *ad hoc* simulation languages will undergo will be presented in the closing section.

11.2. The View Toward Applications

The major theses of this book have been expressed in terms of an Uncertainty Principle of Modeling and of the consequent generality of models of the dynamic, stochastic class. A number of our currently pressing biological, environmental, and societal problems are being recognized in their proper perspective as systemic problems; that simulation will play an immensely important role in their dissolution seems readily apparent.

However, these applications are not without their attendant difficulties. In particular, ecological modeling seems fraught with impediments concerning the appropriate time span over which relevant models should mime biological development. One must select among models that are capable of predicting the immediate effects of a pollution discharge into a stream on the aquatic life therein and those that indicate the long-term, essentially evolutionary, effects of continuous resource development on the diversity of species.

That models of biological ecosystems require stochasticity is perhaps, among many, a moot question. Watt (1970) observes that the ecologist will need to become more attuned to the role of probability in our descriptive models of complicated ecological systems. Nonetheless, any desire to incorporate greater detail in such models will inevitably lead to the requirement for stochasticity in these modeling efforts as well. Effective ecological models will, in all probability, take cognizance of the Uncertainty Principle of Modeling. [See Mihram (1972b).]

Another prime conceptual difficulty in the application of simulation methodology to the modeling of biological, environmental, and societal systems is an apparent lack of agreement regarding the goals these systems are, or should be, striving to achieve. One may recall that a model is constructed in an effort to evaluate the modeled system on the basis of its achievement of certain well-defined goals under alternative environmental conditions. Should an ecosystem be maintained forever in its status quo, or does there exist an appropriate, requisite diversity of species that will allow the system to revive itself, even in the face of geological, meterological, or certain astronomical disasters?

Similarly, models of urban and societal systems require carefully defined goals. Forrester (1969) has indicated some of the essential elements in the life cycle of a city, yet this modeling activity apparently was not undertaken in sufficient detail that stochastic effects were incorporated into the resultant model of urban development.

The success with which business, industrial, and military organizations have been simulated may probably be attributed to our ability to discern for these systems their goals and to make these explicit before undertaking the modeling effort. Stochasticity could often be readily incorporated in these models, because frequently data could be accumulated easily from managers at the appropriate hierarchical levels within the pertinent organization. The construction of models of societal and ecological organizations will also require not only well-defined system goals, but also methods by which statistical data can be efficiently accumulated for use within the symbolic simulation models.

Of interest in this regard is the recent proposal of Simulation Councils, Inc. (see McLeod, 1969). A "World Simulation" model is suggested, to consist of a hierarchy of models that can be used to predict rapidly the effects that social, economic, political, and military forces have in their interactions throughout the world. Whether such an effort will be undertaken remains an important vista in simulation. Once begun, the effort will likely need to incorporate models of biological ecosystems as well.

In addition, considerable emphasis will need to be placed on establishing the credibility of the component modules via proper verification and validation procedures.

11.3. A Glance at Gaming Models

Another simulation-related activity that is likely to become increasingly important is gaming. A gaming model can be either static or dynamic, either physical or symbolic, or deterministic or stochastic. Though gaming models may be subjected to the same categorizations as other models, the more usual conception is that of a dynamic simulation model with the additional feature that the model provides decision points at which one or more human beings interject parametrized decisions. The model then continues as a simulation model, and responses from it can often be analyzed to infer the effects of alternative decision choices on the part of the participants.

Thus, a gaming model is a man–machine simulation. The earliest known gaming models were used as military training devices. Officers could select alternative troop dispositions in accordance with a book of rules; judges or umpires could ensure that the rules were observed, and could evaluate the officers' performance. In passing, one might note that chess is a simplified gaming model of this genre.

Evans *et al.* (1967) note that the more modern symbolic gaming model consists of three phases:

(*a*) a *situation*, or map, *phase*, during which the book of rules and a "fact book" (which may include information regarding previous decisions and their effects) are made available to each player;

(*b*) a *decision phase*, during which parametrized strategies and tactical decisions are specified; and,

(*c*) a *computation phase*, during which the competitive situation is simulated, including the possibility of the inclusion of randomly occurring phenomena.

These symbolic gaming models have become especially useful in training programs for young executives. Implicit in the construction of such models is the development of a simulation model of the competitive situation. Though not necessary, these simulation models are usually of the dynamic, stochastic variety; usually, the detailed structure of the model is not made available to the participants, but some information is made available by means of a "fact book" regarding the gaming situation. At the decision stage, participants are permitted to change their competitive stance by altering prices, reallocating advertising funds, selecting alternative personnel policies, etc.

Thus, gaming models have been more recently of a symbolic variety.

Indeed, Shephard (1965) defines a gaming model as "a simulation of competition or conflict in which the opposing players decide which course(s) of action to follow on the basis of their knowledge about their own situation and intentions, and on their (usually incomplete) information about their opponents." According to this definition, one does not refer to any model that permits human interaction as a game, because the simulation analyst who alters environmental specifications in successive encounters of a simulation model would not normally be considered as engaged in a competitive situation with one or more opposing players. Hence, two important distinguishing characteristics of gaming models are:

(*a*) a model, usually of the simular variety, of a competitive situation; and,

(*b*) human interaction with the model.

A third ingredient of most gaming models is an umpire, who judges the quality of the participants' decisions and who ensures that the rules of the game are maintained (see also Meier *et al.*, 1969).

Such models are of considerable pedagogical value. Indeed, simulation models of urban environments are often readily converted to the gaming format; industrialists may participate in the model by specifying the number and location of factories, citizen representatives may specify the number and location of exhaust-plumed freeways required by the public, and medical–pollution teams may endeavor to impose constraints on these activities by setting public health criteria. In this manner, the competitive nature of daily urban living can be more readily exposed, indicating to all citizenry the need for containment of excessive greed on the part of *any* element of the city. Of course, the credibility of such gaming models will depend on the established validity of the game's kernel—its simulation model.

11.4. Insights in Adaptive Models

One of the most exciting developments in simulation is the effort to simulate the human mind. The literature regarding the subject is now quite extensive, as an examination of the bibliography concluding the excellent collection of papers edited by Feigenbaum and Feldman (1963) will reveal. The present inability to have accomplished the ultimate goal, that of a thinking automaton, has been due to a number of factors, among which may be noted:

(1) our currently inadequate knowledge of the mechanism by which the mind, considered as a physical net of cells (neurons), operates;
(2) an incomplete understanding of efficient methods by which large numbers of possible contributors to a mental decision are disregarded as immaterial to the final decision;
(3) an uncertainty regarding the learning mechanism of the human (as well as many an other) animal;
(4) a speculation that the human mind is a parallel, not a sequential, processor of information made available to it; and
(5) a reticence to admit that the human mind may indeed make some decisions on an essentially random basis.

In the first instance, developments in neurophysiology are rapidly overcoming the "knowledge gap" that appears to exist in our understanding of the mechanistic principles by which the network of neural elements produce the phenomenon known as mind. The interested reader is referred to the expositions of Young (1957) and Rosenblueth (1970) for further details regarding these developments; the latter is of special importance for its optimism regarding the eventual development of a mechanistic, physical theory of the mind.

Nonetheless, the immensity of the human neural net will probably preclude its simulation on a neuron-by-neuron basis (cf. von Neumann, 1958). Even with continually improved computational speeds and costs, the foreseeable future holds little promise for a successful, artificially intelligent, machine constructed on this basis.

Instead, much progress has already been made by viewing the human mind as a transform, an unknown function that maps sensory information into decisions or actions. The distinction between actions that are the result of instinctive, reflexive, or cognitive mental processes are not always apparent, though in many instances brain "centers," each important to certain categories of human decision–action, have been isolated. Unfortunately, the hope of parsing the brain into mutually exclusive centers does not appear too promising either (cf. Young, 1957).

One approach to the reduction of the "dimensionality" of the simulation of the human mind has been the use of heuristics, principles or devices which contribute, on the average, to the reduction of search in problem-solving activities whose combinatorial possibilities are barely finite. Heuristics can be contrasted with algorithms, which provide a straightforward procedure, and which guarantee that the solution to a given problem will be isolated. A typical example of an algorithm is that providing the optimal solution to a linear objective function whose terms are subjected to linear contraints (inequalities and/or equalities); i.e., linear programming. In this instance, the examination of all feasible solutions to the optimization problem can be reduced to a combinatorial problem of (usually) immense size, yet the linear programming algorithm provides a straightforward method by which the particular location (i.e., value of the terms of the objective function) of the optimum can be isolated rather readily.

On the other hand, an exemplary heuristic is a suggested procedure which, if followed, would seemingly lead one to an improved situation, yet with no guarantee that the procedure shall lead to *the* optimum. Typical of a heuristic is the procedure suggested in Section 8.5 for the isolation of important contributors to the simular response.

The proposed use of heuristic procedures in the development of artificially intelligent machines implies the need for describing a sequential decision-making rule to the automaton; indeed, with such heuristics, the machine should be capable of adaptation, or learning, since the sequential decision-making rules shall likely employ records or statistical summaries of the past experiences of the automaton. Some of the typical applications in this regard are concerned with chess-playing machines and with pattern-recognizing devices such as those capable of deciphering handwriting (see, e.g., Sayre, 1965).

The third important difficulty to be overcome in successfully simulating the human mind is concerned with our understanding of the mechanism by which the human mind learns. The importance that culture obtains in this process cannot be overlooked, since the human animal has become highly dependent upon the passage from generation to generation of acquired experiences.

In this regard, studies of neural characteristics and of child development seem to reveal the importance of knowledge reinforcement. Thus, the individual (especially the child) who elects certain courses of action and who finds himself punished tends not to elect the same course of action again, unless conditions are discernably or strikingly different. Indeed, much of the human experience in pedagogy indicates that repetition of material (either in actual verbal repetition, or in providing multiple sources, or in reviews or examination procedures) contributes the kind of reinforcement necessary for transmitting culture between (among) generations. Seen in this light, the distinction between prejudice and insight is not always readily discernible.

In the first chapter, the intelligent use of very elementary machines was observed to be of fundamental importance to scientific discovery (i.e., to scientific modeling). The lens, the microscope, the accurate pendulum clock—each permitted the scientist to observe and record phenomena beyond the scope of man's naturally endowed sensory capabilities. Similarly, our modern machines, with all their relative sophistication, prove even more valuable in extending man's knowledge of the world about him.

Nonetheless, it has become apparent that certain of our machines have provided a mechanism for superimposing knowledge reinforcement on large segments of the human populace. The collective state of Nazi Germany's "mind" was a direct consequence of Propaganda Minister (Dr. Josef) Goebbels' repetitive incantations to the populace of his nation (see Grunberger, 1964). That this concern continues to exist in other highly civilized, sophisticated societies has been noted by several in recent years. Notably, McLeod (1969) writes:

> I believe that much of the strife and unrest in the world today is the result of the improper handling of information. . . . Unfortunately, violence and sensationalism dominate the information that gets through. . . . I believe that the trend in world affairs is now such that our very survival might well depend upon our correcting the bias and making proper use of the information available to us [p. vii].

More to the point, Churchman (1968) notes that

> Certain individuals are in control of the mass media made available to the average citizen. These persons are the politicians themselves and the owners of various television, newspaper, and other types of communication companies. Since they control the mass media, these individuals are capable of directing the way in which the public pays attention to issues. The question therefore is whether those who control what is presented to the citizen are themselves well informed and whether they act on the basis of information or rather a desire to be leaders and controllers of public opinion.

The answer to the question posed by Churchman is one for society at large to answer. Its importance in artificial intelligence is directed to the establishment and specification of the input conditions for a simulation of the human mind. The validity of such a model, once developed, will, like all simulation models, rest on its ability to reproduce human decision-making and inferences; i.e., on its ability to mime the responses of a culturally preconditioned mind. The cultural aspects of the human mind cannot be neglected in its successful modeling.

The fourth of the aforementioned difficulties in simulating the human mind deals with the apparent need to mime a parallel processor. That the mind is apparently such a system is possibly revealed best by self-reflection, but the reports of two leading mathematicians, Poincaré (1913) and Hadamard (1945), each describing the seemingly unusual circumstances often surrounding mathematical innovation and invention, serve to corroborate the conclusion. Though we, as human beings, transmit our mathematical findings to one another in a logical, sequential fashion, it is not at all certain that such a mode of operation is employed in the innovative mind. Thus, the simulation of such a mind (or, of this aspect of the mind, generally referred to as the search for general, theorem-proving machines) may require a parallel processor. Fortunately, the concept of a software-defined simulation clockworks may make possible a better mimicry of the mind via the digital computer.

Another aspect of the search for the "thinking machine" is that concerned with the nature of the human decision and inference processes. The importance of the conditioned mind cannot be denied in this regard, especially when one notes that the restoration of sight to an adult with congenital blindness implies an immensely difficult and time-consuming period during which the affected individual essentially *learns* how to see (Young, 1957). Pasteur, in a statement which provides the epigraph for this book, had noted a century ago that not only is a prepared mind of value to research but also fortuitous circumstances are of prime importance in discovery and innovation. A primary difficulty with the modeling of the mind would seem to center about methods by which these conditional probabilities may be entered meaningfully into such a model's structure.

The eventual modeling of the thought process remains, nonetheless, one of the important vistas on the horizon of simulation methodology. The ability of the dynamic, stochastic simulation model to act as a pseudo-parallel processor, to incorporate with facility random phenomena, and to advance mental states by means of quite general algorithms (and/or heuristics) will improve the likelihood of its use as a fundamental mechanism in the artificially intelligent machine.

11.5. Language Developments

A number of the preceding vistas in simulation methodology will probably have repercussions in *ad hoc* simulation language developments. Noteworthy in this regard will be the need to provide facilities for incorporating heuristics, rather than the more direct algorithms, as event routines in simulation models. Many simulation languages permit a user to define fixed event routines, which are considered as "stationary" operating rules applicable at any point in simular time; perhaps greater flexibility in permitting these operating rules, or event routines, to be altered significantly with the passage of simular time, will be in order. The "process" concept (cf. Blunden and Krasnow, 1967) in simulation languages may prove especially valuable in this regard. Possibly interpretive languages, especially in association with gaming models, will also prove of value here.

Other possible developments include hybrid computation and associated languages for the implementation of simulation models on analog-digital (hybrid) machines. As was noted in the preceding section, such simulation capabilities will be especially useful in the modeling of the human mind; that they should also be of value in environmental and ecological modeling is hardly a moot point.

In this last regard, the validity and acceptability of the responses emanating from these quite scientific models will also rest on the acceptability of the model's implicit assumptions and their builders' use of established scientific facts and knowledge. Thus, every *ad hoc* simulation language should be organised so as to encourage the systems analyst/modeler to provide explicit statements of the assumptions underlying each source language statement (or, at the least, each source language event routine); the means by which computational algorithms and event routines are derived should be made explicit by incorporating scientific references among the source statements of the simulation model. In this way, documentation of a model's structure becomes more automatic and thereby subject to closer scrutiny by the scientific community at large. Though *ad hoc* simulation languages could hardly be structured so as to enforce meaningful documentation and referencing, perhaps more emphasis should be placed on this aspect of a model's credibility.

Other possible language developments and improvements would concern methods for obtaining more pertinent statistical analyses of the response data arising from a single simulation encounter. Many *ad hoc* languages are deficient, for example, in their capability for performing time series analyses. In addition, more languages will need to provide methods by which experimental designs may be readily defined in advance by the analyst; the resulting responses should be automatically analyzed in accordance with the specified experimental design.

Furthermore, methods by which the response surface methodology can be applied in order to isolate the optimal simular response "automatically" would be of immense value to the simulation analyst. The use of heuristics for reducing the dimensionality of the "search space" would also be of eventual concern.

One might note aside the suggested use of *antithetic variates*, a Monte Carlo technique deemed occasionally appropriate to simular experimentation. [See, e.g., Page (1965).] The simular response at simular time t, arising as a result of the environmental specification $\vec{x}$, can be represented generally as

$$Y(t) = \mu_t(\vec{x}) + \varepsilon_t(S; \vec{x}),$$

where $\mu_t(\vec{x}) \equiv E[Y(t)]$ and where $\varepsilon_t(S; \vec{x})$ is an "error" random variable of mean zero, yet is an unknown transformation of the randomly selected value of the juxtaposed seed S. Thus, the essential statistical properties of $Y(t)$ become those of a random process $\{\varepsilon(t)\}$, whose distributional properties depend not only on the simular time t of observation but also on the environmental specification $\vec{x}$. The essence of the antithetic variate technique is the definition of a second random number generator, one to be used in the second of a pair of model encounters so as to yield a response $Y_2(t)$ which would be negatively correlated with the first response (one recorded from the model when employing the original random number generator). The arithmetic mean of the two responses, $Y_1(t)$ and $Y_2(t)$, would then have lower variance than that pertaining to the arithmetic average of the responses arising from a pair of independently seeded encounters, each of which employed the same generator. However, a general procedure, for defining random number generators that will guarantee the desired negative correlations in responses from general simulation models, remains undisclosed. Indeed, it would appear unlikely that a single such technique would be generally applicable to all stochastic simulation model encounters.

11.6. Conclusion

This chapter has hopefully served not only as a review of the book, but also as a vantage point for the applicability of the relatively new scientific modeling capability: the simulation model. The primary theme of this text is that our efforts to model conscientiously and in adequate detail any complex, dynamic system will provide a confrontation with the Uncertainty Principle of Modeling and, consequently, with the need for simulation models of the dynamic and stochastic varieties. Such models are of the most general type currently available to the scientific investigator, so that their acceptability depends heavily upon their builders' ability to represent reality as objectively as possible. The modeler is thus charged with the responsibility of including, in his simulation, appropriate and accurate random effects so that the resulting simular time series may indeed constitute approximate realizations from the stochastic process which would be observed in nature.

The implications of the need for stochasticity in modeling permeate the entire modeling activity. The systems analysis stage requires a constant vigil on the part of the analyst in order that truly stochastic effects not be represented deterministically (and vice-versa). The subsequent system synthesis stage requires that appropriate random variables, in consonance with known probability laws or observational data, be properly embedded into the model's structure. The verification stage may curtail or preclude randomness while model (or submodel) debugging proceeds, but should also include comparisons of one or more simular responses with theoretically applicable random variables. Finally, the Uncertainty Principle of Modeling leads to important considerations in the experimental stages (the validation and model analysis stages) of model development. Primary among these are the Principium of Seeding and the concomitant need to analyze and compare the independent responses via existing statistical methodology: statistical comparisons and tests of hypotheses, factorial experimental designs, multivariate analyses, sequential experimentation, regression analyses, response surface methodology, and time series analysis.

Appendix

STATISTICAL TABLES

TABLE A1

Pseudorandom Digits

04295	41589	08505	45322	40589	72851	13775	16481	20448	92444
13245	21342	82287	89323	36054	27463	73281	46235	30320	47318
15173	68768	08350	64308	34338	35672	98280	90999	52983	54862
24051	68964	88355	59443	85569	69604	56827	28162	60936	05314
29750	64639	02635	10380	37927	19771	49526	00976	71607	91671
26542	73639	72849	87517	53951	51594	67156	81697	38074	38309
31234	54617	65307	87643	43762	46541	71351	49982	16042	10856
07511	03719	49409	01115	75933	31460	16275	76254	55668	50338
11679	58340	91430	55798	72219	27240	66924	88235	09244	86572
34616	81858	53177	85325	23829	05158	55846	29519	99032	52349
47701	68271	90195	95166	96782	88664	67099	54380	66328	03765
79455	00427	17887	68191	34728	42502	56815	05589	35524	15513
67023	82401	48428	24257	31872	12295	26157	54198	88058	25636
54914	08224	09397	88373	48807	78753	67353	04705	63218	14567
65211	87926	48989	91747	92747	33797	19301	48083	98291	80840
51134	90339	25039	91913	93157	33744	08505	41458	01952	83079
81975	42780	78426	14749	86848	99748	26291	69201	34731	17278
04501	13070	18172	54981	95494	80424	16897	58357	61141	52495
96427	51894	08260	85246	45962	28463	35577	44181	52389	19372
84004	55741	57299	79461	62122	34694	93896	71612	68717	96857
50649	85050	84282	16563	58587	71800	53321	38213	49102	35692
29363	01321	79130	58278	04534	88394	70589	96031	95583	55055
60959	33220	08226	51751	11858	24814	51680	96452	72510	13804
25279	07693	44965	57314	49020	57346	47954	45879	59946	52491
26253	50253	46209	05766	02428	80143	40713	03567	17840	89221
46200	31479	59778	10810	13655	66298	21604	58600	45905	94042
92783	76786	48296	63306	25663	73987	98302	33345	75896	25337
55970	26261	63363	77112	87045	42661	50479	38265	20679	50174
84762	93262	13575	04190	02531	20574	42454	10197	40613	51206
96725	87106	52931	51648	93212	40916	78851	65610	84837	08109
60164	98921	85117	78150	53572	81966	80362	54462	35561	94045
51439	63249	46525	84009	76979	69550	71031	71569	39912	09890
01089	63641	09194	00915	79298	70104	18598	33371	68762	53351
14460	10812	46630	95999	94239	42412	68145	21141	07777	49251
98080	49511	43124	93452	56424	27942	68810	89546	75214	13479
54442	07441	13357	47542	41490	26361	26351	04493	86147	49142
37755	49002	46130	36255	09294	86567	33311	68925	56472	41583
04031	00734	06566	47314	08991	07066	45882	82165	74591	91792
53141	36611	37582	60541	65871	45168	04918	19962	76684	67778
60678	12326	06314	54979	91934	44862	50281	81246	55434	23192
46067	80866	56989	48241	57675	70729	65423	86001	24444	94407
32969	69524	70999	71897	43994	42506	25207	89407	63905	03864
41014	13549	10136	62646	73047	64325	17080	62684	99838	31280
16833	86340	42570	67188	07644	96728	76186	43658	31925	27791
79780	03023	35718	81604	23100	90896	31468	42276	36067	03769
36006	65833	58184	36019	05591	55446	14686	60706	39904	81395
16341	28539	76859	55123	29745	19373	50111	16777	15006	30628
31131	45595	77688	36992	27733	52520	31879	05789	60915	64416
21280	02389	91899	59263	95161	70026	21236	61700	86091	18415
31866	58294	86288	05518	97988	41918	69482	46869	31631	44594
55164	36790	88796	68198	62086	15920	07038	72384	47899	69385
96375	29117	81796	90028	55388	03174	47039	91036	34391	68003
20245	02368	17551	16307	24303	35339	45820	74729	01796	49748
52571	82010	05816	07920	33802	40008	55023	50034	24780	96934
49846	75116	05010	72194	96693	62072	03402	82208	37035	15137
25501	76578	28256	68108	74668	43984	73135	31745	89080	97182
44827	18710	66422	96480	04242	30420	98157	21059	56667	40191
85231	47531	44532	57040	57108	45088	23173	04532	65994	46635
16492	99043	78134	05271	99358	61801	34053	95499	03658	49106
99619	68532	94839	35077	79800	21086	15847	31316	16992	87012
45316	77869	45790	11163	66892	89839	26083	14864	96564	94044
26344	12347	64877	40092	78985	87554	00909	20240	79666	90652
14871	82419	52414	63656	26221	70797	52447	54534	34161	78260
28564	29149	77390	45173	16978	40443	79989	88820	88472	64212
30317	97867	20754	60864	89784	76248	17874	72523	78393	70842
48591	14867	33899	67303	25563	73058	91489	88445	97228	61155
80840	79527	74255	54381	87435	14833	62442	53607	75009	09918
23939	91448	16009	73888	38654	39328	26931	86101	87744	24901
55417	31648	31039	19196	15989	80003	00294	02864	21282	41218
80135	70905	05665	84019	98571	85241	88122	50202	98957	34518

TABLE A1 (*continued*)

71264	49688	15271	10513	23343	70618	22597	60509	40159	88864
84665	25033	33716	11325	70344	20311	44517	37389	60957	74828
24364	72939	20284	79365	86539	81263	49145	59889	70247	05261
96418	32664	16190	45314	48382	50971	00147	27193	68762	02800
21437	44377	07831	68892	93148	09166	62966	00514	30982	96957
95026	26328	87626	18057	89917	47741	29494	01409	76734	32108
02731	24613	77848	63155	85349	74604	12310	58013	72371	73373
24453	10219	90858	53102	59562	68075	91695	15087	58491	07729
15006	56827	93103	10367	76082	01645	14398	02865	68685	15235
35218	71308	32620	43489	19395	06731	82427	55989	99218	92456
44093	29543	93099	92411	96639	56129	45302	49802	65467	09631
59629	55506	64340	55743	48914	95572	32871	39401	72221	37191
66392	34141	81606	62524	85084	87743	50332	09753	39235	48528
04394	30738	97516	06715	29252	24643	15124	35171	73620	56920
28697	63965	22214	23020	74858	73076	59309	66198	79259	37648
95000	08802	13021	10155	76029	06413	63419	73854	53055	84208
15711	51904	26251	64925	92966	06525	41092	47809	50779	12568
56211	47902	73743	39879	55206	55070	70558	28820	24263	22120
14635	93335	67477	41403	27101	35927	81743	19262	49047	08926
63087	07709	99916	06423	66328	02709	68896	21232	89923	68428
36198	51284	07894	96844	71089	89793	20690	62082	03564	83597
46877	78844	16513	94034	27519	24326	55288	44733	65130	32978
01513	90691	69076	23494	08036	93018	29286	67404	41882	33724
90173	58643	32098	54907	46613	93467	69343	56746	33877	20104
54125	38651	33377	67496	40533	17945	66116	12526	72362	10476
95699	95095	58488	07501	12794	40423	84383	33248	22915	97951
06628	17510	09396	56223	27323	67650	93438	43119	95239	74408
63100	70812	30616	35852	93113	34829	20461	33890	27385	26584
81229	47681	46079	68774	35769	38336	89138	32985	01393	89294
58127	48915	35959	36726	68271	64561	38839	74354	72563	66771

TABLE A2

PERCENTILES FOR THE STANDARDIZED NORMAL DISTRIBUTION[a]

P	0.00	0.01	0.02	0.03	0.04	0.05	0.06	0.07	0.08	0.09
0.00		-2.3263	-2.0537	-1.8808	-1.7507	-1.6449	-1.5548	-1.4758	-1.4051	-1.3408
0.10	-1.2815	-1.2265	-1.1750	-1.1264	-1.0803	-1.0364	-0.9945	-0.9541	-0.9154	-0.8779
0.20	-0.8416	-0.8064	-0.7722	-0.7388	-0.7063	-0.6745	-0.6433	-0.6128	-0.5828	-0.5534
0.30	-0.5244	-0.4958	-0.4677	-0.4399	-0.4125	-0.3853	-0.3585	-0.3319	-0.3055	-0.2793
0.40	-0.2533	-0.2275	-0.2019	-0.1764	-0.1510	-0.1257	-0.1004	-0.0753	-0.0502	-0.0250
0.50	0.0000	0.0250	0.0502	0.0753	0.1004	0.1257	0.1510	0.1764	0.2019	0.2275
0.60	0.2533	0.2793	0.3055	0.3319	0.3585	0.3853	0.4125	0.4399	0.4677	0.4958
0.70	0.5244	0.5534	0.5828	0.6128	0.6434	0.6745	0.7063	0.7388	0.7722	0.8064
0.80	0.8416	0.8779	0.9154	0.9541	0.9945	1.0364	1.0803	1.1264	1.1750	1.2265
0.90	1.2815	1.3408	1.4051	1.4758	1.5548	1.6449	1.7507	1.8808	2.0537	2.3263

[a] Entries provide values of z for which $P = \int_{-\infty}^{z} \varphi(z; 0, 1)\, dz$.

TABLE A3. Percentiles of Student's t Distributions (1–20 Degrees of Freedom)[a]

P	1	2	3	4	5	6	7	8	9	10
0.50	0.0000	0.0000	0.0000	0.0000	0.0000	0.0000	0.0000	0.0000	0.0000	0.0000
0.51	0.0314	0.0283	0.0272	0.0267	0.0263	0.0261	0.0260	0.0259	0.0258	0.0257
0.52	0.0629	0.0566	0.0545	0.0534	0.0527	0.0523	0.0520	0.0517	0.0516	0.0514
0.53	0.0945	0.0850	0.0817	0.0801	0.0791	0.0785	0.0780	0.0777	0.0774	0.0772
0.54	0.1263	0.1135	0.1091	0.1069	0.1056	0.1047	0.1041	0.1036	0.1033	0.1030
0.55	0.1584	0.1421	0.1366	0.1338	0.1322	0.1311	0.1303	0.1297	0.1293	0.1289
0.56	0.1908	0.1709	0.1642	0.1609	0.1589	0.1575	0.1566	0.1559	0.1553	0.1549
0.57	0.2235	0.2000	0.1920	0.1880	0.1857	0.1841	0.1830	0.1821	0.1815	0.1810
0.58	0.2568	0.2292	0.2200	0.2154	0.2126	0.2108	0.2095	0.2086	0.2078	0.2072
0.59	0.2905	0.2588	0.2482	0.2429	0.2398	0.2377	0.2363	0.2351	0.2343	0.2338
0.60	0.3249	0.2887	0.2767	0.2707	0.2672	0.2648	0.2632	0.2619	0.2610	0.2602
0.61	0.3600	0.3189	0.3054	0.2988	0.2948	0.2922	0.2903	0.2889	0.2878	0.2870
0.62	0.3959	0.3496	0.3345	0.3271	0.3227	0.3197	0.3177	0.3161	0.3149	0.3140
0.63	0.4327	0.3808	0.3640	0.3557	0.3508	0.3476	0.3453	0.3436	0.3423	0.3412
0.64	0.4706	0.4125	0.3939	0.3848	0.3794	0.3758	0.3733	0.3714	0.3699	0.3688
0.65	0.5095	0.4448	0.4242	0.4142	0.4082	0.4043	0.4015	0.3995	0.3979	0.3966
0.66	0.5498	0.4777	0.4550	0.4440	0.4375	0.4332	0.4302	0.4279	0.4262	0.4248
0.67	0.5914	0.5113	0.4864	0.4743	0.4672	0.4625	0.4592	0.4567	0.4548	0.4533
0.68	0.6346	0.5457	0.5184	0.5052	0.4974	0.4923	0.4887	0.4860	0.4839	0.4823
0.69	0.6796	0.5810	0.5510	0.5366	0.5281	0.5226	0.5186	0.5157	0.5135	0.5117
0.70	0.7265	0.6172	0.5844	0.5686	0.5594	0.5534	0.5491	0.5459	0.5435	0.5415
0.71	0.7757	0.6545	0.6186	0.6014	0.5914	0.5848	0.5802	0.5767	0.5740	0.5719
0.72	0.8273	0.6929	0.6536	0.6349	0.6240	0.6168	0.6118	0.6081	0.6052	0.6029
0.73	0.8816	0.7327	0.6896	0.6692	0.6574	0.6496	0.6442	0.6401	0.6370	0.6345
0.74	0.9391	0.7738	0.7267	0.7045	0.6916	0.6832	0.6772	0.6729	0.6695	0.6668
0.75	1.0000	0.8165	0.7649	0.7407	0.7267	0.7176	0.7111	0.7064	0.7027	0.6998
0.76	1.0649	0.8609	0.8044	0.7780	0.7628	0.7529	0.7459	0.7408	0.7368	0.7337
0.77	1.1343	0.9073	0.8454	0.8166	0.8000	0.7893	0.7817	0.7761	0.7718	0.7684
0.78	1.2088	0.9559	0.8879	0.8565	0.8385	0.8268	0.8186	0.8125	0.8079	0.8042
0.79	1.2892	1.0069	0.9322	0.8979	0.8783	0.8655	0.8566	0.8501	0.8450	0.8410
0.80	1.3764	1.0607	0.9785	0.9410	0.9195	0.9057	0.8960	0.8889	0.8834	0.8791
0.81	1.4715	1.1175	1.0270	0.9859	0.9625	0.9474	0.9369	0.9291	0.9232	0.9185
0.82	1.5758	1.1779	1.0780	1.0329	1.0073	0.9909	0.9794	0.9710	0.9645	0.9594
0.83	1.6909	1.2424	1.1318	1.0823	1.0543	1.0363	1.0238	1.0146	1.0075	1.0019
0.84	1.8190	1.3116	1.1889	1.1344	1.1037	1.0840	1.0703	1.0602	1.0525	1.0464
0.85	1.9626	1.3862	1.2498	1.1896	1.1558	1.1342	1.1192	1.1081	1.0997	1.0931
0.86	2.1251	1.4673	1.3150	1.2483	1.2110	1.1872	1.1708	1.1587	1.1494	1.1422
0.87	2.3109	1.5559	1.3853	1.3112	1.2699	1.2437	1.2255	1.2123	1.2021	1.1941
0.88	2.5257	1.6537	1.4616	1.3789	1.3331	1.3041	1.2840	1.2693	1.2581	1.2493
0.89	2.7776	1.7627	1.5452	1.4525	1.4014	1.3691	1.3469	1.3306	1.3182	1.3084
0.90	3.0777	1.8856	1.6377	1.5332	1.4759	1.4398	1.4149	1.3968	1.3830	1.3722
0.91	3.4420	2.0261	1.7413	1.6226	1.5579	1.5172	1.4894	1.4691	1.4537	1.4416
0.92	3.8948	2.1894	1.8589	1.7229	1.6493	1.6033	1.5718	1.5489	1.5315	1.5179
0.93	4.4738	2.3834	1.9950	1.8375	1.7529	1.7002	1.6643	1.6383	1.6185	1.6031
0.94	5.2422	2.6202	2.1562	1.9712	1.8727	1.8117	1.7702	1.7402	1.7176	1.6999
0.95	6.3138	2.9200	2.3534	2.1318	2.0150	1.9432	1.8946	1.8596	1.8331	1.8125
0.96	7.9158	3.3198	2.6054	2.3329	2.1910	2.1043	2.0460	2.0042	1.9727	1.9481
0.97	10.5790	3.8964	2.9505	2.6008	2.4216	2.3133	2.2409	2.1892	2.1504	2.1202
0.98	15.8946	4.8487	3.4819	2.9985	2.7565	2.6123	2.5168	2.4490	2.3984	2.3598
0.99	31.8207	6.9646	4.5407	3.7469	3.3649	3.1427	2.9980	2.8965	2.8214	2.7638

p	11	12	13	14	15	16	17	18	19	20	
0.50	0.0000	0.0000	0.0000	0.0000	0.0000	0.0000	0.0000	0.0000	0.0000	0.0000	0.0000
0.51	0.0256	0.0256	0.0256	0.0255	0.0255	0.0255	0.0254	0.0254	0.0254	0.0254	0.0250
0.52	0.0513	0.0512	0.0511	0.0511	0.0510	0.0509	0.0509	0.0509	0.0508	0.0508	0.0502
0.53	0.0773	0.0769	0.0767	0.0766	0.0765	0.0765	0.0764	0.0763	0.0763	0.0762	0.0753
0.54	0.1028	0.1026	0.1024	0.1023	0.1021	0.1020	0.1019	0.1019	0.1018	0.1017	0.1004
0.55	0.1286	0.1283	0.1281	0.1280	0.1278	0.1277	0.1276	0.1274	0.1274	0.1273	0.1257
0.56	0.1545	0.1542	0.1540	0.1538	0.1536	0.1534	0.1533	0.1531	0.1530	0.1529	0.1510
0.57	0.1806	0.1802	0.1799	0.1797	0.1794	0.1792	0.1791	0.1789	0.1788	0.1787	0.1764
0.58	0.2067	0.2063	0.2060	0.2057	0.2054	0.2052	0.2050	0.2048	0.2047	0.2045	0.2019
0.59	0.2331	0.2326	0.2322	0.2319	0.2316	0.2313	0.2311	0.2309	0.2307	0.2306	0.2275
0.60	0.2596	0.2590	0.2586	0.2582	0.2579	0.2576	0.2573	0.2571	0.2569	0.2567	0.2533
0.61	0.2863	0.2857	0.2852	0.2848	0.2844	0.2841	0.2838	0.2835	0.2833	0.2831	0.2793
0.62	0.3132	0.3125	0.3120	0.3115	0.3111	0.3108	0.3104	0.3102	0.3099	0.3097	0.3055
0.63	0.3404	0.3396	0.3390	0.3385	0.3381	0.3377	0.3373	0.3370	0.3367	0.3365	0.3319
0.64	0.3678	0.3670	0.3663	0.3658	0.3653	0.3649	0.3645	0.3641	0.3638	0.3636	0.3585
0.65	0.3956	0.3947	0.3940	0.3933	0.3928	0.3923	0.3919	0.3915	0.3912	0.3909	0.3853
0.66	0.4236	0.4227	0.4219	0.4212	0.4206	0.4201	0.4196	0.4192	0.4189	0.4186	0.4125
0.67	0.4521	0.4510	0.4502	0.4494	0.4488	0.4482	0.4477	0.4473	0.4469	0.4465	0.4399
0.68	0.4809	0.4798	0.4788	0.4780	0.4773	0.4767	0.4762	0.4757	0.4753	0.4749	0.4677
0.69	0.5102	0.5090	0.5079	0.5071	0.5063	0.5056	0.5051	0.5045	0.5041	0.5037	0.4958
0.70	0.5399	0.5386	0.5375	0.5366	0.5357	0.5350	0.5344	0.5338	0.5333	0.5329	0.5244
0.71	0.5702	0.5688	0.5676	0.5665	0.5656	0.5649	0.5642	0.5636	0.5630	0.5625	0.5534
0.72	0.6010	0.5995	0.5982	0.5971	0.5961	0.5953	0.5945	0.5939	0.5933	0.5927	0.5828
0.73	0.6325	0.6308	0.6294	0.6282	0.6271	0.6262	0.6254	0.6247	0.6241	0.6235	0.6128
0.74	0.6646	0.6628	0.6613	0.6599	0.6588	0.6578	0.6570	0.6562	0.6555	0.6549	0.6433
0.75	0.6974	0.6955	0.6938	0.6924	0.6912	0.6901	0.6892	0.6884	0.6876	0.6870	0.6745
0.76	0.7311	0.7290	0.7272	0.7257	0.7243	0.7232	0.7222	0.7213	0.7205	0.7198	0.7063
0.77	0.7656	0.7633	0.7614	0.7598	0.7583	0.7571	0.7560	0.7550	0.7542	0.7534	0.7388
0.78	0.8012	0.7987	0.7966	0.7948	0.7933	0.7919	0.7907	0.7897	0.7887	0.7879	0.7722
0.79	0.8378	0.8351	0.8328	0.8309	0.8292	0.8277	0.8264	0.8253	0.8243	0.8234	0.8064
0.80	0.8755	0.8726	0.8702	0.8681	0.8662	0.8647	0.8633	0.8620	0.8610	0.8600	0.8416
0.81	0.9146	0.9115	0.9088	0.9065	0.9046	0.9028	0.9013	0.9000	0.8988	0.8977	0.8779
0.82	0.9552	0.9518	0.9489	0.9464	0.9443	0.9424	0.9408	0.9393	0.9380	0.9369	0.9154
0.83	0.9974	0.9937	0.9905	0.9878	0.9855	0.9835	0.9817	0.9802	0.9788	0.9775	0.9541
0.84	1.0415	1.0374	1.0340	1.0311	1.0285	1.0263	1.0244	1.0227	1.0212	1.0198	0.9945
0.85	1.0877	1.0832	1.0795	1.0763	1.0735	1.0711	1.0690	1.0672	1.0655	1.0640	1.0364
0.86	1.1363	1.1314	1.1273	1.1238	1.1208	1.1182	1.1159	1.1138	1.1120	1.1104	1.0803
0.87	1.1876	1.1823	1.1778	1.1739	1.1707	1.1678	1.1653	1.1630	1.1611	1.1593	1.1264
0.88	1.2422	1.2363	1.2314	1.2272	1.2235	1.2204	1.2176	1.2152	1.2130	1.2110	1.1750
0.89	1.3005	1.2940	1.2886	1.2840	1.2800	1.2765	1.2734	1.2707	1.2683	1.2662	1.2265
0.90	1.3634	1.3562	1.3502	1.3450	1.3406	1.3368	1.3334	1.3304	1.3277	1.3253	1.2815
0.91	1.4318	1.4237	1.4170	1.4113	1.4063	1.4021	1.3983	1.3950	1.3920	1.3894	1.3408
0.92	1.5069	1.4979	1.4903	1.4839	1.4784	1.4736	1.4694	1.4656	1.4623	1.4593	1.4051
0.93	1.5906	1.5804	1.5718	1.5646	1.5583	1.5529	1.5482	1.5439	1.5402	1.5369	1.4758
0.94	1.6856	1.6739	1.6641	1.6558	1.6487	1.6425	1.6370	1.6322	1.6280	1.6242	1.5548
0.95	1.7959	1.7823	1.7709	1.7613	1.7531	1.7459	1.7396	1.7341	1.7291	1.7247	1.6449
0.96	1.9284	1.9123	1.8969	1.8875	1.8777	1.8693	1.8619	1.8553	1.8495	1.8443	1.7507
0.97	2.0961	2.0764	2.0600	2.0462	2.0343	2.0240	2.0150	2.0071	2.0000	1.9937	1.8808
0.98	2.3281	2.3027	2.2816	2.2638	2.2485	2.2354	2.2238	2.2137	2.2047	2.1967	2.0537
0.99	2.7181	2.6810	2.6503	2.6245	2.6025	2.5835	2.5669	2.5524	2.5395	2.5280	2.3263

[a] Entries give values τ for which $P = \int_{-\infty}^{\tau} f_T(t; k)\, dt$.

TABLE A4. PERCENTILES OF PEARSON'S CHI-SQUARED DISTRIBUTIONS (1–10 DEGREES OF FREEDOM)[a]

p	1	2	3	4	5	6	7	8	9	10
0.01	0.0002	0.020	0.115	0.297	0.554	0.872	1.239	1.646	2.088	2.558
0.02	0.0006	0.040	0.185	0.429	0.752	1.134	1.564	2.032	2.532	3.059
0.03	0.0014	0.061	0.245	0.535	0.903	1.330	1.802	2.310	2.848	3.412
0.04	0.0025	0.082	0.300	0.627	1.031	1.492	1.997	2.537	3.105	3.697
0.05	0.0039	0.103	0.352	0.711	1.145	1.635	2.167	2.733	3.325	3.940
0.06	0.0057	0.124	0.401	0.788	1.250	1.765	2.320	2.908	3.521	4.157
0.07	0.0077	0.145	0.449	0.862	1.347	1.885	2.461	3.068	3.700	4.353
0.08	0.0101	0.167	0.495	0.931	1.439	1.997	2.592	3.217	3.866	4.535
0.09	0.0128	0.189	0.540	0.999	1.526	2.103	2.716	3.357	4.021	4.705
0.10	0.0158	0.211	0.584	1.064	1.610	2.204	2.833	3.490	4.168	4.865
0.11	0.0191	0.233	0.628	1.127	1.691	2.301	2.945	3.616	4.308	5.018
0.12	0.0228	0.256	0.671	1.188	1.770	2.395	3.054	3.737	4.442	5.163
0.13	0.0268	0.279	0.714	1.249	1.846	2.486	3.158	3.855	4.571	5.304
0.14	0.0311	0.302	0.756	1.308	1.921	2.575	3.260	3.968	4.696	5.439
0.15	0.0358	0.325	0.798	1.366	1.994	2.661	3.358	4.078	4.817	5.570
0.16	0.0408	0.349	0.839	1.424	2.066	2.746	3.455	4.186	4.934	5.698
0.17	0.0461	0.373	0.881	1.481	2.136	2.829	3.549	4.291	5.049	5.822
0.18	0.0518	0.397	0.922	1.537	2.206	2.910	3.642	4.393	5.162	5.943
0.19	0.0578	0.421	0.964	1.593	2.275	2.991	3.733	4.494	5.272	6.062
0.20	0.0642	0.446	1.005	1.649	2.343	3.070	3.822	4.594	5.380	6.179
0.21	0.0709	0.471	1.047	1.704	2.410	3.148	3.911	4.691	5.487	6.294
0.22	0.0780	0.497	1.088	1.759	2.477	3.226	3.998	4.788	5.592	6.407
0.23	0.0855	0.523	1.129	1.814	2.543	3.303	4.084	4.883	5.695	6.518
0.24	0.0933	0.549	1.171	1.868	2.609	3.379	4.170	4.977	5.798	6.628
0.25	0.1015	0.575	1.213	1.923	2.675	3.455	4.255	5.071	5.899	6.737
0.26	0.1101	0.602	1.254	1.977	2.740	3.530	4.339	5.163	5.999	6.845
0.27	0.1191	0.629	1.296	2.031	2.805	3.605	4.423	5.255	6.099	6.952
0.28	0.1285	0.657	1.339	2.086	2.870	3.679	4.506	5.346	6.198	7.058
0.29	0.1383	0.685	1.381	2.140	2.935	3.753	4.589	5.437	6.296	7.163
0.30	0.1485	0.713	1.424	2.195	3.000	3.828	4.671	5.527	6.393	7.267
0.31	0.1591	0.742	1.467	2.249	3.065	3.902	4.754	5.617	6.490	7.371
0.32	0.1701	0.771	1.510	2.304	3.130	3.975	4.836	5.707	6.587	7.475
0.33	0.1816	0.801	1.553	2.359	3.195	4.049	4.918	5.797	6.684	7.578
0.34	0.1935	0.831	1.597	2.415	3.260	4.123	5.000	5.886	6.780	7.681
0.35	0.2059	0.862	1.642	2.470	3.325	4.197	5.082	5.975	6.876	7.783
0.36	0.2187	0.893	1.686	2.526	3.391	4.271	5.164	6.065	6.972	7.886
0.37	0.2321	0.924	1.731	2.582	3.456	4.346	5.246	6.154	7.068	7.988
0.38	0.2459	0.956	1.777	2.639	3.522	4.420	5.328	6.243	7.164	8.090
0.39	0.2602	0.989	1.823	2.696	3.589	4.495	5.410	6.333	7.261	8.193
0.40	0.2750	1.022	1.869	2.753	3.655	4.570	5.493	6.423	7.357	8.295
0.41	0.2903	1.055	1.916	2.811	3.723	4.646	5.576	6.513	7.454	8.398
0.42	0.3062	1.089	1.964	2.869	3.790	4.721	5.660	6.603	7.550	8.501
0.43	0.3227	1.124	2.012	2.928	3.858	4.798	5.743	6.694	7.648	8.605
0.44	0.3397	1.160	2.060	2.987	3.927	4.875	5.828	6.785	7.745	8.706
0.45	0.3573	1.196	2.109	3.047	3.996	4.952	5.913	6.877	7.843	8.812
0.46	0.3755	1.232	2.159	3.107	4.066	5.030	5.998	6.969	7.942	8.917
0.47	0.3944	1.270	2.210	3.169	4.136	5.108	6.084	7.062	8.041	9.022
0.48	0.4139	1.308	2.261	3.231	4.207	5.187	6.170	7.155	8.141	9.128
0.49	0.4341	1.347	2.313	3.293	4.279	5.267	6.258	7.249	8.242	9.235

P	1	2	3	4	5	6	7	8	9	10
0.50	0.4549	1.386	2.366	3.357	4.351	5.348	6.346	7.344	8.343	9.342
0.51	0.4765	1.427	2.420	3.421	4.425	5.430	6.435	7.440	8.445	9.450
0.52	0.4989	1.468	2.474	3.486	4.499	5.512	6.525	7.537	8.548	9.559
0.53	0.5220	1.510	2.529	3.552	4.574	5.595	6.615	7.634	8.652	9.669
0.54	0.5459	1.553	2.586	3.619	4.651	5.680	6.707	7.733	8.757	9.780
0.55	0.5707	1.597	2.643	3.687	4.728	5.765	6.800	7.833	8.863	9.892
0.56	0.5963	1.642	2.701	3.756	4.806	5.852	6.894	7.933	8.971	10.006
0.57	0.6228	1.688	2.761	3.826	4.886	5.940	6.989	8.036	9.079	10.120
0.58	0.6503	1.735	2.821	3.898	4.966	6.029	7.086	8.139	9.189	10.236
0.59	0.6788	1.783	2.883	3.971	5.048	6.119	7.184	8.244	9.301	10.354
0.60	0.7083	1.833	2.946	4.045	5.132	6.211	7.283	8.351	9.414	10.473
0.61	0.7387	1.883	3.011	4.120	5.217	6.304	7.384	8.459	9.528	10.594
0.62	0.7707	1.935	3.076	4.197	5.303	6.399	7.487	8.568	9.645	10.717
0.63	0.8037	1.989	3.144	4.276	5.391	6.496	7.591	8.680	9.763	10.841
0.64	0.8379	2.043	3.213	4.356	5.481	6.594	7.698	8.794	9.883	10.968
0.65	0.8735	2.100	3.283	4.438	5.573	6.695	7.806	8.909	10.006	11.097
0.66	0.9104	2.158	3.355	4.522	5.667	6.797	7.917	9.027	10.131	11.228
0.67	0.9489	2.217	3.430	4.607	5.763	6.902	8.029	9.148	10.258	11.362
0.68	0.9889	2.279	3.506	4.695	5.861	7.009	8.145	9.270	10.388	11.499
0.69	1.0307	2.342	3.584	4.786	5.961	7.119	8.263	9.396	10.521	11.638
0.70	1.0742	2.408	3.665	4.878	6.064	7.231	8.383	9.524	10.656	11.781
0.71	1.1196	2.476	3.748	4.974	6.170	7.346	8.507	9.656	10.795	11.927
0.72	1.1671	2.546	3.834	5.072	6.279	7.465	8.634	9.791	10.938	12.076
0.73	1.2167	2.619	3.922	5.173	6.391	7.586	8.765	9.930	11.084	12.229
0.74	1.2688	2.694	4.014	5.277	6.507	7.712	8.899	10.072	11.234	12.387
0.75	1.3233	2.773	4.108	5.385	6.626	7.841	9.037	10.219	11.389	12.549
0.76	1.3805	2.854	4.207	5.497	6.749	7.974	9.180	10.370	11.548	12.716
0.77	1.4409	2.939	4.309	5.613	6.876	8.112	9.327	10.526	11.713	12.888
0.78	1.5044	3.028	4.415	5.733	7.009	8.255	9.480	10.688	11.883	13.066
0.79	1.5714	3.121	4.526	5.858	7.146	8.404	9.639	10.856	12.059	13.251
0.80	1.6424	3.219	4.642	5.989	7.289	8.558	9.803	11.030	12.242	13.442
0.81	1.7176	3.321	4.763	6.125	7.439	8.719	9.975	11.212	12.433	13.641
0.82	1.7976	3.430	4.890	6.268	7.595	8.888	10.154	11.401	12.632	13.849
0.83	1.8829	3.544	5.025	6.418	7.759	9.064	10.342	11.599	12.840	14.066
0.84	1.9742	3.665	5.167	6.577	7.932	9.250	10.540	11.808	13.058	14.294
0.85	2.0722	3.794	5.317	6.745	8.115	9.446	10.748	12.027	13.288	14.534
0.86	2.1780	3.932	5.477	6.923	8.309	9.654	10.968	12.259	13.531	14.788
0.87	2.2925	4.080	5.649	7.114	8.516	9.875	11.203	12.506	13.799	15.057
0.88	2.4173	4.241	5.833	7.318	8.738	10.112	11.454	12.770	14.066	15.344
0.89	2.5542	4.415	6.033	7.539	8.977	10.368	11.724	13.054	14.363	15.653
0.90	2.7055	4.605	6.251	7.779	9.236	10.645	12.017	13.362	14.684	15.987
0.91	2.8744	4.816	6.491	8.043	9.521	10.948	12.337	13.697	15.034	16.352
0.92	3.0649	5.051	6.759	8.337	9.837	11.283	12.691	14.068	15.421	16.753
0.93	3.2830	5.319	7.060	8.666	10.191	11.660	13.088	14.484	15.854	17.203
0.94	3.5374	5.627	7.407	9.044	10.596	12.090	13.540	14.956	16.346	17.713
0.95	3.8415	5.991	7.815	9.488	11.070	12.592	14.067	15.507	16.919	18.307
0.96	4.2179	6.438	8.311	10.026	11.644	13.198	14.703	16.171	17.608	19.021
0.97	4.7093	7.013	8.947	10.712	12.375	13.968	15.509	17.010	18.480	19.922
0.98	5.4119	7.824	9.837	11.668	13.388	15.033	16.622	18.168	19.679	21.161
0.99	6.6349	9.210	11.345	13.277	15.086	16.812	18.475	20.090	21.666	23.209

[a] Entries give values x for which $P = \int_0^x f_{\chi^2}(y; 2, k/2)\, dy$.

TABLE A5. PERCENTILES OF PEARSON'S CHI-SQUARED DISTRIBUTIONS (11–20 DEGREES OF FREEDOM)[a]

P	11	12	13	14	15	16	17	18	19	20
0.01	3.053	3.571	4.107	4.660	5.229	5.812	6.408	7.015	7.633	8.260
0.02	3.609	4.178	4.765	5.368	5.985	6.614	7.255	7.906	8.567	9.237
0.03	3.997	4.601	5.221	5.856	6.503	7.163	7.832	8.512	9.200	9.897
0.04	4.309	4.939	5.584	6.243	6.914	7.596	8.288	8.989	9.698	10.415
0.05	4.575	5.226	5.892	6.571	7.261	7.962	8.672	9.390	10.117	10.851
0.06	4.810	5.480	6.163	6.859	7.566	8.283	9.008	9.742	10.483	11.231
0.07	5.024	5.710	6.409	7.120	7.841	8.572	9.311	10.058	10.812	11.573
0.08	5.221	5.921	6.634	7.359	8.093	8.836	9.588	10.347	11.112	11.884
0.09	5.405	6.118	6.844	7.581	8.327	9.082	9.845	10.614	11.391	12.173
0.10	5.578	6.304	7.042	7.790	8.547	9.312	10.085	10.865	11.651	12.443
0.11	5.742	6.480	7.228	7.987	8.754	9.530	10.312	11.101	11.897	12.697
0.12	5.899	6.648	7.407	8.175	8.952	9.737	10.528	11.326	12.130	12.939
0.13	6.050	6.809	7.577	8.355	9.142	9.935	10.735	11.541	12.353	13.169
0.14	6.196	6.964	7.742	8.529	9.324	10.125	10.933	11.747	12.567	13.391
0.15	6.336	7.114	7.901	8.696	9.499	10.309	11.125	11.946	12.773	13.604
0.16	6.473	7.259	8.055	8.859	9.669	10.487	11.310	12.139	12.972	13.810
0.17	6.606	7.401	8.205	9.016	9.835	10.659	11.490	12.325	13.165	14.010
0.18	6.737	7.540	8.351	9.170	9.996	10.828	11.665	12.507	13.353	14.204
0.19	6.864	7.675	8.494	9.320	10.153	10.992	11.835	12.684	13.537	14.393
0.20	6.989	7.807	8.634	9.467	10.307	11.152	12.002	12.857	13.716	14.578
0.21	7.111	7.937	8.771	9.611	10.458	11.309	12.166	13.026	13.891	14.759
0.22	7.232	8.065	8.906	9.753	10.606	11.464	12.326	13.193	14.063	14.937
0.23	7.351	8.191	9.039	9.892	10.752	11.615	12.484	13.356	14.232	15.111
0.24	7.468	8.316	9.170	10.030	10.895	11.765	12.639	13.517	14.398	15.283
0.25	7.584	8.438	9.299	10.165	11.037	11.912	12.792	13.675	14.562	15.452
0.26	7.699	8.560	9.427	10.299	11.176	12.058	12.943	13.832	14.724	15.618
0.27	7.812	8.680	9.553	10.432	11.314	12.201	13.092	13.986	14.883	15.783
0.28	7.925	8.799	9.678	10.563	11.451	12.344	13.240	14.139	15.041	15.945
0.29	8.037	8.917	9.802	10.693	11.587	12.485	13.386	14.290	15.197	16.106
0.30	8.148	9.034	9.926	10.821	11.721	12.624	13.531	14.440	15.352	16.266
0.31	8.258	9.151	10.048	10.949	11.855	12.763	13.674	14.589	15.505	16.424
0.32	8.368	9.267	10.170	11.077	11.987	12.901	13.817	14.736	15.657	16.581
0.33	8.477	9.382	10.291	11.203	12.119	13.038	13.959	14.883	15.809	16.737
0.34	8.587	9.497	10.411	11.329	12.250	13.174	14.100	15.029	15.959	16.892
0.35	8.695	9.612	10.532	11.455	12.381	13.310	14.241	15.174	16.109	17.046
0.36	8.804	9.726	10.651	11.580	12.511	13.445	14.381	15.318	16.258	17.199
0.37	8.912	9.840	10.771	11.705	12.641	13.580	14.520	15.463	16.407	17.352
0.38	9.020	9.954	10.890	11.829	12.771	13.714	14.659	15.606	16.555	17.505
0.39	9.129	10.068	11.010	11.954	12.900	13.848	14.798	15.750	16.703	17.657
0.40	9.237	10.182	11.129	12.078	13.030	13.983	14.937	15.893	16.850	17.809
0.41	9.346	10.296	11.249	12.203	13.159	14.117	15.076	16.036	16.998	17.961
0.42	9.455	10.410	11.368	12.328	13.289	14.251	15.215	16.180	17.146	18.112
0.43	9.564	10.525	11.488	12.453	13.419	14.386	15.354	16.323	17.293	18.264
0.44	9.673	10.640	11.608	12.578	13.549	14.520	15.493	16.467	17.441	18.416
0.45	9.783	10.755	11.729	12.703	13.679	14.656	15.633	16.611	17.589	18.569
0.46	9.893	10.871	11.850	12.829	13.810	14.791	15.773	16.755	17.738	18.721
0.47	10.004	10.987	11.971	12.956	13.941	14.927	15.913	16.900	17.887	18.874
0.48	10.116	11.104	12.093	13.083	14.073	15.063	16.054	17.045	18.037	19.028
0.49	10.228	11.222	12.216	13.211	14.206	15.201	16.196	17.191	18.187	19.182

P	11	12	13	14	15	16	17	18	19	20
0.50	10.341	11.340	12.340	13.339	14.339	15.338	16.338	17.338	18.338	19.337
0.51	10.455	11.460	12.464	13.469	14.473	15.477	16.481	17.485	18.489	19.493
0.52	10.570	11.580	12.589	13.599	14.608	15.617	16.626	17.634	18.642	19.650
0.53	10.685	11.701	12.716	13.730	14.744	15.758	16.771	17.783	18.796	19.808
0.54	10.802	11.823	12.843	13.863	14.881	15.899	16.917	17.934	18.951	19.967
0.55	10.920	11.946	12.972	13.996	15.020	16.042	17.065	18.086	19.107	20.127
0.56	11.039	12.071	13.102	14.131	15.159	16.187	17.213	18.239	19.264	20.289
0.57	11.159	12.197	13.233	14.267	15.300	16.333	17.364	18.394	19.423	20.452
0.58	11.281	12.324	13.365	14.405	15.443	16.480	17.516	18.550	19.584	20.617
0.59	11.405	12.453	13.500	14.544	15.587	16.629	17.669	18.708	19.746	20.783
0.60	11.530	12.584	13.636	14.685	15.733	16.780	17.824	18.868	19.910	20.951
0.61	11.657	12.716	13.773	14.828	15.881	16.932	17.982	19.030	20.076	21.122
0.62	11.785	12.851	13.913	14.973	16.031	17.087	18.141	19.193	20.245	21.294
0.63	11.916	12.987	14.055	15.120	16.183	17.244	18.303	19.360	20.415	21.469
0.64	12.049	13.125	14.199	15.269	16.337	17.403	18.466	19.528	20.588	21.646
0.65	12.184	13.266	14.345	15.421	16.494	17.565	18.633	19.699	20.764	21.826
0.66	12.321	13.409	14.494	15.575	16.653	17.729	18.802	19.873	20.942	22.009
0.67	12.461	13.555	14.646	15.732	16.816	17.896	18.974	20.050	21.124	22.195
0.68	12.604	13.704	14.800	15.892	16.981	18.067	19.150	20.230	21.309	22.385
0.69	12.750	13.856	14.958	16.055	17.150	18.241	19.329	20.414	21.497	22.578
0.70	12.899	14.011	15.119	16.222	17.322	18.418	19.511	20.601	21.689	22.775
0.71	13.051	14.170	15.283	16.392	17.498	18.599	19.697	20.793	21.885	22.975
0.72	13.207	14.332	15.452	16.567	17.678	18.784	19.888	20.988	22.086	23.181
0.73	13.367	14.499	15.625	16.745	17.862	18.974	20.083	21.189	22.291	23.391
0.74	13.532	14.670	15.802	16.929	18.051	19.169	20.283	21.394	22.502	23.607
0.75	13.701	14.845	15.984	17.117	18.245	19.369	20.489	21.605	22.718	23.828
0.76	13.875	15.026	16.171	17.311	18.445	19.574	20.700	21.822	22.940	24.055
0.77	14.054	15.213	16.365	17.510	18.651	19.786	20.918	22.045	23.169	24.289
0.78	14.240	15.406	16.564	17.716	18.863	20.005	21.142	22.275	23.404	24.530
0.79	14.432	15.605	16.771	17.930	19.083	20.231	21.374	22.513	23.648	24.779
0.80	14.631	15.812	16.985	18.151	19.311	20.465	21.615	22.760	23.900	25.038
0.81	14.839	16.027	17.207	18.380	19.547	20.708	21.864	23.015	24.162	25.305
0.82	15.055	16.251	17.439	18.620	19.793	20.961	22.124	23.282	24.435	25.584
0.83	15.281	16.485	17.681	18.869	20.051	21.226	22.395	23.559	24.719	25.874
0.84	15.518	16.731	17.935	19.131	20.320	21.502	22.679	23.850	25.016	26.178
0.85	15.767	16.989	18.202	19.406	20.603	21.793	22.977	24.155	25.329	26.498
0.86	16.031	17.262	18.484	19.697	20.902	22.100	23.291	24.477	25.658	26.834
0.87	16.310	17.552	18.783	20.004	21.218	22.425	23.625	24.818	26.007	27.190
0.88	16.609	17.860	19.101	20.333	21.556	22.771	23.979	25.181	26.378	27.569
0.89	16.929	18.191	19.443	20.684	21.917	23.142	24.359	25.570	26.775	27.975
0.90	17.275	18.549	19.812	21.064	22.307	23.542	24.769	25.990	27.204	28.412
0.91	17.653	18.939	20.214	21.478	22.732	23.977	25.215	26.446	27.670	28.887
0.92	18.069	19.369	20.657	21.933	23.199	24.456	25.705	26.947	28.181	29.410
0.93	18.533	19.849	21.151	22.441	23.720	24.990	26.251	27.505	28.751	29.991
0.94	19.061	20.393	21.711	23.017	24.311	25.595	26.870	28.137	29.396	30.649
0.95	19.675	21.026	22.362	23.685	24.996	26.296	27.587	28.869	30.144	31.410
0.96	20.412	21.785	23.142	24.485	25.816	27.136	28.445	29.745	31.037	32.321
0.97	21.342	22.742	24.125	25.493	26.848	28.191	29.523	30.845	32.158	33.462
0.98	22.618	24.054	25.472	26.873	28.259	29.633	30.995	32.346	33.688	35.020
0.99	24.725	26.217	27.688	29.141	30.578	32.000	33.409	34.805	36.191	37.566

[a] Entries give values x for which $P = \int_0^x f_{\chi^2}(y; 2, k/2)\, dy$.

TABLE A6. CRITICAL VALUES OF THE KOLMOGOROV–SMIRNOV STATISTIC[a,b]

N	00	01	02	03	04	05	06	07	08	09
00		**.99500**	**.92929**	**.82900**	**.73424**	**.66853**	**.61661**	**.57581**	**.54179**	**.51332**
		.97500	.84189	.70760	.62394	.56328	.51926	.48342	.45427	.43001
10	**.48893**	**.46770**	**.44905**	**.43247**	**.41762**	**.40420**	**.39201**	**.38086**	**.37062**	**.36117**
	.40925	.39122	.37543	.36143	.34890	.33760	.32733	.31796	.30936	.30143
20	**.35241**	**.34427**	**.33666**	**.32954**	**.32286**	**.31657**	**.31064**	**.30502**	**.29971**	**.29466**
	.29408	.28724	.28087	.27490	.26931	.26404	.25907	.25438	.24993	.24571
30	**.28987**	**.28530**	**.28094**	**.27677**	**.27279**	**.26897**	**.26532**	**.26180**	**.25843**	**.25518**
	.24170	.23788	.23424	.23076	.22743	.22425	.22119	.21826	.21544	.21273
40	**.25205**	**.24904**	**.24613**	**.24332**	**.24060**	**.23798**	**.23544**	**.23298**	**.23059**	**.22828**
	.21012	.20760	.20517	.20283	.20056	.19837	.19625	.19420	.19221	.19028
50	**.22604**	**.22386**	**.22174**	**.21968**	**.21768**	**.21574**	**.21384**	**.21199**	**.21019**	**.20844**
	.18841	.18659	.18482	.18311	.18144	.17981	.17823	.17669	.17519	.17373
60	**.20673**	**.20506**	**.20343**	**.20184**	**.20029**	**.19877**	**.19729**	**.19584**	**.19442**	**.19303**
	.17231	.17091	.16956	.16823	.16693	.16567	.16443	.16322	.16204	.16088
70	**.19167**	**.19034**	**.18903**	**.18776**	**.18650**	**.18528**	**.18408**	**.18290**	**.18174**	**.18060**
	.15975	.15864	.15755	.15649	.15544	.15442	.15342	.15244	.15147	.15052
80	**.17949**	**.17840**	**.17732**	**.17627**	**.17523**	**.17421**	**.17321**	**.17223**	**.17126**	**.17031**
	.14960	.14868	.14779	.14691	.14605	.14520	.14437	.14355	.14274	.14195
90	**.16938**	**.16846**	**.16755**	**.16666**	**.16579**	**.16493**	**.16408**	**.16324**	**.16242**	**.16161**
	.14117	.14040	.13965	.13891	.13818	.13746	.13675	.13606	.13537	.13469

[a] Entries give values d for which $P(D_n > d) = 0.05$ (regular type) or **0.01** (**boldfaced type**).

[b] After L. H. Miller, *J. Amer. Statist. Soc.* **51**, 113–115 (1956). His Table 1. By permission of The American Statistical Society.

GLOSSARY

Activities: *1.* The temporal, or dynamic, structures of a simulation model; *2.* The conceptual processes which occur within the simular boundary, and which require the passage of time, usually demarcated by simular events.

Analysis: The fifth stage of a model's development, concerned with experimentation, by means of independently seeded encounters, with a verified and validated simulation model.

Analyst: A person skilled in the definition of models and in development of procedures for their implementation and use.

Algorithm: A process carried out according to a precise prescription leading from given, possibly variable, data to a proven definite result (Fritz, 1963); should be contrasted with the term Heuristic.

Attribute: Any characteristic, or property, of an entity.

Autocorrelogram: *1.* For a weakly stationary time series, the plot of the correlations of random variables versus their lag (difference in their time indices); *2.* Estimates of this autocorrelation function, as made from a sample record of the time series.

Autocovariance function: For the random variables, $X(t)$ and $X(s)$, in a time series, their covariance as a function of the time indices, s and t; viz., $R(s, t) = \mathrm{Cov}[X(s), X(t)]$.

Boundary (system): A conceptual artifice designating those entities and activities which are essential to the system, yet whose interactions with, and influences upon, elements outside the boundary may be considered either insignificant or not connected with any intrinsic feedback mechanism of the system.

Calendar: *1.* The chronological (in terms of simular time) listing of events yet to transpire in a simular encounter; *2.* Hence, the complete list of known, future events which are to transpire in a simular encounter at any given point in simular time.

Clockworks: That portion, or segment, of the executive routine of a computerized simulation model which maintains the simular clock (simular time).

Consistent estimator: A sequence, T_n, $n = 1, 2, \ldots$, of statistics which, in probability, converges to a parameter θ for which each T_n provides an estimate.

Correlation: The expectation of the product of two jointly distributed random variables, each standardized:

$$\varrho_{x,y} = E\left(\left[\frac{X - \mu_X}{\sigma_X}\right]\left[\frac{Y - \mu_X}{\sigma_X}\right]\right)$$

Covariance: The expectation of the product of two jointly distributed random variables, each of mean zero; viz., $\mathrm{Cov}(X, Y) = E([X - \mu_X][Y - \mu_Y])$.

Cumulative distribution function, $F_X(a)$: The probability that a random variable X shall not exceed the real number, a.

Degenerate random variable: *1.* One having zero variance; *2.* Hence, one assuming its mean value with probability 1.

Deterministic: *1.* Category of models possessing no stochastic effects or elements; *2.* A simulation model requiring no random number seed.

Dynamic model: *1.* One whose properties, features, or attributes are expected to alter with the passage of time; *2.* One which requires time as an explicit variable.

Encounter: *1.* The specification of the environmental conditions and seeds for, and the subsequent observation and recording of the responses of, a simulation model. *2.* A "run" of a simulation model. (Cf. Iteration.)

Endogenous: *See* Event.

Entity: *1.* Any element of a simulated system; *2.* More usually, any element of the simular system; *3.* (Permanent): Any element which may be expected to remain within the simular boundary throughout any given encounter; *4.* (Temporary): Any element which enters a simulation model subsequent to its initiation and/or which may be expected to depart the simular boundary prior to an encounter's completion.

Environment: The collection of elements and/or events which affect measurably the elements within the system boundary, yet which are not significantly affected in return by any of the elements internal to the boundary.

Environmental conditions: *1.* That set of specifications, save the juxtaposed seed, necessary to the definition of a simular encounter; *2.* Input conditions (less the juxtaposed seed) for a simulation model.

Ergodicity: For stationary time series, the equality of time averages determined from any realization in the ensemble and the mean, or expected value, of any one of the random variables constituting the time series.

Event: *1.* Instantaneous modifications of one or more state variables of a simulation model; *2.* An algorithm which states the conditions for, and the extent of, alterations in the state variables of a model; *3.* (Endogenous): The commencement or completion of a simular activity. *4.* (Exogenous): An instantaneous effect produced within the system boundary from an element without (i.e., from the simular environment.)

Event graph: *1.* The directed graph whose nodes are the component events of an analyzed system and whose arcs indicate the nature of the relationships (conditional or not) among events; *2.* A relation graph among the events of a simulation model.

Executive routine: That segment of a computerized simulation model which maintains the simular clockworks and which sequences the successive calls of the event routines in the proper order as simular time evolves.

Exogenous: *See* Event.

Expectation: The first moment of a random variable.

Experiment (nondeterministic): Any activity whose outcome cannot be predicted with certainty in advance.

Experimental error: The failure of two responses to be the same even though the environmental specifications for the two successive, though independently seeded, simulation encounters are identical. (Cf. Replication.)

Feedback: Information concerning the state of the system, at any point in time, which is used in determining the future state of the system (Krasnow and Merikallio, 1964).

Formal models: Symbolic assertions, in logical terms, of idealized, relatively simple situations, the assertions representing the structural properties of the original factual system (Rosenblueth and Wiener, 1945).

Formalization: That class of symbolic models in which none of the physical characteristics of the modeled are reproduced in the model itself, and in which the symbols are manipulated by means of a well-formed discipline such as mathematics or mathematical logic (Sayre and Crosson, 1963).

Heteroscedastic: Of differing variances (scatter).

Heuristic: Description of a procedure that *may* lead to a problem solution, but one that cannot be guaranteed to lead to a solution (Fritz, 1963).

Homoscedastic: *1.* Of equal scatter; *2.* Of equal variance.

Hypothesis: A conjecture regarding the parameter space of a random variable, usually tested by means of a statistic computed from a random sample of the variable.

Independently distributed random variables: Those whose joint cumulative distribution function may be factored as the product of their respective marginal distributions at any point in Euclidean n-space.

Independently seeded encounters: A set of iterations of a stochastic model, the seeds for each of which are selected randomly and without the possibility of their reoccurrence.

Iteration: *1.* A repetition of the specifications for, and the observation of the response emanating from, a model; *2.* For stochastic models, a repetition of all specifications (save possibly the random number seeds), for, and the resulting observation of, the model.

Joint distribution function (cumulative): For a set of n random variables, $X_1, \ldots, X_n$, the probability that $X_1 \leq a_1, \ldots, X_n \leq a_n$, for any point $(a_1, \ldots, a_n)$ in Euclidean n-space.

Marginal distribution: The distribution of one of the members of a random vector, in the complete absence of information regarding the outcomes of any of the remaining members.

Material models: Transformations of original physical objects (Rosenblueth and Wiener, 1945).

Modeling: The act of mimicry.

Monte Carlo method: That branch of experimental mathematics concerned with experiments on random numbers (Hammersley and Handscomb, 1964).

Power: The probability that a null hypothesis will be rejected given that it is false.

Principium of Seeding: In any sequence of n encounters with a verified stochastic simulation model, the n successive random number seeds will be selected randomly and independently, though repetitions will be forbidden.

Pseudorandom numbers: A sequence of numbers which superficially appear random, but which are generated algorithmically (and hence, cyclicly).

Random process: A time series.

Random sample: A set of n independently distributed random variables having the same marginal cumulative distribution functions.

Random variable: A function, or mapping, from the set of elementary outcomes of a nondeterministic experiment into (a subset of) the real numbers. (Associated with any random variable is a cumulative distribution function.)

Random vector: A multivariate function, or mapping, of the elementary outcomes of a deterministic experiment into points in Euclidean n-space.

Realization: One of the time functions in an ensemble constituting a random process, or time series.

Record: An observed realization, or finite portion thereof, of a time series.

Replication: *1.* That class of models which display a significant degree of physical similarity between the model and the modeled (Sayre and Crosson, 1963); *2.* The independent repetition of an experimental design, usually conducted so as to measure, or estimate, experimental error more accurately.

Response: *1.* The output, univariate or multivariate, of a simulation model; *2.* Also, the corresponding observation, or yield, in the simuland.

Response curve: A response surface dependent upon only one environmental condition (input variable).

Response surface: *1.* The geometrical interpretation of the loci of the responses, y, emanating from a deterministic model with environmental specification $(x_1, x_2, \ldots, x_p)$; *2.* The geometrical interpretation of the loci of the *expected* simular responses $E(Y(T))$ as a function of the environmental specifications $(x_1, x_2, \ldots, x_p)$.

Seed: *1.* (Random number): That randomly selected number, $0 \leq U_0 \leq 1$, from which all succeeding random numbers derive by means of an algorithmic, pseudorandom, number generator; *2.* (Random number): That randomly selected, integral, specification which denotes the location of the first random number to be employed in a prerecorded, pretested sequence of random numbers; *3.* (Juxtaposed): The single, randomly selected, number conceptually derived by the juxtaposition of the several seeds required by a simulation model. *4.* (Cf. Independently seeded.)

Simuland: *1.* The system being mimed by a simulation model; *2.* That which is modeled.

Simular: *1.* Having to do with simulation; *2.* Of, or pertaining to a simulation model, as opposed to the simulated system (simuland).

Simulation: *1.* That class of symbolic models whose symbols are not all manipulated by a well-formed discipline (such as mathematics or mathematical logic) in order to arrive either at a particular numerical value or at an analytic solution (Sayre and Crosson, 1963); *2.* (Dynamic, stochastic): A generator of realizations from a time series, or stochastic process, of generally unknown specifications; *3.* A technique for reproducing the dynamic behavior of a system as it operates over time (Kiviat, 1967); *4.* The technique of solving problems by following changes over time of a dynamic model of a system (Gordon, 1969).

Simulator: One who, or that which, simulates.

Spectral density function: The Fourier transform of the autocorrelation function of a stationary time series.

State history: *1.* A multivariate time series describing the entire evolution of every state variable during a simular encounter, from commencement to completion; *2.* Any abridgement of such a state history.

State representation: A possible simultaneous state achievable by the entire collection of state variables at any point in simular time.

State variables: *1.* The static structures of a simulation model; *2.* The entities, their attributes and relationships, including system properties such as simular time.

Static model: *1.* One which need not employ time as an explicit variable; *2.* One whose properties or attributes are not observed to change with the passage of time.

Stationarity: *1.* (First-order): For a time series, the property that each random variable in the collection has the same CDF; *2.* (kth-order): The property that every set of n random variables, $X(t_1), \ldots, X(t_n)$, in a time series has the same joint CDF, for any indices $t_1 < t_2 < \cdots < t_n$, as does $X(t_1 + h), \ldots, X(t_n + h)$, for any $h > 0$; *3.* (Strict): Stationarity for any arbitrary order k; *4.* (Weak, or wide-sense): The property of a time series whose component random variables all have the same mean and whose covariance function, for $X(s)$ and $X(t)$, depends only upon the lag, $\tau = (t - s)$.

Statistic: *1.* Any transformation of a random sample; *2.* Any function of a set of n random variables.

Stochastic: *1.* Random; *2.* Probabilistic.

Stochastic process: A time series.

System: A collection of interdependent and interactive elements which act together in a collective effort to attain some goal.

System synthesis: That stage (the second) of a model's development, during which the symbolic representation of a system's behavior is organized in accordance with the findings of the systems analysis stage.

Systems analysis: *1.* That discipline devoted to the study of systems in order to isolate their salient components and to delineate their interactions, relationships, and dynamic behavior mechanisms; *2.* More broadly, the science of model-building; *3.* The initial phase of a model's development, during which the salient components, interactions, relationships, and dynamic behavior mechanisms of a system are isolated.

Time series: *1.* An infinite collection of random variables, $\{X(t)\}$, usually indexed by a parameter, t, which represents time; *2.* An ensemble of unsystematically fluctuating time functions; *3.* A stochastic, or random, process.

Unbiased estimator (for θ): A statistic whose expectation is θ.

Uncertainty Principle of Modeling: Refinement in modeling eventuates a requirement for stochasticity.

Validation: That stage (the fourth) of a model's development, during which independently seeded encounters of the verified model are defined and the responses emanating therefrom are constrasted with observations made on the simuland in an effort to establish the verisimilitude of the model and the modeled.

Variance: The second central moment of a random variable; its positive square root, the standard deviation, measures the dispersion of a random variable's distribution.

Verification: *1.* That stage (the third) of a model's development, during which the model's responses are compared with those which would be anticipated to appear if indeed the model's structure were programmed as intended; *2.* The collection of techniques, including debugging and controlled Monte Carlo experimentation, which verify the programmed structure of a model.

White noise: A conceptually infinite sequence of independent random variables, each from the same cumulative distribution function. A spectral analysis of such a sequence should reveal no latent periodicities, such as is the case for white light.

BIBLIOGRAPHY

Aitchison, J. and Brown, J. A. C. (1957). "The Log-Normal Distribution." Cambridge Univ. Press, Cambridge.

Anderson, R. L. (1942). Distribution of the Serial Correlation Coefficient, *Ann. Math. Statist.* **13**, 1.

Anderson, T. W. (1958). "An Introduction to Multivariate Statistical Analysis." Wiley, New York.

Bauer, W. F. (1958). The Monte Carlo Method, *J. Soc. Ind. Appl. Math.* **6**, 438–451.

Beer, S. (1965). The World, the Flesh, and the Metal, *Nature* **205**, 223–231.

Bellman, R. and Dreyfus, S. (1962) "Applied Dynamic Programming." Princeton Univ. Press, Princeton, New Jersey.

Berge, C. (1962). "Theory of Graphs and Its Applications." Methuen, London.

Bhat, U. N. (1969). Sixty Years of Queuing Theory, *Management Sci. Appl.* **15**, B280–B294.

Blunden, G. P. and Krasnow, H. S. (1957). The Process Concept as a Basis for Simulation Modelling, *Simulation* **9**, 89–93.

Bose, R. C. and Draper, N. R. (1959). Second Order Rotatable Designs in Three Dimensions, *Ann. Math. Statist.* **30**, 1097–1122.

Box, G. E. P. (1952). Multifactor Designs of First Order, *Biometrika* **39**, 49–57.

Box, G. E. P. (1954). The Exploration and Exploitation of Response Surfaces: Some General Considerations and Examples, *Biometrics* **10**, 16–60.

Box, G. E. P. and Hunter, J. S. (1957). Multifactor Experimental Designs for Exploring Response Surfaces, *Ann. Math. Statist.* **28**, 195–241.

Box, G. E. P. and Hunter, J. S. (1958). Experimental Designs for the Exploration and Exploitation of Response Surfaces. In "Experimental Designs in Industry" (V. Chew, ed.), pp. 138–190. Wiley, New York.

Breiman, L. (1963). The Poisson Tendency in Traffic Distribution, *Ann. Math. Statist.* **34**, 308–311.

Brennan, R. D. and Linebarger, R. N. (1964). A Survey of Digital Simulation: Digital Analog Simulator Programs. In "Simulation" (John McLeod, ed.), pp. 244–258. McGraw-Hill, New York, 1968.

Buffon, Georges Louis Leclerc, Comte de (1733). "Essai d'Arithmétique Morale," Paris, France (1775).

Chorafas, D. N. (1965). "Systems and Simulation." Academic Press, New York.

Churchman, C. W., Ackoff, R. L., and Arnoff, E. L. (1957). "Introduction to Operations Research." Wiley, New York.

Churchman, C. W. (1968). "Challenge to Reason." McGraw-Hill, New York.

Clark, C. E. (1966). "Random Numbers in Uniform and Normal Distribution." Chandler, San Francisco.

Conway, R. W. (1963). Some Tactical Problems in Digital Simulation, *Management Sci.* **10**, 47–61.

Conway, R. W., Johnson, B. M., and Maxwell, W. L. (1959). Some Problems of Digital Systems Simulation, *Management Sci.* **6**, 92–110.

Coveyou, R. R. (1960). Serial Correlation in the Generation of Pseudo-Random Numbers, *J. Assoc. Comput. Mach.* **7**, 72–74.

Coveyou, R. R. and MacPherson, R. D. (1967). Fourier Analysis of Uniform Random Number Generators, *J. Assoc. Comput. Mach.* **14**, 100–119.

Cox, D. R. (1962). "Renewal Theory." Methuen, London.

Cox, D. R. and Smith, W. L. (1961). "Queues." Methuen, London.

Cramér, H. (1946). "Mathematical Methods of Statistics." Princeton Univ. Press, Princeton, New Jersey.

Curtiss, J. H. (1953). "Monte Carlo" Methods for the Iteration of Linear Operators, *J. Math. and Phys.* **32**, 209–232.

Davies, O. L. (ed.) (1954). "The Design and Analysis of Industrial Experiments." Hafner, London.

DeBaun, R. M. (1959). Response Surface Designs for Three Factors at Three Levels, *Technometrics* **1**, 1–8.

DeMoivre, A. (1718). "The Doctrine of Chances." Pearson, London.

Diananda, P. H. (1953). Some Probability Limit Theorems with Statistical Applications, *Proc. Cambridge Philos. Soc.* **49**, 239–246.

Doob, J. L. (1953). "Stochastic Processes," Wiley, New York.

Draper, N. R. (1962). Third Order Rotatable Designs in Three Factors: Analysis, *Technometrics* **4**, 219–234.

Eisenhart, C. (1950). "Tables of the Binomial Probability Distribution" Nat. Bur. Stand., Appl. Math. Ser. No. 6, Washington, D. C.

Evans, G. W., Wallace, G. F., and Sutherland, G. L. (1967). "Simulation Using Digital Computers," Prentice-Hall, Englewood Cliffs, New Jersey.

Feigenbaum, E. A. and Feldman, J. (1963). "Computers and Thought" McGraw-Hill, New York.

Feller, W. (1950). An Introduction to Probability Theory and Its Applications," Vol. I (1950), Vol. II (1965). Wiley, New York.

Fisher, R. A. (1921). On the "Probable Error" of a Coefficient of Correlation Deduced from a Small Sample. *Metron* **1** (Pt. 4), 3.

Fisher, R. A. and Tippett, L. H. C. (1928). Limiting Forms of the Frequency Distribution of the Largest or Smallest Member of a Sample. *Proc. Cambridge Philos. Soc.* **24**, 180–190.

Fishman, G. S. and Kiviat, P. J. (1967). The Analysis of Simulation-Generated Time Series. *Management Sci.* **14**, 525–557.

Forrester, J. W. (1969). "Urban Dynamics." MIT Press, Cambridge, Massachusetts.

Forsythe, G. E. (1948). Generation and Testing of Random Digits at the National Bureau of Standards, Los Angeles. In "Monte Carlo Method," pp. 34–35. Nat. Bur. Stand., Appl. Math. Ser. No. 12 (1951), Washington, D.C.

Fox, B. L. (1963). Sampling from Beta Distributions, *Technometrics* **5**, 269–270.

Fritz, W.B. (1963). Selected Definitions, *Commun. Assoc. Comput. Mach.* **6**, 152–158.

Gafarian, A. V. and Ancker, C. J. (1966). Mean Value Estimation from Digital Computer Simulation, *Operations Res.* **14**, 25–44.

Gardiner, D. A., Grandage, A. H. E., and Hader, R. J. (1959). Third Order Rotatable Designs for Exploring Response Surfaces, *Ann. Math. Statist.* **30**, 1082–1096.

Gauss, K. F. (1809). "Theoria Motus Corporum Cœlestium." *English Translation*: "Theory of Motion of the Heavenly Bodies." Dover, New York, 1963.

Golomb, S. W. (1967). "Shift Register Sequences." Holden-Day, San Francisco.

Gordon, G. (1969). "System Simulation," Prentice-Hall, Englewood Cliffs, New Jersey.

Gossett, W. S. (1908). The Probable Error of a Mean, *Biometrika* **6**, 1–25.

Granger, C. W. and Hughes, A. O. (1968). Spectral Analysis of Short Series—A Simulation Study, *J. Roy. Statist. Soc.* **A131**, 83–99.

Graybill, F. A. (1961). "An Introduction to Linear Statistical Models." McGraw-Hill, New York.

Green, B. F., Smith, J. E. K., and Klem, L. (1959). Empirical Tests of an Additive Random Number Generator, *J. Assoc. Comp. Mach.* **6**, 527–537.

Grunberger, R. (1964). "Germany, 1918–1945." Harper, New York.

Gumbel, E. J. (1958). "Statistics of Extremes." Columbia Univ. Press, New York.

Hadamard, J. (1945). "The Psychology of Invention." Dover, New York.

Hahn, G. J. and Shapiro, S. S. (1967). "Statistical Models in Engineering." Wiley, New York.

Hammersley, J. M. and Handscomb, D. C. (1964). "Monte Carlo Methods." Methuen, London.

Hampton, R. L. (1965). A Hybrid Analog–Digital Pseudo-Random Noise Generator. In "Proceedings, Spring Joint Computer Conference," Vol. 25, pp. 287–301. Spartan Books, Baltimore, Maryland.

Harris, R. (1970). An Application of Extreme Value Theory to Reliability Theory, *Ann. Math. Statist.* **41**, 1456–1465.

Harter, H. L. (1964). "New Tables of the Incomplete Gamma-Function Ratio and of Percentage Points of the Chi-Squared and Beta Distributions." U.S. Govt. Printing Office, Washington, D.C.

Hill, W. J. and Hunter, W. G. (1966). A Review of Response Surface Methodology: A Literature Survey, *Technometrics* **8**, 571–590.

Hull, T. E. and Dobell, A. R. (1962). Random Number Generators, *SIAM Rev.* **4**, 230–254.

Hunter, J. S. (1959). Determination of Optimum Operating Conditions by Experimental Methods, Part II-2, *Ind. Quality Control* **15** (No. 7), 7–15.

Hunter, J. S. and Naylor, T. H. (1970). Experimental Designs for Computer Simulation Experiments, *Management Sci.* **A16**, 422–434.

IBM Corporation (1959). "Random Number Generation and Testing." Reference Manual C20–8011, New York.

Jenkins, G. M. (1961). General Considerations in the Analysis of Spectra, *Technometrics* **3**, 133–136.

Jenkins, G. M. and Watts, D. G. (1968). "Spectral Analysis and Its Applications." Holden-Day, San Francisco.

Johnson, N. L. (1949). Systems of Frequency Curves Generated by Methods of Translation, *Biometrika* **36**, 149.

Karlin, S. (1966). "A First Course in Stochastic Processes." Academic Press, New York.

Kempthorne, O. (1952). "The Design and Analysis of Experiments." Wiley, New York.

Kendall, D. G. (1951). Some Problems in the Theory of Queues, *J. Roy. Statist. Soc.* **B13**, 151–185.

Kendall, M. G. and Babington-Smith, B. (1938). Randomness and Random Sampling Numbers, *J. Roy. Statist. Soc.* **101**, 147–166.

Kendall, M. G. and Stuart, A. (1966). "The Advanced Theory of Statistics," Vol. I (1963), Vol. II (1961), Vol. III (1966). Hafner, London.

Kiviat, P. J. (1967). Development of Discrete Digital Simulation Languages, *Simulation* **8**, 65–70.

Kiviat, P. J. (1969). Digital Computer Simulation: Computer Programming Languages. RAND Corporation Memorandum (RM-5883-PR), January, 1969.

Krasnow, H. S. and Merikallio, R. A. (1964). The Past, Present, and Future of General Simulation Languages, *Management Sci.* **11**, 236–267.

Kunz, K. S. (1957). "Numerical Analysis," McGraw-Hill, New York.

Laplace, Marquis De (1812). "A Philosophical Essay On Probabilities." English edition: Dover, New York (1951).

Lehmer, D. H. (1951). Mathematical Methods in Large-Scale Computing Units, *Ann. Computation Lab., Harvard Univ.* **26**, 141–146.

Leibowitz, M. A. (1968). Queues, *Scientific American* **219**, 96–103.

Lindgren, B. W. (1968). "Statistical Theory." Macmillan, New York.

Lockyer, K. G. (1967). "An Introduction to Critical Path Analysis." Pitman, London.

Malik, H. J. (1968). Exact Distribution of the Product of Independent Generalized Gamma Variables with the Same Shape Parameter, *Ann. Math. Statist.* **39**, 1751–1752.

Martin, F. F. (1968). "Computer Modeling and Simulation." Wiley, New York.

McLeod, J. (1969). A Statement of Personal Beliefs and Intentions, *Simulation* **13**, vii (Editorial).

Meier, R. C., Newell, W. T., Pazer, H. L. (1969). "Simulation in Business and Economics." Prentice-Hall, Englewood Cliffs, New Jersey.

Mendenhall, W. (1968). "Introduction to Linear Models and the Design and Analysis of Experiments." Wadsworth, Belmont, California.

Mesarović, M. D. (ed.) (1964). "Views on General Systems Theory." Proc. 2nd Systems Symp. Case Inst. Technol. Wiley, New York.

Metropolis, N. (1956). Phase Shifts-Middle Squares-Wave Equations. In "Symposium on Monte Carlo Methods" (H. A. Meyer, ed.), pp. 29–36. Wiley, New York.

Mihram, G. A. (1965). "Bivariate Warning-Time/Failure-Time Distributions." Doctoral Dissertation, University Microfilms, Ann Arbor, Michigan.

Mihram, G. A. (1969a). A Cautionary Note Regarding Invocation of The Central Limit Theorem, *American Statistician* **23**, 38.

Mihram, G. A. (1969b). Complete Sample Estimation Techniques for Reparameterised Weibull Distributions, *IEEE Trans. Reliability* **18**, 190–195.

Mihram, G. A. (1969c). A Distended Gamma Distribution, *Sankhya* **B31**, 421–426.

Mihram, G. A. (1970a). A Cost-Effectiveness Study for Strategic Airlift, *Transportation Sci.* **4**, 79–96.

Mihram, G. A. (1970b). An Efficient Procedure for Locating the Optimal Simular Response, *Proc. Fourth Internat. Conf. on Applications of Simulation*, New York.

Mihram, G. A. (1971). Time Series Techniques in the Analysis of Stochastic Simulation Models, *Proc.* 1971 *Summer Comput. Simulation Conference*, Boston, Massachusetts.

Mihram, G. A. (1972a). On Limiting Distributional Forms Arising in Simulation Encounters. In "The Mathematics of Large-Scale Simulation" (Paul Brock, ed.), Simulation Councils, Inc., La Jolla, California.

Mihram, G. A. (1972b). Some Practical Aspects of the Verification and Validation of Simulation Models, *Operational Res. Quarterly*. (To appear.)

Miller, J. C. P. and Prentice, M. J. (1969). Additive Congruential Pseudo-Random Number Generators, *Comput. J.* **11**, 341–346.

Mises, R. von (1957). "Probability, Statistics, and Truth." Macmillan, New York and London.

Molière (1670). "Le Bourgeois Gentilhomme." In "Théâtre Choisi De Molière," pp. 419–516. Garnier, Paris, 1954.

Morrison, D. F. (1967). "Multivariate Statistical Methods." McGraw-Hill, New York.

Muller, M. E. (1958). An Inverse Method for the Generation of Random Normal Deviates on Large Scale Computers, *Math. Comp.* **12**, 167–174.

Muller, M. E. (1959). Comparison of Methods for Generating Normal Deviates on Digital Computers, *J. Assoc. Comput. Mach.* **6**, 376–383.

Naylor, T. H., Balintfly, J. L., Burdick, D. S., Chu, K. (1966). "Computer Simulation Techniques." Wiley, New York.

Neumann, John von (1948). Various Techniques Used in Connection with Random Digits. In "Monte Carlo Method," pp. 36–38. Nat. Bur. Stand. Appl. Math. Ser. No. 12, Washington, D.C. (1951).

Neumann, J. von (1958). "The Computer and the Brain." Yale Univ. Press, New Haven, Connecticut.

Oplinger, J. D. (1971). "Generation of Correlated Pseudo-Random Numbers for Monte Carlo Simulation." University of Pennsylvania Thesis (No. 1971.38), Philadelphia, Pennsylvania.

Page, E. S. (1965). On Monte Carlo Methods in Congestion Problems: Simulation of Queueing Systems, *Operations Res.* **12**, 300–305.

Papoulis, A. (1965). "Probability, Random Variables, and Stochastic Processes." McGraw-Hill, New York.

Parzen, E. (1961). Mathematical Considerations in the Estimation of Spectra, *Technometrics* **3**, 167–190.

Parzen, E. (1962). "Stochastic Processes." Holden-Day, San Francisco.

Pearson, K. (1948). "Tables of the Incomplete Beta-Function." *Biometrika* Office, University College, London.

Pearson, K. (1957). "Tables of the Incomplete Γ-Function." *Biometrika* Office, University College, London.

Phillips, D. T. (1971). Procedures for Generating Gamma Variates with Non-integer Parameter Sets, *Proc. Fifth Conf. on Applications of Simulation*, New York.

Poincaré, H. (1913). "Mathematics and Science: Last Essays," Dover, New York.

RAND Corporation (1955). "A Million Random Digits with 100,000 Normal Deviates." Free Press, Glencoe, Illinois.

Riordan, J. (1952). "Stochastic Service Systems." Wiley, New York.

Romig, H. G. (1953). "50–100 Binomial Tables." Wiley, New York.

Rosenblueth, A., and Wiener, N. (1945). The Role of Models in Science, *Philos. Sci.* **12**, 316–321.

Rosenblueth, A. (1970). "Mind and Brain." MIT Press, Cambridge, Massachusetts.

Saaty, T. L. (1957). Résumé of Useful Formulas in Queuing Theory, *Operations Res.* **5**, 162–187.

Sayre, K. M., and Crosson, F. J. (eds.) (1963). "The Modeling of Mind: Computers and Intelligence." Simon and Schuster, New York.

Sayre, K. M. (1965). "Recognition: A Study in the Philosophy of Artificial Intelligence." Univ. of Notre Dame Press, Notre Dame, Indiana.

Scheuer, E. and Stoller, D. S. (1962). On the Generation of Normal Random Vectors, *Technometrics* **4**, 278–281.

Schuster, A. (1898). On the Investigation of Hidden Periodicities with Application to a Supposed 26-Day Period of Meteorological Phenomena, *Terrestrial Magnetism* 3, 13–41.

Shephard, R. W. (1965). War Games and Simulations. In "Digital Simulation In Operations Research" (S. H. Hollingdale, ed.), pp. 210–217. American Elsevier, New York.

Shreider, Yu. A. (1964). "Method of Statistical Testing: Monte Carlo Method." Elsevier, Amsterdam.

Shreider, Yu. A. (ed.) (1966). "The Monte Carlo Method: The Method of Statistical Trials." Pergamon, Oxford.

Stacy, E. W. (1962). A Generalization of the Gamma Distribution, *Ann. Math. Statist.* **33**, 1187–1192.

Stacy, E. W. and Mihram, G. A. (1965). Parameter Estimation for a Generalised Gamma Distribution, *Technometrics* **7**, 349–358.

Teichroew, D. (1965). A History of Distribution Sampling Prior to the Era of the Computer and Its Relevance to Simulation, *J. Amer. Statist. Assoc.* **60**, 27–49.

Teichroew, D. and Lubin, J. F. (1966). Computer Simulation-Discussion of the Technique and Comparison of Languages, *Commun. Assoc. Comput. Mach.* **9**, 723–741.

Tintner, G. (1940). "The Variate-Difference Method," Cowles Commission Monograph No. 5, Bloomington, Indiana.

Tocher, K. D. (1965). Review of Simulation Languages, *Operations Res. Quart.* **15**, 189–217.

Watt, K. E. F. (1970). Details versus Models, *Science* **168**, 1079.

Weibull, W. (1951). A Statistical Distribution Function of Wide Applicability. *J. Appl. Mech.* **18**, 293–297.

Wilde, D. J. (1964). "Optimum Seeking Methods." Prentice-Hall, Englewood Cliffs, New Jersey.

Williamson, E. and Bretherton, M. H. (1963). "Tables of the Negative Binomial Probability Distribution." Wiley, New York.

Wold, H. (1955). "Random Normal Deviates," London Univ. Tracts for Computers, No. 25, Cambridge Univ. Press, London and New York.

Young, J. Z. (1957). "Doubt and Certainty in Science," Oxford Univ. Press, Oxford.

Bibliographic Sources

A number of extensive bibliographies on simulation have been compiled and published in various journals and textbooks. The following is a partial listing of these bibliographic sources:

Cragin, S. W., Jr., Fernald, P. J. *et al.* (1959). "Simulation: Management's Laboratory." Harvard Univ. Press, Cambridge, Massachusetts.

Cruiskshank, D. R. and Broadbent, F. W. (1970). "Simulation in Preparing School Personnel: *A Bibliography*." ERIC Clearinghouse on Teacher Education, Washington, D.C.

Deacon, A. R. L., Jr. (1961). A Selected Bibliography: Books, Articles, and Papers on Simulation, Gaming, and Related Topics. In "Simulation and Gaming, a Symposium," pp. 113–131. Amer. Management Assoc. Rept. No. 55.

Dutton, J. M. and Starbuck, W. H. (1971). Computer Simulation Models of Human Behavior: A History of an Intellectual Technology, *IEEE Trans. Systems, Man, and Cybernetics* **1**, 128–171.

Feigenbaum, E. A. and Feldman, J. (eds.) (1963). "Computers and Thought." McGraw-Hill, New York.

Hammersley, J. M. and Handscomb, D. C. (1964). "Monte Carlo Methods," Methuen, London.

IBM Corporation (1966). "Bibliography on Simulation," IBM Corp. Rept. No. 320-0924-0, White Plains, New York.

Malcolm, D. G. (1960). Bibliography on the Use of Simulation in Management Analysis, *Operations Res.* **8**, 169–177.

Martin, F. F. (1968). "Computer Modeling and Simulation." Wiley, New York.

Mihram, G. A. (1971). Time Series Techniques in the Analysis of Stochastic Simulation Models. *Proc. 1971 Summer Computer Simulation Conference*, Boston, Massachusetts.

Mize, J. H. and Cox, J. G. (1968). "Essentials of Simulation." Prentice-Hall, Englewood Cliffs, New Jersey.

Morgenthaler, G. W. (1961). The Theory and Application of Simulation in Operations Research. In "Progress in Operations Research" (R. L. Ackoff, ed.), Vol. I, pp. 361–413. Wiley, New York.

Naylor, T. H., Balintfly, J. L., Burdick, D. S., and Chu, K. (1966). "Computer Simulation Techniques." Wiley, New York.

Naylor, T. H. (1969). Simulation and Gaming. Bibliography Number 19 in *Comput. Rev.* **10**, 61–69.

Shreider, Yu A. (1964). "Method of Statistical Testing: Monte Carlo Method." Elsevier Amsterdam.

Shubik, M. (1960). Bibliography on Simulation, Gaming, Artificial Intelligence, and Allied Topics, *J. Amer. Statist. Assoc.* **55**, 736–751.

Shubik, M. (1970). "A Preliminary Bibliography on Gaming," Inst. Management Sci., College on Simulation and Gaming, Providence, Rhode Island.

Sisson, R. L. (1968). Simulation: Uses. In "Progress in Operations Research" (J. Aronofsky, ed.). Wiley, New York.

Teichroew, D. (1965). A History of Distribution Sampling Prior to the Era of the Computer and Its Relevance to Simulation, *J. Amer. Statist. Assoc.* **60**, 27–49.

Teichroew, D. and Lubin, J. F. (1966). Computer Simulation: Discussion of the Technique and Comparison of Languages, *Commun. Assoc. Comput. Mach.* **9**, 723–741.

SUBJECT INDEX

V

W